Springer-Verlag Berlin Heidelberg GmbH

EBO – Experimental Biology Online Annual 1996/97

C.R. Bridges • A. Curtis • D. Sanders

Springer

Prof. Dr. Christopher R. Bridges
Institut für Zoophysiologie
Lehrstuhl für Stoffwechselphysiologie
Heinrich-Heine-Universität
D-40225 Düsseldorf
Germany

Professor Dale Sanders
Department of Biology
University of York
Heslington
York YO1 5DD
United Kingdom

Professor Adam Curtis
Division of Infection and Immunity
Joseph Black Building
University of Glasgow
Glasgow G12 8QQ
United Kingdom

ISSN 1435-3237

ISBN 978-3-662-00934-5 ISBN 978-3-662-00932-1 (eBook)
DOI 10.1007/978-3-662-00932-1

Cataloging-in-Publication Data applied for

Die Deutsche Bibliothek - CIP-Einheitsaufnahme
Experimental Biology Online : annual. - Berlin ; Heidelberg ; New York ;
Barcelona ; Budapest ; Hong Kong ; London ; Milan ; Paris ;
Santa Clara ; Singapore ; Tokio : Springer
Erscheint jährl. - Aufnahme nach Vol. 1. 1996/97 (1998)
Vol. 1. 1996/97 (1998) -

Additional material to this book can be downloaded from http://extras.springer.com

Cover design: design & production GmbH, Heidelberg

SPIN 10669335 31/3190 – 5 4 3 2 1 0 – Printed on acid-free paper

PREFACE

In March 1996 the Society of Experimental Biology (UK) together with two other international scientific societies, the Australian and New Zealand Society for Comparative Physiology and Biochemistry and the Society for Integrative and Comparative Biology (formally the American Society for Zoology) joined forces with Springer-Verlag, Heidelberg to produce one of the first fully electronic online, peer-reviewed biological journals, **Experimental Biology Online**. The present product represents the fruits of this joint venture and encapsulates Volumes 1 and 2 of the journal. This will be an ongoing series such that an archival version of the journal will be available to all libraries as well as the on-line version.

At the outset this was "new land" for all concerned but the launching of a journal which would cover experimental biology in terms of Animal , Cell and Plant topics was daunting but we all felt that the use of electronic media and the internet would be ideally suited to this purpose.

In the current two volumes you will find a wealth of examples of the benefits of the use of the WWW and the advantages of what an "on-line" can offer above and beyond a normal print journal. Multimedia presentations are possible with both quicktime movies, videos and animated GIFs, colour can be used through out, not only for graphics but also for illustrations and some of the remarkable pictures taken through the confocal laser microscope really make this evident. A whole new range of techniques have now become available for the author and reader including sound, particularly important in behavioural studies and the use of demonstration software.

With the increasing pressure on library resources and scientists the need to publish rapidly and at low costs have become more important. EBO has certainly proved to be a success with 29 papers from an international forum of scientists spread around the globe with the shortest publication time of 10 days for one paper. Total accesses to the EBO web site have reached over 12000 hits /month with the first paper being accessed >1000 times within the first 6 months of appearing.

Thanks to the use of electronic communications such as Email, FTP servers etc and automated report forms we are able to provide our authors and readers with **rapid, refereed, rigorous and readable** papers.

We would like to take this opportunity to thank our authors, editorial board members, guest referees and the Journal Production staff of Springer for making this all possible.

Editors In Chief

Christopher R. Bridges
Adam Curtis
Dale Sanders

Table of Contents Vol. 1 (1996)

Table of Contents Vol. 2 (1997)

Exp. Biol. Online 1:1 (1996)

Activation of neuron adhesion and neurite extension by tethered neurotrophins and adhesion molecules

Adam Curtis, Russell Quinn and Michael McGrath

Cell Engineering Centre, Joseph Black Building, University of Glasgow, Glasgow G12 8QQ, Scotland, UK

Correspondence to: A.Curtis@bio.gla.ac.uk

Received: 7 March 1996 / Accepted: 31 March 1996
Published: April 12, 1996

Summary. Laminin, neural cell adhesion molecule (N-CAM) and the neurotrophins, ciliary neurotrophic factor (CNTF) and brain-derived neurotrophic factor (BDNF) play important roles in the formation and maintenance of the axons and dendrites of vertebrate neurons. The conventional explanation of the action of N-CAM and laminin has been that they provide suitable substrata on which the cells can adhere, spread and extend. We show by applying these molecules immobilised to small beads (tethered) that they activate adhesion and cell extension on substrata where no special adhesion molecules have been placed. This activation occurs even when the beads are attached to the side of the cell not in contact with the substratum, implying that the signal is transduced to activate parts of the cell remote from the vicinity of the bead.

Key words: Neurite, neurotrophin, activation

Introduction

Many reports have appeared in recent years in which neurons have been shown to have a requirement for various specific substratum molecules if neuronal extension of axons or dendrites is to occur (Edgar et al. 1988; Ford-Holevinski et al. 1988; Asou et al. 1992; Payne et al. 1992; Rutishauser et al. 1988). In these tests neurons were shown to extend neurites (axons or dendrites) extensively on substrata of such molecules compared with controls. The neurotrophins can stimulate neurite extension, axon sprouting and axon branching (Lindsay et al. 1985; Loughlin and Fallon 1993; Cohen-Cory and Fraser 1995). Such findings have led to the suggestion that these or similar molecules act during animal development to control the development of the innervation pattern (Harrelson 1982; Hammarback and Letourneau 1986; Harrelson and Goodman 1988; Parnavelas et al. 1988; Harrelson et al. 1989). These and similar experiments on other cell types have been interpreted as demonstrations that the molecules concerned act directly as part of the adhesive system used by the cell.

However, the results might also be explained as the need for an activation of the adhesive processes by a signalling molecule. Doherty et al. (1991) have shown that a region of the neural cell adhesion molecule (N-CAM) molecule resembles other signalling sequences, although N-CAM has been described as a molecule to which neurons and other cells adhere. The same group (Harper et al. 1993) have also shown that embryonic chick spinal neurons cultured on sections of foetal muscle respond in their outgrowth to N-CAM, N-cadherin and to the beta-1 subunit of the integrin. We (Curtis et al. 1992) showed that fibronectin could activate cell adhesion changes from one part of the cell to another if a bead bearing immobilised fibronectin was attached to one small area on a cell. Both these reports suggest that some molecules may stimulate cell adhesion by signalling processes. If this view is right observing adhesion of cells to a substratum of such molecules is an ambiguous result that could be due either to the activation of signalling systems or to cell adhesion. We describe a test between the two possibilities for neuronal extension using tethered adhesion molecules and neurotrophins and discuss the light the results throw on how cell adhesions may be produced. Other workers such as Seiler et al. (1992) have shown that tethered molecules may initiate cell signalling events.

Materials and methods

Beads

Magnetic beads, tosyl form, 4.5 μm diameter (Dynal, Warrington, UK), were derivatised with N-CAM (a mixture of the 180 kDa, 140 kDa and 120 kDa isoforms from mouse brain) (a gift from Prof. F. Walsh), or BDNF (a gift from Regeneron), Laminin. (Calbiochem, Nottingham, UK), Interleukin-8 (IL-8) (British Biotechnology, Oxford, UK) or CNTF (a gift from Amgen, Inc.) or blocked by reaction with tris-hydroxy methyl aminomethane (Tris). IL-8 was regarded as a control molecule.

The beads were activated by raising the pH to 9.5 in an 0.01 M borate buffer. The appropriate protein was then added to the reaction system at levels of 1 to 10 ng per 1×10^8 beads. The presence of signal molecule on the bead was established by quantitative immunofluorescence using the method described by Kinn et al. (1990) for IL-8, laminin, N-CAM, CNTF and BDNF.

Neurons and bead attachment

Neonatal (4-6 day) rat (Sprague-Dawley strain) nodose ganglia were dissected and dissociated in a mixture of 1 $mg.ml^{-1}$ of each of collagenase (Clostridiopeptidase A, Sigma Chemical, Poole, UK) trypsin (Gibco BRL, U.K.), and 0.1 $mg.ml^{-1}$ of DNAase Type 1 (Sigma) dissolved in Hepes-buffered Hanks balanced salt solution. The tissue fragments were incubated at 37°C for 20 min, then dissociated by gentle trituration with a Pasteur pipette. Cells were washed once in culture medium to remove enzymes.

Embryonic chick dorsal root ganglia (DRG) were obtained from 8-day embryos and dissociated with trypsin solution 300 Benzzoyl-arginine ethyl ester (BAEE) units. ml^{-1} in 1 mM EDTA (ethylene diamine tetra-acetic acid) dissolved in phosphate-buffered saline. The tissue fragments were then dissociated and washed as described for nodose ganglion.

Neurons were recognised by their reaction with anti-neurofilament antibody and by the presence of a growth cone. The nodose ganglia were also recognised by their very large soma, which are about 20-30 μm in diameter.

Bead attachment

The beads were allowed to attach to suspensions of freshly dissociated nodose ganglion neurons or to chick dorsal root ganglion cells. The beads were reacted with the cells for 30 min at ratios of from 2:1 to 10:1 at 37°C. Unattached beads were removed by gentle dilution and washing of the cells with culture medium.

Culture

The cells were then plated out on glass coverslips for the nodose cells or on poly-lysine (or laminin for controls)-coated glass coverslips for the DRG cells and cultured in Ham's F12 medium with 3% foetal calf serum, N2 supplement with glutamine and antibiotic supplements with Hepes buffering as recommended by Payne et al. (1992) for 24 h (DRGs) or 48 h (nodose cells).

Double stimulation (bead attachment to the neurites as well as to the soma)

In these experiments BDNF or control beads were applied to the cells at the start of culture, and also after 24 h the attached cells were exposed to a second lot of BDNF-derivatised beads . They were grown a further 24 h and the neurite length measured and bead attachment on soma and on neurites measured.

Measurement of neurite length

The cells were then fixed, stained and the neurite length measured with ocular micrometers and the number of attached beads counted.

Statistical methods

Standard linear regression analysis methods were used to examine the significance of dose (bead number)-response(neurite length) plots.

Interference reflection microscopy (IRM)

This was carried out by the procedure described by Curtis (1989). Because of the very small depth of focus of the method at NA = 1.0, the position of a bead relative to the contact side of a neuron can be established with very few cases of ambiguity.

Results

Adhesion molecules or activators ?

The basic experimental approach has been to take molecular species which have been shown to affect cell adhesion and to present them to the cell immobilised on small (4.5 μm diameter) beads.

Four different signal molecules have been tethered to beads, laminin, N-CAM , ciliary neurotrophic factor (CNTF) and brain-derived neurotrophic factor (BDNF). Interleukin-8 tethered to beads and Tris-blocked beads were used as controls.

In the first experiment the beads carrying the immobilised proteins were mixed with the cells in suspension, allowed to attach at various cell/bead ratios and then allowed to settle onto glass coverslips. They were then cultured for 24 or 48 h and fixed. Cells in both control (no beads or control beads) or those bearing tethered BDNF, CTNF, etc. on beads extended neurite in this period, see Fig. 1.

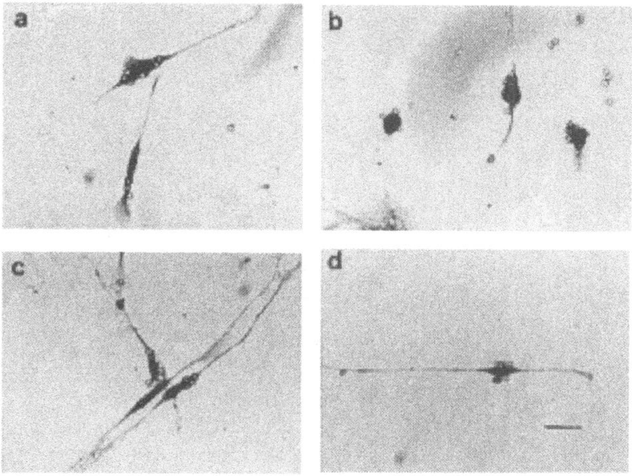

Fig. 1a-d. Appearance of rat nodose neurons with attached beads. Fixed and stained with Coomassie blue preparations. **a, b** Controls. Nodose neurons on glass with attached control (tris-blocked) beads. **c, d** Rat nodose neurons with BDNF beads. *Scale bar* = 30 μm

There is a close positive relationship between neurite length and the number of beads bound which bear N-CAM, laminin, CNTF or BDNF; see Figs. 2-5 (Fig. 2, Fig. 3, Fig. 4, Fig. 5). Control beads (IL-8 or Tris) had no effect on neurite extension; see Fig. 2 and Fig. 6. The experiments on rat nodose cells were done with plain glass surfaces, and there was no need to use poly-lysine or laminin surfaces to obtain good neurite extension if the cells were bead-stimulated. Subsidiary experiments showed that a poly-lysine substratum changed neurite outgrowth pattern with BDNF beads to one with shorter but more numerous neurites. Figure 1 show that the majority of beads are attached to the cell body, but some are found far down the neurites. Similar results were obtained with the DRG cells. Time-lapse video records suggest that these beads are not translocated along the extending neurites, but are adventitious beads that were carried over unattached to cells when the cultures were set up and attached later.

Fig. 2a

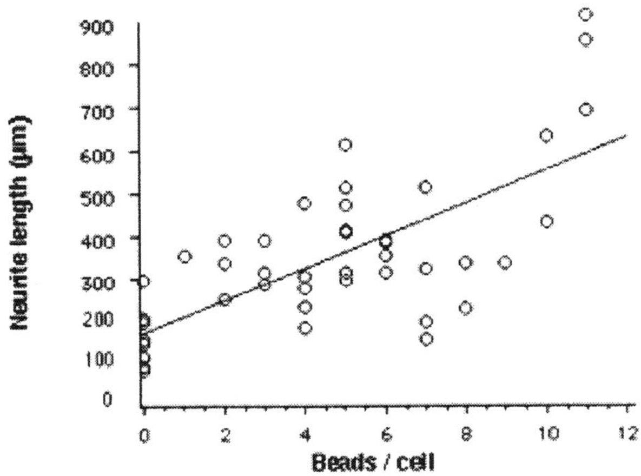

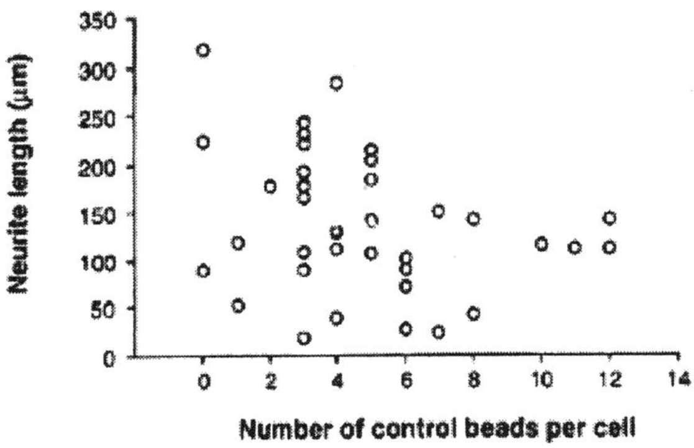

Fig. 2b

Fig. 2. a Effect of numbers of BDNF beads of nodose neurite extension. The regression $y = 167 + 36.7x$. The 95% confidence limits of the regression coefficient are 24 and 51. The regression coefficient is significantly different from zero slope ($t = 5.3$, $P = 0.001$, $df = 47$). **b** shows the reaction of nodose cells to control beads (Tris-blocked). The regression (omitted in the figure for clarity) $y = 169 - 6.96x$. The 95% confidence limits of the regression coefficient are -13.8 and -0.3x; regression coefficient is significantly less than zero slope ($t = 11.09$). Nodose cells treated with BDNF in solution at 10 $ng.ml^{-1}$ had a mean length of 168.6 µm (SD 54.4, $n = 31$). This is significantly longer than control cells ($t = 4.3$, df 56

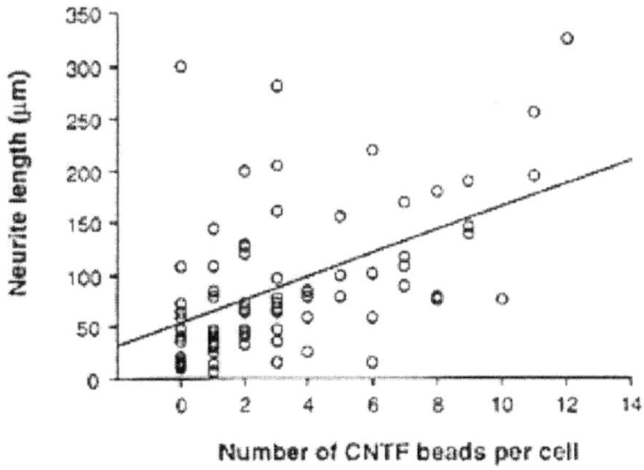

Fig. 3. Effect of numbers of CNTF beads of DRG neurite extension. The regression $y = 11.1x + 54$. The 95% confidence limits on the regression coefficient are 7.1 and 15.2. The regression coefficient is significantly different from zero slope ($t = 5.3$ $P = 0.001$ $df = 94$). Control beads (Tris-blocked) $y = 0.169\text{-}6.96x$. DRG cells treated with CNTF in solution at 10 ng.ml^{-1} had a mean length of 168.6 μm (SD 54.4 $n = 31$); this is significantly longer than control cells without beads ($t = 4.3$, df 56)

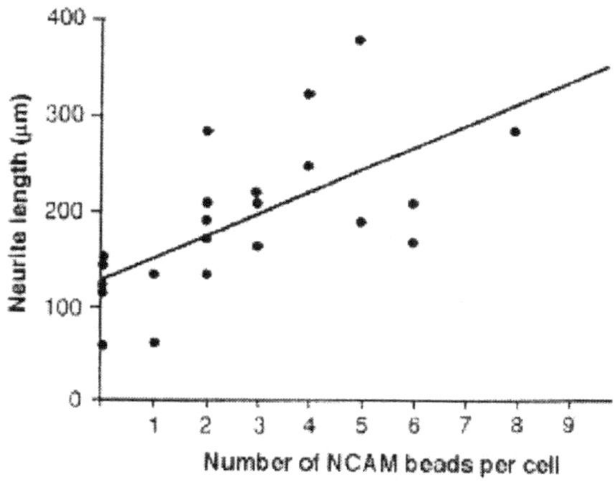

Fig. 4. The reaction of DRG neurons to N-CAM beads. The 95% confidence limits of the regression coefficient are 3.4 and 9.0. The regression of neurite length on bead number $y = 61.0+19.9x$ is significant ($t = 7.5$, $P<0.001$, $df = 29$)

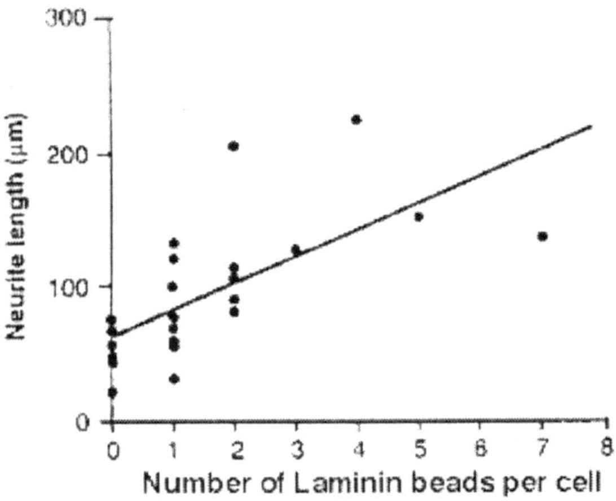

Fig. 5. DRG cells to laminin beads. The regression $y = 60.9+119.9x$, $t = 5.1$ is significant ($P = 0.001$, $df = 29$). The 95% confidence limits of the regression coefficient are 11.9 and 27.9. Tris-beads on DRG cells give a non-significant regression $y = 84.2\text{-}9.08x$ ($P = 0.35$, $df = 36$), i.e. the regression (not shown) is not significantly different from a slope of 0. Control DRG cells without treatment had neurites of mean length 118.5 μm (SD 60, $n = 40$) and control nodose cells neurite lengths of 110.4 μm (SD = 48.1, $n = 30$). DRG cells on laminin substrates and on N-CAM substrate had neurites of 117.8 μm (SD = 25.2, $n = 44$) and of length 122.1 μm (SD 27.2, $n = 45$). Adsorptions of laminin were from 10 μg.ml $^{-1}$ solutions of laminin

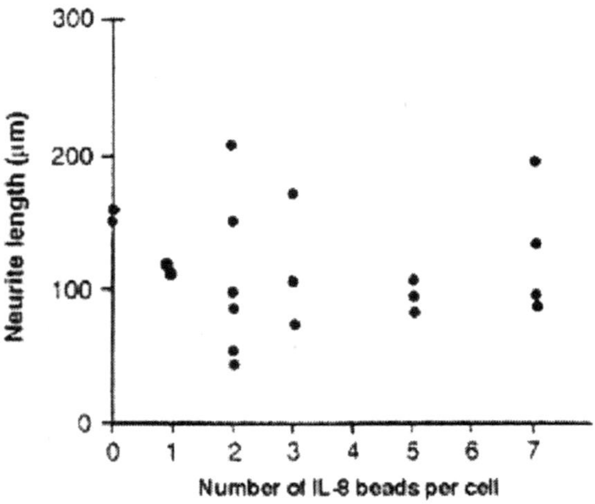

Fig. 6. Reaction of DRG neurons to IL-8 beads. The regression $y = 130 \text{-}2.2x$ has a slope not significantly different from zero ($t = 0.57$, $df = 44$)

Medium taken from the cultures after 24 h had no effect on neurite outgrowth of freshly plated-out neurons. This suggests that it is unlikely that the beads are releasing neurotrophic proteins. The quantity of protein that could be released would in any event be extremely small since it would be related to the density of tosyl groups of the original bead stock.

Do the beads attach the cells or activate attachment?

A test of this possibility arose from the chance discovery in another system by myself and colleagues (Curtis et al. 1992) that cells could be made to adhere to otherwise inappropriate substrates, such as adsorbed haemoglobin, by beads which were on the medium side of the cells and not on the substratum side. This observation made with interference reflection microscopy (IRM) was repeated for the neurons and laminin and BDNF beads. If the beads were sticking the cells to the substrate directly, the beads and cell contacts would be simultaneously in contact, but in fact, as Fig. 7 shows, when the cell contacts are in focus, the beads are not. Table 1 shows the counts of cell position.

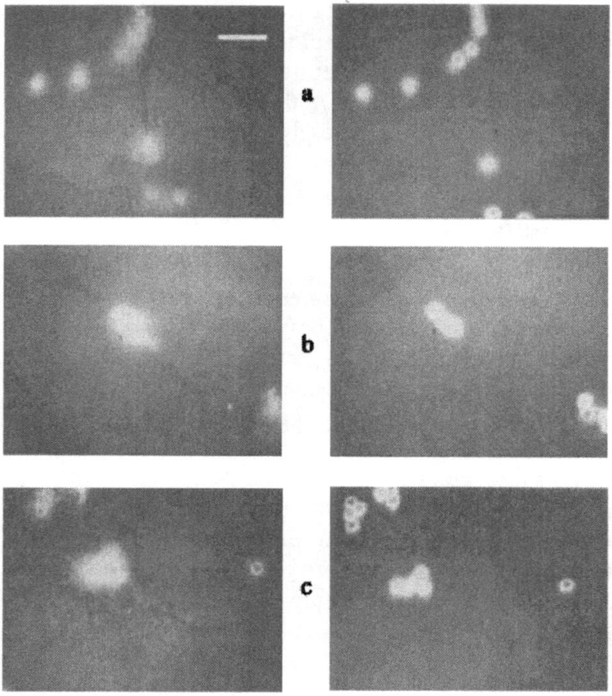

Fig. 7a-c. IRM of bead attachment. Paired photographs of beaded chick DRG neurons attached to poly-lysine coated glass. **a, b** Tris-blocked, i.e. control beads **c** Laminin beads. The left-hand pictures are in focus on the substratum surface and the contact zone of the cell; the right-hand pictures are in focus on the 2.8 μm diameter beads. Note that beads attached to cells are on the non-contact side of the cell. Note also that the bead on the left-hand side of the middle pair and on the right hand side of the bottom pair, which are not in contact with cells, are in focus in the left hand pictures, i.e. on the substratum surface. *Scale bar* = 10 μm

Table 1. Position of beads on cells as detected by IRM counts of instances

	Substratum side	Medium side	Uncertain
Laminin beads, glass subst	3	159	1
Laminin beads, PolyLysine substr	14	48	0
Tris beads, glass substr	14	42	0
Tris beads, PolyLysine substr	26	64	0

These results show that the proteins used do not have to be presented to the cells as a substratum on which the cells spread and extend. The IRM results show that they can be presented to the free surface of the cells. Such results support the alternative explanation that they are acting as signalling molecules to activate adhesion. Tanaka et al. (1993) have shown that interleukin 1 is probably presented to cells immobilised on a proteoglycan, which is a mode of presentation analogous to the use of beads.

Effects of double stimulation

In a second experiment the neurite outgrowth obtained in 24 h was tested to discover whether a second round of bead attachment (mainly to the neurites themselves) would cause additional extension, The results, (Fig. 8) show that application of beads bearing signal molecules to the neurites caused additional extension. This is perhaps not unsurprising, but it suggests that the lateral surface of the neurites may be an appropriate point for signal reception as well as the cell body. It also suggests that 'guidepost' cells described in insects by Harrelson and Goodman (1988) might act in a similar manner. The results should also be compared with those experiments in which beads with polyethyleneimine derivatised on their surfaces were placed on *Aplysia* neurons (Forscher et al. 1992). These induced F-actin assembly radiating from the bead site, which suggests that the beads may be signalling to the cells.

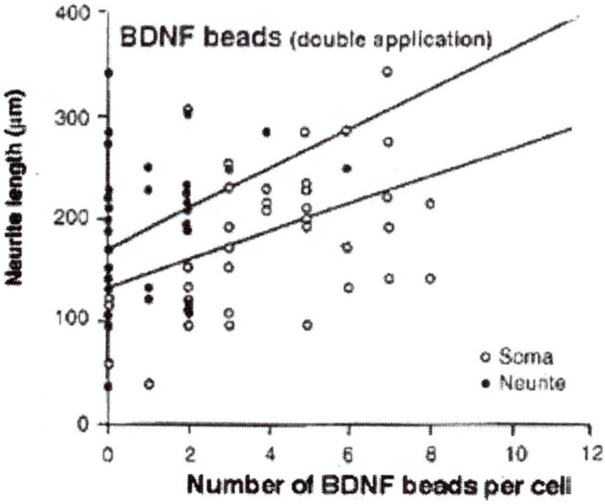

Fig. 8. In these experiments BDNF beads were applied to the cells at the start of culture, and after 24 h the attached cells were exposed to a second lot of beads, BDNF-derivatized. They were grown a further 24 h and the neurite length measured and bead attachment on soma and on neurites recorded. For beads applied at 0 and at 24 h soma-bound beads are predominantly those applied at 0 h, neurite bound beads almost entirely those bound at 24 h. For controls see Fig. 2a. Regression to soma-bound beads, $y = 129.0 + 13.3x$, the regression is significant ($t = 3.3,5$, $P = 0.002$, df 42), and to neurite-bound beads $y = 165,1 + 10.45x$, the regression is significant ($t = 2.7$, $P = 0.01$, df 42). This suggests that BDNF can act to promote neurite extension when applied to the neurite

Discussion

In all these experiments beads with CNTF, BDNF, N-CAM or laminin tethered to their surfaces produced an increase in growth cone activity even though the growth cone might be (at least at the end of the experiment) remote from the bead(s). Our results suggest that laminin, N-CAM and BDNF may do so by providing an appropriate activating signal to the cell rather than being adhesive substrates. Such an interpretation is consistent with the finding (Hammarback and Letourneau 1986) that fine filopodia can extend across 'non-adhesive' surfaces in an exploratory manner. Later they form more permanent adhesions-often at the nearest edge of substratum covered with adhesion-promoting molecules. These results also bear on the question of the effects of prolonged receptor occupancy because the immobilised ligands are likely to remain on the cell's receptors for a long while. In this connection it is of interest that continuous occupancy has been found important for maintenance of effects such as protein phosphorylation (Meisenhelder et al. 1989).

It could be suggested that the factors are slowly eluting from the beads in culture and working as soluble signals. This is unlikely because of the very irreversible nature of the bond formed by the tosyl groups and also because the molecules remain on the beads during the preparation and examination phases of the stages of staining with fluorescent immune reagents (see Materials and methods).

A related question of importance is the size of the signal that can stimulate a cell. The beads have contact areas with the cell not exceeding one-third of the bead surface (see Curtis et al. 1992 for figures): in other words, 20 μm^2. Estimates of the density of surface binding of fibronectin and of laminin based on immunofluorescent data give values of ca. 1 $\mu gm\ cm^{-2}$, i.e. 20×10^{-14} g per bead, or about 45,000 molecules per bead. This in excess of the minimum required occupancy of fibronectin receptors to produce cell spreading (Aplin and Hughes 1981).

One apparent problem raised by the results with interference microscopy on bead location is how even in controls there should be a preponderance of beads on the non-contact side of the surface rather than holding the cells in contract with the surface. It is possible that if a cell settles with a bead on the underside this raises the cell so far off the surface that it cannot easily attach, and either rolls over through chance disturbance or is removed as an unattached cell.

No evidence was found that BDNF increased axon branching, but this may be a consequence of the cell types used.

Acknowledgements. We thank Regeneron Inc. for making BDNF and an antibody to it available to us and Prof F.Walsh for a gift of N-CAM. We also thank Amgen for providing CNTF and an antibody. We also thank Drs. I. Jakob and A.J. Lawrence for critical review of the typescript. The work was supported by the General Funds of the University of Glasgow.

References

Aplin, J.D. and Hughes, R.C. (1981) Cell adhesion on model substrata: threshold effects and receptor modulation. *J. Cell Sci.* **50**, 89-103

Asou, H., Miura, M., Kobayashi, M., Uyemura, K. and Itoh, K. (1992) Cell adhesion molecule L1 guides cell migration in primary reaggregation cultures of mouse cerebellar cells. *Neurosci. Lett.* **144**, 221-224

Cohen-Cory, S. and Fraser, S.E. (1995) Effects of brain-derived neurotrophic factor on optic axon branching and remodelling in vivo. *Nature* **378**, 192-196

Curtis, A.S.G. (1989) Interference reflection microscopy. In: Lacey. AJ, ed. *Light microscopy in biology: a practical approach.* Oxford: IRL Press, 45-49 (part of Chap. 2)

Curtis, A.S.G., McGrath, M. and Gasmi, L. (1992) Localised application of an activating signal to a cell: experimental use of fibronectin bound to beads and the implications for mechanisms of adhesion. *J. Cell Sci.* **101**, 427-436

Doherty, P., Ashton, S.V., Moore S.E. and Walsh, F.S. (1991) . Morphoregulatory activities of NCAM and N-cadherin can be accounted for by G protein-dependent activation of L- and N-type neuronal Ca++ channels. *Cell* **67**, 21-33

Edgar, D., Timpl, R. and Thoenen, H. (1988) Structural requirements for the stimulation of neurite outgrowth by two variants of laminin and their inhibition by antibodies . *J. Cell Biol.* **106**, 1299-1306

Ford-Holevinski, T.S., Hopkins, J.M., McCoy, J.P. and Agranoff, B.W. (1986) Laminin supports neurite outgrowth from explants of axotomized adult rat retinal neurons. *Dev. Brain Res.* **28**, 121-126

Forscher, P., Lin, C.H. and Thompson, C. (1992) Novel form of growth cone motility involving site-directed actin filament assembly. *Nature* **357**, 515-518

Hammarback, J.A. and Letourneau, P.C. (1986) Neurite extension across regions of low cell-substratum adhesivity: implications for the guidepost hypothesis of axonal pathfinding . *Dev. Biol.* **117**, 655-662

Harper, S.J., Walsh, F.S. and Doherty, P. (1993) Neurite outgrowth of spinal neurons on tissue sections of embryonic muscle is largely integrin independent. *Neurosci. Lett.* **159**, 202-206

Harrelson, A.L. (1992) Molecular mechanisms of axon guidance in the developing insect nervous system. *J. Exp. Zool.* **261**, 310-321

Harrelson, A.L. and Goodman, C.S. (1988) Growth cone guidance in insects: Fasciclin II is a member of the immunoglobulin superfamily . *Science* **242**, 700: 531-70

Harrelson, A.L, Bastiani M.J, Snow P.M., and Goodman, C.S. (1989) From cell ablation experiments to surface glycoproteins: Selective fasciculation and the search for axonal recognition molecules . From *Cell ablation axperiments to surface glycoproteins*, Chap. 7, pp. 96-109

Kinn, S.R, Curtis, A. and Dow, J.A.T. (1990) Detection and quantitation of HLA class II molecules on keratinocytes by quantitative immunofluorescence. *J. Immunol. Methods* **134**, 243-251

Lindsay, R.M., Thoenen, H. and Barde, Y.-A. (1985) Placode and neural crest-derived sensory neurones are responsive at early developmental stages to brain derived neurotrophic factor. *Dev.l Biol.* **112**, 319-328

Loughlin, S.E. and Fallon, J.H. (eds.) (1993) *Neurotrophic factors.* San Diego: Academic Press.

Meisenhelder J., Suh P.-G., Rhee, S.G. and Hunter, T. (1989) Phospholipase C-y is a substrate for the PDGF and EGF receptor protein-tyrosine kinases in vivo and in vitro *Cell* 57, 1109-1122

Parnevelas, J.G., Stern, C.D. and Stirling, R.V. (eds.) (1988) *The making of the nervous system.* Oxford: Oxford University Press

Payne, H.R, Burden, S.M., and Lemmon, V. (1992) Modulation of growth cone morphology by substrate-bound adhesion molecules. *Cell Motil. Cytoskel.* **2**, 65-73

Rutishauser, U., Acheson, A., Hall, A.K., Mann, D.M. and Sunshine, J. (1988) The neural cell adhesion molecule (NCAM) as a regulator of cell-cell interactions. *Science* **240**; , 53-57

Seiler, S.M., Michel, I.M. and Fenton, J. (1992) Involvement of the tethered ligand receptor in thrombin inhibition of platelet adenylate-cyclase. *Biochem. Biophys. Res. Commun.* **182**, 1296-1302

Tanaka Y, Adams D. H, Hubscher S, Hirano, H, Siebenlist, U and Shaw, S. (1993) T-cell adhesion induced by proteoglycan-immobilized cytokine MIP-1ß.*Nature* **361**, 79-82

Exp. Biol. Online 1:2 (1996)

Morphogenetic guidance cues can interact synergistically and hierarchically in steering nerve cell growth

Stephen Britland[1,2], Cameron Perridge[1], Morgan Denyer[1], Hywell Morgan[2], Adam Curtis[1], and Chris Wilkinson[2]

[1]Centre for Cellular Engineering, University of Glasgow, Glasgow G12 8QQ, UK
[2]Department of Electronics and Electrical Engineering, University of Glasgow, Glasgow G12 8QQ, UK

Correspondence to: S. Britland (Tel: 0141 339 8855 x5979 or x2074; Fax: 0141 330 4600; e-mail: S.Britland@elec.gla.ac.uk)

Received: 13 March 1996 / Accepted: 13 May 1996
Published: May 13, 1996

Abstract. Nerve cell growth is influenced by guiding properties of its substratum. Microfabricated cell culture substrata were used to determine whether rat dorsal root ganglia (DRG) nerve cells could detect and integrate simultaneous model adhesive and topographic guidance cues. Interference reflection microscopy demonstrated strips of surface contact under the marginal zone of growth cones on planar surfaces which were coincident with actin immunostaining at the periphery of the C-domain. Clusters of focal contacts below the growth cone C-domain delineated the track edges on adhesive gratings. Neurite extension was guided most effectively by adhesive gratings of 25-μm period where highly aligned cells were typically bipolar. Nanometric steps and differences in surface texture between the adhesive tracks was detected using atomic force microscopy (AFM). Neurites did not align to 12- to 100-μm pitch grooves which were less than 1 μm deep. The proportion of aligned neurites increased with groove depth. Maximum neurite alignment was seen when 6-μm-deep, 25-μm-wide grooves contained superimposed parallel adhesive tracks of matched pitch. Neurites aligned preferentially to adhesive tracks superimposed orthogonally over shallow grooves (1 μm deep). Primary neurites aligned increasingly to grooves with orthogonal adhesive tracks as their depth increased. These neurites frequently had highly branched terminal arbours aligned to the orthogonal adhesive tracks. We conclude that morphogenetic guidance cues can interact synergistically and hierarchically to steer nerve cell growth.

Key words: Nerve cell – Microelectronics – Development – Regeneration

Introduction

The long-distance targeting that is characteristic of nerve cells suggests that axon growth and regeneration might be steered by adhesive pathways and topographic channelling (Ebendal 1982; Hynes and Lander 1992). Adhesive pathways can take different forms; columns of cells (Rakic 1990), pre-existing neural tracks (Hynes et al. 1986) and aligned extracellular matrix (Leisi 1990). Suspected examples include pathways of differential adhesiveness in the dorsal columns directing frog DRG afferents rostrally and caudally after the root transitional zone (Holder et al. 1987) and preferential routing of motor axons towards the ventral roots (Lance-Jones and Dias 1991). Tracks of laminin and fibronectin are known to appear transiently in developing nervous tissues (Rogers et al. 1986;Riggott and Moody 1987). Pathways of neural cell adhesion molecule (NCAM) are also thought to guide axons during chick limb (Tang et al. 1992) and mouse cochlea development (Silver and Rutishauser 1984). Antibodies which interfere with fibronectin-binding membrane receptors in neural crest cells (Duband1990) and antisense integrin RNA disruption of integrin function in the optic tectum (Galileo et al. 1992) can disrupt neuroblast migration. The shape of cellular microenvironment and compartmentalisation is also considered to influence the direction of nerve cell growth in vivo (Ebendal and Jakobsen 1977;Lofberg et al. 1980; Newgreen 1989; Nakata and Nakatsuji 1991).

Experimental tissue culture substrata have been fabricated using techniques adapted from the microelectronics industry (Connolly et al. 1991). This has allowed the adhesive and topographic properties of cell culture substrates to be controlled, and varied, at the micrometer (cellular) scale in isolation from incidental 'unwanted' guiding effects. Geometric and functional aspects of both adhesive (Vielmetter et al. 1990; Clark et al. 1993; Soekarno et al. 1993) and topographic guidance (Hirono et al. 1988; Clark et al. 1990, 1991) of neurite outgrowth from cultured nerve cells have been examined in this way. The orienting response of different nerve cell types to very similar cues (Nakataet al. 1993), and even the same cue (Rajnicek et al. 1994), has sometimes proved inconsistent. One explanation for this could be that different cell types are predisposed to favour particular cues in certain circumstances. Single guidance cues might provide only part of the navigational information required by growth cones and are therefore unable to guide cells accurately on their own. Identification of dominant guidance principles could have implications for the design of prosthetic implants to aid tissue repair (Rolfe 1995).

In this paper we have examined the ability of rat dorsal root nerve cells to detect and respond to two model morphogenetic guidance cues simultaneously. Adhesive tracks and topographic anisotropy were presented to nerve cells either individually, or together in parallel or orthogonally opposed. Particular attention was paid to evidence of possible synergistic and hierarchical interactions between the cues and their effect on neurite orientation and growth cone motility.

Materials and methods

Substratum fabrication

Fused-silica microscope slides were cleaned using 7:1 conc. sulphuric acid:hydrogen peroxide at 80°C for 10 min. After rinsing in RO water and drying, the slides were spin-coated with 1400–31 +ve photoresist (Shipley Chemicals, Lancashire, UK) at 4000 rpm for 30 s and mask containing gratings of 100, 50, 25 and 12 μm pitch. Exposed resist was developed using 1:1 Microposit developer in water for 75 s. The patterned resist was used as a mask for reactive ion etching (RIE80, Plasma Technology, UK) in CHF_3 plasma at 29 sccm flow rate and 15 mTorr and at 100 W rf at 13.6 kHz (etch rate approx. 50 nm/min.). The six slides were etched to depths of 50, 100 and 500 nm, and 1, 3 and 6 μm. Etch depth was measured using a Dektak surface profiler (Veeco Sloan Technology Division, California, USA). The remaining resist was removed by sonication in acetone and the slides blanket-etched in CHF_3 for 30 s as before.

Patterning surface chemistry

The slides were cleaned (as above) and spin-coated with 1400–31 +ve photoresist at 4000 rpm for 30 s. After baking at 90°C for 30 min, the slides were again exposed to UV light through the photolithographic mask that was used to fabricate the topographic gratings. A mask aligner (HTG, San Jose, USA) was used to place the adhesive grating either parallel to the topographic grating (methylsilane on ridges), or perpendicular to the orientation of the topographic grating (Fig. 1). The slides were then exposed to dimethyldichlorosilane (Sigma) vapour for 30 s, before being immersed in acetone for 3×5 min periods (+5 second ultrasonication) to remove the remaining resist. The patterned surface chemistry consisted of alternating strips of methylsilane (a hydrophobic surface) and plain silica. The slides were then blown dry using compressed nitrogen and placed in 6-cm Petri dishes, covered with 5 μg/ml laminin in PBS (Sigma, UK) and incubated at 37°C for 2–3 h. Finally the excess laminin solution was removed and the cell suspension added.

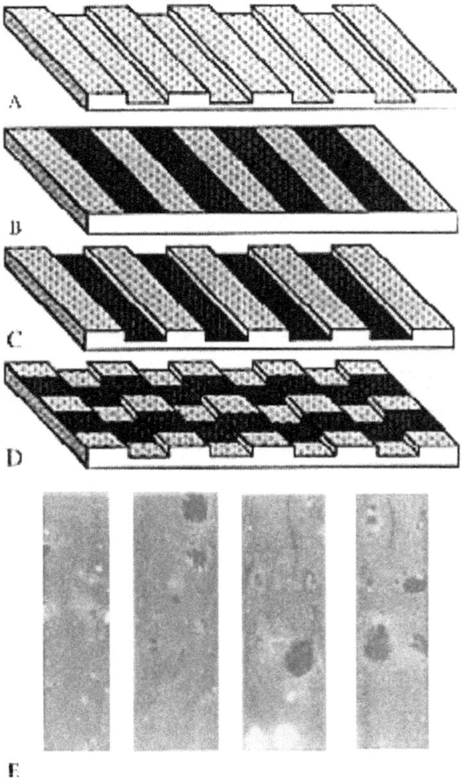

Fig. 1A–E. Microfabricated culture substrata. **A** Grooves and ridges only. **B** Adhesive tracks only. **C** Grooves and ridges with superimposed parallel adhesive tracks. **D** Grooves and ridges with superimposed orthogonal adhesive tracks. **E** Immunofluorescently labelled vitronectin adsorbed to the patterned culture substrate (*bright strips*)

Cell culture

P1 Sprague-Dawley rats were killed by an overdose of intraperitoneal Sagatal (0.6 ml Na^+ phenobarbital) and the skin opened to expose the vertebral column. The spinal ganglia were then exposed by laminectomy and approximately 50 removed. These were finely minced in divalent-free Hank's balanced salt solution (HBS) and then transferred to a solution of 0.2 mg/ml EDTA (Flow Labs, UK) containing 1 mg/ml trypsin, 1 mg/ml collagenase and 0.1 mg/ml DNAase type II (all Sigma, UK) at 37°C for 30 min. Trypsinisation was stopped by the addition of serum-containing medium. The cell suspension was then centrifuged at 2000 rpm for 5 min and resuspended by trituration in serum-containing medium. The culture medium was 100 ml Hepes-buffered Hams F12 supplemented with 3% fetal calf serum, 3 mM L-glutamine, 1% N2 (Gibco, UK), 100 units/ml penicillin, 100 μg/ml streptomycin, 2.5 μg/ml ampherotericin-B and 10 ng/ml 7S NGF (Sigma, UK). The cells were plated onto the fabricated surfaces at a concentration of approx. 5×10^3 cells cm^{-2} and maintained at 37°C in humidified air for 24 h. Laminin and FCS-coated culture substrates were routinely tested for their ability to support nerve cell growth.

Quantification of neurite alignment

The orientation (line of best fit from the cell body to the growth cone) of 100 primary neurites per grating to the underlying patterned adhesive or topographic substrata was measured using an eyepiece protractor graticule and inverted phase-contrast optics. Orientations were recorded and grouped into 10° intervals from 0 to 90°. Those neurites within 10° either side of 0° and 180° were scored as aligned; all other orientations were scored non-aligned. The cells were scored twice for the *perpendicular experiment*, once for alignment to the adhesive tracks and once for alignment to the topographic grating. The significance of measured differences in the degree of alignment between different substrates was determined by calculating chi^2 from 2×2 contingency tables, and association between alignment responses to adhesive tracks and grooves and ridges of matched pitch was made using Pearson's product-moment correlation test. In both analyses statistical significance was taken at $P<=0.05$.

Immunocytochemistry

Preferential adsorbtion of laminin and serum adhesion molecules to different parts of the culture substrates was examined immunocytochemically. The slides were rinsed 3 times in PBS, blocked with fraction V BSA (Sigma) at 10 μg/ml in PBS for 1 h and then immersed in rabbit anti-laminin for 1 h at room temperature (1:40; Sigma, UK), rabbit anti-bovine vitronectin for 24 h at 4°C (1:9000 in BSA/PBS, Chemicon) and monoclonal anti-human fibronectin (1:1000 in BSA/PBS, Calbiochem) for 2 h at room temperature. The slides were then washed 3 times in PBS before immersion in TRITC-labelled sheep anti-mouse IgG (1:200 in PBS, Nordic), TRITC-labelled donkey anti-rabbit IgG (1:200 in PBS, Amersham) for 1 h in the dark in humidified air. Finally the slides were rinsed 3 times in PBS and mounted in Citifluor.

Cells for actin staining were rinsed in PBS at 37°C and then fixed using 4% formaldehyde in 0.1 M PBS for 15 min. The membrane was permeabilised using 0.5% Triton X100 for 15 min. After a further rinse in PBS the cells were immersed in 5-ml aliquots of TRITC-phalloidin (Sigma) at 1 μg/ml in PBS for 45 min at room temperature in the dark. The cells were then rinsed 3 times in PBS and mounted in Citifluor.

Microscopy and photography

Neurite alignment was measured using a Leica DMIRB inverted phase-contrast light microscope, set to Kohler illumination and equipped with a long working distance ×20, 0.4NA, infinity-corrected Neoplan objective. Focal contacts and other regions of close growth cone membrane-substratum apposition in live growth cones were examined using a Vickers M17 fluorescence (interference-reflection) microscope (IRM) (Curtis 1964). The cells were grown on type-0 coverglass (Chance Propper, UK) with a chemically patterned surface, mounted in medium on a depression slide, and examined through a stopped-down 100× oil-immersion, 1.3NA Neofluor objective and a Panasonic video camera. Illumination was provided by a 100 W mercury lamp through a green band-pass filter (546 nm), intensity-reducing neutral density filters, a centrable field iris and a neutral beam splitter. Images were gathered using an image analysis program (Semper, Synoptics, Cambridge) running on a Compaq 386 computer and photographed directly from a VDU screen. Immunofluorescence preparations were examined using a 60× oil-immersion objective on a Nikon inverted microscope with a Tracor Odyssey laser scanning confocal attachment (model no. VSM-LSM, Noran Instruments, Milton Keynes, UK).

Scanning electron micrographs were obtained using a Hitachi SEM 800. Cells were fixed in 2% glutaraldehyde in 0.1 M PBS for 30 min and then rinsed 3 times in PBS with 2% sucrose, before post-fixation in 1% w/v osmium tetroxide for 30 min. The cells were then rinsed thoroughly but gently in RO water and dehydrated in graded concentrations of ethanol and water to 100% ethanol. The ethanol was gradually substituted for graded concentrations of hexamethyldisilazane (Sigma, UK) in ethanol to 100% HMDS. The specimens were then air dried in a fume hood for 30 min before being sputter-coated with 20 nm gold/palladium (Polaron E5000, Polaron, UK). The samples were finally mounted onto SEM stubs, earth-strapped using colloidal graphite and examined at 20 kV.

Fabricated patterned adhesive culture substrata and nerve cell growth cones were imaged using a Nanoscope III Atomic Force Microscope (Digital Instruments, Calif., USA) in tapping mode in air. The cells were processed as for SEM but not coated. Growth cones were located using an optical microscope and simultaneous AFM height mode and error signal images recorded. Tapping-mode silicon cantilevers (Digital Instruments) with resonant frequencies of approx. 300 kHz were used. Images of 256×256 data points were collected at scan rates of 1–2 Hz. No image processing or enhancement was performed on the final images.

Results

Model guidance cues

The reader should refer to Fig.1 for graphical illustrations of the experimental culture substrata described in the following results sections.

Adhesive tracks only

After 24 h exposure to serum, vitronectin but not fibronectin was preferentially adsorbed to the plain glass tracks separating the laminin tracks. An AFM was used in contact mode and with a high imaging force to selectively remove a small region of adsorbed protein (Fig.2A). Re-imaging using tapping mode allowed the protein layer to be measured at 15–20 nm in thickness (Fig.2B). The step height between the adhesive tracks varied across the grating between 0 and 20 nm. The vitronectin tracks had non-uniform surface structure/topography, with many crystallites of protein present. This led to the appearance of a much rougher or uneven surface than laminin tracks, as shown in Fig.2A and C.

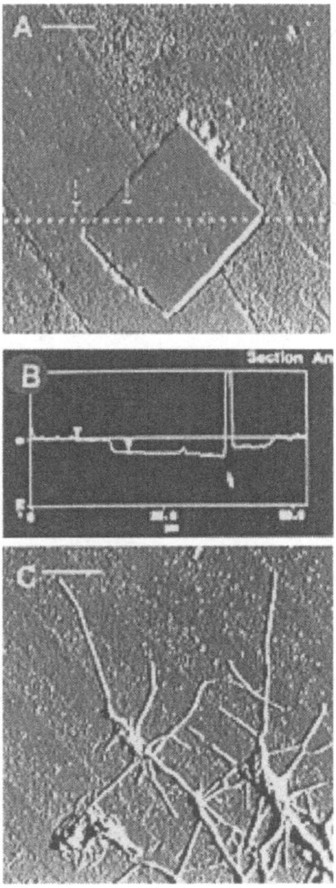

Fig. 2A-C. AFM error-signal images of the patterned adhesive substrate. **A** Laminin tracks appeared smoother in texture than adjacent vitronectin tracks (see Fig. 1B, E). *Scale bar* = 2 µm. **B** The adsorbed protein layer was approx. 15--20 nm thick (between the arrows in **A** and **B**). **C** Primary neurites were not aligned to the adhesive tracks after 4 h in vitro. *Scale bar* = 7 µm

DRG neurons growing on a plain serum-treated surface were typically multipolar and highly branched with many growth cones (Fig.4A). Neurite outgrowth on patterned adhesive gratings did not appear aligned after 4 h in culture (Fig. 4B). After 24 h, quantitation showed that neurites were highly aligned on 25-µm pitch tracks (Fig.3). These cells were typically bipolar and predominantly occupied vitronectin tracks. Confocal fluorescence microscopy of growth cones on plain surfaces revealed positive actin staining in the filopodia and actin condensations in the C-domain (Fig.5B). IRM showed that the periphery of the actin staining in the C-domain was coincident with strips of close apposition between the underside of the growth cone and the substrate immediately below the marginal zone (Fig.5A, *black arrows*). It is our opinion that these darks bands are consistent with surface contact and not artefact resulting from thickness changes in the growth cone body. The strips of contact were not always continuous around the entire perimeter of the C-domain, but often appeared opposite filopodia with focal adhesions at their tips (Fig.5B, *white arrow*). There were occasionally concentric strips of close contact under the marginal zone which had a period of around 0.25 μm. Actin accumulated eccentrically within growth cones on 5-μm adhesive gratings, paralleling accompanying filopodial asymmetry and the direction of neurite extension (Fig.5C). On homogenous surfaces growth cones were between 5 and 20 μm in width and protruded numerous filopodia (Fig.6A). An AFM height-mode image revealed that the growth cone P-domain was approx. 0.5 μm in thickness and the raised C-domain around 1.0 μm in thickness (Fig.6A, B). An error signal AFM image highlighted the morphology of the same growth cone with respect to the lanes of the underlying 5 μm period grating (Fig.6B). A live growth cone viewed using IRM and aligning on a 12-μm adhesive track had multiple focal adhesions beneath the C-domain which appeared to delineate the borders of the underlying tracks (Fig.6C).

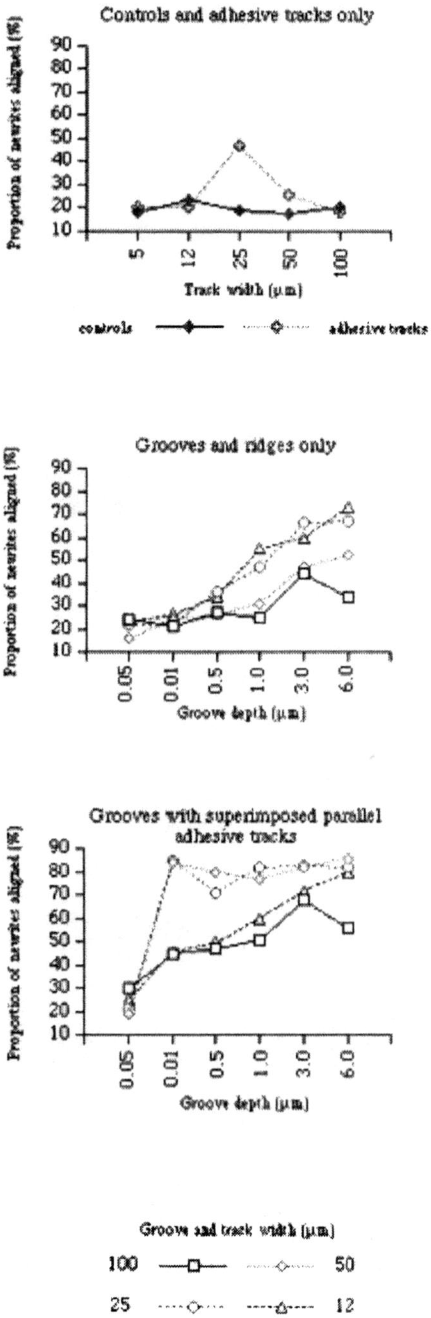

Fig. 3. Line graphs showing the proportion of neurites aligned by different microfabricated substrates

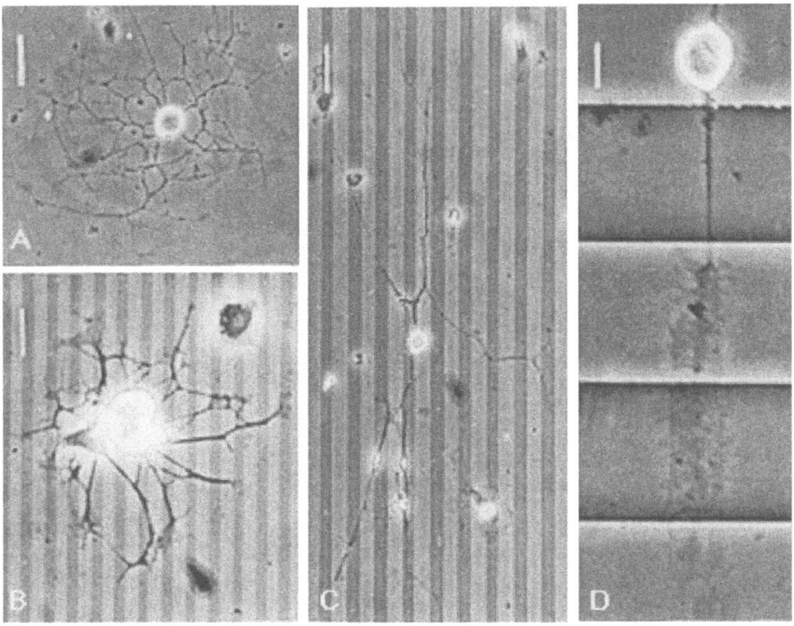

Fig. 4. A A phase-contrast photomicrograph of a DRG nerve cell after 24 h growth on a planar surface. The cell is typically multipolar, highly branched and has many growth cones. *Scale bar* = 30 μm. **B** A DRG cell after 4 h growth on a 5-μm pitch, 1.0-μm-deep topographic grating with superimposed parallel adhesive tracks (see Fig. 1C). There is no evidence of neurite alignment. *Scale bar* = 15μm. **C** A DRG cell after 24 h aligning to 12-μm pitch, 1.0-μm-deep grooves with superimposed parallel adhesive tracks (see Fig. 1C). *Scale bar* = 48 μm. **D** A cell preferentially aligning to adhesive tracks superimposed orthogonally over grooves (see Fig. 1D). Note the elongated growth cone spanning the width of a groove by following the adhesive track. *Scale bar* = 10 μm

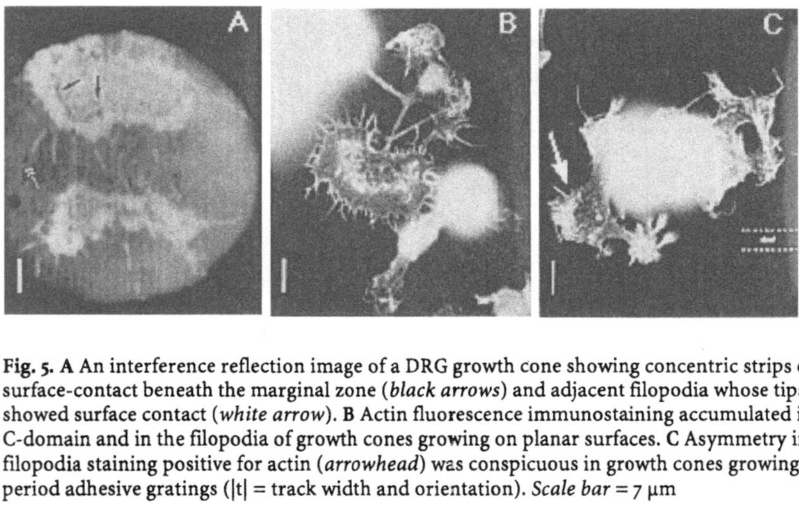

Fig. 5. A An interference reflection image of a DRG growth cone showing concentric strips of surface-contact beneath the marginal zone (*black arrows*) and adjacent filopodia whose tips also showed surface contact (*white arrow*). **B** Actin fluorescence immunostaining accumulated in the C-domain and in the filopodia of growth cones growing on planar surfaces. **C** Asymmetry in filopodia staining positive for actin (*arrowhead*) was conspicuous in growth cones growing on 5-μm period adhesive gratings (|t| = track width and orientation). *Scale bar* = 7 μm

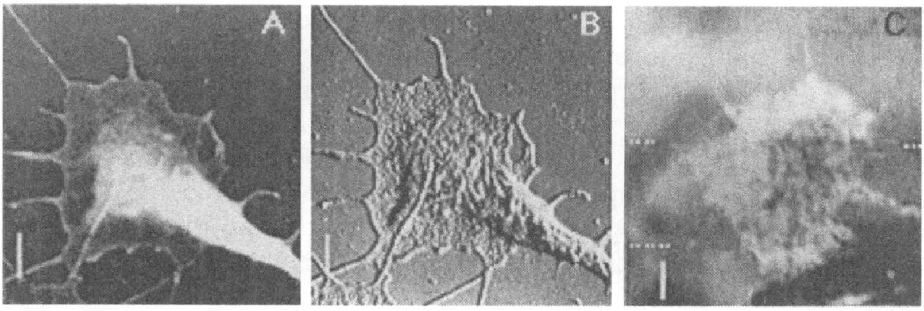

Fig. 6. A An AFM height-mode image of a growth cone. The filopodia, flattened P-domain and raised C-domain are clearly recognisable. *Scale bar* = 2 μm, *height bar* = 0–2 μm. **B** An AFM error-signal image showing the position and morphology of a growth cone relative to the underlying adhesive tracks after 4 h in culture. *Scale bar* = 2 μm. **C** An IRM image of a growth cone aligned to an adhesive track after 24 h in culture. Clusters of focal contact have accumulated beneath the C-domain. These appeared to delineate the edges of the underlying tracks (*dotted lines*). *Scale bar* = 5 μm

Grooves and ridges only

Neurites did not align to topographic gratings of any pitch until groove depth exceeded 1 μm (Fig.3). Cells growing in 12-μm-wide grooves were consistently more highly aligned than those in 25-, 50- or 100-μm pitch grooves at 1, 3 and 6 μm depth. Maximum neurite alignment was seen in 12-μm-pitch, 6.0-μm-deep grooves (Fig.7A).

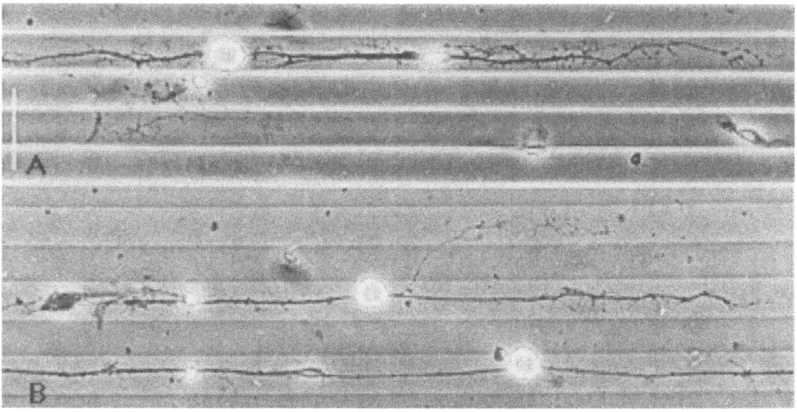

Fig. 7. A A photomicrograph of a highly aligned DRG nerve cell growing in a 12-μm-wide, 6-μm-deep groove (no superimposed adhesive tracks). **B** Two nerve cells growing in adjacent 12-μm-wide, 3-μm-deep grooves with superimposed parallel adhesive tracks (see Fig. 1C). Note the typical bipolar linear morphology that is characteristic of guided cells. *Scale bar* = 24 μm

Grooves and ridges with superimposed parallel adhesive tracks

Neurites did not appear to be aligned to these substrates after 4 h in culture (Fig.4B). After 24 h, the neurites were highly aligned on substrates comprised of grooves and ridges with superimposed parallel adhesive tracks of matched pitch

(Fig.4C), except on the shallowest gratings (Fig.3). There was no difference between the proportion of neurites aligned to grooves and ridges only and to superimposed parallel topographic and adhesive gratings at the shallowest groove depth (50 nm). Neurites were consistently more highly aligned on 25-μm and 50-μm pitch gratings than on 12-μm and 100-μm gratings for grooves deeper than 100 nm. Equal proportions of neurites were aligned on 12-, 25- and 50-μm wide, 6-μm-deep grooves. Aligned cells were usually positioned in the centre of the groove/adhesive track (Fig.7B), and even these cells occasionally had branched primary neurites (Fig.9B).

Grooves and ridges with superimposed orthogonal adhesive tracks

The proportion of neurites aligning to superimposed orthogonal adhesive tracks was greatest on the shallowest grooves and lowest on the deepest grooves (Fig.8). There was significant negative correlation ($r_p = -0.963$; $P<0.01$) between the proportion of neurites aligned to the tracks and to the grooves with increasing groove depth. Cells aligning to the orthogonal adhesive tracks often spanned numerous grooves and ridges ;Fig.9A). Bipolar cells were again most frequently found on 25- and 50-μm-wide tracks (Fig.10D), but cells with branching neurites were also seen (Fig.10C). A conspicuous feature of these bipolar cells was very large, elongated and flattened structures which strongly resembled growth cones and which occupied the full width of 12- and 25-μm-wide tracks (Fig.10D). These were often elongated along the tracks and spanned consecutive grooves and ridges (Fig.4D). Cells growing on the 6-μm-deep substrate, which appeared under phase-optics to have primary neurites aligned to the grooves, also had a highly branched terminal neurite arbour which was revealed under SEM to be oriented to the orthogonal adhesive tracks (Fig.10A, B).

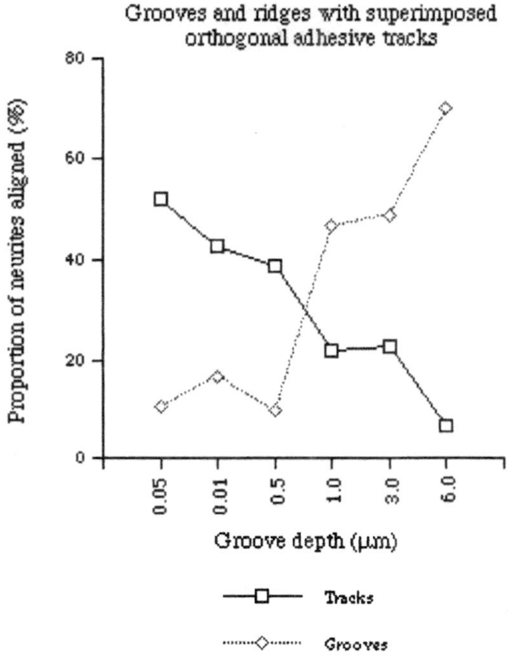

Fig. 8. Graphs showing the proportion of neurites aligned to adhesive tracks superimposed orthogonally over grooves of matched pitch

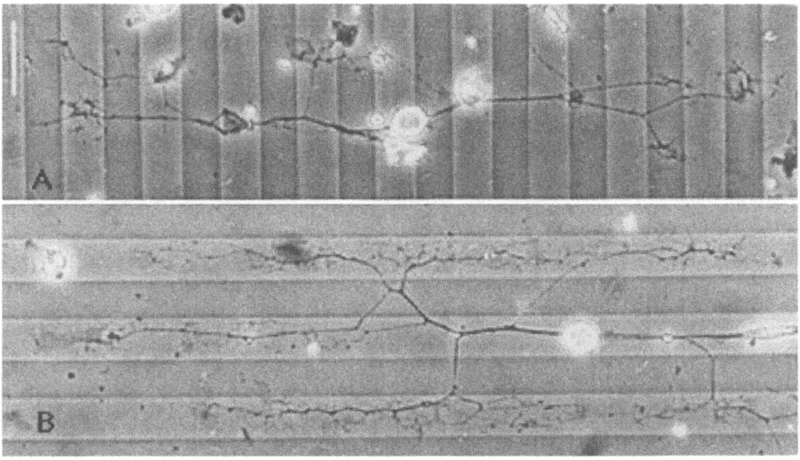

Fig. 9. A A DRG nerve cell aligned on a 25-µm-wide adhesive track superimposed orthogonally over 25-µm-wide, 3-µm-deep grooves. **B** A cell aligned to 25 µm wide, 3-µm-deep groove with a superimposed parallel adhesive track. *Scale bar* = 20 µm

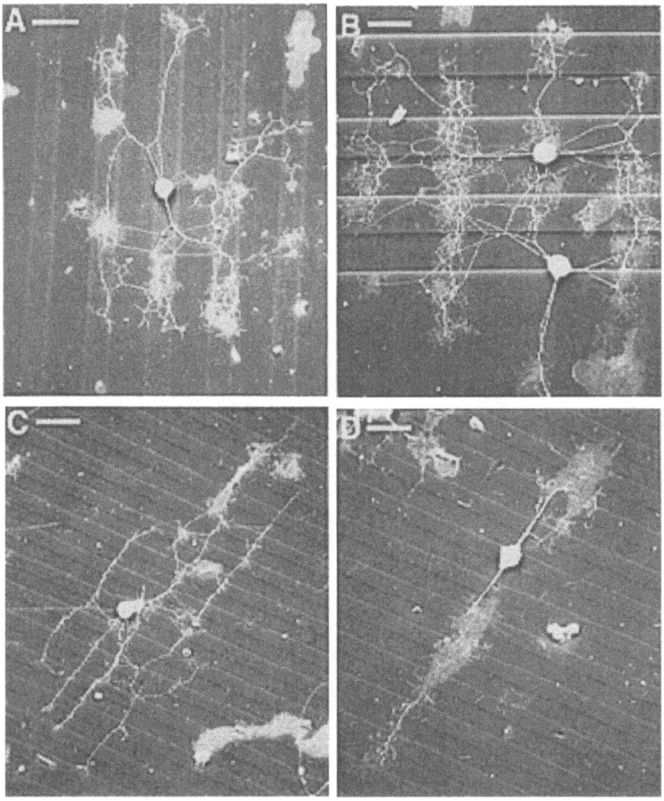

Fig. 10. A An SEM of a DRG nerve cell growing on a 12 μm period adhesive grating. Note the highly branched terminal neurite arbours aligned to vitronectin lanes. Primary neurite branching is unusually infrequent. *Scale bar* = 16 μm. **B** A cell aligned to an adhesive grating superimposed orthogonal to 3-μm-deep grooves of matched width. The terminal neurites are aligned to the adhesive grating and the primary neurites to the grooves. *Scale bar* = 10 μm. **C** A cell on a similar surface as in **B**, but with shallower grooves. Primary neurites are now primarily oriented to the adhesive track. *Scale bar* = 25 μm. **D** Cells aligned to adhesive tracks orthogonal to grooves often had elongated growth cones which spanned successive grooves and ridges. *Scale bar* = 20 μm

Discussion

Directional information is supplied to nerve cell growth cones by multiple extrinsic guidance cues. These cues are likely to act in concert to produce a resultant directional response from a cell. This study has examined the orienting effects of two model morphogenetic guidance cues on rat primary DRG nerve cell growth. In order to identify possible synergistic or hierarchical interaction in their effects, adhesive tracks and topographic anisotropy were presented to cells independently, simultaneously in parallel and simultaneously but orthogonally opposed.

The patterned adhesive surfaces consisted of tracks of laminin separated by tracks of vitronectin. Vitronectin-mediated cell adhesion to serum-treated surfaces is known to predominate after several hours in culture and focal contacts which form contain a vitronectin receptor that co-localises with talin, vinculin and stress fibre termini (Fath et al. 1989). Qualitative observations on DRG cells indicated that neurites did not orient to adhesive tracks of any pitch after 4 h. This is in agreement with previous observations that patterned laminin substrata have little effect on the position of neuritogenesis from nerve cell soma (Britland and McCaig 1994). Sensory neurons degrade fibronectin culture substrata during early neurite outgrowth (McGuire and Seeds 1990). This type of degradation process may have destroyed any directional information in the form of patterned proteins. It is possible that directional signals may not be utilised until growth cone motility is well under way. In effect, nerve cells may have to come to terms with their microenvironment before guidance can be induced.

Neurites were aligned to 25-μm-wide vitronectin tracks after 24 h. The neurites of highly aligned cells were frequently positioned in the centre of the vitronectin lanes, not at the edges. This implies that geometric constraints were imposed on growth cone advance by the adhesive pattern. Growth cones, viewed using phase-optics, extended filopodia over to adjacent laminin tracks and IRM confirmed that surface contact was made. Gomez and Letourneau (1994) have suggested that spatially localised ECM components, laminin/fibronectin borders in their experiments, direct axon outgrowth through mechanisms involving intracellular signalling through filopodial contact with alternating substrata. Here the contacts made by lateral filopodia were frequently transient and were unaccompanied by the strips of contact under the marginal zone (see Rochlin et al. 1995) which were characteristic of growth cones exploring homogenous surfaces. Clusters of focal contacts were observed underneath the C-domain which delineated the edges of the tracks. The route of surface-signal transduction through integrins to the cytoskeletal machinery might therefore be housed within the body of the growth cone. Matrix proteins are known to induce phosphorylation of intercellular proteins involved in focal adhesion (Burridge et al. 1992). The growth cones may have been receiving additional stimuli from adjacent tracks. Studies using neuroblastoma cells have shown that filopodia form contacts equally on more and less adhesive regions of chemically patterned substrates (Soekarno et al. 1993). These authors suggested that cell guidance on such surfaces is due to preferential adhesion to more adhesive areas. Transient contact with laminin has been shown to provide directional information and maintain the velocity of growth cone translocation on fibronectin (Kuhn et al. 1995). Therefore laminin might have affected lamellipodial advance along the vitronectin lanes by signalling through lateral filopodia (Kleitman and Johnson 1989). This effect would reinforce the tendency for the linear neurite to advance straight ahead (Katz 1985). Growth cones aligning to adhesive borders (Clark et al.1993), or edges of very wide tracks, may do so by utilising primarily filopodial steering. Premature detachment of filopodia can redirect growth cone advance towards remaining attached filopodia (Wessels and Nuttall 1978).

AFM revealed nanometric steps between the lanes of the adhesive grating.

Whether this topography contributed to the orienting response of DRG cells on adhesive grating was not established. Two observations suggest that these nanometric steps may not have played a part. Firstly, there was dimensional mismatch between the widths of grooves and tracks which induced maximum neurite alignment. Secondly, the cells had already exhibited little response to gratings of 50 and 100 nm depth.

Tension develops in extending neurites resulting from exploratory activity of the preceding growth cone (Lamoureuxetetetet et al. 1989). For growth cones to pull on the neurite they must obtain purchase from the substrate. IRM showed that growth cones may pull from transient anchorages in the form of an adhesion belt located underneath the marginal zone. This contact was coincident with the periphery of actin staining in the C-domain. Myosin IIB has been found to be concentrated in the marginal zone and also co-localised with F-actin bundles throughout the C-domain of growth cones (Rochlin et al. 1995). This myosin IIB/actin/surface contact interaction may comprise the motor that generates the tension required to extend the neurite.

Neurites did not align to topographic gratings of 12, 25, 50 and 100 μm pitch (with *uniform* surface chemistry) when groove depth was less than 1 μm. This is consistent with the orienting behaviour of adult mouse DRG cells (Hirono et al. 1988) and chick embryo cerebral hemisphere cells (Clarket al. 1990) on similar surfaces. Rat optic nerve oligodendrocytes and hippocampal neurons (Webb et al. 1995), and *Xenopus* spinal cord neurons (Rajnicek et al. 1994), all align to 1- to 4-μm-wide, less than 1-m-deep grooves. The orienting response of nerve cells to topography may be dependent on cell type (see Rajnicek et al. 1995). Chick and rat cerebellar hemisphere cells (Clark et al. 1991) and rat hippocampal cells (Webb et al. 1995) are not aligned by 100- to 400&-nm-deep ultrafine gratings. This variability in orienting response to very similar features may be a product of at least two separate developmental influences, cell genotype and extrinsic morphogenetic factors.

Guidance of neurite extension on topographic gratings was enhanced by superimposed parallel adhesive tracks. Virtually all of the neurites were aligned in 2- μm-wide and 6-μm-deep grooves containing 25-μm-wide vitronectin tracks. Two lateral obstacles must have prevented turning of the growth cone, the groove wall and an adhesive boundary. The orienting effect of the combined cues sometimes appeared to be greater than the sum of the individual cues. Exposure to one cue may have altered the cells' sensitivity to a subsequent cue. It is also possible that previously unresponsive cells may have been recruited under the combined influence of both cues. Whether adhesive and topographic guidance is affected through common or separate pathways within the growth cone remains to be established. Adhesive and topographic guidance cues are therefore capable of interacting synergistically to enhance the orienting behaviour of cells.

During morphogenesis, certain cues may be weighted at appropriate points during gestation. This theory was tested by superimposing adhesive tracks orthogonally over grooves. The model cues appeared to interact hierarchically in steering neurite extension, but neither was consistently dominant. Neurites aligned preferentially to adhesive tracks positioned perpendicularly over shallower grooves, but aligned increasingly to grooves with increasing depth. Neurite alignment to adhesive tracks was negatively correlated with groove depth. A point was reached at groove depths of between 0.5 and 1.0 μm where the cues were statistically balanced. That is, similar numbers of primary neurites aligned to grooves and to tracks. Some cells appeared to have 'balanced morphologies' with primary neurites arranged in a cruciform configuration around the cell soma.

A remarkable feature of this experiment was extremely elongated growth cones which were aligned to the perpendicular adhesive tracks and bridged successive grooves and ridges. Traversing these obstacles must have required severe cytoskeletal distortion within the C-domain of growth cones. Bends in the growth cone cytoskeleton did not appear to restrict their motility, nor were growth cones observed to accumulate at ridge edges. Cells which appeared under phase-optics to be aligned to deep grooves had peripheral regions of intense neuritogenic activity supported by primary neurites with very few branches. Differential effects of concanavalin-A and laminin on branching in different parts of the neuritic tree of the same cell have been described before (Grumbacher-Reinert 1989), but not simultaneous divergent orientation responses and branching patterns. There remained a small number of cells which showed no orienting response. DRG neurons can be subdivided on the basis of their neurotransmitter phenotype and neurotrophic dependency. The same may be true of their response to guidance cues.

These experimental cues were models which could, in some senses, be regarded as physiologically inaccurate. However, this study has demonstrated that growth cones can detect and integrate directional information presented in the form of model adhesive and topographic morphogenetic guidance cues.

Acknowledgements. This work was funded, in part, by the EPSERC.

References

Britland, S., McCaig, C.D. (1994) The response of cultured *Xenopus* neurites to simultaneous electrical and adhesive guidance cues. *Exp. Cell Res.* (in press)

Burridge, K., Petch, L.A., Romer, L.H. (1992) Signals from focal adhesions. *Curr. Biol.* **2**: 537–540

Clark, P., Connolly, P., Curtis, A.S.G., Dow, J.A.T., Wilkinson, C.D.W. (1990) Topographical control of cell behaviour: II. Multiple grooved substrata.*Development* **108**: 635–644

Clark, P., Connolly, P., Curtis, A.S.G., Dow, J.A.T., Wilkinson, C.D.W. (1991) Cell guidance by ultrafine topography in vitro. *J. Cell Sci.* **99**: 73–77

Clark, P., Britland, S.T., Connolly, P. (1993) Growth cone guidance and neuron morphology on micropatterned laminin surfaces. *J. Cell Sci.* **105**: 203–212

Connolly P, Britland ST, Clark P, Monaghan W, Moores GM (1991) Microelectronic and nanoelectronic interfacing techniques for biological systems. In: Seyfried, P., Kunzmann, H., McKeown, P., Weck, M. (eds.) *Progress in precision engineering.* Springer, Berlin Heidelberg New York, pp. 225–230

Curtis, A.S.G. (1964) The mechanism of adhesion of cells to glass.*J. Cell Biol.* **20**: 199–215

Duband, J.L. (1990) Extracellular matrix and embryonal morphogenesis: role of fibronectin in cell migration. *Reprod. Nutr. Dev.* **30**: 379–395

Ebendal, T. (1982) Orientational behaviour of extending neurites. In: Bellairs, R., Curtis, A.S.G., Dunn, G. (eds.) *Cell behaviour* (Eds) Cambridge University Press, Cambridge, pp. 281–297

Ebendal, T., Jakobsen, C,-O. (1977) Tissue explants affecting extension and orientation of axons in cultured chick embryonic ganglia. *Exp. Cell Res.* **105**: 379–387

Fath, K.R., Edgell, C.J., Burridge, K. (1989) The distribution of distinct integrins in focal contacts is determined by the substratum composition. *J. Cell Sci.* **92**: 67–75

Galileo, D.S., Majors, J., Horwitz, A.F., Sanes, J.R. (1992) Retrovirally induced antisense integrin RNA inhibits neuroblast migration in vivo. *Neuron***9**: 117–1131

Gomez, T.M., Letourneau, P.C. (1994) Filopodia initiate choices made by sensory neuron growth cones at laminin fibronectin borders in vitro. *J. Neurosci.* **14**: 5959–5972

Grumbacher-Reinert, S. (1989) Local influence of substrate molecules in determining distinctive growth-patterns of identified neurons in culture.*Proc. Natl. Acad. Sci. USA* **86**: 7270–7274

Hirono, T., Torimitsu, K., Kawana, A., Fukada, J. (1988) Recognition of artificial microstructures by sensory nerve fibres in culture. *Brain Res.* **446**: 189–194

Holder, N., Clarke, J.D.W., Tonge,D. (1987) Pathfinding by dorsal column axons in the spinal cord of the frog tadpole. *Development* **99**: 577–587

Hynes, R.O., Patel, R., Miller, R.H. (1986) Migration of neuroblasts along pre-existing axonal tracks during prenatal cerebellar development. *J. Neurosci.* **6**: 867–876

Hynes, R.O., Lander, A.D. (1992) Contact and adhesive specificities in the associations, migrations, and targeting of cells and axons. *Cell* **68**: 303–322

Katz, M.J. (1985) How straight do axons grow?. *J. Neurosci.* **5**: 589–595

Kleitman, N., Johnson, M.I. (1989) Rapid growth cone translocation on laminin is supported by lamellipodia not filopodial structures. *Cell Motil. Cytoskel.* **13**: 288–300

Kuhn, T.B., Schmidt, M.F., Kater, S.B. (1995) Laminin and fibronectin guideposts signal sustained but opposite effects to passing growth cones. *Neuron* **14**: 275–285

Lamoureaux, P., Buxbaum, R.E., Heidemann, S.R. (1989) Direct evidence that growth cones pull. *Nature* **340**: 159–162

Lance-Jones, C., Dias, M. (1991) The influence of presumptive limb connective tissue on motorneuron axon guidance. *Dev. Biol.* 143:92–110

Leisi, P. (1990) Extracellular matrix and neuronal movement. *Experientia* **46**: 900–907

Lofberg, J., Ahlfors, K., Fallstrom, C. (1980) Neural crest migration in relation to extracellular matrix organisation in the developing axolotl trunk. *Dev. Biol.* **75**: 148–167

McGuire, P.G., Seeds, N.W. (1990) Degradation of underlying extracellular matrix by sensory neurons during neurite outgrowth. *Neuron* 4: 633–642

Nakata, I., Nakatsuji, N. (1991) Rodent CNS neuroblasts exhibit both perpendicular and parallel contact guidance on the aligned parallel neurite bundle. *Development* **112**:581–590.

Nakata, I., Kawana, A., Nakatsuji, N. (1993) Perpendicular contact guidance of CNS neuroblasts on artificial microstructures. *Development* **117**: 401–408

Newgreen, D.F. (1989) Physical influences on neural crest cell migration in avian embryos: contact guidance and spatial restriction. *Dev. Biol.* **131**: 136–148

Rajnicek, A.M., Britland, S., McCaig, C. (1994) Differential orientation of *Xenopus* neurites and rat hippocampal neurites growing on grooved surfaces in vitro. *J. Physiol. (Lond.)* **481**: 25P

Rajnicek, A.M., Robinson, K.R., McCaig, C.D. (1994) The influence of the growth surface on the orientation of cultured *Xenopus* neurites exposed to weak electric fields. *J. Physiol. (Lond.)* **481**: 26P

Rajnicek, A.M., Wallace, M.N., Britland, S., McCaig, C.D. (1995) CNS neurones from frog and rat grow orthogonally on the same micropatterned substrates. *Soc. Neurosci. Abstr.* **21**: 322.1

Rakic, P. (1990) Principles of neural cell migration. *Experientia* **46**: 882–891

Riggott, M.J., Moody, S.A. (1987) Distribution of laminin and fibronectin along peripheral trigeminal axon pathways in the developing chick. *J. Comp. Neurol.* **258**:580–596

Rochlin, W., Itoh, K., Adelstein, R.S., Bridgman, P.C. (1995) Localisation of myosin II A and B isoforms in cultured neurons. *J. Cell Sci.* **108**: 3661–3670

Rogers, S.L., Edson, K.J., Letourneau, P.C., McLoon, S.C., (1986) Distribution of laminin in the developing peripheral nervous system of the chick. *Dev. Biol.* **113**: 429–435

Rolfe, P. (1995) Editorial. *J. Cell. Eng.* **1**: 1–2

Silver, J., Rutishauser, U. (1984) Guidance of optic axons in vivo by a preformed adhesive pathway on neuroepithelial endfeet. *Dev. Biol.* **106**: 485–499

Soekarno, A., Lom, B., Hockberger, P.E. (1993) Pathfinding by neuroblastoma cells in culture is directed by preferential adhesion to positively charged surfaces. *Neuroimage* **1**: 129–144

Tang, J., Landmesser, L., Rutishauser, U. (1992) Polysialic acid influences specific pathfinding in avian motor neurons. *Neuron* **8**: 1031–1044

Vielmetter, J., Stolze, B., Bonhoeffer, F., Stuermer, C.A.O. (1990) In vitro assay to test differential substrate affinities of growing axons and migratory cells. *Exp. Brain Res.* **81**: 283–287

Webb, A., Clark, P., Skepper, J., Compston, A., Wood, A. (1995) Guidance of oligodendrocytes and their progenitors by substratum topography. *J. Cell Sci.* **108**: 2747–2760

Wessels, N.K., Nuttall, R.P. (1978) Normal branching, induced branching, and steering of cultured parasympathetic motor neurons. *Exp. Cell Res.* **115**: 111–122

Exp. Biol. Online 1:3 (1996)

Review

Physiological states and intracardiac shunting in non-crocodilian reptiles

James W. Hicks[1] and Egle Krosniunas[2]

[1] Department of Ecology and Evolutionary Biology, University of California, Irvine, Irvine, CA 92717, USA (e-mail: JHICKS@UCI.EDU; WWW: http://compphys.bio.uci.edu)

[2] Department of Biologic and Materials Science, School of Dentistry, University of Michigan, Ann Arbor, MI 48109-1078, USA (e-mail: egle@umich.edu)

Correspondence to: J. Hicks (e-mail: JHICKS@UCI.EDU; WWW: http://compphys.bio.uci.edu)

Received: 16 April 1996 / Accepted: 13 May 1996
Published: May 21, 1996

Abstract. Historically, the evolution of the vertebrate heart was portrayed as a continuous progression from the two-chambered heart of fish to the completely divided, four-chambered heart of birds and mammals. The incompletely divided ventricle of non-crocodilian reptiles and the resulting central mixing of oxygen-rich and oxygen-poor blood (cardiac shunting) was viewed as an intermediate phylogenetic step that was functionally inefficient compared to the circulation of "higher" vertebrates. This perception changed when studies showed that the degree of mixing within the reptilian heart is influenced by the ventilatory state of the animal and is regulated by the autonomic nervous system. The ability to control cardiac shunts is hypothesized to provide several unique physiological advantages to reptiles and has given rise to the notion that cardiac shunting is a highly derived condition, representing a successful adaptation for animals with low metabolic rates and intermittent breathing patterns. The modern view of the reptilian circulation is based, to a large measure, on a relatively small number of observations during very specific types of behaviors or physiological states, being primarily restricted to the description of cardiovascular changes associated with intermittent lung ventilation. The goals of this review are to summarize the studies that have described cardiac shunting patterns in non-crocodilian reptiles during a variety of physiological states and to emphasize the need for additional research on this important group of vertebrates.

Key words: Reptiles – Cardiovascular – Cardiac shunts – Diving – Exercise – Respiration

Introduction

The evolution of the vertebrate heart was historically portrayed as a continuous progression from the two-chambered heart of fish to the completely divided, four-chambered heart of birds and mammals (Burggren 1985). Consequently, the incompletely divided ventricle of non-crocodilian reptiles and the resulting central mixing of oxygen-rich and oxygen-poor blood (cardiac shunting) was viewed as an intermediate phylogenetic step that was functionally inefficient compared to the circulation of "higher" vertebrates (Foxon 1955). This perception changed when several studies showed that the degree of mixing within the reptilian heart (cardiac shunting) is influenced by the ventilatory state of the animal and is probably regulated (White and Ross 1966; White 1968, 1970; Shelton and Burggren 1976; Burggren and Shelton 1979; Burggren et al. 1989; White et al. 1989; Hicks and Malvin 1992; Comeau and Hicks 1994; Hicks and Comeau 1994; Hicks 1994). The ability to control cardiac shunts is hypothesized to provide several unique physiological advantages to reptiles (White 1976; Johansen and Burggren 1980; Burggren 1985; Table 1)

Table 1. Physiological function of cardiac shunts in non-crocodilian reptiles

Physiological function	Type of shunt	Reference
Saves cardiac energy	R-L	Burggren 1987
"Meters" lung oxygen stores	R-L	Burggren and Shelton 1979
Reduces plasma filtration into the lungs	R-L	Burggren 1982
Reduces CO_2 flux into the lungs	R-L	White 1985
Facilitates warming	R-L	Baker and White 1970
Minimizes V/Q mismatching	L-R	Wood 1984
Facilitates CO_2 elimination into the lung	L-R	White 1985

These apparent physiological functions have given rise to the opinion that the cardiovascular system of reptiles is a highly derived condition, representing a successful adaptation for animals with low metabolic rates and intermittent breathing patterns (Burggren and Warburton 1994). While the proposed physiological functions of cardiac shunts are intuitively appealing and often appear in standard textbooks of comparative physiology (e.g., Withers 1992), most of the them either lack direct experimental support, have not been evaluated in detail or are based on a limited number of behavioral observations or physiologcial states. Based on this background, this review will briefly describe the anatomy of the heart of non-crocodilian reptiles, discuss the mechanisms that result in intracardiac shunts, the factors that may regulate cardiac shunting and discuss the effects of various physiological stimuli on cardiac shunting patterns.

Cardiac anatomy

The cardiac anatomy of turtles, snakes and lizards has been extensively studied for over 100 years (see Griel 1903, for review of 19th century work; O'Donoghue 1918; Rau 1924; Mathur 1946; White 1959, 1968; Webb et al. 1971, 1974; Van Mierop and Kutsche 1981, 1985). First, we will briefly review the basic cardiac anatomy of non-crocodilian reptiles and provide a brief conceptual overview of cardiac shunting. Detailed descriptions of cardiac anatomy and cardiac shunting can be found in papers by Van Mierop and Kutsche (1981, 1985) or a recent review by Hicks (1993). The heart consists of two atrial chambers (RAt and LAt), separated by a complete septum and a single ventricle. There are two atrioventricular ostia with each ostia having a set of cup-shaped valves (AV valves). A distinctive feature of the ventricular anatomy is the presence of a septum-like structure, called the *Muskelleiste* (Brücke 1852) or muscular ridge (MR). The MR originates from the ventral ventricular wall, runs from apex to base and divides the ventricle into two main chambers; a smaller right ventro lateral chamber, the cavum pulmonale (CP) and a larger left dorso lateral chamber, the cavum dorsale (Van Mierop and Kutsche 1981, 1985). The dorsolateral border of the MR is free, resulting in the potential for blood flow between the CP and cavum dorsale. Some reptiles have a prominent vertical septum originating from the dorsal aspects of the MR and subdividing the cavum dorsale into the cavum arteriosum (CA) and the cavum venosum (CV) (Van Mierop and Kutsche 1985; Fig. 1). A small passage over the anterior border of the vertical septum and just below the atrio-ventricular valves is termed the interventricular canal (White 1959). The vertical septum is highly developed in *Python* and in *Varanus* (Webb et al. 1971). However, not all reptiles have a recognizable vertical septum. The hearts of *Chelonia mydas, Pseudemys floridana, Kinosternon sp., Iguana iguana, Boa constrictor*, and *Crotalus adamanteus* do not show a discrete vertical septum (Van Mierop and Kutsche 1985). Rather the cavum dorsale of these species contains many sagittally oriented septa, similar to the ventricular septa observed in anurans (Van Mierop and Kutsche 1985).

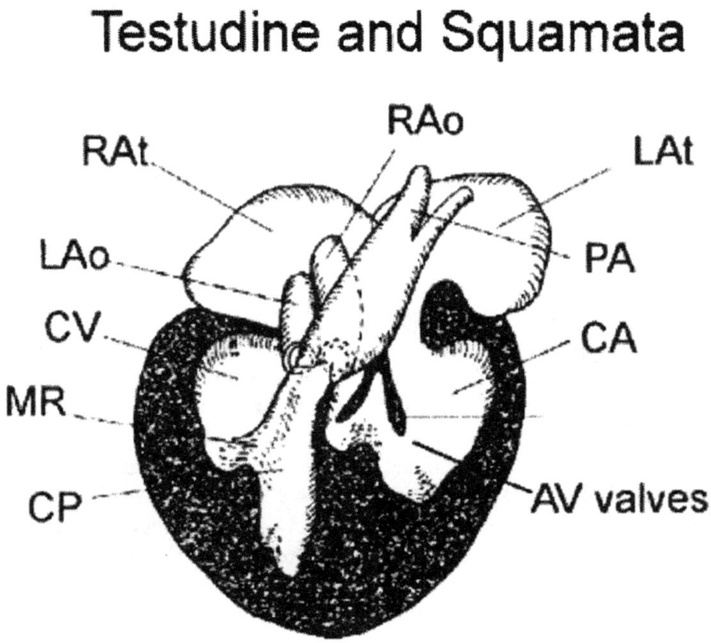

Fig. 1. Schematic diagram of a reptilian heart: *RAt*, right atrium; *LAt*, left atrium; *RAo*, right aortic arch; *LAo*, left aortic arch; *PA*, pulmonary artery; *AV valves*, atrioventricular valves; *CV*, cavum venosum; *CP*, cavum pulmonale; *CA*, cavum arteriosum; *MR*, muscular ridge. Schematic is redrawn from original by F.N. White

In all reptiles, three great vessels arise from the ventricle: the pulmonary artery (PA), the right aortic arch (RAo) and left aortic arch (LAo). The PA emerges to the left of the two aortic arches and originates from the CP. The RAo and LAo arise from the CV. Immediately after emerging from the heart, the RAo divides into subclavian and carotid arteries, and a third branch of the RAo (brachiocephalic) unites with the LAo mediocaudally to form the dorsal aorta.

The spatial relationship of the three cava and the dynamics of the MR in the heart of *Trachemys scripta* are shown in the attached Quick Time(R) movie (Fig. 2). In this movie, a 5 MHz pediatric ultrasonic probe was placed against the ventricular wall on the ventral side near the apex of the heart and the ventricle was imaged for several seconds. A single sweep of the bar, located on the bottom of the image, is equal to 1 s. Note that during systole the MR adpresses against the dorsal wall of the ventricle and separates the CP from the CV/CA.

Anatomical locations were verified by the injection of microbubbles.

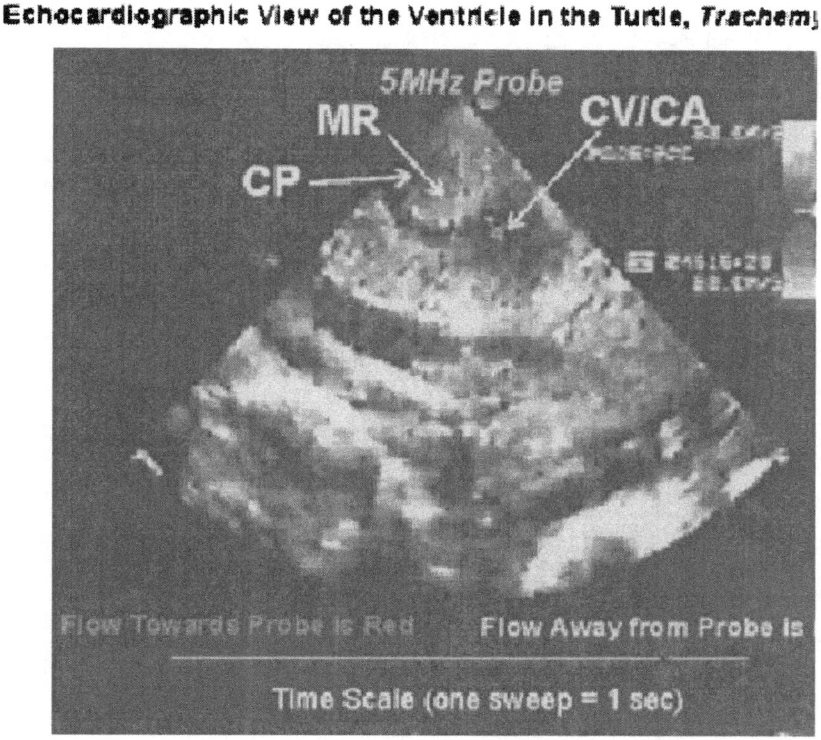

Fig. 2. Echocardiographic view of the ventricle in the turtle, *Trachemys scripta*. Ultrasonic flow probe (5 MHz; located at the top of the figure) was placed on the ventral surface of the heart, near the apex. *CP*, cavum pulmonale; *CV/CA*, cavum venosum and arteriosum; *MR* muscular ridge. Accompanying QT movie (5 s ~ 900 Kbytes) shows the dynamic relationship between the three ventricular subchambers and the muscular ridge during the cardiac cycle. A single sweep of the time bar (located on the bottom of the image) is equal to 1 s

Cardiac shunting

The anatomy of the heart and great vessels results in the potential for cardiac shunts. The term cardiac shunt denotes blood flow that bypasses the lung or the systemic circuit and represents alternative pathways to the "normal circulation." Cardiac shunts are defined by their direction as right-to-left (R-L) and left-to-right (L-R). A R-L shunt represents recirculation of systemic venous blood (oxygen poor) back into the systemic arterial circulation. Since this blood bypasses the lungs, the shunted portion has no chance of being oxygenated. In contrast, a L-R shunt represents the recirculation of pulmonary venous blood (oxygen rich) to the pulmonary circulation (Fig. 3). In the presence of unidirectional cardiac shunts, pulmonary (Q_{pul}) and systemic (Q_{sys}) blood flows are not equal. During conditions of L-R shunt, Q_{pul} exceeds Q_{sys}, while during R-L shunting Q_{sys} exceeds Q_{pul}. In both cases, the difference between Q_{pul} and

Q_{sys} (Q_{pul}-Q_{sys}) represents the net shunt flow (Q_{shunt}). The level of shunting can thus be expressed either as net shunt, the ratio of pulmonary to systemic blood flow (Q_{pul}/Q_{sys}) or shunt fraction (Q_{shunt}/Q_{pul} or Q_{sys}) (Hicks 1994). In non-crocodilian reptiles bidirectional shunting can occur (Heisler and Glass 1985; Ishimatsu et al. 1988; Hicks and Comeau 1994). In hearts with significant bidirectional shunting, one must be cautious in interpreting net shunt flows since they tend to underestimate the actual shunt flows (Hicks 1994). Bidirectional shunting can be quantified by measurements of blood oxygen content from arterial and cardiac sites (Ishimatsu et al.1988).

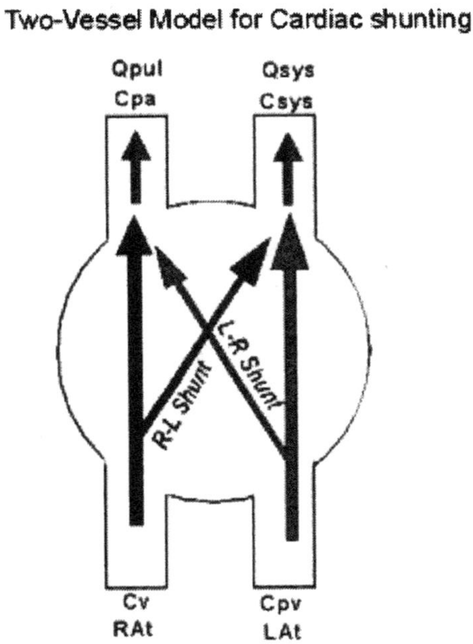

Fig. 3. The conventional two-vessel model for analysis of intracardiac shunting. *C*, O_2 content; *Q*, blood flow rate; pul, pulmonary; sys, systemic; *pa*, pulmonary artery; *pv*, pulmonary vein; *a*, arterial; *LAt*, left atrium; *RAt*, right atrium. *Blue arrows*, deoxygenated blood; *red arrows*, oxygenated blood, *combination blue-red*; mixed blood. (Diagram redrawn from Ishimatsu et al. 1988)

Cardiac shunts affect the blood oxygen levels in the systemic and pulmonary circulations. A R-L shunt lowers the arterial oxygen content (C_aO_2) and a L-R shunt raises the pulmonary arterial oxygen content ($C_{pa}O_2$) (Fig. 3). Recent studies quantified the effects of Q_{pul}/Q_{sys} on systemic arterial oxygen levels (Hicks and Comeau 1994). As Q_{pul}/Q_{sys} increases to values greater than 1.0, the C_aO_2 and $C_{pa}O_2$ increase. A large L-R shunt can potentially reduce C_aO_2 through changes in the *V*/*Q* distributions or through the development of regional pulmonary edema (Lister and Pitt 1983). However this has not been demonstrated in turtles under such hemodynamic conditions (Hopkins et al., in press).

The changes in the blood oxygen content can be used to calculate the size of cardiac shunting (see Ishimatsu et al. 1988; Hicks 1993). It should be emphasized that a reduction in C_aO_2 does not always indicate an increase in R-L shunt. Reduction in C_vO_2 or $C_{pv}O_2$ can reduce systemic oxygen levels even at a constant shunt level. In addition, the potential for the differential distribution of systemic venous blood into the two aortic arches (Fig. 4) complicates the interpretations of arterial blood gases within the context of the standard shunt equation (Ishimatsu et al. 1988; Hicks and Wood 1988; Hicks, 1994; Ishimatsu et al. 1996).

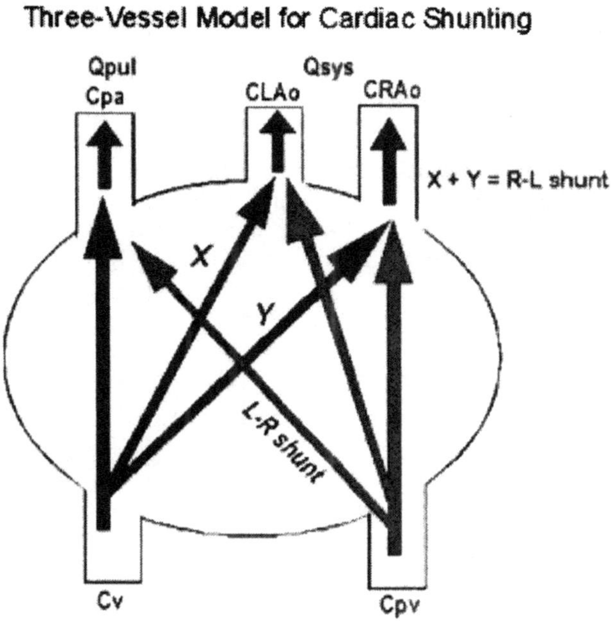

Fig. 4. The three-vessel model for the description of intracardiac flow patterns in non-crocodilian reptiles. *PA*, pulmonary artery; *RAt*, right atrium; *LAt*, left atrium; *RAo*, right aortic arch; *LAo*, left aortic arch. This model shows the potential for differential distribution of deoxygenated blood (*blue arrows*) into the aortic arches. *Blue arrows* deoxygenated blood, *red arrows* oxygenated blood, *combination blue-red* mixed blood (redrawn from Ishimatsu et al. 1988)

Regulation and mechanisms of cardiac shunt

The factors that control cardiac shunting have been studied in only few representatives of extant non-crocodilian reptiles, with two species, the monitor lizard *Varanus exanthematicus* and turtle*Trachemys scripta* receiving the most attention (Steggarda and Essexs 1957; White and Ross 1966; Shelton and Burggren 1976; Burggren and Johansen, 1982; Hicks and Malvin 1992; Hicks and Comeau 1994; Hicks 1994; Hicks et al. 1996). In turtles, the size and direction of the cardiac shunt is primarily determined by factors that control the vascular resistance of the pulmonary and systemic circulations. These factors include adrenergic and cholinergic mechanisms (Luckhardt and Carlson 1921; White 1976; Burggren 1977; Milsom et al. 1977; Berger and Burnstock 1979; Nilsson 1983; Burggren 1985; Lillywhite and Donald 1989; Comeau and Hicks 1994; Hicks and Comeau 1994; Hicks 1994). There is growing evidence that non-adrenergic non-cholinergic (NANC) factors may also affect the vascular resistance in reptiles and thus influence the size and direction of cardiac shunts (Lillywhite and Donald 1989; Conlon et al. 1990; Comeau et al. 1992). In addition to factors that control peripheral vascular resistances, cardiac shunts are also determined by the end-systolic and end-diastolic volumes of blood within the ventricular chambers (Heisler et al. 1983; Heisler and Glass 1985; Hicks and Malvin 1992; Hicks and Comeau 1994; Hicks and Malvin 1996; Hicks et al. 1996; Ishimatsu et al. 1996).

Cardiac shunts and physiological states

We will briefly review what is known about cardiac shunting patterns during various physiological states and, where appropriate will discuss preliminary results from our laboratory. Our discussion will focus primarily on turtles, since the vast majority of work has been on this group.

Cardiac shunting patterns and ventilatory state

There are distinct cardiovascular changes associated with intermittent ventilation in turtles. These changes are referred to as cardiorespiratory synchrony (White et al. 1989).

Apnea. During apnea there is a bradycardia and an increase in the pulmonary vascular resistance (R_{pul}) which leads to a reduction in Q_{pul} and the development of a net R-L shunt (Millen et al. 1964; White and Ross 1966; Shelton and Burggren 1976; White et al. 1989; Wang and Hicks 1996a,b). A small L-R shunt also occurs during apnea (Glass and Heisler 1985; White et al. 1989). This is not surprising since bidirectional shunting occurs within the turtle heart (Hicks and Comeau 1994). The small L-R shunt is detectable only by analysis of blood gases and cannot be measured by blood flow probes (see Hicks 1993 for details). In turtles the size of the R-L shunt is periodically diminished during the apneic period, resulting in a transfer of O_2 from the lung to blood and an increase in arterial O_2 levels (Burggren and Shelton 1979; Burggren et al. 1989). This type of regulation has been termed "oxy-stat" (Burggren et al. 1989).

It has been suggested that a R-L shunt facilitates O_2 uptake from the lung, sequesters CO_2 away from the lung, enhances tissue CO_2 stores and subsequently facilitates O_2 delivery to the tissues (White 1985; Grigg and Johansen 1987). We have recently questioned these benefits (Hicks and Wang 1996) for the following reasons. First, if the length of apnea is within the aerobic limits, then a R-L shunt provides few additional benefits to O_2 exchange, since the transfer of O_2 from lung to blood and blood to tissue must ultimately be equal (Hicks and Wang 1996). Second, sequestration of CO_2 away from the lung does not require a R-L shunt. For example in man, voluntary breath holding results in changes in the pulmonary CO_2 and gas exchange ratio (R) that are qualitatively similar to those observed during apnea in reptiles (Douglas and Haldane 1922; Otis et al.1948; Mithoefer 1959). The pattern of CO_2 exchange during breath holding, in all air breathing vertebrates, is the result of the physicochemical properties of tissue, blood and air and does not depend on cardiac shunting (Hicks and Wang 1996). Third, although increasing the tissue CO_2will acidify the perfusing blood and thus right-shift the O_2 equilibrium curve, a R-L shunt also reduces systemic arterial O_2 levels. Thus the benefits of a CO_2 retention are only apparent when the reduction in blood O_2 affinity is greater than the reduction in arterial PO_2 resulting from R-L shunt (Hicks and Wang 1996). Based on the measured changes in blood pH during breath holding within aerobic limits (White et al. 1989; Hicks and White 1992), the magnitude of the Bohr shift is not large enough to offset the reductions in arterial O_2 content and PO_2 due to shunt. It is difficult, therefore, to reconcile the notion that reductions in arterial O_2 levels improve arterial O_2 delivery (Hicks and Wang 1996).

Ventilation. During ventilatory periods, the cardiovascular changes are the reciprocals of those occurring during apnea. For example, in the turtle, *Trachemys scripta*, the HR doubles, the Rpul decreases by more than 50%, leading to a threefold increase in the pulmonary blood flow (Burggren 1975; Shelton and Burggren 1976; Wang and Hicks 1996a,b). Blood flow measurements from selected arteries estimate that a net L-R shunt develops during ventilation (White and Ross 1966; Shelton and Burggren 1976). In the turtle, *Trachemys scripta*, analysis of the blood PO_2 from the LAt, RAt and PA confirms a L-R intracardiac shunt during ventilation (White et al. 1989). A recent study indicates that in quiet, resting turtles, a net L-R shunt does not always develop during the short ventilatory periods, although the ratio, Q_{pul}/Q_{sys} approaches 1.0 (Wang and Hicks 1996b).

It has been suggested that the developement of a L-R shunt is important for CO_2 elimination in intermittent lung breathers (Ackermann and White 1979; White 1985). However, other periodic lung breathers (diving mammals, crocodilians) have not retained the ability to L-R shunt. Recently, we have found that decreasing the size of the L-R shunt had no significant effect on CO_2 elimination (Wang and Hicks, submitted). It has also been suggested that L-R shunting improves *V*/*Q* distributions in the lung. In turtles, recent experiments confirms that a L-R shunt reduces the *V*/*Q* heterogeneity within the lung and subsequently improves gas exchange (Hopkins et al.1995). However, it is doubtful that the regulation of L-R shunting has evolved to improve *V*/*Q* matching. The improvement of *V*/*Q* distributions is, in part, dependent on increasing the pulmonary blood flow. Thus it is important to have mechanisms that increase pulmonary blood flow and this does not necessarily require a L-R shunt.

Results from this laboratory suggests that cardiac shunting under quiet resting conditions provides few, if any, unique physiological benefits to O_2 or CO_2 exchange. Obviously, in the resting state, cardiac shunts are variable and appear to be a function of ventilatory state. However, during ventilation, it is pulmonary blood flow that is being adjusted so as to improve gas transport during brief ventilatory periods and the level of shunting follows rather than leads to these events.

Diving and cardiac shunts

The relationship between ventilatory state and shunt was first described in studies of diving, where submergence was associated with the development of a R-L shunt in turtles (Millen et al. 1964; White and Ross 1966). R-L shunts in reptiles were thus interpreted as part of the dive reflex, a suite of adaptive cardiovascular responses to breath holding and submergence seen in many air-breathing vertebrates. In recent decades, however, the general concept of the dive reflex has been increasingly challenged, based on differences in observed response to forced versus voluntary dives. Unlike earlier studies which forcibly submerged animals, subsequent research in reptiles and other divers has found diving bradycardia reduced or absent in voluntary dives (review in Kooyman 1989).

In this laboratory, we have found qualitative differences between forced and voluntary dives in the turtle, *Trachemys scripta,* both in heart rate and cardiac shunt pattern (Krosniunas and Hicks, in preparation). Forcibly submerging turtles leads to significant bradycardia, reduced cardiac output and the development of a sizable net R-L shunt. This response is similar to, though often more extreme than, the cardiovascular reactions shown during voluntary submerged rest (see Fig. 5). In contrast, voluntary diving (Kooyman 1989) or the departure from the surface to depth by active swimming efforts, is characterized by tachycardia, an increase in cardiac output and a net L-R shunt. These are cardiovascular changes seen during exercise generally (see below). Therefore, while voluntary and forced dives share the same ventilatory state, apnea, each elicits the opposite shunt pattern. Such uncoupling of shunt from ventilation may occur during submerged exercise, and may be due to the overriding effects of activity on the direction of shunt (see below).

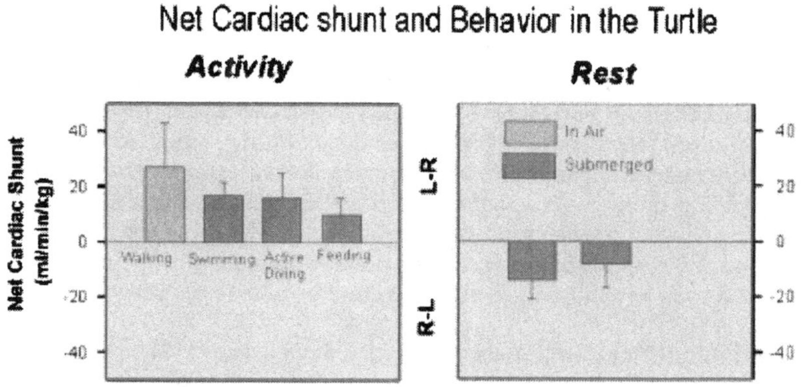

Fig. 5. Preliminary analysis of net cardiac shunting and different voluntary aquatic behaviors in the turtle *Trachemys scripta*. See text for details

Activity

The effect of activity on the size and direction of cardiac shunting remains fragmentary. In the turtles, *Trachemys scripta* and *Terrapene ornata*, a significant reduction in the oxygen saturation of LAo blood, from 90% at rest to 50% during activity, has been reported (Gatten 1975). These results are probably not indictive of the events that occur during exercise, since the activity was induced by electric shock to the shoulder and thigh musculature. In the lizards, *Varanus exanthematicus* and *Iguana iguana*, the O_2 content and PO_2 of the RAo remains constant during treadmill exercise, up to the VO_2 max (Gleeson et al. 1980). This suggests that the shunt fraction may decrease or, at least, remain constant during exercise (Mitchell et al. 1981a,b).

Previous studies on turtles have noted that during activity there are increases in Q_{pul} that are independent of ventilation (Shelton and Burggren 1979; Butler et al. 1984). Recently, West et al. (1992), measured left pulmonary artery blood flow (Q_{Lpul}) and left aortic arch blood flow (Q_{LAo}) during rest and swimming in the green sea turtle, *Chelonia mydas*. During swimming the HR increases and the Rpul decreases, leading to an increase in the Q_{Lpul}. The total increase in cardiac output is accounted for, primarily, by an increase in Q_{Lpul} suggesting the development of a L-R shunt during exercise. Similar cardiovascular changes occur in the the turtle *Trachemys scripta* following intravenous infusion of epinephrine or during sympathetic nerve stimulation (Comeau and Hicks 1994; Hicks 1994; Hicks et al. 1996). These hemodynamic changes result in a reduction in R-L shunt and a subsequent increase in the arterial O_2 levels (Hicks and Malvin 1992; Hicks and Comeau 1994; Ishimatsu et al. 1996). However, the development of a L-R shunt may not be a universal response of reptiles to exercise. Recent studies in the lizards, *Varanus exanthematicus*, indicate that a signficant L-R shunt does not develop during treadmill exercise (Hopkins et al. 1995).

Preliminary research in our laboratory suggests that the pattern of cardiac shunting during voluntary behavior is primarily associated with activity level (Krosniunas and Hicks 1994, 1995). Chronically instrumented turtles (*Trachemys scripta*) were tested in a naturalistic laboratory setting, with access to an aquatic or terrestrial environment. Six categories of voluntary behavior were examined in spontaneously behaving turtles: walking, rest in air, rest in water, active diving (movement into water from air), feeding and swimming. The latter four types of behavior were operationally defined as those in which turtles were completely submerged and, therefore, unable to breathe.

Our results indicate that rest is associated with a net R-L shunt (Fig. 5), while exercise leads to a reduction or elimination of net R-L cardiac shunt in turtles. Longer bouts of exercise, those lasting one minute or more, resulted in the development of a net L-R shunt (Fig. 6). Shunt direction for both rest and activity were found to be independent of submergence and, hence, ventilatory state. These preliminary results also suggest that the magnitude of shunt may be related to metabolic demand, since longer-duration high energy activities (i.e., walking) had the highest L-R values, while the size of R-L shunt tended to increase over time during rest.

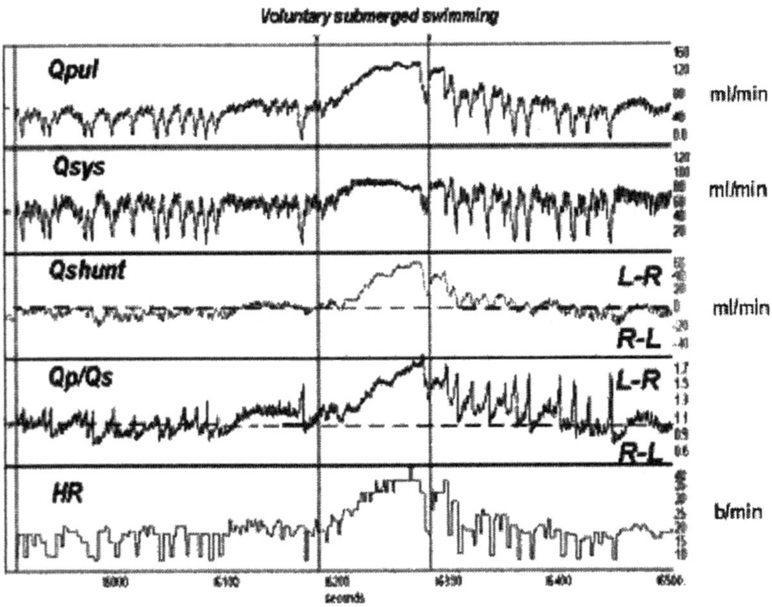

Fig. 6. The cardiovascular response to voluntary submergence swimming. The panels are, starting from the top, total pulmonary blood flow (Q_{pul}), total systemic blood flow (Q_{sys}), net cardiac shunt (Q_{shunt}), pulmonary/systemic flow ratio (Q_{pul}/Q_{sys}) and heart rate (*HR*). Note that Q_{shunt} values that are negative represent a net R-L shunt and positive values represent a net L-R shunt. Voluntary swimming while submerged is designated by the *solid line brackets*. Tracing represents raw, on-line measurements, using the Bio-Pac data acquisition system

Thermoregulation

Reptiles exhibit differential rates of heating and cooling: in general, an increased rate during warming and a decreased rate during cooling (Templeton 1970). This hysteresis enables reptiles to remain at or near their preferred body temperature for longer periods (Tempelton 1970). These differential rates of heating and cooling result from changes in the cutaneous, deep muscle and visceral blood flow in response to elevation of core body temperature or in response to local heating of the skin (see White 1976).

Few studies have addressed the effects of heating and cooling on cardiac shunting patterns. In the lizards, *Iguana iguana* and *Tupinambis nigropunctatus*, the arterial-venous O_2 difference decreases during heating, indicating the development of a R-L intracardiac shunt (Baker and White 1970). It is suggested that the R-L shunt may facilitate warming by reducing heat loss from the lung (Baker and White 1970). However, direct measurements of cardiac output and, more importantly, cardiac shunting during heating and cooling have been scarce. In the turtle, the average R-L shunt fraction during apnea is 60% at 15°C and is reduced to 27% during apnea at 30°C (Heisler and Glass 1985). A recent study examined the effects of temperature on cardiac shunting in the lizard *Varanus niloticus* (Ishimatsu et al. 1988). In this animal, as the body temperature increases from 25 to 35°C, the average R-L shunt fraction is reduced. At 25°C, the R-L shunt is 33% of the cardiac output and decreases to 8.5% at 35°C. It should be noted that the shunt values reported in this study represent measurements made after the animals were at the higher temperature for at least 12 h and the dynamic response remains unknown.

Preliminary experiments from this laboratory confirm the development of a R-L shunt during heating. This net R-L shunt may be as large as 60% of the systemic output. During cooling, a net L-R shunt develops, which is not consistent with the hypothesis that R-L shunting retards heat loss (Fig. 7). In this figure, RAo/LAo represents the ratio of right aortic to left aortic arch blood flow, Q_{sub} is the blood flow in the right subclavian artery. Preliminary evidence suggests that the cardiovascular response to heating and cooling, shown in Fig. 7, remains intact after the infusion of atropine. This indicates that the local vasodilation of peripheral vessels, in response to a heat load, is primarily responsible for the development of the R-L shunt.

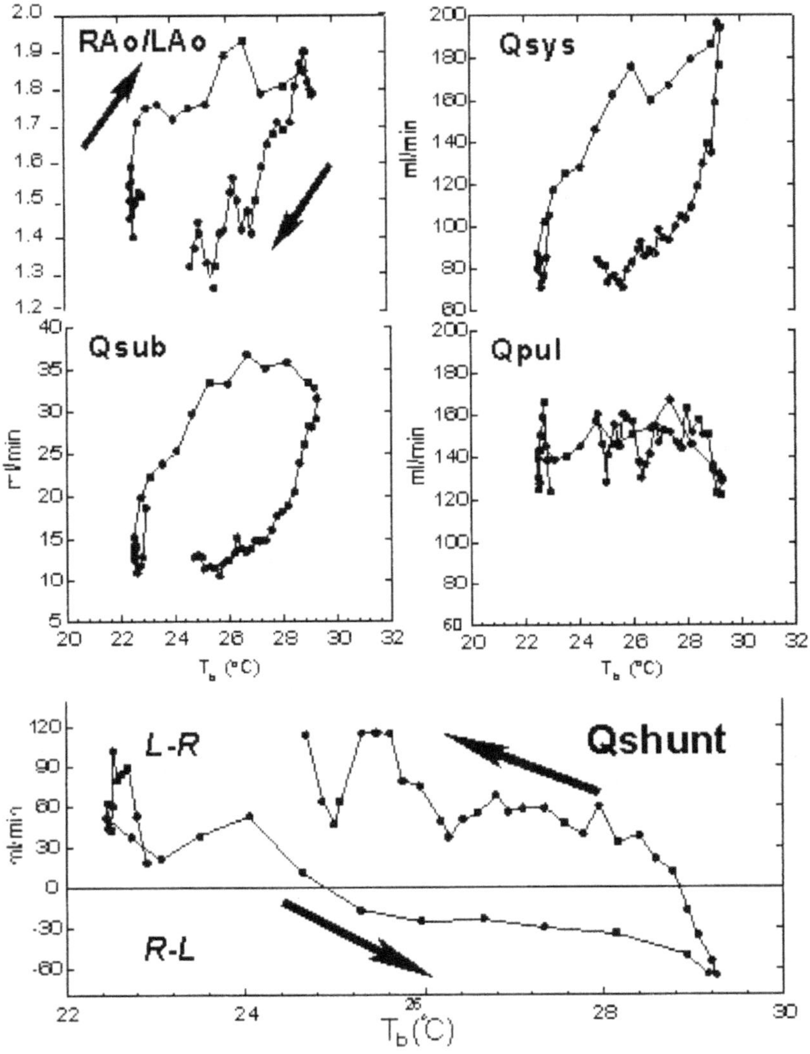

Fig. 7. Preliminary experiment on the dynamic cardiovascular response of a turtle (*Trachemys scripta*) during heating and cooling. Note the development of a R-L shunt during heating and a reversal of this trend during cooling. Also, note lack of hysteresis in Q_{pul}. Q_{sub}, subclavian blood flow, *RAo/LAo*, the ratio of blood flow in the right aortic and left aortic arches; Q_{shunt}, net cardiac shunt. Negative values represent a net R-L shunt and positive values represent a net L-R shunt. Turtle = 1.85 kg

Other physiological states or environmental challenges

Hypoxia and hypercapnia. The ventilatory response of reptiles to hypoxia and hypercapnia has been well studied (see Shelton et al. 1986). However the effects of inspiratory gases on the shunt direction and size are virtually unknown.

Millard and Johansen (1973) examined the effects of hypoxia and hypercapnia on ventricular outflow dynamics in the Nile monitor lizard. In response to hypoxia and hypercapnia, the pulmonary and systemic blood pressures increased, but were never superimposable. It was concluded that intracardiac shunting was small and did not change in response to either hypoxia or hypercapnia.

Preliminary evidence from this laboratory indicates that the level of R-L cardiac shunting is sensitive to progressive hypoxia. In the turtle, 5% inspired O_2 at 15°C results in signficant reductions in the R-L shunt determined from microspheres (Hicks, Ishimatsu and Heisler, unpublished observations). In *Trachemys scripta*, progressive reductions in the PIO_2 result in an average increase in Q_{pul}, heart rate and a reduction in R-L shunting, with the subsequent development of L-R shunting (Wang and Hicks 1994;Fig. 8). It should be noted that at inspired levels below 10% O_2, cardiorespiratory synchrony was diminished and Q_{pul} remained elevated during apnea.

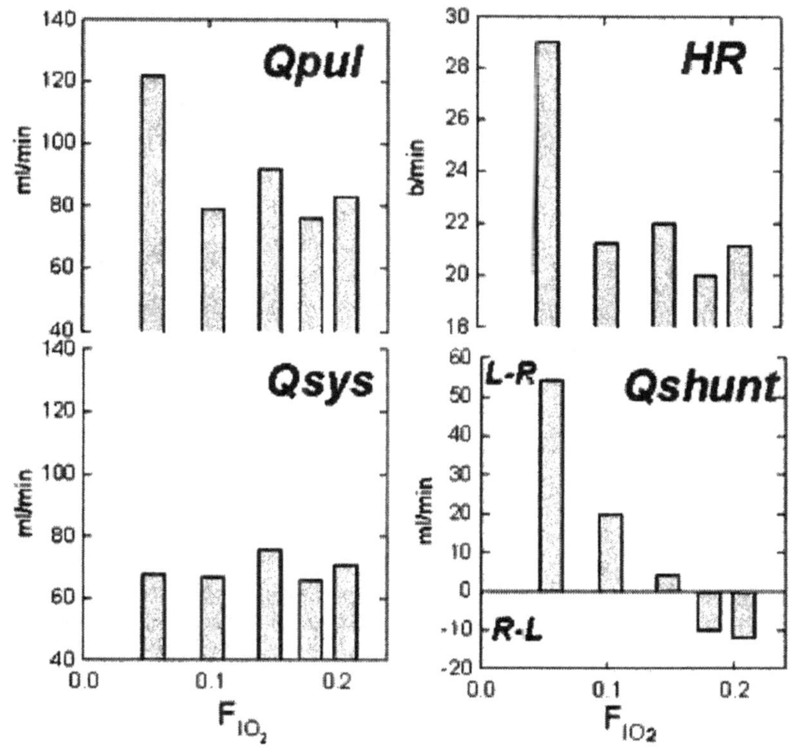

Fig. 8. Preliminary data on the cardiopulmonary response to progressive hypoxia in the turtle, *Trachemys scripta*. Q_{pul}, pulmonary blood flow; Q_{sys}, total systemic blood flow; Q_{shunt}, net cardiac shunt (negative values represent net R-L shunt, positive values represent net L-R shunt); *HR*, heart rate. Turtle = 1.9 kg

The postprandial response in reptiles. Increments in post-feeding metabolic rate have also been found in both vertebrate and invertebrate ectothermic animals (reviewed in Houlihan 1991). Specific dynamic action, strictly defined, is very similar, if not identical, in magnitude in reptiles and mammals, being in excess of 25% for high protein meals and about 10% for fat (Benedict 1932; Secor et al. 1994). However, the impact on reptilian metabolic rate is substantially greater, principally because of the low initial metabolic plane of these animals and the huge meal sizes that can be ingested. The increment in metabolic requirements varies from 1.7 to 24 times the resting rate. A postprandial increase in O_2 consumption has been measured in turtles (Sievert et al. 1988), lizards (Hicks and Bennett, unpublished observation; Secor et al., personal communication) and snakes (Benedict 1932; Secor and Diamond 1995).

Except for measurements of organ weights and blood gases (Secor and Diamond 1995), little is known about the cardiopulmonary response during the postprandial period. In the crocodilians, it has been suggested that a R-L shunt may develop during digestion (Jones and Shelton 1993), because the delivery of shunted acidic blood, preferentially to the gut, may be important to the mechanisms involved in HCl secretion during the postprandial period. Recent evidence has shown that blood flow in the LAo and the celiac artery increases 8–6 h following feeding in *Crocodylus porosus* (Axelsson et al. 1991), however, the development of a R-L shunt could not be directly determined. Recently, Secor and Diamond (1995) measured reductions in blood PO_2 during the postprandial period in the Burmese python. However, since the blood samples were acquired by cardiac puncture and because no other cardiopulmonary measurements were made, it is difficult to determine whether the reduction in blood PO_2 was due to cardiac shunt, hypoventilation, changes in V/Q distributions or resulted from the mixing of blood at the sampling site.

Conclusion

For almost 200 years, biologists have been fascinated by the cardiovascular system of reptiles; however, understanding the physiological function of their unique cardiovascular "design" remains a major challenge. Little is known about the extent of cardiac shunting in most species of lizards, turtles and snakes. The vast majority of studies have concentrated on two genera of turtles (*Trachemys* and *Chrysemys*) and a single genus of lizard (*Varanus*). More importantly, there is little information on the cardiac shunting patterns associated with basic physiological states, other than the detailed descriptions of the cardiovascular changes that occur during intermittent ventilation in turtles. Consequently, interpretations of the physiological function of cardiac shunts have focused primarily on the potential advantages to pulmonary gas exchange. In addition the cardiovascular changes that occur during intermittent ventilation have often be translated to the events that might occur during "natural" diving. However, this is probably overly simplistic. Our preliminary evidence suggests that the physiological response to voluntary diving, in recovered animals, is highly dependent on the animal's behavioral state. We believe to fully understand the functional significance of cardiac shunting it will be important to measure the various shunting patterns that occur in fully recovered animals under a wide variety of behavioral states and physiological challenges. Recent advances in blood flow measuring devices and computer-assisted data acquisition systems are making these types of experiments possible.

References

Ackerman, R., White, F.N. (1979) Cyclic carbon dioxide exchange in the turtle, *Pseudemys scripta*. *Physiol. Zool.* **52**: 378–389

Axelsson, M., Fritsche, R., Holmgren, S., Grove, D.J., Nilsson, S. (1991) Gut blood flow in the esturine crocodile,*Crocodylus porosus*. *Acta Physiol. Scand.* **142**:(2) 509-516

Baker, L.A., White, F.N. (1970) Redistribution of cardiac output in response to heating in *Iguana iguana*. **35**: 253–262

Benedict, F. G. (1932) *The physiology of large reptiles*. Carnegie Institute Washington Publ. 425

Berger, P.J., Burnstock, G. (1979) Autonomic nervous system. In: Gans, C., Northcutt, G., Ulinksi, P. (eds.) *Biology of the reptilia., Vol. 10. Neurology.* Academic Press, New York, pp.1–47

Brücke, E. (1852) Beiträge zur vergleichenden Anatomie und Physiologie des Gefäßsystems. *Denkschr Akad Wien* **3**:335-367

Burggren, W. (1975) A quantitative analysis of ventilation tachycardia and its control in two chelonians, *Pseudemys scripta* and *Testudo graeca*. *J. Exp. Biol.* **63**: 367–380

Burggren, W. (1977) The pulmonary circulation of the chelonian reptile: morphology, pharmacology and hemodynamics. *J. Comp. Physiol. B* **116**: 303–324

Burggren, W. (1985) Hemodynamics and regulation of central cardiovascular shunts in reptiles. In: Johansen, K., Burggren, W. (eds.) *Cardiovascular shunts: phylogenetic, ontogenetic and clinical aspects*. Munksgaard, Copenhagen, pp. 121–142

Burggren, W., Johansen, K. (1982) Ventricular hemodynamics in the monitor lizard, *Varanus exanthematicus*, pulmonary and systemic pressure separation. *J. Exp. Biol.* **96**: 343–354

Burggren,W., Shelton, G. (1979) Gas exchange and transport during intermittent breathing in chelonian reptiles. *J. Exp. Biol.* **82**: 75–92

Burggren, W.W., Warburton, S.J. (1994) Patterns of form and function in developing hearts: contributions from non-mammalian vertebrates. *Cardioscience* **5**: 183–191

Burrgen, W., Smits, A., Evans, B. (1989) Arterial O2 homeostasis during diving in the turtle *Chelodina longicollis*. *Physiol. Zool.* **62**: 668--686

Butler, P., Milsom, W.K, Woakes, A.J. (1984) Respiratory, cardiovascular and metabolic adjustments during steady state swimming in the green turtle, *Chelonia mydas*. *J. Comp. Physiol. B* **154**: 167–74

Comeau, S.G., Hicks, J.W. (1994) Regulation of central vascular blood flow in the turtle. *Am. J. Physiol.* **36**: R569–R578

Comeau, S., Lance, V.A., Hicks, J.W., Conlon, J.M. (1992) Purification and biological activity of alligator bradykinin. *Am. J. Physiol.* **263**: R400–R404

Conlon, J.M., Hicks, J.W., Smith, D.D. (1990) Isolation and biological acitivty of a novel kinin ([Thr6]) bradykinin from the turtle, *Pseudemys scripta*. *Endocrinology* **126**: 985–991

Douglas, C.G., Haldane, J.S. (1922) The regulation of the general circulation rate in man. *J. Physiol. (Lond)* **56**:69–100

Foxon, G.E.H. (1955) Problems of the double circulation in vertebrates. *Biol. Rev.* **30**:196–228

Gatten, R.E. (1975) Effects of activity on blood oxygen saturation, lactate and pH in the turtles *Pseudemys scripta* and *Terrapene ornata*. *Physiol. Zool.* **48**: 24–35.

Gleeson, T.T., Mitchell, G.S., Bennett, A.F. (1980) Cardiovascular responses to graded activity in the lizards *Varanus* and*Iguana*. *Am. J. Physiol.* **239**: R174–R179

Griel, A. (1903) Beiträge zur vergleichenden Anatomie und Entwicklungsgeschicte des Herzens und des Truncus arteriosis der Wirbelthiere. *Morphol. Jahrb.* **31**: 123–310

Grigg, G.C., Johansen, K. (1987) Cardiovascular dynamics in *Crododylus porosus* breathing air and during voluntary aerobic dives. *J. Comp. Physiol. B* **157**: 381–392

Heisler, N., Glass, M.L (1985) Mechanisms and regulation of central vascular shunts in reptiles. In: Johansen, K., Burggren, W. (eds.) *Cardiovascular shunts: phylogenetic, ontogenetic and clinical aspects*. Munksgaard, Copenhagen, pp. 334–353

Heisler, N., Neumann, P., Maloiy, G.M.O. (1983) The mechanism of intracardiac shunting in the lizard *Varanus exanthematicus*. *J. Exp. Biol.* **105**: 15–31

Hicks, J.W. (1993) Regulation of intracardiac shunting in reptiles: anatomic vs effective shunting. In: Bicudo, J.E., Glass, M.L. (eds.) *The vertebrate gas transport cascade: adaptations to environment and mode of life.* CRC Press, Boca Raton, pp. 252–265

Hicks, J.W. (1994) Adrenergic and cholinergic regulation of intracardiac shunting in chelonias. *Physiol. Zool.* **67**: 1325–1346

Hicks, J.W., Comeau, S.G. (1994) Vagal regulation of intracardiac shunting in the turtle *Pseudemys scripta. J. Exp. Biol.* **186**: 109–126

Hicks, J.W., Malvin, G.M. (1992) Mechanism of intracardiac shunting in the turtle *Pseudemys scripta. Am. J. Physiol.* **262**: R986–R992

Hicks, J.W., Malvin, G.M. (1996) Mechanism of intracardiac shunting in reptiles: pressure vs washout shunting. In: Heisler, N. (ed.) *Advances in comparative and environmental physiology, mechanisms of systemic regulation. Vol I. Respiration and circulation.* Springer, Berlin Heidelberg New York, pp. 137-155

Hicks, J.W., White, F.N. (1992) Ventilation and gas exchange during intermittent ventilation in the American alligator, *Alligator mississipiensis. Respir. Physiol.* **88**: 23–36

Hicks, J.W., Wang, T. (1996) The functional role of cardiac shunts in reptiles. *J. Exp. Zool.* (in press)

Hicks, J.W., Wood S.C. (1989) Oxygen homeostasis in lower vertebrates: the impact of external and internal hypoxia. In: Wood, S.C. (ed.) *Lung biology in health and disease-comparative pulmonary physiology : current concepts.* Marcel Dekker, New York, pp. 311-341

Hicks, J.W., Ishimatsu, A., Molloi, S., Erashin, A., Heisler, N. (1996) The mechanism of cardiac shunting in reptiles: a new synthesis. *J. Exp. Biol.* (in press)

Hopkins, S., Hicks, J.W., Cooper, T., Powell, F. (1995) Ventilation-perfusion relations in the savannah monitor during treadmill exercise. *J. Exp. Biol.* **198**:1783–1789

Hopkins, S.R., Wang, T., Hicks., J.W. (1995) Ventilation-perfusion heterogeneity improves with increased pulmonary blood flow in anaesthetized turtles. *FASEB J.* **9**: (4) A641, abstr. no. 3716

Hopkins, S.R., Wang, T., Hicks, J.W. (1996) The effects of altering pulmonary blood flow on pulmonary gas exchange in the turtle *Trachemys (Pseudemys) scripta. J. Exp. Biol.* (in press)

Houlihan, D. F. (1991) Protein turnover in ectotherms and its relationships to energetics. *Adv. Comp. Environ. Physiol.* 7: 1–43

Ishimatsu, A., Hicks, J.W., Heisler, N. (1988) Analysis of intracardiac shunting in the lizard, *Varanus niloticus*: a new model based on blood oxygen levels and microsphere distribution. *Respir. Physiol.* **71**: 83–100

Ishimatsu, A., Hicks, J.W., Heilser, N. (1996) Analysis of intracardiac shunting in the turtle, *Trachemys (Pseudemys) scripta*: application of the three outflow vessel model. *J. Exp. Biol.* (in press)

Johansen, K., Burggren, W.W. (1980) Cardiovascular function in lower vertebrates. In: *Hearts and hart like organs.* Academic Press, New York, pp. 61–117

Jones, D.R., Shelton, G. (1993) The physiology of the alligator heart: left aortic flow patterns and right-to-left shunts. *J. Exp. Biol.* **176**: 247–269

Kooyman, G.L. (1989) *Diverse divers, physiology and behavior.* Springer, Berlin Heidelberg New York

Krosniunas, E., Hicks, J.W. (1994) Cardiovascular correlates of behavior in the turtle (abstract). *The Physiologist* A-95, abstr. no. 58.24

Krosniunas, E., Hicks, J.W. (1995) Intracardiac shunts during voluntary activity and rest in the turtle (abstract). *Am. Zool.* **35**: A63, abstr. no. 253

Lillywhite, H.B., Donald, J.A. (1989) Pulmonary blood flow regulation in an aquatic snake. *Science* **245**: 293–295

Lister, G., Pitt, B.R. (1983) Cardiopulmonary interactions in the infant with congenital cardiac disease. *Clin. Chest Med.* **4**: 219–232

Luckhardt, A.B., Carlson, A.J. (1921) Studies on the visceral sensory nervous system. VIII. On the presence of vasomotor fibres in the vagus nerve to the pulmonary vessels of the amphibian and reptilian lung. *Am. J. Physiol.* **56**: 72–112

Mathur, P.N. (1946) The anatomy of the reptilian heart. II. Serpentes, Testudinata and Loricata. *Proc. Indian Acad. Sci.* **20**: 1–29

Millard, R.W., Johansen, K. (1973) Ventricular outflow dynamics in the lizard, *Varanus niloticus:* responses to hypoxia, hypercarbia and diving. *J. Exp. Biol.* **60**: 871–880

Millen, J.E., Murdaugh, H.V., Bauer, C.B., Robin, D. (1964) Circulatory adaptation to diving in the freshwater turtle. *Science* **145**: 591–593

Milsom, W.K., Langille, B.L., Jones, D.R. (1977) Vagal control of pulmonary vascular resistance in the turtle *Chrysemys scripta. Can. J. Zool.* **55**: 359–367

Mitchell, G. S., Gleeson, T. T., Bennett, A. F. (1981a) Ventilation and acid-base balance during graded activity in lizards. *Am. J. Physiol.* **240**: R29–R37

Mitchell, G. S., Gleeson, T. T., Bennett, A. F. (1981b) Pulmonary O_2 transport during activity in lizards. *Respir. Physiol.* 43: 365–375

Mithoefer, J.C. (1959) Mechanism of pulmonary gas exchange and CO_2 transport during breath holding. *J. Appl. Physiol.* 14:706–710

Nilsson, S. (1983) *Autonomic nerve function in the vertebrates.* Springer, New York Berlin Heidelberg

O'Donoghue, C.H. (1918) The heart of the leathery turtle, *Dermochelys (Sphargis) coriacea.* With a note on the septum ventriculorum in the Reptilia. *J. Anat. Lond.* **52**: 823–890

Otis, A.B., Rahn, H., Fenn, W.O. (1948) Alveolar gas exchange during breath holding. *Am. J. Physiol.* **152**: 674–686

Rau, A.S. (1924) Observations of the anatomy of the heart of *Tiliqua scincoides* and *Eunectes murinus. J. Anat. Lond.* **59**: 60–71

Secor, S. M., Diamond, J. (1995) Adaptive responses to feeding in Burmese pythons: pay before pumping *J. Exp. Biol.* **198**: 1313–1325

Secor, S.M. , Stein, E.D., Diamond, J. (1994) Rapid upregulation of snake intestine in response to feeding: a new model of intestinal adaptation. *Am. J. Physiol.* **266**: G695–G705

Shelton, G., Burggren, W.W. (1976) Cardiovascular dynamics of the Chelonia during apnea and lung ventilation. *J. Exp. Biol.* **64**: 323–343

Shelton, G., Jones, D.R., Milsom, W.K. (1986) Control of breathing in ectothermic vertebrates. In: *Handbook of physiology. Section 3: The respiratory system. Vol. II.* American Physiological Society, Bethesda, pp. 857–909

Sievert, L.M., Sievert, G.A., Cupp, P.V. (1988) Metabolic rate of feeding and fasting juvinile midland painted turtles, *Chrysemys picta marginata. Comp. Biochem. Physiol.***90A**: 157-161

Steggerda F.R., Essex, H.E. (1957) Circulation and blood pressure in the great vessels and heart of the turtle *Chelydra serpentina. Am. J. Physiol.* **190**: 320--326.

Templeton, R. (1970) *Comparative physiology of thermoregulation.* Whittow, G.C. (ed.). Academic Press, New York, Chapter 5, pp. 167–221

Van Mierop, L.H.S., Kutsche L.M. (1981) Comparative anatomy of the ventricular septum. In: Wenink, A.C.G. (ed.) *The ventricular septum of the heart*. Martinus Nijhoff, The Hague, pp. 35–46

Van Mierop, L.H.S., Kutsche, L.M. (1985) Some aspects of comparative anatomy of the heart. In: Johansen, K., Burggren, W. (eds.) *Cardiovascular shunts: phylogenetic, ontogenetic and clinical aspects*. Munksgaard, Copenhagen, pp. 38--56

Wang, T., Hicks, J.W. (1994) Hypoxia and reduced blood hemoglobin concentration elicits left-to-right cardiac shunting in the turtle (abstract). *The Physiologist* A-95, abstr. no. 58.21

Wang, T., Hicks, J.W. (1996a) The interaction of pulmonary ventilation and cardiac shunts on arterial oxygen levels. *J. Exp. Biol.* (in press)

Wang, T., Hicks, J.W. (1996b) Cardiorespiratory synchrony in the turtle. *J. Exp. Biol.* (in press)

Webb, G., Heatwole, H., Bavay, J. de (1971) Comparative cardiac anatomy of the Reptilia. I. The chambers and septa of the varanid ventricle. *J. Morphol.* **134**: 335–350

Webb, G., Heatwole, H., Bavay, J. de (1974) Comparative cardiac anatomy of the Reptilia. II. A critique of the literature on the Squamata and Rhynchocephalia. *J. Morphol.* **142**: 1–20

West, N. , Butler, P.J., Bevan, R.M. (1992) Pulmonary blood flow at rest and during swimming in the green turtle, *Chelonia mydas*. *Physiol. Zool.* **65** : 287–310

White, F.N. (1959) Circulation in the reptilian heart (Squamata). *Anat. Rec.* **135**: 129–134

White, F.N. (1968) Functional anatomy of the heart of reptiles. *Am. Zool.* **8**: 211–219

White, F.N. (1970) Central vascular shunts and their control in reptiles. *Fed. Proc. Fed. Soc. Exp. Biol.* **29**: 1149–1153

White, F.N. (1976) Circulation. In: Gans, C., Dawson, W.R. (eds.) *Biology of the reptilia*. Academic Press, New York, pp. 275–334

White, F.N. (1985) Role of intracardiac shunts in pulmonary gas exchange in chelonian reptiles. In: Johansen, K., Burggren, W. (eds.) *Cardiovascular shunts: phylogenetic, ontogenetic and clinical aspects*. Munksgaard, Copenhagen, pp. 296–309

White, F.N., Ross, G. (1966) Circulatory changes during experimental diving in the turtle. *Am. J. Physiol.* **211**: 15–18

White, F.N., Hicks, J.W., Ishimatsu, A. (1989) Respiratory states and intracardiac shunts in turtles. *Am. J. Physiol.* **256**: R240-R247

Withers, P.C. (1992) *Comparative animal physiology*. Saunders, Philadelphia

Exp. Biol. Online 1:4 (1996)

Distribution and dynamics of mitochondrial nucleoids in animal cells in culture

Jürgen Bereiter-Hahn and Monika Vöth

Cinematic Cell Research Group, Biocentre, Johann Wolfgang Goethe University Frankfurt am Main, Germany

Correspondence to: J. Bereiter-Hahn (e-mail: bereiter-hahn@zoology.uni-frankfurt.de)

Received: 8 May 1996 / Accepted: 14 June 1996
Published: June 24, 1996

Abstract. Mitochondrial (mt) nucleoids were visualized in living cells in culture by staining with the fluorochrome picoGreen. The cell types included a line derived from *Xenopus* heart endothelial cells (XTH-2), 3T3 cells, SV40-transformed 3T3 cells and primary cultures of *Xenopus* tadpole epidermis cells. In the permanent cell lines 6-60% of the mitochondria were found to be devoid of DNA. The peaks of the frequency distribution of mtDNA content, as revealed by microfluorometry, were not very distinct, indicating the presence of a high amount of aneuploid mt nucleoids. The maximum size of nucleoids (as derived from fluorescence intensity) was 10-12 times that of the minimum peak value in proliferating cell cultures. A linear ratio was found between the volume of the nucleoids and their DNA content, which is interpreted as a uniform package density. In terminally differentiating tadpole epidermis cells mitochondria form large bodies containing giant nucleoids, while in mitotic cells the mt nucleoids are small and of uniform size. Fusion and fission of the nucleoids were observed to occur either for no visible reason or in connection with fusion and fission events of the mitochondria.

Key words: Mitochondrial nucleoids – *Xenopus laevis*

Introduction

The presence of DNA in mitochondria has been well established since the pioneering studies of Chévremont (1963), Swift (1964), Borst (1972) and Nass (1977). For a review of the early literature see for example, Roodyn and Wilkie (1968). The length and information content of mt DNA varies considerably among different organisms.

Mitochondria of *Physarum* and of plant cells contain sufficient DNA (mtDNA) to follow the behaviour of the mitochondrial (mt) nucleoid using DAPI staining and to correlate the appearance of the nucleoid with the changes in shape of mitochondria seen by phase-contrast or Nomarski optics in vivo (Kuroiwa 1982). Mitochondria of animal cells (in culture or in tissue sections) may not contain sufficient DNA to be identified unequivocally by DAPI fluorescence and to relate it to the changes in shape and division cycles of the mitochondria. Satoh and Kuroiwa (1991) were the first to visualize mt nucleoids in human A2780 cells after fixation. They confirmed observations made with the electron microscope that in thin tissue sections stained with osmium tetroxide the mtDNA appeared in a compact state in the matrix and provided evidence of the presence of 1-15 copies of mtDNA per mitochondrium.

In a variety of mammalian cells Robin and Wong (1988) determined a mean of 2.6±0.3 SE DNA molecules per virtual mitochondrion. The term "virtual mitochondrion" takes into account the fact that mitochondria may fuse or divide and the size of an isolated mitochondrium may differ considerably from that in a living cell. Furthermore, it is not known whether all mitochondria in a given cell contain DNA and how the mt nucleoids are distributed and may interact with each other.

Using a new nucleic acid selective probe, "picoGreen", we succeeded in staining the nucleoids of mitochondria in living endothelial cells derived from *Xenopus laevis* tadpole hearts (line XTH-2) and a series of other cells, in culture. The cells were transiently permeabilized either by electroporation or by glycerol treatment to allow staining. Identification of organelles as mitochondria is based on the appearance in phase contrast (in many cases unequivocal identification is possible, cf. Bereiter-Hahn 1978, 1990) or on staining with the mitochondria selective fluorochrome DASPMI (Bereiter-Hahn 1976) or using Mito-tracker (Whitaker et al. 1991), a fluorescent probe developed by Molecular Probes. mtDNA forms distinct nucleoids of varying DNA content, which seems to be related to the proliferative activity of the cells. Some of the mitochondria, the very long ones in particular, contain several nucleoids, while other mitochondria seem to be devoid of DNA. Often the mt-nucleoid is found in specific zones of the mitochondria, and mt nucleoids may become displaced in branching or moving mitochondria. In addition, fusion and fission of mitochondria and the behaviour of mt nucleoids during these processes have been observed.

Materials and methods

Cells and cell cultures

XTH-2 is a line of *Xenopus* tadpole heart endothelial origin. These cells spread extremely flat on glass slides (Schlage et al. 1981). These cells were kept in minimum essential medium (MEM) supplemented with 5% fetal calf serum (FCS). Swiss 3T3 cells and SV40-transformed 3T3 cells were kept in Dulbecco's MEM supplemented with 5% FCS. 10T1/2 cells were originally derived from mouse mesoderm cells and were maintained in basal medium (Eagle's BME) plus 10% FCS. The line HaCaT was derived from human basal epidermis cells (Boukamp et al. 1988) and was propagated in the Gibco keratinocyte medium, either at the original low calcium concentration or at an elevated concentration (1.2 mM) for 10 days without replacement of the medium.

All the culture media were obtained from Gibco (Glasgow, UK).

Staining procedures

PicoGreen (Molecular Probes Europe, Leiden) does not penetrate into living cells. Therefore the cell membranes had to be permeabilized. This was done either by electroporation following the method of Kwee et al. (1990) or – as was the case in most of our experiments – by glycerol treatment (see below for details).

The first set of experiments using Bio-Rad Gene pulser with pulse controller and capacitance extender (Bio-Rad, Munich, Germany) was performed on cell cultures exposed to 20 pulses at 760 V / cm (32 ms each) to permeabilize the membrane. The staining solution (picoGreen diluted 1:50 in saline containing a high level of potassium) was applied for 30 min. After this time the cells recovered from electroporation in normal culture medium for about 30 min. Then cells were either treated with mitotracker or picoGreen fluorescence was immediately observed.

Permeabilization by glycerol and staining were achieved by dissolving picoGreen 1:200 in culture medium containing 5% glycerol. After 30 min in this solution, cells were transferred to normal culture medium without glycerol and then observed with a true confocal laser scanning microscope (Leica TCS, Leica, Bensheim, Germany) equipped with a 100× Fluotar lens. In this case large pinholes (150 nm projected diameter) were used to increase focal depth. The advantage of the TCS microscope for the present study was not its high depth resolution but rather the possibility of synchronous acquisition of fluorescence and phase contrast images. This was a prerequisite for the comparison of mt nucleoid behaviour and motility events of the whole mitochondrion.

Some cultures were also stained after fixation to prove the specificity of picoGreen staining. In these cases cultures were washed with saline and then fixed according to Karnovsky (1965). After elution with acetone at –20°C for 30 min, cells were either stained directly or pretreated with 1mg /ml RNAase (DNAase-free) in saline at 37°C (from bovine pancreas; Serva, Heidelberg, Germany) for 30 min. DNAase treatment followed the same protocol. The DNAase (450 units/mg) was also from bovine pancreas (Serva). PicoGreen was used at a 1:2000 dilution of the purchased solution (1 mg/ml).

Results

Shape of mitochondria and distribution of mt nucleoids

In the well-spread XTH-2 cells mitochondria may vary from small spherical bodies to long filamentous structures that branch and remain either single or interconnect each other (Fig. 1), thus forming a huge mitochondrial reticulum (Fig. 2) (for review on intracellular motility and identification see Bereiter-Hahn 1990). Unequivocal identification in living cells is possible using either phase contrast microscopy (in well-spread cells) or fluorescence microscopy after selective staining with such fluorochromes as, for instance, DASPMI (Bereiter-Hahn 1976), rhodamine 123 (Johnson et al. 1980), mitotracker (Whitaker et al. 1991) or, more recently, by specifically targetted green fluorescent protein (Rizzutto et al. 1995). Mitochondria are able to bend, branch, fuse and divide within a few seconds or minutes. In other cells, like HaCaT cells of human epidermis origin or tadpole epidermal cells, mitochondria preferentially exhibit a rod-like appearance. During mitosis short rods are more frequent than long ones and no reticular organization is present.

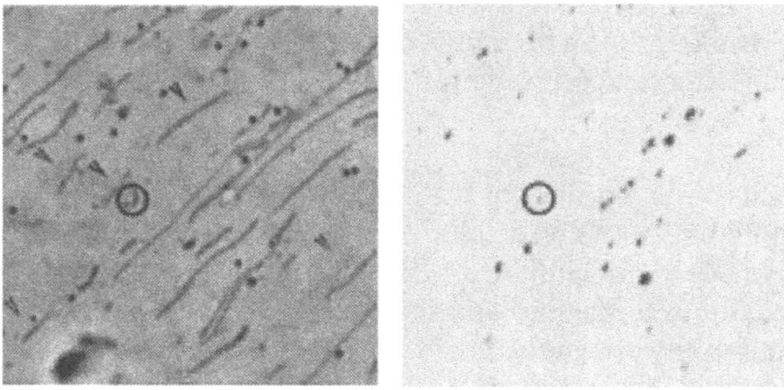

Fig. 1a,b. Confocal laser scan micrographs (Leica TCS System). **a** Phase-contrast image of a part of an XTH-2 cell showing the distribution of mitochondria. In the encircled mitochondrion and the mitochondrion above it, the nucleoid is situated in the centre of the organelle, a zone of preferred bending (small mitochondrion). **b** The same area as in **a**. Fluorescence of picoGreen indicates mitochondrial DNA (negative image of the fluorescence). The weak background staining is assumed to result from mtRNA. The encircled area marks a mitochondrion with the nucleoid at a central bending site. Mitochondria devoid of DNA are marked by *black arrowheads* in **a**. *Image width*: 24.7 µm

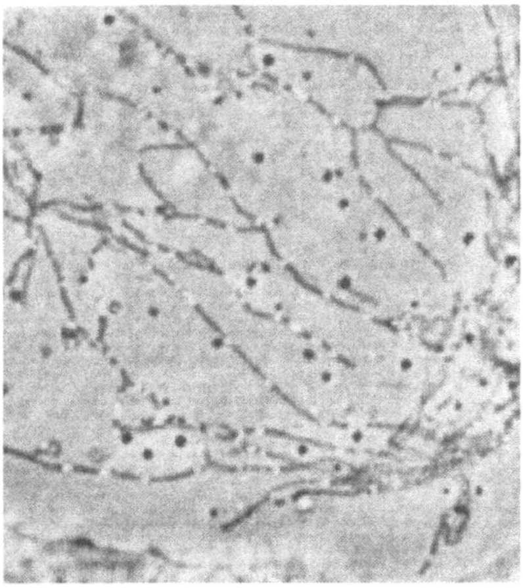

Fig. 2. Confocal laser scan images of the periphery of a XTH-2 cell: phase contrast and picoGreen fluorescence have been overlaid. The mt nucleoids appear bright on the darker phase image of the mitochondria. In this well-spread XTH-2 cell mitochondria form long filamentous and interconnected structures giving rise to a huge mitochondrial reticulum. *Image width*: 25 μm

mtDNA appears in the form of distinct small bodies, about 0.4 μm in diameter, which were revealed by staining with picoGreen. This dye has been reported to bind stoichiometrically to DNA (information leaflet from the producer) and also with much lower affinity to RNA also. Therefore it was necessary to prove whether the fluorescent structures in mitochondria are indeed composed of DNA. Cells were treated with RNase and/or DNase after fixation with glutaraldehyde. RNase treatment abolished all the background staining outlining mitochondria (see, for example Fig. 1) due to mtRNA. Also cytoplasmic staining of the endoplasmic reticulum disappeared. No change was observed in the presence or distribution of mt nucleoids. DNase treatment on the other hand, destroyed all the structures identified as mt nucleoids thus containing DNA.

In long filamentous mitochondria, mt nucleoids of various size can be located throughout the entire length like pearls on a collar. For instance mt nucleoids have no specific location inside a mitochondrium; however, some preferred locations can be identified; for example the central part of a long mitochondrion, if it contains only one mt nucleoid. Nucleoids have often been found in the very tips of mitochondria (Fig. 1b) and in zones of bending (Fig. 1, circle). Almost spherical mitochondria very often contain one particularly large mt nucleoid; swollen areas of mitochondria under stress also contain large mt nucleoids. Sometimes in phase-contrast microscopy the nucleoid-bearing locus may appear either with higher density or brighter; no unequivocal identification is possible with this method.

Some mitochondria are devoid of nucleoids (white arrowheads in Fig. 1a). In XTH-2 cells and in 3T3 cells this fraction is between 6% and up to 60% of all the mitochondria identifiable in a single cell (Fig.3). In other cell types not as flat as these two types, individual mitochondria are not distinguished either; therefore the fraction devoid of DNA cannot be determined.

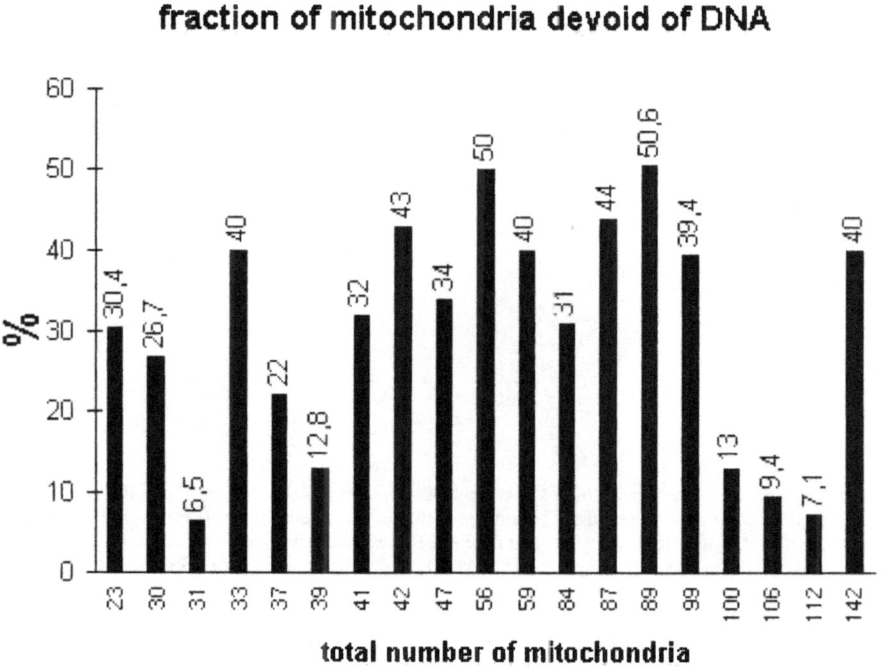

Fig. 3. Percentage of mitochondria devoid of mt nucleoids in XTH-2 cells. Each *bar* represents one cell (or a part of a cell where all mitochondria have been identified). The *numbers* above the bars indicate the numbers of mitochondria on which the percentage has been based

Size of mt nucleoids in various cell types

The number of mt nucleoids per mitochondrion varies, as does the size of the mitochondria. In most cases 1-10 mt nucleoids are found in a single mitochondrium; however, because of frequent fusion and fission processes this number is of no significance. It may be more important to estimate the distribution of DNA . The relative amount of DNA per mt nucleoid has been quantified by measuring the total fluorescence of single nucleoids. To eliminate factors which may influence staining intensity only those nucleoids were compared with each other which were located in the same cell. Calibration with DNA from *Escherichia coli* kept under minimal conditions failed because of the wide variation in *E. coli* DNA content; therefore only relative values for the mt nucleoids can be given. In growing cultures of XTH-2 cells or 3T3 cells and SV40-transformed 3T3 cells the total fluorescence per mt-nucleoid is linearly related to the volume of the mt nucleoid (Fig. 4), indicating a constant package density of DNA, independent of the size of the nucleoid.

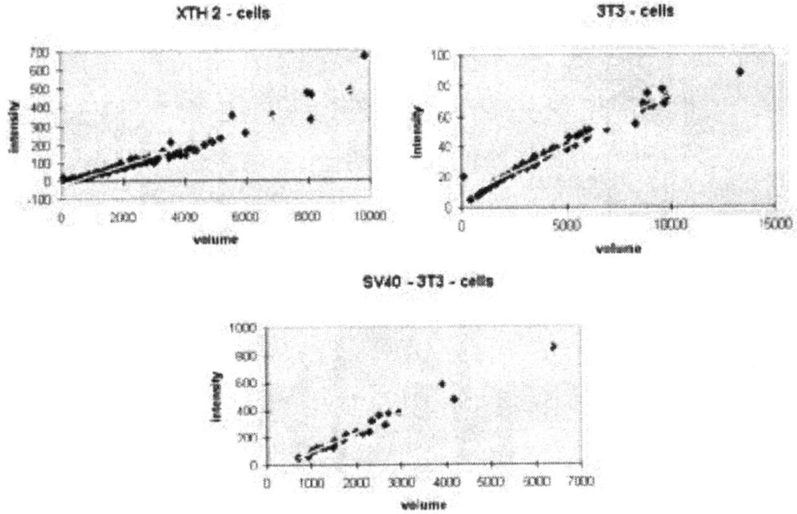

Fig. 4. Fluorescence/volume ratio of single nucleoids in various cell types. The total fluorescence of single mt nucleoids was determined by multiplication of the pixels with their grey value. Volume was calculated based on the assumption of spherical shape and circular projection areas. The correlation coefficients (*R*) are very high: 0.9376 (XTH-2 cells), 0.9709 (3T3 cells), 0.9713 (SV40-3T3 cells)

Distribution of fluorescence of mt nucleoids depends on the cell type under investigation

In XTH-2 cells, initially an almost continuous decrease in frequency with increasing DNA amount is observed. The mt nucleoid with the highest picoGreen fluorescence contains about 38 times the amount of DNA of that with the lowest fluorescence, which corresponds to about 11 times the value of the first peak value (300). Multiples of about 300 fluorescence units (300, 600, 1200 units) indicate ploidy values, although no clear maximum is obvious in the frequency distribution (Fig.5). 3T3 cells exhibit a maximum at about 2, 3 and 5 times the mean of the lowest class value (ca. 600, 1200, 1800, 3000); the maximum value is 10 times the DNA content of the mean of the first peak and about 16 times that of the smallest nucleoid.

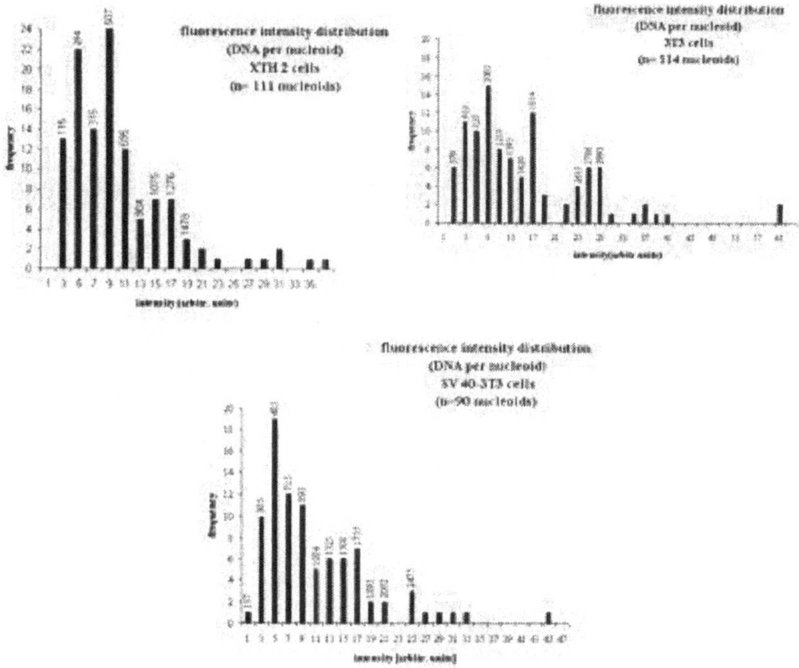

Fig. 5. Frequency distribution of fluorescence of single mt nucleoids in various cell types to estimate ploidy situation. For further description see text

In the fast-growing SV-40 transformed 3T3 cells, ploidy is equal to that of 3T3 cells. The maximum ploidy observed was 10 times the mean of the first peak value, and mitochondria showed a clear peak around 450 units with some increase at 900 and 1800 units.

Very clear differences in DNA content of mt nucleoids were found in the epidermis of tadpoles of the clawed toad, *X. laevis* Daudin. This epidermis is composed of two layers of keratocytes, a mitotic basal layer and a superficial one undergoing terminal differentiation (Fox 1987). This arrangement is maintained in primary cultures of tadpole epidermis (Strohmeier and Bereiter-Hahn 1991). The nuclear chromatin of the surface cells appears dense and stains more intensively with picoGreen than the nuclei of the basal cells. These cells contain only a few mitochondria (less than 30). mt nucleoids are small in basal cells and preferentially distributed in the perinuclear area. In superficial cells, large mitochondrial bodies have been formed and only a few very large mt nucleoids are present (Fig. 6).

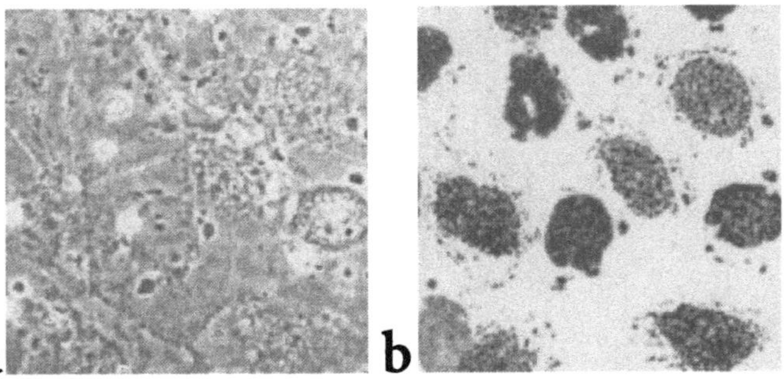

Fig. 6a,b. Bilayered *Xenopus laevis* tadpole epidermis in primary culture (Confocal laser scan images, "all-max"-projection). **a** Phase contrast image. The superficial cells are delineated by almost straight adhesion zones which are easily identified by phase contrast microscopy. The large mitochondrial bodies appear black. **b** The same area as shown in **a** after staining with picoGreen (negative image of the fluorescence). The nuclear chromatin of the surface cells is more heavily and more homogeneously stained than that of the basal cells. In the vicinity of the superficial cell nuclei, large dark spots mark the huge mt nucleoids; many smaller mt nucleoids are found around basal cell nuclei. *Image width*: 71.8 μm

Mitochondria during mitosis

During mitosis long filamentous mitochondria were missing in all the cell types examined. Mitochondria are homogeneously distributed throughout the cytoplasm with the exception of the mitotic spindle, which is free of mitochondria. mt nucleoids are more homogenous than in interphase cells, i.e. the relative brightness of the mt nucleoids is almost uniform (Fig. 7 and Fig. 8). In some HaCaT cells mt nucleoids during mitosis contain considerably less DNA than interphase cells (Fig. 7), while in others the amount of DNA is almost equal to that in interphase cells (Fig. 8).

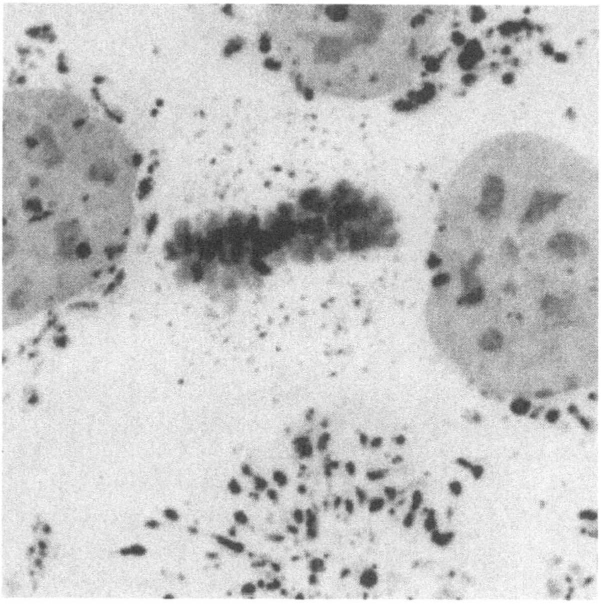

Fig. 7. Epidermis cells from *X. laevis*. mt nucleoids are more homogeneous during mitosis (here a metaphase cell) than in interphase cells, i.e. the relative brightness of the mt nucleoids is almost uniform. Only the basal cells remain proliferative. The interphase cells surrounding the mitotic cell show a broad range of mt nucleoids preceding the formation of the giant nucleoids shown in Fig. 6. *Image width*: 42μm

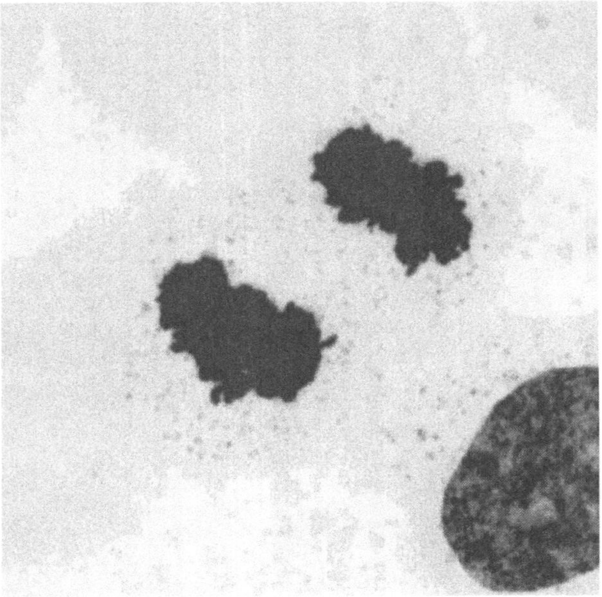

Fig. 8. PicoGreen staining of a telophase cell in an epidermis culture from *Xenopus* tadpole. The mt nucleoids become evenly distributed to the daughter cell. In the adjacent interphase cell mt nucleoids equal those of the mitotic cell. *Image width*: 50 μm

Fusion and fission of mitochondria

Fusion and fission of mitochondria require that the contents of the fusing organelles intermingle with each other or become separated. This is valid for the role of mt nucleoids in fusion and fission processes in particular, because genetic recombination is assumed to represent the functional significance of these processes. Therefore we investigated the behaviour of mt nucleoids and of the inner mt compartment during fusion and fission. Obviously it is not possible using light microscopy to decide whether genetic recombination takes place. The packaging of mtDNA is the only feature which can be followed in living mitochondria in situ.

The series of images in Fig. 9 shows a fusion process. The fusion site, a thickened part of a long mitochondrion containing a large nucleoid, is marked with an arrowhead. The mitochondrion on the right develops a branch emanating at about 90°. The nucleoid at the branching site splits into two while the branching occurs. One of the daughter nucleoids is found in the advancing tip of the branch (images 6-8 in the series); then it fuses with the nucleoid of the fusion partner (image 9 at the position marked by the arrowhead). Later non the nucleoids seem to separate again (image 14 = last image). During this process the newly formed branch separates from its original mitochondrion (images 11 - 14 in Fig. 9).

During fission of mitochondria the reverse processes occur (Fig. 10). The mitochondrial parts separate from each other (white arrow in Fig. 10); then the two peripheral nucleoids fuse with each other (image 7 and 8 in Fig. 10).

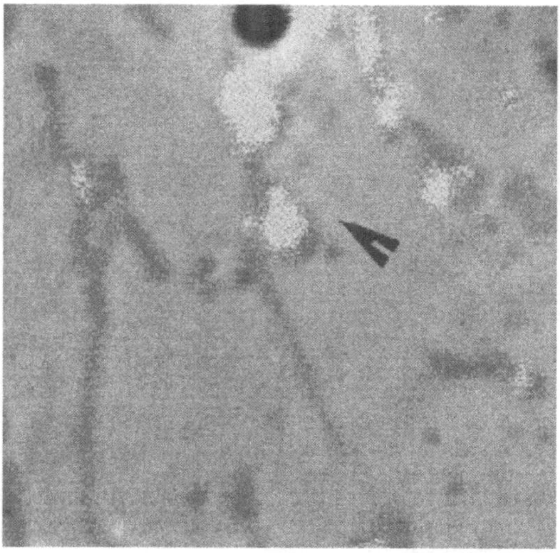

Fig. 9. Branching and fusion of a mitochondrion in a XTH-2 cell. Time series in phase contrast (*black and white*) and picoGreen fluorescence (*green*). Misinterpretation of overlapping of mitochondria as a fusion event can be excluded due the the very thin cytoplasm in the area under observation. In addition, electron micrographs of serial sections through this cell type support our interpretation (Bereiter-Hahn and Vöth 1994)

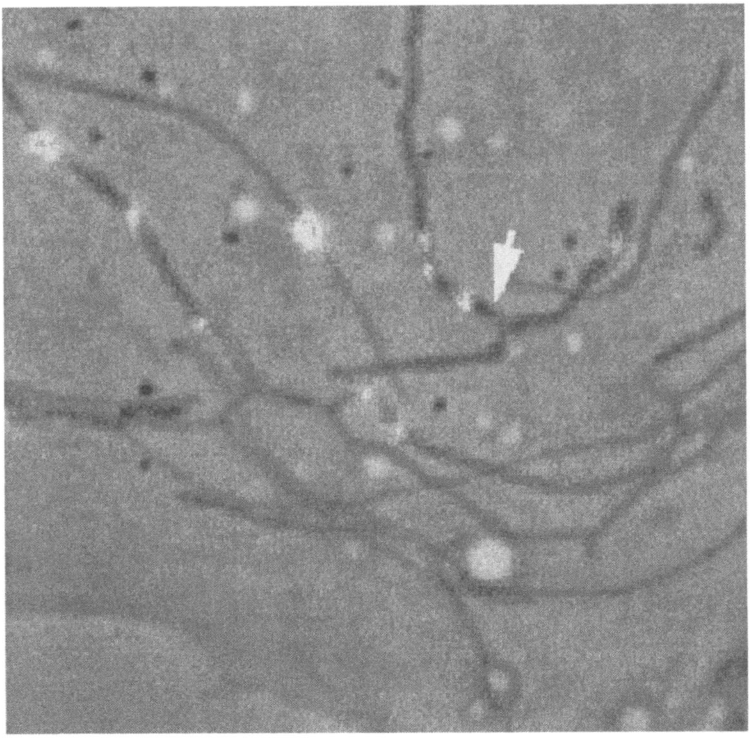

Fig. 10. Fission of a mitochondrion in a XTH-2 cell. Time series in phase contrast (*black and white*) and picoGreen fluorescence (*green*). Time difference between single images: 7s. *Image width*: 8 μm

A CLSM series of mitochondria stained selectively with the membrane potential sensitive dye DASPMI (Bereiter-Hahn 1976; Fuchs et al. 1987) in an XTH-2 cell illustrates the enormous activity of mitochondria in changing shape, fusing and dividing (Fig. 11). In addition this series provides some insight into the dislocations of inner membrane material. Pseudocoloring reveals a pattern of fluorescent patches that move along single mitochondria. In principle local changes in membrane potential cannot be distinguished from variations in inner membrane densities. Comparison of the behaviour revealed by fluorescence and electron micrographs of mitochondria which have been seen to fuse or divide before fixation strongly suggests that areas of higher fluorescence be interpreted as zones with a higher density of crista membranes. At the site numbered 1, the upper mitochondrion bends toward the flank of the long one and fuses for a short time "head to side"; crista accumulation at the fusion site is evident. At the same time the long mitochondrion contracts towards the region of fusion and segregates from its right-hand end (arrowhead) in 1 min. This cleavage remains for the next 30 s while the head-to-side fusion is dissolved (1.5 min). Mitochondrion 2 gives another example of short-lived fusion (0.5, 1, 2, 2.5 min) and fission cycles (1.5, 3, 3.5 min). The fissions taking place in mitochondrion 3 (1-1.5 min, 1.5-2.5 min) are preceded by reduction of fluorescence in the division regions, a very common phenomenon. Crista density increases again when the free ends are withdrawn and thickened (1.5/2 min). In this mitochondrion another interesting kind of behaviour can be observed. At 0.5 min a small mitochondrion fuses with the long one and travels down along its side (1-1.5 min) as is indicated by a highly fluorescing streak running the length of the long mitochondrion. Later on the streak disappears in the fusion product (2 min). Also, electron micrographs of freshly fused mitochondria indicate that the rearrangement of the cristae in the fusion product may take several minutes. This difficulty in rearrangement of cristae might be the reason why many of the fusions are not stable and dissolve. The mitochondria at positions 4 are good examples of unstable fusion. Crista accumulations at the fusion sites in both partners are indicated by the very strong fluorescence in pictures at 2.5 and 3 min, then the mitochondria separate again. At least 80% of the fusions we observed in these cell cultures were disunited again. The reason for believing that fusion had occurred was that the fusion partners exhibited similar movements and changes in shape.

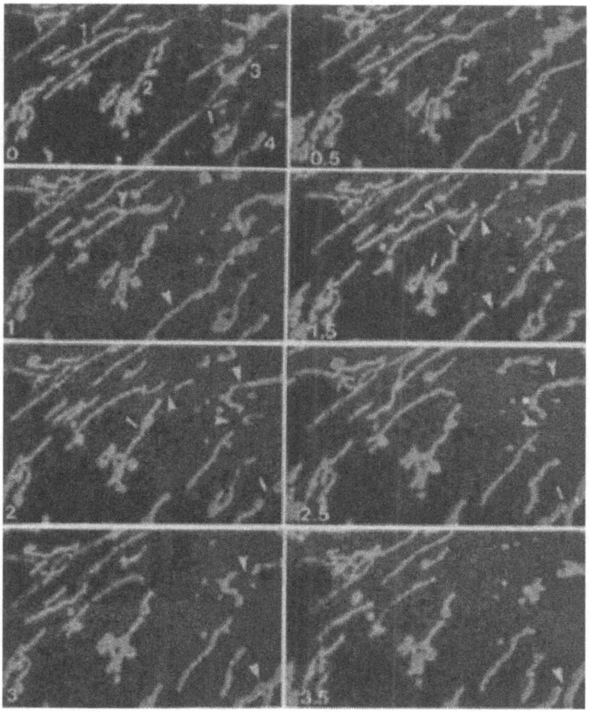

Fig. 11. CLSM image series of fusion and fissions in mitochondria in XTH-2 cells stained with the mitochondria-selective fluorochrome DASPMI. Pseudocolouring marks the intensity distributions of the fluorescence. The images were taken at 30-s intervals by E. van Spronsen (Amsterdam). For detailed description see text. *Image width*: 20 μm

The interpretation of this series is supported by the direct observation of the movements of mt nucleoids in a larger group of mitochondria, as shown in Fig. 12. Here several association and dissociation processes of mt nucleoids are observed, together with more or less extensive displacement of nucleoids inside the mitochondria.

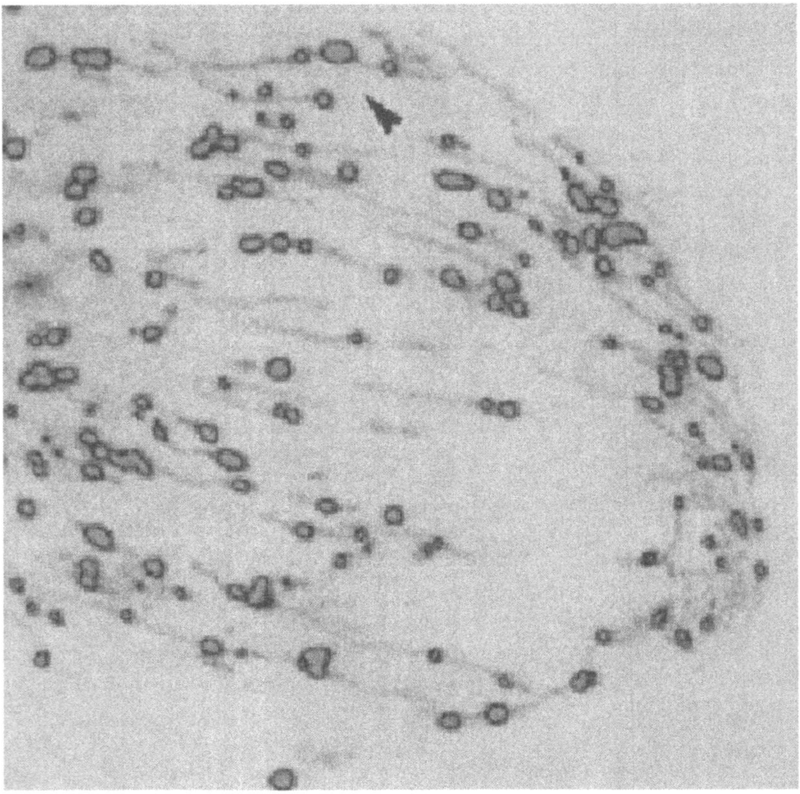

Fig. 12. Time series of picoGreen stained mt nucleoids (*red*) at the periphery of a XTH-2 cell showing intramitochondrial movements of the nucleoids, association and dissociation of mt nucleoids and the behaviour of mitochondria devoid of nucleoids. The weak staining of RNA outlined the whole mitochondria. Time interval between single images: 7s. *Image width*: 26.6 µm

Discussion

For a long time mitochondria have been known to be highly motile structures (for review see Bereiter-Hahn 1990; Bereiter-Hahn and Vöth 1994) which are able to fuse with each other and to divide again. The physiological significance of these processes remains unclear. However, two functions may be involved:

1. Immobilization of mitochondria by fusions, thus forming a large chondriom which supports all parts of a cell equally with metabolites (i.e. in cardiomyocytes, *Chlamydomonas*, some yeast cells).
2. Fusion may provide the structural basis for recombinations among different mt nucleoids. Mitochondrial fusion is considered to be a basic mechanism for spreading mobile genetic elements through

mitochondrial populations within species against a background of almost uniparental mtDNA transmission (Kawano et al.1991).

On the other hand, during mitotic divisions mitochondria must be distributed equally to the daughter cells. In this case many small mitochondria guarantee an approximately equal distribution without the need for specific structures. At the moment we are a long way from being able to explain either fusion or fission on the basis of molecular interactions. Fusion-inducing plasmids have only been identified in *Physarum* (Kawano et al. 1991).

In vertebrate cells it was by no means clear whether in a living cell all mitochondria contain DNA or whether this is restricted to one part of a population only. rho° mutants of HeLa cells entirely lacking mtDNA have been described. If these mutants are fused with cytoplasts of wild type cells rapid and homogeneous mixing of the DNA throughout all mitochondria has been shown by the distribution of ethidium bromide and DAPI staining (Hayashi et al. 1991), cytochromoxidase activity has been transferred to mitochondria lacking this enzyme. Although the appearance of these stainings does not allow conclusions regarding the distribution of DNA in HeLa cells, spreading of cytochromoxidase activity and stainability with ethidium bromide through the whole mitochondrial population revealed extensive recombination among the mitochondria.

Our observations indicate for the first time that a considerable fraction of mitochondria in a living cell in culture might be devoid of DNA. These mitochondria behave like all the others containing DNA. No differences in the frequency of fusion and fission have been observed. This coincides with the observation that fusion and fission processes may take place on the same mitochondrion at a given time, but at different sites. The half-life of mtRNA has been estimated to be several hours (Zylber et al. 1969; Gelfand and Attardi 1981) an hourly fusion of an mt nucleoid-less mitochondrion with another one could be sufficient to maintain a given mRNA level. In XTH-2 cells we observed about 40 fusion and fission processes per hour, corresponding to at least one such event per hour and mitochondrion.

In trypanosomes, due to the specific organization of the mitochondrion (kinetoplast), precise orchestration of nuclear and mitochondrial genome positioning is required to ensure constant segregation during division (Matthews et al. 1995). This is achieved by microtubule-dependent positioning of kinetoplast and nucleus. In this protist all the mtDNA is concentrated in the kinetoplast. This situation differs considerably from the metazoan cells investigated in the present study. In the latter mt nucleoids become assembled with giant nucleoids in cells which are not about to undergo mitosis, in particular during terminal differentiation in keratocytes. But also XTH-2 cells, if they are not immediately abiout to undergo mitosis, exhibit increased multivalency of the mt nucleoids. In contrast, the fast-dividing 3T3 cells and the SV40-transformed 3T3 cells (doubling time about 22 h) exhibit more pronounced peaks at multiples of a first maximum of DNA content and the maximum DNA content of single mt nucleoids is 10 times that of the first frequency peak. The lack of easily identifiable multiples of mitochondrial genome units could be a result of the high variability of some segments of the mtDNA, as has been reported for different teleost fish species (Lee et al. 1995). Similarly, selective amplification and increase of genome heterogeneity during cell culturing has been reported to occur in mitochondria of cultured wheat callus (Weigel et al. 1995).

The linear relationship between total fluorescence intensity (as a measure of DNA content) and mt nucleoid volume indicates that packing of the DNA does not change by compaction of mtDNA into large nucleoids. In addition this relationship indicates that diffusion is not a limiting factor for penetration and binding of the dye to mtDNA. Glycerol treatment and electroporation render the plasma membranes sufficiently permeable for picoGreen. The mitochondrial membranes seem to be highly permeable to a series of dyes. This can be shown with fluorescein, which does not penetrate the plasma membrane but, after being injected into the cytoplasm, accumulates in the mitochondria in a few seconds. In spite of the large nuclear pores, the nuclear membrane in many cells is not easily penetrated by picoGreen, as can be concluded from the varying staining intensity of the nuclei of different cell types (cf. Fig. 6, Fig. 7, Fig. 8). The conditions for cell permeabilization had to be as mild as possible for all the endomembranes to remain intact. This was obviously achieved by mild glycerol treatment and by thoroughly controlled electroporation as well. Mitochondrial motility remained unchanged and the cells appeared in good condition as judged by their appearance on phase-contrast microscopy.

Mitochondria in most eukaryotic cells exist in a flexible state which forms a branching network that ramifies throughout the peripheral cytoplasm; it branches constantly, fuses, pinches apart and changes location. The membranes, enzymes and nucleic adic components are thus continually redistributed within a "fluid" mitochondrial system. The behaviour of mt nucleoids during these changes in shape corresponds broadly to that described for other intramitrochondrial inclusions (Bereiter-Hahn 1978; Bereiter-Hahn and Morawe 1972). They seem to follow the deformations of mitochondrial cristae. The locations of the mt-nucleoids represent zones of higher motility (i.e. bending occurs preferentially around centrally located mt nucleoids, Fig. 1). In addition, fusion and fission (association/dissociation) of the nucleoids has been observed (Fig. 12). The parts derived from fission of a mt nucleoid are about the same size; therefore we conclude that the whole organelle divides rather than separating single copies of mtDNA. Fission of mt nucleoids seems to be induced close to a site of mitochondrial fusion (Fig. 9).

Further investigations are needed to understand the control of fusion and fission of mitochondria as well as their relationship to the behaviour of mtDNA. The present observations open a first window for the direct observation of these events.

References

Bereiter-Hahn, J. (1976) Dimethylaminostyrylmethylpyridiniumiodine (DASPMI) as a fluorescent probe for mitochondria in situ. *Biochim. Biophys. Acta* **423**: 1–14

Bereiter-Hahn, J. (1978) Intracellular motility of mitochondria: role of the inner compartment in migration and shape changes of mitochondria in XTH-cells. *J. Cell Sci.* **30**: 99–115

Bereiter-Hahn, J. (1990) Behavior of mitochondria in the living cell. *Int. Rev. Cytol.* **122**, 1–63

Bereiter-Hahn, J., Morawe, G. (1972) Stoffwechselabhängige mitochondriale Bewegungen in epithelialen Kaulquappenherzzellen in Gewebekulturen. *Cytobiology* **6**: 447–467

Bereiter-Hahn, J., Vöth, M. (1994) Dynamics of mitochondria in living cells: shape changes, dislocations, fusion and fission of mitochondria. *Microsc. Res. Tech.* **27**: 198–219

Borst, B.P. (1972) Mitochondrial nucleic acids. *Annu. Rev. Biochem.* **41**: 333–376

Boukamp, P., Petrussevska, R.T., Breitkreutz, D., Hornung, J., Markham, A. (1988) Normal keratinization in a spontaneously immortalized aneuploid human keratinocyte cell line. *J. Cell Biol.* **106**: 761–771

Chévremont, M. (1963) Cytoplasmic DNAs. The mitochondrial localisation and synthesis in somatic cells under experimental conditions and during the normal cell cycle in relation to the preparation for mitosis. *Symp. Int. Soc. Cell Biol.* **2**

Fox, H. (1987) The skin of amphibia. In: Bereiter-Hahn, J., Matoltsy, A.G., Richards, K.S. (eds.) *Biology of the integument*, vol. II. Springer, Berlin Heidelberg New York, pp. 78–149

Fuchs, J., Zimmer, G., Bereiter-Hahn, J. (1987) A multiparameter analysis of the perfused rat heart: responses to ischemia, uncouplers and drugs. *Cell Biochem. Funct.* **5**: 245–253

Gelfand, R., Attardi, G. (1981) Synthesis and turnover of mitochondrial ribonucleic acid in HeLa cells: the mature ribosomal and messenger ribonucleic acid species are metabolically unstable. *Mol. Cell Biol.* **1**:497–511

Hayashi, J.-I., Ohta, S., Kikuchi, A., Takemitsu, M., Goto, Y. (1991) Introduction of disease-related mitochondrial DNA deletions into HeLa cells lacking mitochondrial DNA results in mitochondrial dysfunction. *Proc. Natl. Acad. Sci. USA* **88**: 10614–10618

Johnson, L.V., Walsh, M.L., Chen, L.B. (1980) Localization of mitochondria in living cells with rodamine 123. *Proc. Natl. Acad. Sci. USA* **77**: 990–994

Karnowski, M.J (1965) A formaldehyde-glutaraldehyde fixative of high osmolality for use in electron microscopy. *J. Cell Biol.* **27**: 137A–138A

Kawano, S., Takano, H., Mori, K., Kuroiwa, T. (1991) A mitochondrial plasmid that promotes mitochondrial fusion in *Physarum polycephalum. Protoplasma* **160**: 167–169

Kuroiwa, T. (1982) Mitochondrial nuclei. *Int. Rev. Cytol.* **75**: 1–59

Kwee, S., Nielsen, H., Celis, J. (1990) Electropermeabilization of human cultured cells grown in monolayers incorporation of monoclonal antibodies. *Bioelectrochem. Bioenerg.* **23**: 65–80

Lee, W.J., Conroy, J., Howell, W.H., Kocher, T.D. (1995) Structure and evolution of teleost mitochondrial control regions. *J. Mol. Evol.* **41**: 54–66

Matthews, K.R., Sherwin, T., Gull, K. (1995) Mitochondrial genome repositioning during the differentiation of the African trypanosome between life cycle forms is microtubule mediated. *J. Cell Sci.* **108**: 2231–2239

Nass, M.M.K. (1977) Studies on the synthesis and structure of mitochondrial DNA in cells infected by Rous sarcoma viruses and on the occurrence of intramitochondrial virus-like particles in certain RSV-induced tumor cells. *Mol. Cell Biochem.* **14**: 121–128

Rizzuto, R., Brini, M., Pizzo, P., Murgia, M., Pozzan, T. (1995) Chimeric green fluorescent protein as a tool for visualizing subcellular organelles in living cells. *Curr. Biol.* **5**: 635–642

Robin, E.D., Wong, R. (1988) Mitochondrial DNA molecules and virtual number of mitochondria per cell in mammalian cells. *J. Cell Physiol.* **136**: 507–513

Roodyn, D.B., Wilkie, D. (1968) The biogenesis of mitochondria. Methuen´s monographs on biological subjects. Methuen, London

Satoh, M., Kuroiwa, T. (1991) Organization of multiple nucleoids and DNA molecules in mitochondria of a human. *Cell Exp. Cell Res.* **196**: 137–140

Schlage, W., Kuhn, C., Bereiter-Hahn, J. (1981) Established *Xenopus* tadpole heart endothelial (XTH) cells with selected properties of primary cells. *Eur. J. Cell Biol.* **24**:342

Strohmeier, R., Bereiter-Hahn, J. (1991) Emigration of bilayered epidermal cell sheets from tadpole tails (*Xenopus laevis*). *Cell Tissue Res.* **266**: 615–621

Swift, H. (1964) Nucleic acids of mitochondria and chloroplasts. *II. Congr. Int. Histochem.* **82**. Springer, Berlin Heidelberg New York

Weigel, R., Wolf, M., Hesemann, C.U. (1995) Mitochondrial DNA variation in plants regenerated from embryogenic callus cultures of CMS triticale. *Theor. Appl. Genet.* **91**: 1237–1241

Whitaker, J.E., Moore, P.L., Haugland, R.P., Haugland, R.P. (1991) Dihydrotetramethylrosamine: a long wavelenth fluorogenic peroxidase substrate evaluated in vitro and in model phagocyte. *Biochem. Biophys. Res. Commun.* **175**: 387–393

Zylber, E., Vesco, C., Penman, S. (1969) Selective inhibition of the synthesis of mitochondria associated RNA by ethidium bromide. *J. Mol. Biol.* **44**:195–204

Exp. Biol. Online 1:5 (1996)

ConcanavalinA receptors and the chemosensory behaviour of *Tetrahymena thermophila*

Vagn Leick[1], Thorkild C. Bøg-Hansen[2], Søren T. Christensen[1], and Stephen J. Kaufman[3]

[1] Department of Medical Biochemistry and Genetics Biochemistry Laboratory B, The University of Copenhagen, The Panum Institute, Blegdamsvej 3C, DK-2200 Copenhagen N, Denmark
[2] The Protein Laboratory, The University of Copenhagen, The Panum Institute, Blegdamsvej 3C, DK-2200 Copenhagen N, Denmark
[3] Department of Cell and Structural Biology, University of Illinois, 506 Murrill Hall, 505 South Goodwin Avenue, Urbana, IL 61801, USA

Correspondence to: V. Leick (Tlf. +45 35 32 77 66; Fax: +45 31 39 60 42; e-mail: imbgvl@pop.kubism.ku.dk)

Received: 6 June 1996 / Accepted: 30 July 1996
Published: August 2, 1996

Abstract. The relationship between concanavalin A (ConA) receptors and the chemosensory behaviour of the ciliated protozoan *Tetrahymena thermophila* was studied using the peptide chemoattractants proteose peptone and fibroblast growth factor. Studies on the chemosensory behaviour in semisolid methylcellulose showed that 50 μg/ml ConA selectively inhibited the persistent element of swimming behaviour by reducing time runs of cells responding to proteose peptone from 12.2±4.5 min to 0.8±0.3 min.
Methyl-alpha-D-mannoside, but not methyl-alpha-D-galactoside, abolished the inhibitory effect of ConA, suggesting that mannoside-containing ConA receptors are involved in maintaining a persistent swimming behaviour. Control experiments, carried out in liquids where persistent swimming is less important for cellular behaviour, showed that ConA did not affect proteose-peptone-induced chemoattraction under these conditions as measured by a two-phase assay for chemoattraction. Also, no inhibitory effect of ConA could be found on swimming rates when individual velocities of ConA-treated cells were determined. When tested in liquid chemoattraction assays, ConA was found to be a weak but significant chemoattractant. Studies of the cellular location of ConA receptors on the plasma membrane of starved cells showed an unequal distribution. A preferential clustering of receptors at the anterior end of

the cell was observed when determined at high concentrations (100 µg/ml) of fluorescent ConA. Methyl-alpha-D-mannoside but not methyl-alpha-D-galactoside abolished the fluorescent ConA labelling, indicating a preferential clustering of these mannoside-containing receptors at the anterior part of the plasma membrane and cilia. At lower concentrations (25 µg/ml), FITC-ConA produced more general labelling of the entire cell membrane. The results suggest that ConA receptors are necessary for the persistent element of swimming and that binding of ConA to its receptors interferes with processes related to signal transduction rather than by limiting the free movement of cilia required for locomotion. The gradient of receptors seen at high FITC-ConA concentrations may be important for a putative spatial chemosensory mechanism, i.e. chemotaxis.

Key words: Concanavalin A receptors – Recognition – Chemosensory behaviour – *Tetrahymena*

Introduction

The ciliated protozoa offer an interesting opportunity to study the relationship between chemoreception, cell behaviour and cell structure. The membrane ultrastructure of ciliates resembles that of the chemosensory neurons and the olfactory epithelium in mammals (Hufnagel 1992; Menco 1992). As in higher organisms, the cilia may have adapted a sensory role. Ciliates may therefore be viewed as "swimming receptors" where correlations between cell behaviour, cell structure and molecular events can be studied experimentally more easily than in higher cells because many ciliates are easy to grow and handle in the laboratory (Leick and Hellung-Larsen 1985; Wheatley et al. 1994; Wolfe 1993)

Previous studies of the chemosensory behaviour of *Tetrahymena* have shown that this ciliate is chemoattracted to a range of different proteins, peptides and amino acids. They include mammalian signal compounds such as fibroblast growth factor (FGF) and platelet-derived growth factor (PDGF), single amino acids like L-phenylalanine, L-cysteine, L-histidine, and L-leucine and complex peptide mixtures like proteose peptone (Almagor et al. 1981; Leick et al. 1996 (review)). However, although these molecules all act as chemoattractants, they promote different response intensities. Firstly, the single amino acids are "weak" attractants compared to proteose peptone, FGF and PDGF, and they are active only at 1 mM, whereas FGF and PDGF have effects in the nano- and picomolar concentration range. Secondly, only "strong" attractants like PDGF and proteose peptone have been shown to produce a chemosensory response that enables the cells to overcome physical hindrances. For example, when cells are migrating through semisolid gelatine in the Zigmond chamber only proteose peptone, FGF and PDGF are strong enough chemoattractants to induce oriented cell "gliding" under these high viscosity conditions (Leick et al. 1995). Moreover, in a filter paper assay (Boyden-type assay) developed for *Tetrahymena* only proteose peptone and PDGF were able to induce altered locomotion strong enough to make the cells swim through (penetrate) the filter paper membrane (Leick et al. 1990). The high-intensity responses have recently been further characterised as a persistence of cellular swimming caused by intensive ciliary beating and to much longer time runs for cells swimming positively towards an attractant than for cells swimming negatively away from an attractant (Leick et al. 1995).

Studies in other laboratories have shown that if two different mating types of starved *Tetrahymena thermophila* are mixed, a mating-type-dependent swimming is induced (costimulation), leading to the formation of cell pairs (conjugants). It has also been shown that the lectin concabavalin A (ConA) inhibits the onset of pair formation as well as inducing complete separation of early pairs during the conjugation process (Casci and Hufnagel 1988; Frisch and Loyter 1977; Watanabe et al. 1981; Wolfe et al. 1986). In this paper evidence is provided to show that the plant lectin ConA also interferes with the persistent element of swimming during chemoattraction towards peptide chemoattractants. It will be discussed how this phenomenon might relate to the effect of ConA on cell/cell recognition (pair formation) during the conjugation process and/or to its impact on cellular survival and proliferation.

Materials and methods

Growth of cells

Tetrahymena thermophila (strain B7) was grown at 21-28 °C to early or mid-stationary phase in complex growth medium: 0.75 % proteose peptone (Difco) containing 0.75 % yeast extract (Difco) and 1.5 % glucose, 0.001 M $MgSO_4$, 0.00005 M $CaCl_2$, and 0.0001 M ferric citrate, in thin-layer (1-2 cm) cultures without agitation in Fernbach flasks. For chemoattraction experiments, cells were collected by centrifugation at 500 *g* for 2 min, washed twice and re-suspended in 10 mM HEPES (*N*-[2-hydroxyethyl] piperazine-*N′*-[2-ethanesulphonic acid]), pH 7.4 (neutralised with KOH) to a cell concentration of $1.5\text{-}2 \times 10^{-5}$ cells/ml.

Assays of cell chemosensory behaviour

The two-phase population assay (liquid assay). The two-phase assay is a liquid assay for chemoattraction of a cell population of *Tetrahymena*. The cell suspension, layered on the top of a high density metrizamide solution of chemoattractant, responds to the gradient of the attractant established in the interphase between the two solutions.

The two-phase assay was carried out at 28 °C as described (by Koppelhus et al. (1994). The chemoattractant was dissolved in 1 ml 5% (w/v) metrizamide (2-[3-acetamido-5-*N*-methyl-acetamido2,4,6-triiodobenzamido]-2-deoxyglucose) from Nycomed, Oslo, Norway in 10 mM HEPES, pH 7.2 in a disposable plastic cuvette. A 1.5-ml suspension of starved cells was carefully layered on the top of the metrizamide solution. The increase in cell concentration in the lower phase was monitored by following the optical density (OD600) of the metrizamide phase monitored automatically in a thermostated spectrophotometer (Shimadzu UV-160). Methyl-alpha-D-mannopyranoside, methyl-alpha-D-galactopyranoside and basic FGF from bovine pituitary glands were from Sigma, St. Louis, Mo., USA.

The filter paper population assay. In the filter paper assay the cells migrate through the filter paper membrane in response to a gradient of chemoattractant established by diffusion through the membrane of the filter paper.

The filter paper assay for chemoattraction was carried out at 28°C as described in detail earlier (Leick et al. 1990). A Whatman 3MM filter paper membrane (Whatman no. 3030917) divided a 2-ml chamber into two 1-ml compartments. One ml cell suspension was placed in one compartment and 1 ml 10 mM HEPES buffer, pH 7.2 containing the attractant in the other. The appearance of cells (increase in cell concentration) in the attractant-containing well (compartment) was determined by a Coulter electronic cell counter .

The field assay (gel assay); motion analysis of single cells. Analysis of the migration (motion analysis) of individual cells in semisolid media with a gradient of chemoattractant was carried out by video microscopy or by computer analysis of digitised video images.

The field assay (chemoattractant gradient gel chamber) was set up as described earlier (Leick et al. 1995).Two 1-ml liquid wells were drilled in Plexiglas to make a Zigmond chamber. The two wells were separated by 1 mm semisolid (gel) placed under a microscope coverslip. The gradient of chemoattractant through the 1 mm gel was established by diffusion (15-20 min) by dissolving the attractant in one of the wells. The thickness of the gel was 100-200 µm. In most experiments 5-10 % methyl cellulose (w/v) (Sigma, No. M-0512) was used as the solidifying substance. The methylcellulose was suspended in 10 mM HEPES buffer pH 7.2. The highly viscous solution was allowed to become homogeneous at 4°C for 2-3 days, with occasional stirring with a spatula. The gel was placed on the bridge of the Zigmond chamber with a spatula, and a coverslip was used to distribute carefully the semisolid evenly on the top of the bridge, avoiding inclusion of air bubbles. Solutions of proteose peptone (1 mg/ml) or FGF (300 ng/ml) containing human serum albumin (50 µg/ml) as carrier were used as attractants. The attractant was placed in one well of the gradient gel chamber and the cell suspension in the other. The cells had to migrate along the gradient of the chemoattractant in the methylcellulose gel in order to arrive at the solution containing the attractant. For motion analysis, measurements of the microscopic images were carried out using a video camera (Philips CCD camera, type LDH070/10) connected to a video tape recorder or an Apple Macintosh II computer equipped with a Video Spigot frame grabber board. The duration of cell "time runs" between "stops" was determined by following cell migration on the computer screen as described in detail previously (Leick et al. 1995). By placing the screen cursor at the very tip of a migrating cell, transitions between the two different cell morphological stages associated with "runs" and "stops" could be determined with a time range of error of ±5 s. Each experiment was repeated at least three times. The time runs are expressed as means and standard deviations of the mean (SD). Significance of differences was determined by Student's *t*-test and analysis of variance.

Determinations of swimming speeds

Swimming speeds of individual cells in free solution were determined by video microscopy using the same chamber as above for the field assay, but without solidifying the substance on the bridge. Two-seconds swimming tracks were transferred onto transparencies from the screen of video-taped recordings of *Tetrahymena thermophila* pretreated as indicated in each experiment. The length (mm) of individual tracks was monitored using an electronic graphics calculator (Numonics Corp., USA) and the distribution of swimming speeds under various conditions was calculated as described.

Fluorescence microscopy

Living cells were prepared for microscopy by preincubation for 20 min with FITC-ConA ± methyl-alpha-D-mannopyranoside. Fixed cells were prepared by preincubation of cells in 2% glutaraldehyde in HEPES buffer for 10 min followed by two washes with HEPES buffer and labelling with FITC-ConA as above.

Fluorescence microscopy was carried out using a Leitz Diaplan epifluorescence microscope using the oil emersion objectives PL Fluotar 100 and NPL Fluotar 40. For FITC-ConA fluorescence microscopy an A2 (ultraviolet) excitation filter was used and the microscopic image was monitored using the video camera connected to a video tape recorder. Prints were made from still frames of the video recordings using a Sony video printer or by photomicrography using a Wild/Leitz photoautomat MPS 45.

Results

Inhibition of persistent cell locomotion by ConA

Initially we observed that cells treated with ConA (1-50 µg/ml) were unable to enter (burrow into) semisolid media even in the presence of a strong chemoattractant. These observations suggested that an element of persistence in the swimming behaviour, which is consistently observed in control cells (Leick et al. 1995), is lost in the presence of this lectin.

To investigate this observation in detail, quantitative measurements of the swimming behaviour ("gliding") were therefore carried out on individual cells using the field assay and the filter paper assay. The field assay is used for detection of cell "gliding" in a semisolid methylcellulose gel by measuring time runs of individual cells using computer-aided motion analysis.

In Film no.1, examples of "gliding" as well as idling cells (stops) are shown. In the filter paper assay, the physical hindrance is a membrane of filter paper that the cells must overcome in order to be scored as responders to an attracting chemical signal.

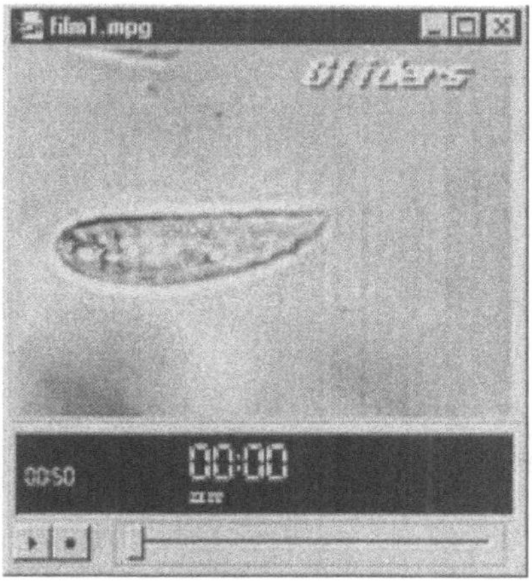

Video recording in real time of swimming ("gliding") of *Tetrahymena thermophila* in semisolid 4% methylcellulose in the field assay chamber. In the film, examples of both "gliding" and "idling" cells are shown as well as transitions between the two stages. A time run of a "gliding" cell was operationally defined as the time spent in this behavioural stage before a stop (idling) lasting for more than 10 s. The metachronal ciliary waves become most conspicuous if the film is played on a high performance computer (24 frames/s) or if it proceeds manually (frame by frame) on the computer screen.

Using the field assay, cells "gliding" in chemoattractant gradients were induced to perform prolonged runs using proteose peptone as stimulant (Leick et al. 1995). When we looked at the effect of ConA and cycloheximide on time runs of indiviual cells (10 cells scored at each condition in Table 1), we saw that they were reduced dramatically and significantly by ConA or cycloheximide compared to control cells (from 12.2±4.5 min to 0.8±0.3 and 0.7±0.2 min, respectively; $P = 0.0002$). Methyl-mannoside abolished most of the inhibitory effect of ConA: time runs were increased from 0.8±0.3 min to 9.4±1.8 min when methyl-mannoside was present together with ConA ($P = 0.19$ when comparing 12.2±4.5 min with 9.4±1.8 min). In contrast, the addition of 50 mM methyl-alpha-D-galactopyranoside did not significantly alter the effect of ConA as there was a significant difference between 12.2±4.5 min and 2.3±0.7 min ($P = 0.0006$) .

Table 1. Effect of concanavalin A (ConA) on time runs of individual cells "gliding" in semisolid medium (5% methylcellulose) in gradients of the attractant proteose peptone in the field assay. Cells were starved for 3 days and allowed to migrate into the semisolid medium in the Zigmond chamber as described in Materials and methods. Time runs were measured by following individual cells using the cursor on the computer screen. Time runs towards the chemoattractant were measured on cells migrating in the 180° angle towards the attractant. The chemotactic gradient in the Zigmond chamber was from 0 mg/ml (*left well A*) to 1 mg/ml proteose peptone (*right well B*) in the presence and absence of ConA (50 µg/ml) ±methyl-alpha-D-mannopyranoside (50 mM) or ±methyl-alpha-D-galactopyranoside (50 mM) placed in both wells as indicated. When present, cycloheximide (20 µg/ml) was also added to both wells. Ten (10) individual cells were scored under each condition.

Conditions in field assay chamber	**Average time runs of individual cells in min ±SD 10 runs were scored under each condition**
Unstimulated control (no gradient of proteose peptone)	1.9±1.2
Stimulated, adapted control (uniform conc. of proteose peptone)	2.0 ±0.5
*Gradient of proteose peptone in gel (0-1 mg/ml)	12.2 ±4.5 (towards attractant)
*Gradient of proteose peptone + cycloheximide (uniformly 20 µg/ml)	0.7 ± 0.2
*Gradient of proteose peptone + ConA (uniformly 50 µg/ml)	0.8 ± 0.3
*Gradient of proteose peptone + ConA (uniformly 50 µg/ml)+ methyl-alpha-D-mannopyranoside (uniformly 50 mM)	9.4 ± 1.8
*Gradient of proteose peptone + ConA (uniformly 50 µg/ml) + methyl-alpha-D-galactopyranoside (uniformly 50 mM)	2.3±0.7

When using the filter paper assay procedure, the chemosensory response of the cell population, as a whole, is monitored by cell counting. This method works with strong peptide chemoattractants like proteose peptone and PDGF (Leick et al. 1990). ConA has a clear inhibitory effect as seen in Table 2. The inhibitory effect is abolished by methyl-alpha-D-mannopyranoside, but not by methyl-alpha-D-galactopyranoside, suggesting that mannoside-containing receptors are important for the cell's potential for a high-intensity response necessary for migration through the Whatman 3MM filter paper. Also, a significant stimulatory effect was consistently seen when this mannoside derivative was added, as seen in Table 2. As methyl-mannoside is not a chemoattractant, it can be explained by an immediate stimulatory effect on the swimming speed induced by methyl-mannoside (see Fig. 2).

Table 2. Effect of ConA on chemoattraction towards proteose peptone in the filter paper assay. Cells starved for 3 days, responding at 28°C to proteose peptone in the presence and absence of ConA (50 µg/ml) and methyl-alpha-D-mannopyranoside (50 mM) or methyl-alpha-D-galactopyranoside (50 mM). The assay was carried out as described earlier (Leick et al. 1990) and in Materials and methods. Starved cells (2x105 cells/ml) were placed on one side (*A*) of the Whatman 3MM filter paper membrane and the chemoattractant on the other side (*B*) in the assay chamber. The concentration of cells in well B was determined in the Coulter counter at 30 min after set-up.The experiment shown, which is representative, was repeated three times

Conditions	**Cells/ml in B-comp. at 30 min**	**% of stimulated control**
Unstimulated control (no attractant)	1.8 x 10e3	2.8
Stimulated control (Proteose peptide, 1mg/ml in B)	6.4 x 10e4	100
Stimulated control + ConA (50 µg/ml in A and B)	2.8 x 10e4	44
Stimulated control + ConA (50 µg/ml) + methyl-alpha-D-mannopyranoside (50 mM) in A and B	9.6 x 10e4	150
Stimulated control + ConA (50 µg/ml) + methyl-alpha-D-galactopyranoside (50 mM) in A and B	3.1 x 10e4	48

We also tested the effect of ConA on the chemoattraction induced by proteose peptone and FGF in the liquid two-phase assay system (Fig. 1A, 1B). In these experiments adaptation is the major cellular mechanism responsible for accumulation toward a chemoattractant. As seen in the figures, ConA seems to have no (or little) inhibitory effect on either of the compounds. These results suggest that ConA (50 µg/ml) does not block any cell surface receptors that are crucial for the cell's overall potential to respond to the chemoattractants proteose peptone and FGF in liquids. In fact, it was observed that in the presence of methyl-mannoside the response was slightly (but significantly) stimulated. This stimulation may also be explained by the stimulatory effect of methyl-mannoside on swimming speeds, as described in Fig. 2.

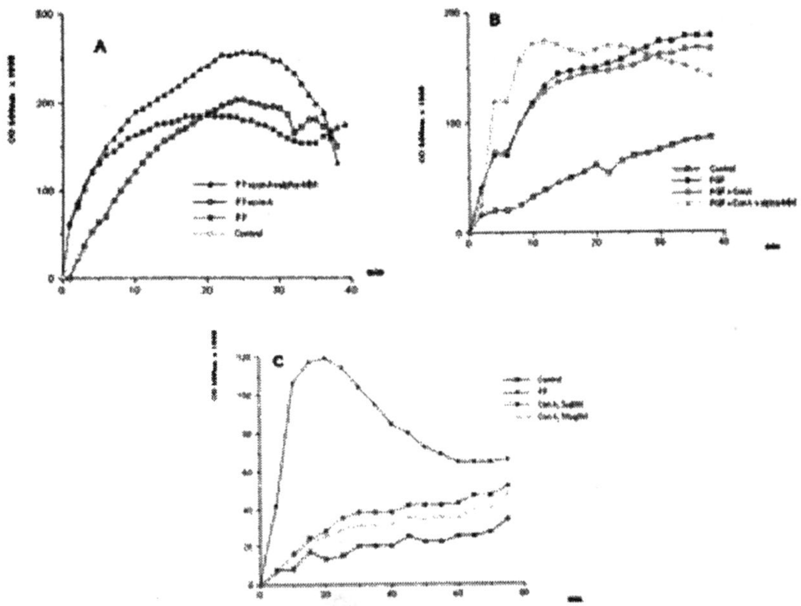

Fig. 1A,B. Time course of the effect of concanavalin A (ConA) on chemoattraction towards proteose peptone and fibroblast growth factor (FGF) in the two-phase (liquid) assay using **A** proteose peptone (1 mg/ml) and **B** FGF (300 ng/ml). Cells starved for 3 days in HEPES buffer were incubated with ConA and immediately layered carefully on the top of a 5% solution of metrazamide also containing other additions, as stated below. The appearance of cells in the metrizamide solution (lower phase) containing the attractant was monitored automatically at 600 nm in a recording spectrophotometer at 28°C. **C.** Time course of the chemoattractant effect of ConA in the two-phase (liquid) assay. Experiment was carried out as in **A** and **B** except that ConA was used as attractant in the lower phase at the concentrations indicated

Stimulatory effect of ConA on swimming speed

Recently we have observed that ConA may act as a positive signal for cell proliferation and survival and can prevent cell death in *T. thermophila* (Christensen and Leick, unpublished observations). Therefore we performed measurements of the swimming speeds in the presence of ConA, proteose peptone and methyl-mannoside. The results are depicted in Fig. 2 and show that, in all cases and combinations of these compounds, a stimulatory effect was seen on the immediate swimming speeds as measured 5 min after the addition to starved (control) cells. ConA induced a small but significant increase in swimming speed compared to control cells and an additive effect was observed when proteose peptone and/or methyl-mannoside was added to ConA.

Taken together with the fact that ConA can act as a positive signal for cell survival and proliferation in lymphocytes and thymocyte cells, possibly through binding to mannoside residues in the plasma membrane (Calder et al. 1991; Grassili et al. 1992), it may be suggested that putative mannoside-containing receptors are involved in signalling processes related to cell proliferation/survival as well as to chemosensory responses in *Tetrahymena*.

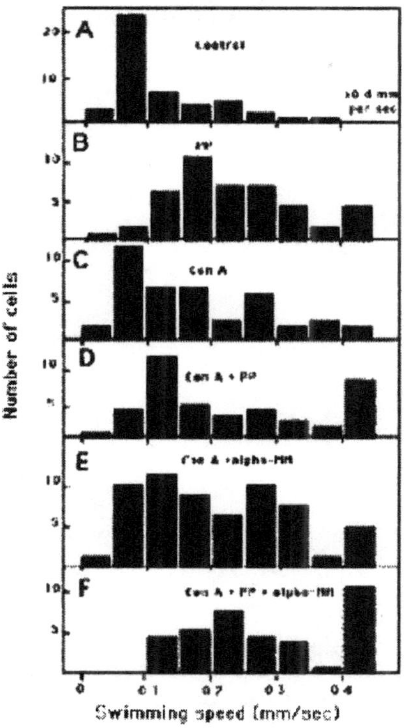

Fig. 2. Distribution of ConA-treated cells swimming at various speeds. Starved cells were treated for 5 min with ConA, proteose peptone, methyl-alpha-D-mannopyranoside and combinations of these compounds (same concentrations as in Fig.1). Swimming speeds of individual cells were determined as described in Materials and methods

Fluorescence microscopy of FITC-ConA-labelled cells

To determine the location of ConA receptors, fluorescent (FITC-labelled) ConA was added to a suspension of living starved cells and to cells fixed before staining and followed by observation by epifluorescence microscopy. In line with previous findings we observed that FITC-ConA at 25 µg/ml induced a general labelling of the plasma membrane (Kovacs et al. 1995; Watanabe et al.1981; Wolfe and Feng 1988) which could be abolished in the presence of 50 mM methyl-mannoside. However in the presence of a higher concentration of FITC-ConA (100 µg/ml) we observed a preferential anterior labelling of cells which was also completely abolished with 50 mM methyl-alpha-D-mannopyranoside as seen in Fig. 3B but not when the same concentration of methyl-alpha-D-galactopyranoside was added (results not shown). The anterior labelling of living cells at 100 µg/ml FITC-ConA (which is also demonstrated on living cells in Film no. 2) is not the result of migration of the ConA/receptor complex in the membrane leading to receptor capping as even glutaraldehyde-prefixed cells show a very similar pattern of labelling, as seen in Fig. 3C and D. Figure 3D also shows that FITC-ConA labelling can be found in the anterior cilia.

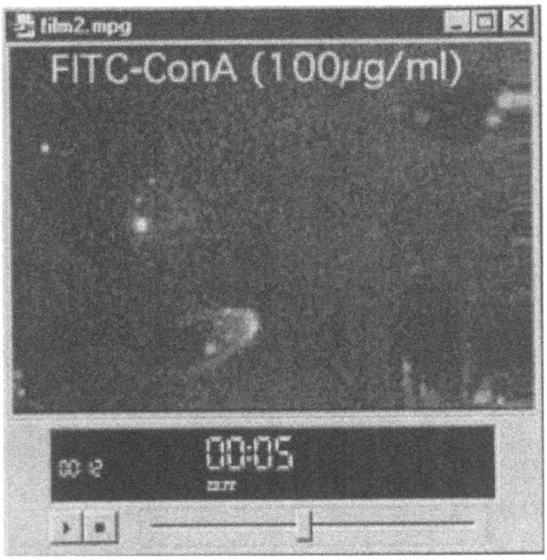

Video recording in real time of free swimming *T. thermophila* labelled with FITC-ConA (100 µg/ml) ± methyl-alpha-D-mannopyranoside (50 mM) as described in the legend to Fig. 3A

* Gradient of proteose peptone attractant through the gel on the bridge in the Zigmond chamber (1 mm width) was established by diffusion, which takes approx. 15-20 min

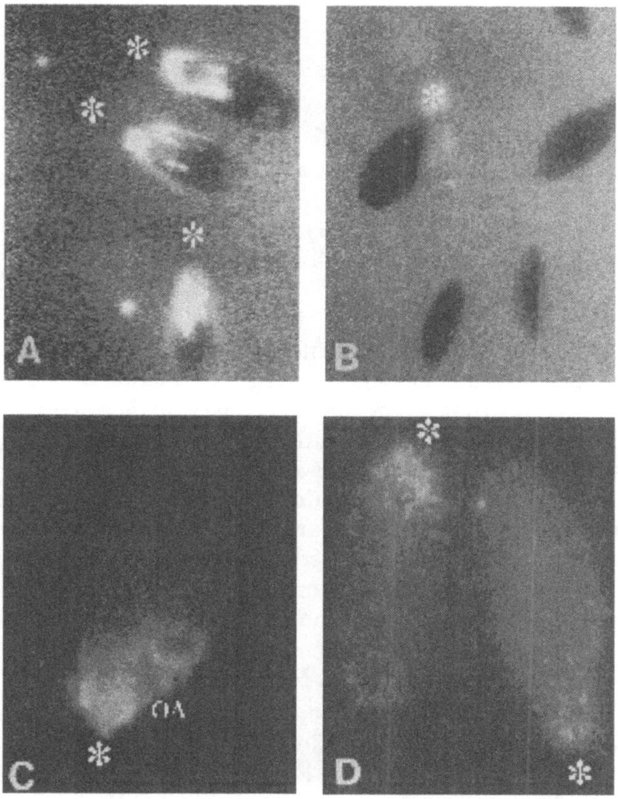

Fig. 3. Fluorescence video- and photomicrographs of the binding of FITC-ConA (100 µg/ml) to *Tetrahymena thermophila* in the absence (**A**), and presence (**B**) of methyl-alpha-D-mannopyranoside (50 mM) and to glutaraldehyde-prefixed cells with the microscopic image focused on the anterior plasma membrane (**C**) and on the anterior cilia (**D**). Conditions of the FITC labelling, prefixation and video-and photomicroscopy were as described in Materials and methods. **A, B** are Video print images; **C, D** photomicrographs. Anterior tip of the cells is indicated by (*); *OA*, oral apparatus

It can be concluded that anteriorly capped ConA binding receptors are present in starved single cells since they can be detected at high FITC-ConA concentrations (Film no.2). These receptors could either have a rather low affinity towards ConA or be partially masked. Watanabe and co-workers have suggested that masked ConA receptors may exist in the anterior surface region of starved cells and be exposed by the collision of opposite mating types during the co-stimulation period leading to "capping" of anterior ConA receptors (Casci and Hufnagel 1988; Watanabe et al. 1981; Wolfe et al. 1986).

Discussion

ConA has a preferential inhibitory effect on the persistent element of swimming of *T. thermophila.* This effect can be monitored on single cell behaviour by measuring the decrease of time runs in semisolids. In the presence of ConA, time runs of cells responding to proteose peptone are reduced from 12.2±4.5 min to 0.8±0.3 min. Methyl-alpha-D-mannoside, but not methyl-alpha-D-galactose, abolishes the inhibitory effect of ConA, suggesting that mannoside-containing receptors are involved in maintaining a persistent swimming behaviour.

In contrast, ConA did not have any pronounced effect on the adaptative element, i.e. on proteose peptone- and FGF-induced chemoattraction in liquids. Thus, adaptation which is supposed to be the major cell mechanism responsible for cell accumulation in liquids, is affected very little by the presence of this lectin. Also, swimming rates of ConA-treated cells are not reduced, but are rather stimulated supporting the hypothesis that ConA interferes with processes related to signal transduction involved in chemotaxis (and growth) rather than interfering with the free movement of cilia required as motor organelles for cell locomotion. When ConA was tested as a chemoattractant in the two-phase (liquid) assay it appeared to act as a weak but significant attractant in concentrations (1-50 µg/ml) where, at the same time, it had a clear inhibitory effect on the persistent element of cell swimming. Moreover, in line with previous findings that autocrine signalling may regulate cell survival in *T. thermophila* (Christensen and Rasmussen 1992; Christensen et al. 1995) experimental evidence of the impact of ConA and methyl-alpha-D-mannopyranoside points to the conclusion that at least some cell signal transduction (and intercellular communication) may act through mannoside-containing receptors (Christensen and Leick, unpublished).

The experiments in which living cells were labelled with fluorescent ConA showed that, at low FITC-ConA concentration (25 µg/ml), fluorescence was found equally distributed over the cell surface and in the food vacuoles, similar to that observed previously in other laboratories (Kovacs et al. 1995; Watanabe et al. 1981). At high FITC-ConA concentrations (100 µg/ml), living cells (as well as cells prefixed with 2% glutaraldehyde) showed a preferential enhancement of the binding of FITC-ConA at the cellular tip anterior to the oral apparatus (see also Film no.2). In all cases 50 mM methyl-alpha-D-mannopyranoside, but not methyl-alpha-D-galactopyranoside, blocked FITC-ConA binding indicating that the anterior receptor structures may contain glycoproteins with this sugar attached. A number of ciliary and membrane ConA binding proteins have been identified in *Tetrahymena.* (Dentler 1992; Wolfe et al. 1993)). In particular, FITC labelling of fixed cells showed a clear fluorescence associated with cilia positioned anteriorally to the oral structure (Fig. 3C, D) also supporting the notion that the anterior part of the cell (and the cilia) might participate in food selection and prey recognition of, for example, bacteria having mannoside-recognising lectins on their surface in a similar way to polarised neutrophils (Bar-shavit et al. 1977; Weinbaum et al. 1980).

The observed gradient of receptors, revealed at high FITC-ConA concentration, showing a preferential anterior location at the cellular tip and of the anterior cilia, may provide an explanation for the selective effect of ConA on the persistent element of cellular swimming. ConA binding to receptors on cilia and the anterior plasma membrane may interfere with the signal transduction pathway required for long time runs. It would be of interest to know how this effect on cell behaviour relates to the ConA inhibition of pair formation during conjugation (Wolfe and Feng 1988). Since the vegetative (growth) cycle and sex cycle are distinct cell cycles having two distinctly independent genetic programs it is possible that ConA has two different effects. On the one hand, it prevents conjugation and pair formation by mechanisms that might be associated with persistent swimming. On the other, it may also induce signal transduction, inducing chemoattraction and other events pushing cells into the vegetative growth cycle.

The gradient of receptors may also be important for a putative spatial chemosensory mechanism, i.e. chemotaxis. In a gradient of attractant, the direction of migration may be determined by the number of hits by chemotactic factors at the cell surface recognition sites. If the hits are more frequent at the anterior end, the cell will continue to move forward (be more persistent) than when the hits are distributed randomly over the whole surface.

Acknowledgements. This work was supported by Carlsbergfondet and Dir. Henriksens Foundation and the Danish Natural Science Research Council. The excellent technical assistance provided by Mrs. Charlotte Iversen is gratefully acknowledged.A preliminary report of this work was given at the Second European Congress of Protistology Clermont-Ferrand, 1995 (Preston and Grebecki 1995).

References

Almalgor, M.,Ron, A., Bar-Tana, J. (1981) Chemotaxis in *Tetrahymena thermophila. Cell Motil.* **1**:261- 268

Bar-shavit, Z., Ofek, I., Goldman, R., Mirelman, D., Sharon, D. (1977) *Biochem Biophys. Res. Commun.* **78**:455-460

Calder, P.C., Bond, J.A., Hunt, S.V., Newsholme, E. (1991) Effect of fatty acids on the proliferation of concanavalin A-stimulated rat lymph node lymphocytes. *Int. J. Biochem.* **23**: 579-588

Casci, R.J., Hufnagel, L.A. (1988) Cell pairing during mating in *Tetrahymena.* Does phagocytosis or cell surface receptor participate in ConA block? *J. Protozool.* **35**: 424-430

Christensen S.T., Rasmussen, L. (1992) Evidence for growth factors which control cell multiplication in *Tetrahymena thermophila*. *Acta Protozool.* **31**: 215-219

Christensen, S.T., Wheatley, D.N., Rasmussen, L. (1995) Mechanisms controlling death, survival and proliferation in a model unicellular eukaryotic cell system, *Tetrahymena thermophila*. *Cell Death Differ.* **2**: 301-308

Dentler, W. L.(1992) Identification of *Tetrahymena* ciliary proteins labeled with sulfosuccinimidyl 6-(biotinamido) hexanoat and Concanavalin A and fractionated with Triton X-114. *J. Protozool.* **39**: 368-378

Frisch, A., Loyter, A. (1977). Inhibition of conjugation in *Tetrahymena pyriformis* by ConA. *Exp. Cell Res.* **110**: 337-346

Grassili, E., Prati, C. de, Monti, D., Troiano, L., Menegazzi, M., Barbieri, D., Francesci, C., Suzuki, H. (1992) Studies on the relationship between proliferation and cell death. II. Early gene expression during concanavalin A-induced proliferation or dexamethasone-induced apoptosis of rat thymocytes. *Biochem. Biophys. Res. Commun.* **188**: 1261-1266

Hufnagel, L.A. (1992) Cortical ultrastrucure and chemoreception in ciliated protists (ciliophora). *Microsc. Res. Tech.* **22**: 225-264

Koppelhus, U., Hellung-Larsen, P., Leick, V. (1994). An improved quantitative assay for chemokinesis in *Tetrahymena*. *Biol. Bull.* **187**: 8-15

Kovacs, P., Sunderman, C., Estridge, B.H., Csaba, G. (1995) A concofocal microscopic evaluation of the effects of insulin imprinting on the binding of concanavalin A by *Tetrahymena pyriformis*. *Cell Biol. Int.* **19**: 973-978.

Leick, V., Hellung-Larsen, P. (1985) Chemosensory responses in *Tetrahymena*; the involvement of peptides and other signal substances. *J. Protozool.* **32**: 550-553.

Leick, V., Frederiksen, K., Lyhne, I., Hellung-Larsen, P. (1990) A paper membrane filter assay for ciliate chemoattraction. *Anal. Biochem.* **184**: 63-66

Leick, V., Koppelhus, U., Rosenberg, J. (1995) Cilia-mediated oriented chemokinesis in *Tetrahymena thermophila*. *J. Eukaryot. Microbiol.* **41**: 546-553

Leick, V., Grave, M., Hellung-Larsen, P. (1996) In: Csaba, G., Muller, W.E.G. (eds) Signal peptide-induced chemosensory behaviour in free cililiates: bioassays and cellular mechanisms. *Progress in molecular and subcellular biology*,vol 17. Springer, Berlin Heidelberg New York, pp. 61-66

Menco, B. Ph. M. (1992). Ultrastructural studies on membrane, cytoskeletal, mucous, and protective compartments in olphactation. *Microsc. Res. Tech.* **22**: 215-224

Preston, T.M., Grebecki, A. (1995) Motility and behaviour in protists. In: Brugerolle, G., Mignot, J.-P. (eds.) *Protistological actualities (Proceedings of the Second European Congress of Protistology).* Université Blaise Pascal de Clermont-Ferrand, Laboratoire de Biologie de Protistes, Aubiere, pp. 185-193

Watanabe, S., Toyohara, A., Suzaki, T., Shigenaka, Y. (1981) The relation of ConA receptor distribution to the conjugation process in *Tetrahymena thermophila. J. Protozool.* **28**: 171-175

Weinbaum, D.L., Sullivan, J.A., Mandell, G.L. (1980) Receptors for ConA cluster at the front of polarized neutrophils. *Nature* **286**: 725-726

Wheatley, D.N., Rasmussen, L., Tiedtke, A. (1994) *Tetrahymena*: a model for growth, cell cycle, and nutritional studies, with biotechnological potential. *Bioessays* **16**: 367-371

Wolfe, J. (1993) The ciliary membrane and its engagement in conjugation. In: Plattner, H. (ed.) *Membrane traffic in protozoa.(Adv. Cell and Mol. Biol. of Membr.*, Vol. 2A, pp. 149-179) JAI Press, Greenwhich, CT

Wolfe, J., Feng, S. (1988) ConA receptor "tipping" in *Tetrahymena* and its relationship to cell adhesion during conjugation. *Development* **102**: 699-708

Wolfe, J., Pagliaro, L., Fortune, H. (1986) Coordination of concanavalin-A-receptor distribution and surface differentiation in *Tetrahymena. Differentiation* **31**: 1-9

Wolfe, J., Mpoke, S., Tirone, S.F. (1993) Cilia, ciliary proteins and mating recognition in *Tetrahymena thermophila. Exp. Cell Res.***209**: 342- 349

Exp. Biol. Online 1:6 (1996)

Scaling effects on muscle function in fast and slow muscles of *Xenopus laevis*

John D. Altringham, Terry Morris, Rob S. James and Chris I. Smith

Department of Biology, University of Leeds, Leeds, LS2 9JT, UK

Correspondence to: J. D. Altringham (J.D.Altringham@Leeds.ac.uk)

Received: 15 July 1996 / Accepted: 7 October 1996
Published: October 11, 1996

Abstract. Fibre bundles or whole muscles from *Xenopus laevis*, ranging in size from 0.5-60g, were studied. Maximum power output of predominantly fast (sartorius) and slow (adductor magnus) muscles was measured at cycle frequencies between 0.5 and 20Hz, using the work loop technique. Power output was highly dependent on cycle frequency, and in 50-60g adults was maximal at 6 Hz for fast fibres (65 Wkg^{-1}), and 2 Hz for slow fibres (14 Wkg^{-1}). The cycle frequency for maximum power output was dependent on body mass (M_b), and decreased as a function of $M_b^{-0.07}$ in fast fibres, and $M_b^{-0.23}$ in slow fibres. The functional significance of these differences is discussed.

Key words: Allometry - Scaling - Muscle - Work loops - Power output

Introduction

Several recent studies have investigated the relationship between body size and the mechanical properties of vertebrate skeletal muscle (e.g. Altringham and Johnston, 1990b; Archer et al., 1990; Rome et al., 1990; Altringham and Young, 1991; Seow and Ford, 1991; Anderson and Johnston, 1992). These studies show that as animal size increases, muscles become slower. They have slower twitches, lower maximum contraction velocities, and in work loop studies, generate maximum power at lower cycle frequencies. These changes are broadly in line with the observed changes in the respiration rate or frequency of the locomotory cycle which the muscles are involved in (e.g. Young et al., 1992b). Rome *et al.* (1990) compared their own data on horse muscle with those from rabbit and rat derived in other laboratories. Maximum contraction velocity (V_{max}) of fast locomotory muscle fibres was found to be much less dependent upon body mass (M_b) than that of slow muscles (<alpha> $M_b^{-0.07}$ and $M_b^{-0.18}$ respectively; but see Seow and Ford, 1991). The allometric equation relating V_{max} to body mass for slow locomotory fibres was very similar to that obtained by Altringham and Young (1991) for frequency for maximum work loop power output (f_{opt}) of mammalian diaphragm:

$$V_{max} = 0.99 M_b^{-0.18} \qquad f_{opt} = 4.42 M_b^{-0.16}$$

In turn, these equations are very similar to those relating body mass to stride frequency at the trot-gallop transition in mammals (f_s) and preferred galloping speed (f_p):

$$f_s = 4.19 M_b^{-0.15} \qquad f_p = 4.44 M_b^{-0.16}$$ (Heglund and Taylor, 1988)

It has been suggested that this striking similarity may have some functional significance. Evolution towards optimal design should give muscles mechanical properties which maximise their performance (in terms of power, efficiency or other relevant parameters) when conditions match those under which they operate in life. Running mammals link respiration and locomotion, breathing once in each stride (Young et al., 1992a;b). If respiratory, and some locomotory, muscles operate to produce maximum power output during running, then it is perhaps not surprising that their properties have a similar dependence on body size as preferred galloping speed (Altringham and Young, 1991; Young et al. 1992b). However, this relation holds only for the slow muscles studied by Rome et al. (1990) and the diaphragm (Altringham and Young, 1991). V_{max} for the fast muscle studied by Rome et al. is much less dependent upon body mass. If this is a property of other fast muscles, it suggests that the selective pressures which determine the properties of fast and slow muscles are somewhat different. In this paper we describe experiments which measure power output in relation to work loop cycle frequency and body size in fast and slow muscle fibres from the aquatic toad, *Xenopus laevis*. In a paper in preparation, these results will be related to a kinematic analysis of swimming in *Xenopus* of a similar size range. Post-metamorphosis *Xenopus*, weighing only 0.5g, have-well developed muscles and swim strongly, and mature adults can weigh well in excess of 50g. This 100-fold size range makes *Xenopus* an ideal animal for such studies: experiments cited above involved inter-specific comparisons.

Materials and methods

Xenopus (body mass 0.5-60g) were killed by decapitation and pithing, the skin was removed from the legs, and the latter immersed in Ringer solution at room temperature (19-22 °C). Ringer composition in mmol l^{-1} = NaCl, 115; KCl, 2.5; Na_2HPO_4, 2.15; NaH_2PO_4, 0.85; Na pyruvate, 5.0; $CaCl_2$, 1.8; pH 7.2 at 20 °C. The legs were pinned out on a transparent base (Sylgard 184, Dow Corning) and either the sartorius or adductor magnus muscle was dissected under dark field illumination with frequent changes of Ringer solution. In small animals it was usual to use the entire muscle for mechanical studies, but muscles from animals >7g were pared down to an approximate diameter of 1.5mm. Where possible slow fibres were removed from sartorius and fast fibres from adductor magnus, based on the known fibre distributions from histochemical studies. Tendons and/or a slice of bone were clamped in aluminium foil clips for attachment of the muscle to the apparatus. The oblique insertion of the adductor magnus into the periostium of the femur meant that care had to be taken in paring and clipping this muscle, to ensure that the preparation consisted of a bundle of parallel fibres of similar length.

Muscles or fibre bundles were transferred to a chamber through which Ringer solution flowed at 20±0.2 °C. One end of the preparation was attached to an isometric force transducer (AE801, SensoNor, Horten, Norway), the other to a servo motor. The preparation was adjusted to approximately its *in vivo* length (leg mid-way between flexion and extension), and left for a minimum of 30min before experimentation. Supramaximal, 2ms duration stimuli were used to elicit the maximum twitch force, and the length of the muscle adjusted to place the preparation on the plateau of the length-force relation, defined as l_0. Twitch parameters were then determined for each preparation: time from stimulus to peak force (twitch rise time), and time from peak force to half maximum force (half relaxation time). Short (15-30 stimuli) tetani were used to determine the stimulation frequency for maximum tetanic force, and this frequency was used in all subsequent experiments on that preparation.

Maximum power output was then determined at a range of cycle frequencies using the work loop technique (Josephson, 1985; Altringham and Johnston, 1990a,b). Briefly, preparations are subjected to sinusoidal strains symmetrical about lo, and stimulated during part of each strain cycle. A plot of muscle length against force yields a hysteresis loop for each cycle, the area of which is the work performed in the cycle. Initial experiments established that over the range of cycle frequencies yielding close to absolute maximum power output, the strain yielding maximum power output at a given frequency was ±6% l_0 (12% peak to peak), and this was used in all subsequent experiments. The number of stimuli, and the phase shift between the onset of stimulation and the strain cycle, were manipulated to maximise power at each frequency. Phase shift is defined from the strain cycle: 0/360° = muscle at l_0 and lengthening, one complete cycle =

360°. In each experimental run, a preparation performed 2-4 work loops, and was then allowed to recover for 8min before the next run. The power output was calculated from the first loop of each run: power typically fell by 0-5% over a four loop run.

Some muscles were dissected and rapidly frozen in isopentane cooled in liquid nitrogen for histochemical analysis. Thin sections (10μm) were cut on a cryostat and stained for succinate dehydrogenase (Sperry, 1981), a mitochondrial marker and thus an indicator of slow muscle fibres with a high oxidative capacity. No attempt was made to differentiate fibres further, along the lines described by Lannergren and colleagues (e.g. Lannergren, 1979).

Results

Morphology

Muscle length scaled with body mass with exponents of 0.31 and 0.25 for adductor magnus and sartorius muscles tively (Table 1). Both values are close to the value of 0.33 expected from geometric similarity.

Histochemistry

Sartorius and adductor magnus muscles were examined from two animals in each of three size ranges of *Xenopus*: 0.5-1g, 5-10g and >40g. Staining for succinate dehydrogenase indicated that sartorius contained <10% slow twitch fibres and adductor magnus <5% fast twitch fibres in all animals.

Table 1 Allometric relations of some morphological and physiological properties of adductor magnus and sartorius muscles (mass = kg, length = m, stress = kN m^{-2}, time = s, power = W kg^{-1}, frequency = Hz). *P* value, r^2 and allometric equation determined from least-squares linear regression (SigmaPlot and SigmaStat, Jandel, San Diego); *n* = no. of observations. If *P*>0.05 (n.s.), then the parameter is not statistically dependent on body mass. Where a significant relationship with body mass was found, the allometric equation is shown in **bold**, and the range of values obtained for that parameter, over the size range of animals used, is given. Where the relationship was not significant, then the mean SE are given

	Adductor magnus				Sartorius			
	Equation	r^2	*P*<	*n*	Equation	r^2	*P*	*n*
Length Range (mm)	**0.055$M_b^{0.31}$** 6-25	0.84	0.001	9	**0.048$M_b^{0.25}$** 6-22	0.92	0.001	20
Stress Mean SE (kN m^{-2})	189$M_b^{0.19}$ 82, 12	0.28	n.s.	8	301$M_b^{0.06}$ 215, 12	0.11	n.s.	14
Twitch rise time Range (ms)	**0.10$M_b^{0.16}$** 35-65	0.51	0.05	9	**0.037$M_b^{0.07}$** 21-39	0.28	0.05	20
Half twitch relaxation time Range (ms)	0.065$M_b^{0.03}$ 37-81	0.01	n.s.	9	**0.075$M_b^{0.19}$** 11-65	0.30	0.05	18
Maximum power output Mean SE (W kg^{-1})	15.3$M_b^{0.05}$ 14, 26	0.02	n.s.	8	71.5$M_b^{0.02}$ 65, 43	0.01	n.s.	13
Frequency for maximum power output Range (Hz)	**1.11$M_b^{-0.23}$** 5-2	0.89	0.01	7	**4.89$M_b^{-0.07}$** 9-6	0.80	0.01	7

Mechanical properties

Table 1 summarises the major findings of the study. In both muscles, maximum isometric stress was statistically independent of body size (Table 1). Twitch rise time for the slow adductor magnus showed a greater dependence on size than that of the fast sartorius (Table 1): both increased with increasing body mass. Relaxation time also showed a significant increase with increasing body mass in sartorius, but not in adductor magnus, possibly due to the greater variability seen between adductor magnus preparations in this parameter.

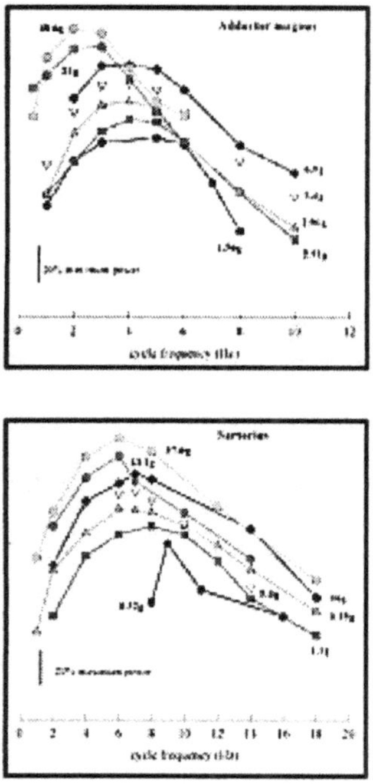

Fig. 1 Relative power output plotted against cycle frequency for fibre bundles from the adductor magnus (**a**) and sartorius (**b**) muscles from *Xenopus* of different masses. The mass of each animal is given by the relevant curve. Curves have been displaced on the vertical axis (by increments of 10% max. power) for clarity

Figure 1 shows power frequency curves for both adductor magnus and sartorius muscles from animals of a wide range of sizes. Power output is expressed relative to the maximum obtained, and each curve is displaced by 0.1 maximum power for clarity. There is a significant increase in the frequency for maximum power with decreasing body mass in both muscles, but the effect is more marked in the adductor magnus. This is seen more clearly in Fig. 2, which shows the relationship between the frequency for maximum power output and body mass. The points were derived by (a) plotting the experimentally tested frequencies which gave the maximum power, and (b) by fitting third order polynomials to the curves shown in Fig. 1, and taking the frequency at the maximum of each curve. Slopes of -0.23 ± 0.037 ($\pm$ standard error of the slope) and -0.20 ± 0.041 were obtained for the adductor magnus for methods (a) and (b) respectively, and -0.07 ± 0.042 and -0.071 ± 0.045 for sartorius. The 95% confidence intervals for both curves overlapped almost completely. Only data from the first method are shown. The vertical lines indicate the range of frequencies over which the muscles develop >90% maximum power. Maximum power of both muscles was independent of body mass (Table 1, Fig.3).

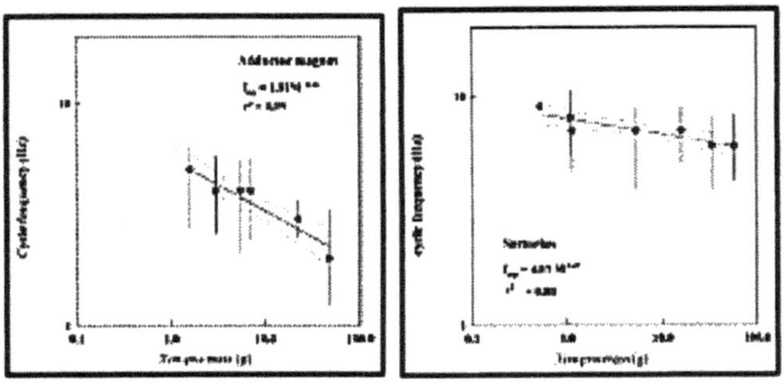

Fig. 2 Log. v. log. plots of cycle frequency for maximum power output against body mass for fibre bundles from the adductor magnus (**a**) and sartorius (**b**) muscles. Mass is expressed in grams on the plot, but the allometric equation uses SI units. *Vertical bars* indicate the frequency range over which more than 90% maximum power is generated. *Dotted lines* show 95% confidence limits

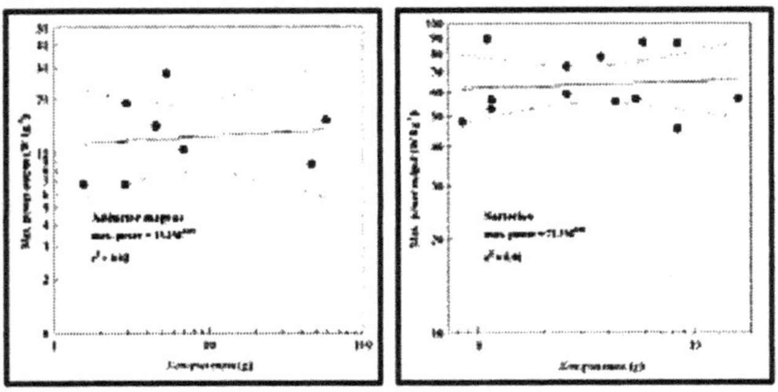

Fig. 3 Log. v. log. plots of maximum power output against body mass for fibre bundles from the adductor magnus (**a**) and sartorius (**b**) muscles. Mass is expressed in grams on the plot, but the allometric equation uses SI units. *Dotted lines* show 95% confidence limits

Discussion

Maximum isometric stresses are similar to those reported for other fast and slow vertebrate muscles at the same temperature, and work loop power outputs are comparable to those reported at similar temperatures and operating frequencies (summarised in James et al., 1995).

Over the size range studied (0.5-50g), and the swimming speeds achieved (I.S. Young and R.S. James, in preparation), the Reynold's number during swimming will be in the range 10^2 - 10^5. Muscle length scales geometrically with body size, as does cross-sectional area (Sperry, 1981). Unpublished measurements (I.S. Young and R.S. James, in preparation) show that the body and limbs also scale geometrically. Any changes in swimming performance or kinematics with size must therefore have their origins in changes in the properties of the locomotory muscles, or the way in which they are used. The proportions of fast and slow fibres in both were comparable in all sizes, and in agreement with a study carried out by Sperry which included the sartorius muscle (1981). Changes in muscle properties are therefore presumably due largely to changes within the fast and slow fibre populations.

The frequency which yielded f_{opt} decreased with increasing body mass for both muscles, but f_{opt} for the adductor magnus decreased more rapidly (<alpha>$M_b^{-0.23}$) than f_{opt} for the sartorius (<alpha>$M_b^{-0.07}$). The exponent for the allometric equation for sartorius was identical to that reported for mammalian fast twitch fibres (Rome et al., 1990), and that for the adductor magnus even greater than that reported for f_{opt} of mammalian diaphragm (-0.16, Altringham and Young, 1991) and V_{max} of mammalian slow locomotor fibres (-0.18, Rome et al., 1990). Rome et al. (1990), Altringham and Young (1991) and Young et al. (1992b) have all argued that the most likely driving force behind size-related changes in slow fibre properties is the maintenance of near-maximum efficiency. The observed changes will ensure that muscles will operate under conditions which yield close to maximum power and efficiency. Curtin and Woledge (1993a,b) have confirmed that fast and slow dogfish muscle fibres are maximally or near-maximally efficient when operating at frequencies which generate maximum power, and similar results have been obtained for fast and slow muscles from the mouse (Barclay, 1994). The kinematics of the power and recovery strokes at different speeds have not yet been determined, so it is not yet possible to relate slow muscle properties to locomotory function.

Rome et al. (1990) suggested that fast muscles may sacrifice efficiency for a large mechanical power output. This power would be necessary for the rapid movements involved in predator avoidance or prey capture. The results from *Xenopus* fast fibres are consistent with this hypothesis. It could be argued that the rapid contraction kinetics present in the fast fibres immediately post-metamorphosis are preserved as much as possible during growth, to retain the potential for a high mechanical power output. The power stroke of the escape response of *Xenopus* also shows a weak dependence on body mass (I.S. Young and R.S. James, in preparation). However, Johnson *et al.* (1993) found that fast muscle f_{opt}, and limb cycle frequency during burst running, showed a marked dependence on body size in the desert iguana (<alpha>$M_b^{-0.24}$). There is clearly no simple explanation, and systematic study of a wider range of animals and muscles is required to resolve this question.

Although f_{opt} decreased with increasing body mass, power output itself was independent of body mass in both muscles. This may in part be due to the finding that muscle stress was also independent of body mass. Power is the rate at which work is performed, and work is force multiplied by distance. The changes in f_{opt} and twitch kinetics with size are relatively small, and any effect they may have had on muscle power may have been masked by the variation in stress and hence power observed between preparations from similar-sized animals.

The size dependence of twitch kinetics can explain much of the size-dependent change in f_{opt}, particularly in the case of the sartorius (Table 1). However, changes in the force-velocity relation are also likely to be important (Rome et al., 1990), and possibly other dynamic muscle properties such as shortening-induced deactivation, which leads to a more rapid relaxation phase in muscle performing work loops (Altringham and Johnston, 1990b).

At this stage we can only speculate on the causes of the changes in muscle kinetics. Since both activation and relaxation times change in isometric contractions, there are probably changes in both the contractile proteins and the calcium cycling proteins. (e.g. Greaser et al., 1988; Sweeney et al., 1988).

Acknowledgements This work was supported in part by grants from the Royal Society to J.D.A and an undergraduate bursery from the Nuffied Foundation to T.M. R.S.J was supported by a SERC-CASE studentship. Thanks to John Oughton for technical support and Iain Young for useful discussions.

References

Altringham, J.D, Johnston, I.A. (1990a) Modelling muscle power output in a swimming fish. *J. Exp. Biol.* **148**:395-402

Altringham, J.D, Johnston, I.A. (1990b) Scaling effects in muscle function: power output of isolated fish muscle fibres performing oscillatory work. *J. Exp. Biol.* **151**:453-467

Altringham, J.D, Young, I.S. (1991) Power output and the frequency of oscillatory work in mammalian diaphragm muscle: the effects of animal size. *J. Exp. Biol.* **157**:381-389

Anderson, M.E, Johnston, I.A. (1992) Scaling of power output in fast muscle fibres of the Atlantic cod during cyclical contractions. *J. Exp. Biol.* **170**:143-154

Archer, S.D., Altringham, J.D, Johnston, I.A. (1990) Scaling effects on the neuromuscular system, twitch kinetics and morphometrics of the cod, *Gadus morhua. Mar. Behav. Physiol.* **17**:137-146

Barclay, C.J. (1994) Efficiency of fast- and slow-twitch muscles of the mouse performing cyclic contractions. *J. Exp. Biol.* **193**:65-78

Curtin, N.A, Woledge, R.C. (1993a) Efficiency of energy conversion during sinusoidal movement of white muscle fibres from the dogfish, *Scyliorhinus canicula. J. Exp. Biol.* **183**:137-147

Curtin, N.A, Woledge, R.C. (1993b) Efficiency of energy conversion during sinusoidal movement of red muscle fibres from the dogfish, *Scyliorhinus canicula*. *J. Exp. Biol.* **185**:195-206

Greaser, M.L., Moss, R.L, Reiser, P.J. (1988) Variations in contractile properties of rabbit single muscle fibres in relation to troponin T isoforms and myosin light chains. *J. Physiol. (Lond)* **406**:85-98

Heglund, N.C, Taylor, C.R. (1988) Speed, stride frequency and energy cost per stride: how do they change with body size and gait. *J. Exp. Biol.* **138**:301-318

James, R.S., Altringham, J.D, Goldspink, D.F. (1995) The mechanical properties of fast and slow skeletal muscles of the mouse in relation to their locomotory function. *J. Exp. Biol.* **198**:491-502

Johnson, T.P., Swoap, S.J., Bennett, A.F, Josephson, R.K. (1993) Body size, muscle power output and limitations on burst locomotor performance in the lizard *Dipsosaurus dorsalis*. *J. Exp. Biol.* **174**:199-213

Josephson, R.K. (1985) Mechanical power output from striated muscle during cyclic contractions. *J. Exp. Biol.***114**:493-512

Lannergren, J. (1979) An intermediate type of muscle fibre in *Xenopus laevis*. *Nature* **279**:254-256

Rome, L.C., Sosnicki, A.A, Goble, D.O. (1990) Maximum velocity of shortening of three fibre types from horse soleus muscle: implications for scaling with body size. *J. Physiol. (Lond)* 431:173-185

Seow, C.Y, Ford, L.E. (1991) Shortening velocity and power output of skinned muscle fibres from amammals having a 25,000-fold range of body mass. *J. Gen. Physiol.* **97**:541-560

Sperry, D.G. (1981) Fibre type composition and postmetamorphic growth of anuran hindlimb muscles. *J. Morphol.* **170**:321-345

Sweeney, H.L., Kushmerick, M.J., Mabuchi, K., Sreter, F.A, Gergely, J. (1988) Myosin alkali light chain and heavy chain variations correlate with altered shortening velocity of isolated skeletal muscle fibres. *J. Biol. Chem.* **263**:9034-9039

Young, I.S., Alexander, R.McN., Woakes, A.J., Butler, P.J, Anderson, L. (1992a) The synchronisation of ventilation and locomotion in horses (*Equus caballus*). *J. Exp. Biol.* **166**:19-31

Young, I.S., Warren, R.D, Altringham, J.D. (1992b) Some properties of the mammalian locomotory and respiratory systems in relation to body mass. *J. Exp. Biol.* **164**:283-294

Exp. Biol. Online 1:7 (1996)

Cardiovascular drift can occur without a concomitant increase in skin blood flow

Michael J. Buono, Jay H. Heaney, Julia A. Cline, Sandra G. Leichliter

Departments of Biology and Exercise and Nutritional Sciences, San Diego State State University, San Diego, CA 92182, USA

Naval Health Research Center, San Diego, CA 92186, USA

Correspondence to: Michael J. Buono (mbuono@mail.sdsu.edu)

Received: 23 August 1996 / Accepted: 17 October 1996
Published: October 18, 1996

Abstract. Previous studies have suggested that increases in skin blood flow (SkBF) are the primary physiological mechanism responsible for cardiovascular drift during exercise in the heat. Most of these studies, however, used exercise bouts of 60 min in duration or less. The purpose of this study was to explore the possibility that during prolonged (> 60 min) exercise in the heat, cardiovascular drift can occur without a concomitant increase in SkBF. The subjects were five heat-acclimated female volunteers. Each subject completed a 6-h heat exposure (38°C, 62% RH). Heart rate (HR), stroke volume (SV), and two independent measures of SkBF were obtained each hour. Cardiovascular drift occurred, as evidenced by a significant (p<0.05) 19% increase in HR and a significant 21% decrease in SV. Interestingly, however, SkBF plateaued by hour 2 and showed no further increase. Such results suggest that during prolonged exercise in the heat, when SkBF has reached very high values (>20 ml / 100 ml per min) and plateaued, cardiovascular drift can still occur.

Key words: Cardiovascular drift – Exercise – Heart rate – Stroke volume

Introduction

Next to severe hemorrhage, the greatest stress ever imposed on the human cardiovascular system is the combination of exercise and hyperthermia. Together these stresses can present a life-threatening situation (Rowell, 1986).

This is particularly evident during prolonged, constant-load exercise in the heat, which results in a downward drift in stroke volume (SV), while simultaneously there is a compensatory rise in heart rate (HR). This phenomenon has been termed "cardiovascular drift" (Ekelund, 1967).

It is generally assumed that the physiological mechanism responsible for cardiovascular drift is related to the thermoregulatory need to increase skin blood flow (SkBF). Specifically, it is believed that a progressive rise in SkBF (Johnson and Rowell 1975) leads to a fall in central venous pressure (Ekelund and Holmgrem 1964), which ultimately decreases SV.

Although this hypothesis seems quite plausible, most of the studies that have examined the relationship between SkBF and cardiovascular drift have used exercise bouts of 60 min in duration or less (Ekelund 1967; Johnson and Rowell 1975). Recently, several studies (Freund et al. 1987; Nose et al. 1990) have shown that SkBF plateaus after approximately 30–60 min of exercise in the heat. Such a finding suggests that during prolonged (> 60 min) exercise in the heat, cardiovascular drift may occur without a concomitant increase in SkBF. The purpose of this study was to explore such a possibility.

Methods

The subjects for the study were five healthy, heat-acclimated, female volunteers. The study was approved by a Human Subject Committee, and written informed consent was obtained prior to the start of data collection. Their mean (±SD) age, height and weight was 22 ± 2 years, 163 ± 4 cm, and 67 ± 6 kg, respectively.

Each subject completed a 6-h heat exposure (38 °C, 62% RH) in an environmental chamber. Specifically, each hour the subjects walked the first 20 min on a motorized treadmill set at 3 mph and 3% grade, followed by 40 min of seated rest. The workload produced a mean (± SD) oxygen uptake of 1.0 ± 0.1 l/min which represented approximately 30% of each subject's maximum oxygen uptake. HR, SV, mean skin temperature, and core temperature were measured during each of the six 20-min exercise sessions. HR was measured using a trans-sternal Polar HR monitor. Skin temperature was measured on the right shoulder, chest, thigh, and calf using Y.S.I (series 400) temperature probes. Mean skin temperature (T_{sk}) was calculated using the Ramanathan (1964) formula. Core temperature (T_{re}) was recorded using a Sheridan rectal temperature probe inserted 10 cm past the anal sphincter. Cardiac output (Q) was measured via CO_2 rebreathing using the equilibration method originally developed by Collier (1956) and later modified by Jones and Campbell (1982). A fast responding Beckman LB-2 CO_2 analyzer was used to measure end-tidal and equilibrium gas tensions. SV was calculated by dividing cardiac output by HR.

To eliminate movement artifact produced by walking on the treadmill, forearm blood flow (FBF) and SkBF were measured immediately upon completion of each 20-min exercise session. FBF was measured using venous occlusion plethysmography with a mercury-in-silastic strain gauge (Whitney, 1953). During the measurement of FBF the arm was suspended and stabilized above the venostatic pressure point, using a support collar around the wrist. Also, the hand was occluded using a blood pressure cuff inflated to 250 Torr. The venous occlusion cuff was inflated to 50 Torr and placed around the bicep. SkBF was measured using a laser-Doppler right-angle probe attached to the flexor surface of the left forearm. During the 6-h heat exposure, water was allowed on an ad lib. basis, and each subject was encouraged to consume enough to prevent dehydration. This was successfully accomplished since the group as a whole lost less than 0.5% of their body weight over the course of the 6-h heat exposure.

Data collected for each dependent variable were analyzed using a repeated measures MANOVA. Significance was set at the $p<0.05$ level.

Results

The mean (±SD) values for all seven measured dependent variables (HR, SV, Q, T_{re}, T_{sk}, SkBF, FBF) across all 6 h of the study are presented in Table 1. HR significantly ($p<0.05$) increased 24 bpm, or 19%, from hour 1 to hour 6. Conversely, SV significantly decreased during the 6 h of heat exposure by 21%, or 14 ml/beat. T_{re} increased continuously throughout the study, rising from 37.2 °C during the first exercise bout to 38.3 °C during the sixth and last exercise bout.

Table 1 Mean (± SD) values for physiological variables collected during 6 h of exercise in the heat

	Hour 1	Hour 2	Hour 3	Hour 4	Hour 5	Hour 6
Heart rate (bpm)	125 ± 21	135 ± 19	140 ± 23	140 ± 19	143 ± 19	149 ± 17
Stroke volume (ml/beat)	73 ± 14	68 ± 11	70 ± 10	63 ± 10	60 ± 6	59 ± 10
Cardiac output (l/min)	8.9 ± .9	8.9 ± .9	9.4 ± 1.1	8.8 ± .7	8.6 ± 1.4	8.7 ± .8
Core temperature (°C)	37.2 ± .1	37.6 ± .2	37.9 ± .1	38.0 ± .1	38.1 ± .1	38.3 ± .1
Mean skin temperature (°C)	36.5 ± .2	37.0 ± .3	37.1 ± .2	37.2 ± .4	37.3 ± .6	37.4 ± .7
Forearm blood flow(ml / 100ml per min)	8.4 ± 3.3	20.1 ± 4.9	20.6 ± 2.5	19.9 ± 3.8	19.2 ± 2.8	20.4 ± 1.5
Skin blood flow (flow units)	4.6 ± 3.1	8.0 ± 1.7	10.6 ± 3.9	9.6 ± 2.5	10.6 ± 6.2	9.4 ± 2.4

In contrast to the response of the above variables, SkBF, FBF, Q, and T_{sk} all demonstrated plateaus by hour 2, as evidenced by non-significant ($p>0.05$) differences across time. Specifically, the mean FBF was 20.1 ml/100 ml/min at hour 2 and 20.4 ml / 100 ml per min at hour 6. Likewise, mean SkBF was 8 flow units at hour 2 and 9.4 flow units at hour 6. Lastly, T_{sk} was 37.0 °C at hour 2 and 37.4 °C at hour 6.

Mean resting FBF and SkBF was 2.7 ml / 100 ml per min and 1.5 flow units, respectively. These values were obtained in a thermoneutral environment prior to the start of exercise.

Discussion

The most important finding of the current study is that during prolonged exercise in the heat, cardiovascular drift can occur without a concomitant increase in SkBF. This is evidenced by the fact that both HR and SV demonstrated significant drift throughout the 6-h heat exposure, yet SkBF and FBF both plateaued by hour 2 and showed no further increase.

Evidence of a plateau in SkBF is strengthened by the fact that two independent techniques both showed the same temporal pattern. Traditionally, venous occlusion plethysmography has been considered the method of choice by which to measure SkBF quantitatively. Although this measurement may be influenced by changes in both skin and muscle blood flow, forearm muscle blood flow is not significantly changed during leg exercise or body heating. Therefore, alterations in FBF may be attributed to changes in SkBF alone (Saumet, et al. 1988). More recently, the use of laser-Doppler flowmetry has become a popular technique qualitatively to measure relative changes in SkBF (Johnson et al. 1984). As presented in Table 1, although the two techniques produced quantitatively different values, both clearly demonstrated that SkBF plateaued by hour 2.

The current study agrees with and extends the findings of Montain and Coyle (1992). They found that cardiovascular drift occurred in trained cyclists during the last 60 min of a 2-h exercise bout in the heat, despite the fact that FBF had plateaued at approximately 13-15 ml / 100 ml per min. Their findings have been questioned, however, because the high air flows (via fanning) they used to cool the subjects may have reduced the potential rise in FBF. The current study supports their findings and shows that the same results can occur with FBFs that are more consistant with the peak values previously reported in the literature (Nadel et al. 1980; Kenney 1988).

Since the data in the current study would suggest that there are circumstances when cardiovascular drift can occur without a concomitant increase in SkBF, alternate physiological mechanisms must be explored. It could be argued that although SkBF plateaued by hour 2, this does not mean that skin blood volume

(SkBV) did not increase. The two primary physiologic determinants of SkBV are SkBF and vascular compliance. Both are directly related to T_{sk}; thus warming the skin increases SkBV (Roberts and Wenger 1979). However, in the current study since both SkBF and T_{sk} did not increase after hour 2, it is difficult to imagine that SkBV could have increased significantly during that time period. Therefore, the current results would not support the hypothesis that increases in SkBV caused the cardiovascular drift observed from hour 2 to hour 6. Additionally, it has previously been shown that cardiovascular drift is linearly related to whole-body dehydration (Montain and Coyle 1992). However, in the current study dehydration was essentially prevented by a vigorous fluid replacement regimen, yet significant cardiovascular drift still occurred. Thus, these findings would suggest that dehydration is not a mandatory requirement, which is absolutely needed to elicit cardiovascular drift during prolonged exercise in the heat.

Lastly, the results of the current study agree with the work of others (Nadel et al. 1980; Kenney 1988) who have reported that SkBF plateaus at approximately 20 ml / 100 ml per min during exercise in the heat. In all of these studies, however, core temperature continued to rise after the plateau in SkBF occurred. Such results would suggest that the plateau in SkBF may result from baroreflex control via either reduced cutaneous vasodilator gain to a given T_{re} elevation, or baroreceptor-induced vasoconstriction (Patterson et al. 1994).

In summary, it must be emphasized that we are not suggesting that cardiovascular drift is not related to SkBF and SkBV during exercise bouts lasting less than 60 min. In fact, there are numerous studies that have supported such a hypothesis. The current results do, however, suggest that during prolonged exercise in the heat, when SkBF has reached very high values ($\geq$ 20 ml / 100 ml per min) and plateaued, cardiovascular drift can still occur. The physiological mechanism responsible for controlling cardiovascular drift despite a plateau in SkBF is unknown at the current time. Clearly, further work on this topic is warranted.

References

Collier, C.R. (1956) Determination of mixed venous CO_2 tensions by rebreathing. *J. Appl. Physiol.* **9**:25-29

Ekelund, L.G. (1967) Circulatory and respiratory adaption during prolonged exercise. *Acta Physiol. Scand.* **70**:1-38

Ekelund, L.G., Holmgren, A. (1964) Circulatory and respiratory adaptations during long-term, non-steady state exercise in the sitting position. *Acta Physiol. Scand.* **62**:240-255

Freund, B.J. Joyner, M.J., Jilka, S.M., Kalis, J. Nittolo, J.M., Taylor, J.A., Peters, H., Freese, G., Wilmore, J.H. (1987) Thermoregulation during prolonged exercise in heat: alterations with beta-adrenergenic blockade. *J. Appl. Physiol.* **63**:930-936

Johnson, J.M., Rowell, L.B. (1975) Forearm skin and muscle vascular responses to prolonged leg exercise in man. *J. Appl. Physiol.* **39**:920-924

Johnson, J.M., Taylor, W.F., Shepherd, A.P., Park, M.K. (1984) Laser-Doppler measurement of skin blood flow: comparison with plethysmography. *J. Appl. Physiol.* **56**:798-803

Jones, N.L., Campbell, E.J.M. (1982) *Clinical exercise testing.* Saunders, Philadelphia, pp. 130-151

Kenney, L.W. (1988) Control of heat-induced cutaneous vasodilatation in relation to age. *Eur. J. Appl. Physiol.* **57**:120-125

Montain, S.J., Coyle E.F. (1992) Influence of graded dehydration on hyperthermia and cardiovascular drift during exercise. *J. Appl. Physiol.* **73**:1340-1350

Nadel, E.R., Fortney, S.M., Wenger, C.B. (1980) Effect of dehydration state on circulatory and thermal regulation. *J. Appl. Physiol.* **49**:715-721

Nose, H., Mack, A.W., Shi, X., Morimoto, K., Nadel, E.R. (1990) Effect of saline infusion during exercise on thermal and circulatory regulations. *J. Appl. Physiol.* **69**:609-616

Patterson, M.J., Warlters, D., Taylor, N.A.S. (1994) Attenuation of the cutaneous blood flow response during combined exercise and heat stress. *Eur. J. Appl. Physiol.* **69**:367-369

Ramanathan, N.L. (1964) A new weighing system for mean surface temperature of the human body. *J. Appl. Physiol.* **19**:531-533

Roberts, M.F., Wenger, C.B. (1979) Control of skin circulation during exercise and heat stress. *Med. Sci. Sports* **11**:36-41

Rowell, L.B. (1986) *Human circulation: regulation during physical stress.* Oxford University Press, New York, pp. 363-406

Saumet, J.L., Kellogg, D.L., Taylor, W.F., Johnson, J.M. (1988) Cutaneous laser-Doppler flowmetry: influence of underlying muscle blood flow. *J. Appl. Physiol.* **65**:478-481

Whitney, R.J. (1953) The measurement of volume changes in human limbs. *J. Physiol. Lond.* **121**:1-27

Exp. Biol. Online 1:8 (1996)

Modulation of Na^+ transport across leech skin by pesticides and heavy metals

Helga Milde, Wolfgang Clauss and Wolf-Michael Weber

Institute for Animal Physiology, Justus-Liebig-University Giessen, Wartweg 95, D-35392 Giessen, Germany

---

Correspondence to: W.-M. Weber (Tel: ++641 702 5894; from 20.12.96: ++641 99 35255; Fax: ++641 702 5878; from 20.12.96: ++641 99 35059; e-mail: Michael.Weber@physzool.bio.uni-giessen.de)

Received: 13 August 1996 / Accepted: 12 November 1996
Published: November 22, 1996

Abstract. The aim of the present study was to investigate the effects of environmental pollutants, such as heavy metals and pesticides on ion transport across the skin of the leech (*Hirudo medicinalis*). We wanted to examine the suitability of this epithelium as a model system for studies concerning the mechanisms of toxic action caused by environmental pollutants. For this purpose we performed Ussing chamber experiments to test three representative heavy metals and pesticides, respectively, for their effects on current flow across leech dorsal integument. Two representatives of each substance class showed distinct effects on ion transport across this epithelium. The heavy metal ions Pb^{2+} and Hg^{2+} produced a significant inhibition of amiloride-sensitive Na^+ transport across leech skin in concentrations below or close to their limiting values in waste water. Therefore, it seems feasible to use leech skin for future investigations of the toxic actions of these heavy metals. The fact that Pb^{2+} and Hg^{2+} exerted their effects only when applied apically points to a specific action of these divalent cations on ion channels in the apical membrane. However, this inhibition does not seem to be a general feature of divalent cations because Cd^{2+} did not influence ion transport across leech skin at all. Since current flow through amiloride-sensitive Na^+ channels in typical vertebrate tight epithelia is stimulated by numerous divalent cations, the pronounced inhibition of amiloride-sensitive Na^+ channels in leech skin by Pb^{2+} and Hg^{2+} might lead to a further differentiation of amiloride-sensitive Na^+ channels. The two widespread pesticides lindane and promecarb exerted their effects only at comparativ high concentrations. This low sensitivity restricts the usefulness of leech skin as a subject for further analysis of toxicity mechanisms, at least for these two pesticides.

Key words: Leech skin - Tight epithelium - *Hirudo medicinalis* - Pesticides - Heavy metals - Pollutants

Introduction

The leech *Hirudo medicinalis* lives in freshwater habitats which are strongly hypoosmotic to its body fluids. Therefore, the animal is faced permanently with the problem of water penetrating the relatively thin, semipermeable integument and subsequently entering the hyperosmotic body fluids. In addition, salts that are at much higher concentrations in the body fluids than in the surrounding water leave the leech body continuously. Thus, mechanisms exist to reabsorb salts in the nephridium before final urine elimination (Zerbst-Boroffka and Wenning 1986) and to take up salts from fresh water through the skin (Sawyer 1986). Although it has been known for quite some time that leech skin is capable of actively transporting certain ions, e.g. Na^+ and K^+, into the animal against substantial concentration gradients (Cobbold 1971), only limited information about the underlying ion transport systems has been available up to now (Weber et al. 1993, 1995). From these studies, which established leech dorsal integument as a valuable model to investigate Na^+ transport and its regulation in invertebrates, it can be concluded that leech skin is a typical tight epithelium which takes up Na^+ mainly through apical amiloride-sensitive channels with high selectivity for Na^+ over other monovalent cations. These channels seem to belong to the class of low amiloride affinity channels which differ from the typical high amiloride affinity channels with regard to their kinetic properties and pharmacological profile (Smith and Benos 1991). A smaller amount of Na^+ is transported by an amiloride-insensitive mechanism, possibly a second type of Na^+ channels or, as reported for leech caecal epithelium (Mildeet *al.* 1996), unspecific cation channels. Basolateral Na^+ extrusion is mediated by ouabain-sensitive Na^+/K^+-ATPases. Elevation of intracellular cAMP levels increased ion transports across the leech integument (Weber et al. 1993), although the physiological stimulus for this signal cascade has not been identified up to now.

In the present study we wanted to investigate if heavy metals or pesticides influence Na^+ absorption across leech skin and if this tissue preparation can be used to analyse the mechanisms underlying the toxic action of certain environmental pollutants. Additionally, a modulation of Na^+ channel characteristics by heavy metals or one of the other compounds might provide new insight into the basic architecture of this transport protein.

Sodium fluoroacetate, lindane (hexachlorocyclohexane) and promecarb

(3-isopropyl-5-methylphenyl-methylcarbamate) were chosen as examples of frequently used pesticides. Sodium fluoroacetate is well known for inhibiting the Krebs cycle, thereby blocking cellular ATP synthesis and increasing intracellular levels of citrate, which acts as a Ca^{2+} chelator (Hayes and Laws 1991). Lindane reportedly blocks ligand- and voltage-dependent Cl^- channels (Bloomquist 1993) as well as the Na^+/K^+-ATPase (Hayes and Laws 1991) in neurons. Effects on corresponding transport systems in non-excitable cells might be possible. Promecarb inhibits cholinesterase of central and peripheral neurons.

For this study we selected the heavy metals cadmium, lead and mercury. Heavy metal ions are known for their ability to react with sulfhydryl groups (SH groups) in proteins. The amiloride-sensitive Na^+ channel includes several SH groups which are exposed on the extracellular side of the apical membrane. Interactions of heavy metal ions with SH groups on the channel protein would presumably change its transport properties.

Materials and methods

Animals and tissue preparation

The methods and materials used for the present study corresponded to those described in detail by Weber et al. (1993). Briefly, they were as follows. Leeches (*Hirudo medicinalis*) purchased from ZAUG (Giessen, Germany) were kept in tap water at room temperature (22-25 °C) without feeding. Before tissue preparation the animals were hypothermally anaesthetized (10-15 min at -20 °C). The integument was dissected by two lateral incisions and the dorsal skin subsequently separated from the intestines. By careful scraping with a scalpel the internal muscle layers were removed until the diagonal muscle layers became visible. Further dissection was impossible without damaging the integument. A lucite ring was then glued onto the basolateral side of the tissue using tissue adhesive (Histoacryl blau; Braun, Melsungen, Germany) and was mounted in an Ussing chamber with an aperture of 0.5 cm^2 which was specially designed to minimize damage to the edges. The edges on both sides of the tissue were sealed with Silicone grease, so ion exchange between the two half-chambers could only occur through the tissue mounted between them. The perfusion rates for the Ringer solutions were about 7 ml/min in the apical half-chamber and 3 ml/min in the basolateral half-chamber. All electrical parameters are given normalized to an area of 1 cm^2. All experiments were performed at room temperature.

Solutions and chemicals

The composition of the basolateral NaCl-Ringer solution was (in mM): 115 NaCl, 4 KCl, 1.8 $CaCl_2$; in the apical Ringer KCl was replaced by tetramethylammonium chloride (TMA-Cl). Nonpermeant TMA^+ in equimolar concentrations was also substituted for Na^+ in the Na^+-free solution. All solutions were adjusted to pH 7.4 using 1 mM trishydroxymethyl-aminomethane (Trizma base) and were buffered with 5 mM hydroxyethylpiperazine-ethanesulfonic acid (HEPES). All these compounds were obtained from Sigma (Deisenhofen, Germany) and were of analytical grade, unless otherwise stated.

Sodium fluoroacetate was dissolved in distilled water and added to the Ringer solution in the respective concentrations. Lindane was used as a saturated aqueous solution instead of distilled water for the preparation of the Ringer solutions used in the experiments testing the effects of lindane on Na^+ transport across leech skin. Its maximal solubility at room temperature (20 °C) is 10 mg/ml. This corresponds to a concentration of 34 µM. Promecarb was dissolved in dimethylsulfoxide (DMSO). The final concentration of DMSO in the Ringer solutions never exceeded 0.1 %, a concentration which had no effect on bioelectrical properties of leech skin in control experiments. To ensure that current changes were really due to promecarb itself, the promecarb-free solutions contained the same DMSO concentration that was used to dissolve promecarb in the test solutions. Lead and mercury were used as acetates, cadmium as chloride salt. Stock solutions of 100 µM and 10 mM were prepared in distilled water and added to the Ringer solutions to give the final concentrations mentioned in the text. The concentrations of heavy metals used were in the range of the limiting values for waste water contamination in Germany.

All test substances were added unilaterally to the apical or basolateral Ringer solution. In the course of apical application the experimental supplements were also included in amiloride-containing and Na^+-free solutions.

Electrical measurements

Current- and voltage-measuring electrodes were prepared by placing Ag-AgCl wires in a 1 M KCl solution and connecting them to the bathing compartments through 1 M KCl agar bridges. During the transepithelial measurements the tissue was voltage clamped to 0 mV using a low-noise voltage clamp (Van Driessche and Lindemann 1978). The short-circuit current (I_{SC}) was recorded continuously by a stripchart recorder and a computer (Apple IIsi) with a MacLab interface and a chart recorder program (Analog Digital Instruments). The transepithelial resistance (R_T) was calculated according to Ohm's law from the changes in I_{SC} induced by short 10 mV pulses.

Statistical analysis

All data are given as means ± standard error of the mean (SEM); *n* is the number of experiments. Although the dorsal skin of one leech is large enough to obtain several tissue preparations for Ussing chamber experiments, only one preparation from each animal was used to analyse the experiments. Statistical significance was tested using Student's *t*-test (Sigma-Plot, Jandel Scientific); a significance level of $p < 0.05$ was accepted to be sufficient.

Results

Basic electrophysiological parameters

Pieces of leech skin could be preserved for several hours in Ussing chambers with apical and basolateral NaCl Ringer on the respective sides of the epithelium. Transepithelial potential difference (V_T), I_{SC} and R_T stabilized within ca. 60 min after mounting the epithelium and remained stable for several hours. Control values for electrophysiological parameters exhibited considerable differences for epithelia from different animals but varied only slightly between preparations from different regions of one animal's dorsal integument. Responses to altered experimental conditions were comparable for all tissue preparations under examination, irrespective of the divergent initial values. Electrophysiological parameters of leech dorsal integument measured under control conditions in the present study are summarized in Table 1. These values are comparable to those determined by Weber et al. (1995).

Table 1 Control values for short-circuit current (I_{SC}), transepithelial potential difference (V_T), and transepithelial resistance (R_T) of leech skin. The data are given as means ± SEM from a pool of 78 experiments

I_{SC} (μA/cm²)	15.0 ± 0.5
V_T (mV)	- 22.3 ± 1.2
R_T (k*cm²)	1.8 ± 0.1

After the equilibration period 100 μM amiloride was added to the apical Ringer solution in order to determine the amiloride-sensitive current; the difference in I_{SC} before and during amiloride application is termed I_{amil}. Subsequently, a Na^+-free solution was given from the apical side and the transepithelial Na^+ transport rate (I_{Na}) was calculated as the difference between the values for I_{SC} with 115 mM Na^+ and 0 mM Na^+ in the apical Ringer solution. I_{Na} made up ca. 62 % of total I_{SC}. It is not known which ions mediate the remaining 38 % of I_{SC}. Amiloride-sensitive Na^+ channels are responsible for about 78 % of I_{Na}; the remaining 22 % of Na^+ current must be mediated by as yet unidentified amiloride-insensitive transport molecules. Apical reapplication of 115 mM Na^+ following the Na^+-free period generated a large transient overshoot in I_{SC} that can be ascribed to activation of silent Na^+ channels during the Na^+-free period and their subsequent rapid downregulation after readdition of Na^+ (Weber et al. 1995). Figure 1 shows the typical effects of apical application of amiloride and Na^+-free Ringer solution. The present study was restricted to effects of the test substances on the amiloride-sensitive Na^+ transport, since these channels are the predominant apical ion conductances in leech dorsal integument.

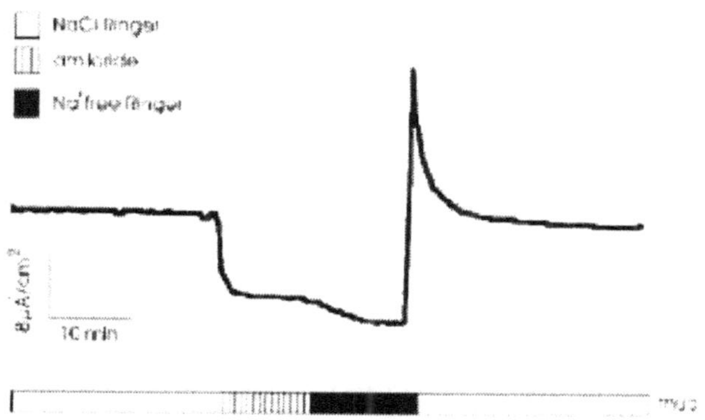

Fig. 1 Typical time-course of a control experiment. Transepithelial currents are inhibited by apical application of amiloride (100 μM) and after apical Na^+ depletion. A typical trace representative of 78 experiments is shown

Pesticides

Up to 100 μM sodium fluoroacetate did not influence ion transport across leech skin during short-term exposure (15 min). Neither apical nor basolateral application of this pesticide induced any changes in bioelectrical activity of leech dorsal integument.

Apical application of 34 μM lindane had no effect on ion transport across leech skin. However, from the basolateral side this pesticide concentration increased I_{SC} significantly by 28 ± 9 % (n=4). This reversible rise in transepithelial current was not mediated by increased Na^+ transport, since I_{Na} and I_{amil} were not affected (Fig. 2).

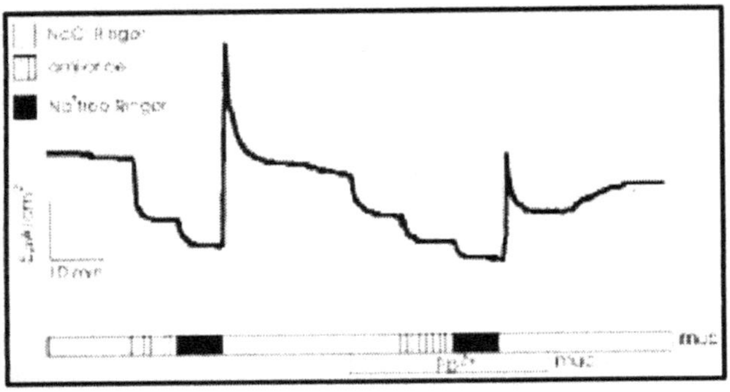

Fig. 2 Effect of lindane. Basolateral application of lindane (34 µM) increased I_{SC} significantly without changing transepithelial Na^{+} transport. The trace shown here is representative of 4 experiments

Apical and basolateral application of 100 µM promecarb reduced I_{SC} by 21 ± 5 % (n=6) and 15 ± 6 % (n=4), respectively. In both cases the inhibition of I_{SC} was only partially due to a decrease in amiloride-sensitive currents: from the apical side promecarb reduced I_{amil} by 12 ± 3 %, while basolateral application caused an inhibition by 11 ± 1 %. All effects of promecarb were reversible by washout with promecarb-free Ringer solutions. The effects of lindane and promecarb are summarized in Table 2.

Table 2 Percentage changes of I_{SC} and I_{amil} caused by apical and basolateral application of the pesticides lindane (34 µM) and promecarb (100 µM); $^{*}P<0.05$

	Percentage changes		
Experiment	I_{SC}	I_{amil}	n
Lindane, apical	0	0	4
Lindane, basolateral	+ 28 ± 9 *	0	4
Promecarb, apical	- 21 ± 5 *	- 12 ± 3 *	6
Promecarb, basolateral	-15 ± 6 *	- 11 ± 1 *	4

Heavy metals

In the present study Cd^{2+} was used in concentrations ranging from 100 nM to 10 µM. Even at the highest concentration, short-term application (15 min) of Cd^{2+} had no significant effects on ion transport across leech skin, irrespective of the side of application.

Lead and mercury were used as acetates. Control experiments with 200 µM sodium acetate added to the apical or basolateral Ringer solution demonstrated that at this concentration acetate itself had no influence on current flow across leech dorsal integument.

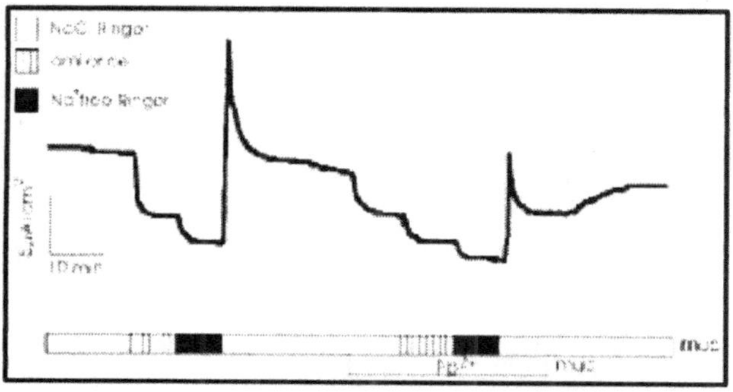

Fig. 3 Inhibition of Na^+ transport by Pb^{2+}. Apical application of lead acetate (100 μM) reduced I_{SC} and I_{amil} by about 42 % and 62 %, respectively. The trace shown is representative of 4 experiments

Pb^{2+} was added to the Ringer solution in final concentrations from 1 μM up to 100 μM. Immediately after addition of 100 μM Pb^{2+} to the apical Ringer solution I_{SC} decreased by 42 ± 5 % (n=4); this reduction was completely mediated by an inhibition of amiloride-sensitive Na^+ currents, which were diminished by 62 ± 7 % (Fig. 3). The effects of Pb^{2+} were reversible by washout with Pb^{2+}-free Ringer solution. In contrast to these results, Pb^{2+} did not change ion transport across leech skin when added to the basolateral Ringer solution (n=3). In order to obtain a dose-response relationship, we subsequently added increasing concentrations (1, 2, 5, 10, 20, 50 and 100 μM) of Pb^{2+} to the apical Ringer solution. This treatment resulted in a stepwise inhibition of I_{SC} (n=4), which is shown in Figure 4a. The data were fitted to the Michaelis-Menten equation; Figure 4b shows the resulting dose-response curve. A small but significant inhibition of I_{SC} by 13 ± 7 % could already be observed during application of 1 μM Pb^{2+}.

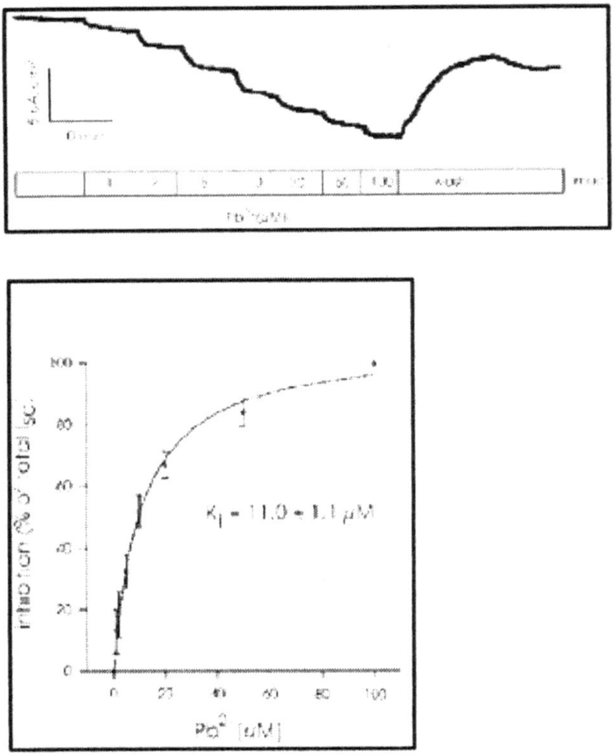

Fig. 4a,b Effects of different Pb^{2+} concentrations. **a** Increasing concentrations of Pb^{2+} in the apical Ringer solution produced a stepwise decrease of I_{SC}. The trace shown here is representative of 4 experiments. **b** Dose-response curve demonstrating the effects of apical Pb^{2+} on I_{SC} across leech skin; data given as means ± SEM. The *solid line* fits the Michaelis-Menten equation (K_I = 11.0 ± 1.1 µM)

Hg^{2+} was used in concentrations between 100 nM and 10 µM. Immediately after addition of 5 µM Hg^{2+} to the apical Ringer solution I_{SC} decreased by 38 ± 3 % (n=4). This effect was not entirely mediated by the inhibition of I_{amil} by 62 ± 9 %; amiloride-insensitive Na^+ transport and currents not carried by Na^+ were diminished as well. For a dose-response curve we used successively increasing Hg^{2+} concentrations from the apical side of the epithelium. This procedure evoked graduated decreases of I_{SC} (n=5), which are shown in Fig. 5a. Only the effects of up to 5 µM Hg^{2+} were always fully reversible by washout with Hg^{2+}-free Ringer solution. Even the lowest concentration used (0.1 µM) brought about a distinct inhibition of I_{SC} by 27 ± 7 %. Figure 5b shows the dose-response curve for the inhibition of I_{SC} by apical Hg^{2+}. Applying Michaelis-Menten kinetics resulted in a K_I of 0.38 ± 0.07 µM. Basolateral application of Hg^{2+} had no influence on ion transport across leech skin. The current changes induced by Pb^{2+} and Hg^{2+} are summarized in Table 3.

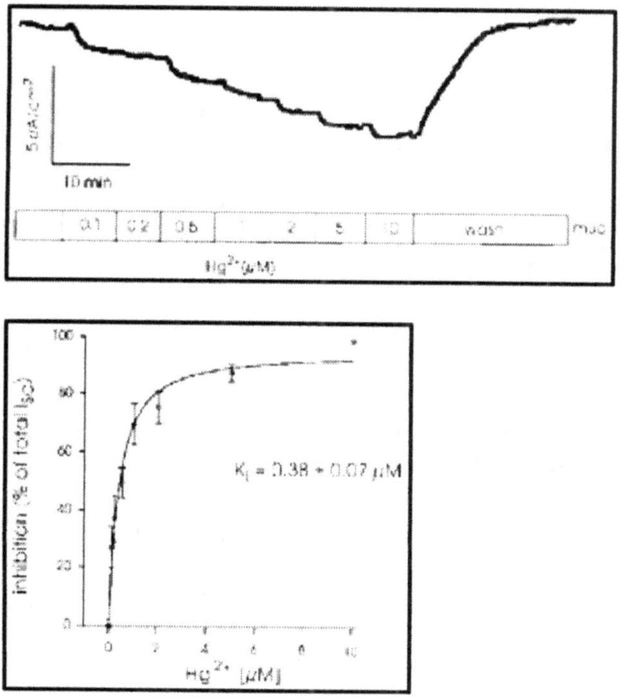

Fig. 5a,b Effects of increasing Hg^{2+} concentrations. **a** Successive application of increasing Hg^{2+} concentrations in the apical Ringer solution caused progressive reductions of I_{SC}. The trace shown is representative of 5 experiments. **b** Dose-response relationship of the effects of apical Hg^{2+} on I_{SC} across leech integument. Data are given as means ± SEM. The solid line was fitted using the Michaelis-Menten equation ($K_I = 0.38 \pm 0.07$ µM)

Table 3 Percentage changes of I_{SC} and I_{amil} caused by apical and basolateral application of the heavy metals Pb^{2+} (100 µM) and Hg^{2+} (5 µM); $^{*}P<0.05$

	Percentage changes		
Experiment	I_{SC}	I_{amil}	n
Pb^{2+}, apical	- 42 ± 5 *	- 62 ± 7 *	4
Pb^{2+}, sersosal	0	0	3
Hg^{2+}, apical	- 38 ± 3 *	- 62 ± 9 *	5
Hg^{2+}, basolateral	0	0	4

Discussion

Pesticides

In the present study we chose one substance each from three different pesticide categories: Sodium fluoroacetate as an example of fluoroacetates, lindane for polychloridated hydrocarbons and promecarb as a typical carbamate. These substances were tested in Ussing chamber experiments for their potency to influence ion transport across leech skin. Emphasis was placed on amiloride-sensitive Na^+ absorption, which plays a predominant role in transepithelial electrolyte transport in this tissue.

Sodium fluoroacetate firstly causes ATP depletion and secondly leads to an accumulation of citrate inside the cells (Hayes and Laws 1991). Shortage of ATP, on one hand, reduces the activity of the basolateral Na^+/K^+-ATPase, thereby destroying the Na^+ concentration gradient across the apical membrane, which is the driving force in transepithelial Na^+ transport. On the other hand, citrate is a Ca^{2+} chelator. Since during Na^+ autoregulation amiloride-sensitive Na^+ channels are inhibited by increasing cytosolic Ca^{2+} concentrations (Chase 1984) it could be speculated that a reduction of intracellular Ca^{2+} levels due to an intracellular Ca^{2+} chelator might increase amiloride-sensitive Na^+ currents. Nevertheless, in our experiments we could not observe any changes in ion transport across leech skin after apical or basolateral application of sodium fluoroacetate. This may be due to the short period of application (15 min); effects like ATP depletion probably need more time to become effective. However, we were mainly interested in direct effects on transport proteins and long-term application of pesticides is very likely to have general cytotoxic effects that would influence ion transport indirectly via an impaired viability of the cells.

Apical application of lindane had no effect on leech skin, while application from the basolateral side of the epithelium increased I_{SC}. However, this reaction was not mediated by an augmentation of amiloride-sensitive or amiloride-insensitive Na^+ currents. Other cations are rather unlikely to underlie the observed Na^+-independent current increase, since besides Na^+ only low concentrations of Ca^{2+} and/or K^+ are contained in the Ringer solutions. Therefore, the observed increase of I_{SC} must be mediated by stimulation of Cl^- transport from the basolateral to the apical side of the epithelium or by inhibition of Cl^- transport in the opposite direction, because Cl^- is the only anion in the Ringer solutions. Such a mode of action is not unlikely, because lindane in concentrations comparable to those used in the present study (Ogata *et al.* 1988; Wafford et al. 1989) has been shown to act on several types of Cl^- channels, but up to now only in neurons. However, the mechanisms of Cl^- transport across leech dorsal integument have not been investigated to date.

Therefore we cannot make any statement, as to if and how lindane influences Cl^- flow through this tissue.

Promecarb evoked a reduction in current flow across leech skin. During apical application amiloride-sensitive and amiloride-insensitive Na^+ transport as well as Na^+-independent currents were affected. This generalized transport inhibition argues against a specific effect on a particular transport system; non-specific cytotoxicity seems to be a more likely explanation. After addition of promecarb to the basolateral Ringer solution only Na^+-dependent currents were diminished. This result suggests an inhibitory effect on the Na^+/K^+-ATPase, which is responsible for basolateral export of Na^+ taken up through apical Na^+-permeable channels. Alternatively the effects of basolateral application of promecarb could be explained by an inhibition of cholinesterase in neurons that may be left in the partially stripped preparations that were used in the present study. The inhibition of these neurons could in turn reduce ion transport across leech skin.

Heavy metals

Pb^{2+} and Hg^{2+} caused a pronounced inhibition of amiloride-sensitive Na^+ transport and were exclusively effective when applied from the apical side. Therefore an effect on SH groups in the extracellular parts of amiloride-sensitive Na^+ channels could be assumed. However, Garty and Benos (1988) found that apical application of many divalent ions caused an increase of amiloride-sensitive Na^+ currents in model vertebrate tight epithelia, while in leech skin we observed a reduction of the same currents by Pb^{2+} and Hg^{2+}. On the other hand, Weber et al. (1995) demonstrated that apical Ca^{2+} also significantly reduced amiloride-sensitive Na^+ transport across leech skin in concentrations similar to the concentrations of Pb^{2+} and Hg^{2+} which were effective in the present study. Taken together these results suggest a fundamental difference between invertebrate and vertebrate mechanisms of epithelial Na^+ absorption. One could postulate that ions with a double positive charge generally inhibit amiloride-sensitive Na^+ transport. This assumption is questioned by the fact that Cd^{2+}, another divalent heavy metal ion, had no effect on ion transport across leech skin, independent of the side of application. Alternatively the Pb^{2+}-induced inhibition of amiloride-sensitive Na^+ transport could be due to uptake of Pb^{2+} into the cytoplasm and a subsequent intracellular mode of action. This conjecture is substantiated by the report of Goldstein (1993) that Pb^{2+} can act as an intracellular Ca^{2+} substitute and the observation that in the course of Na^+ feedback inhibition intracellular Ca^{2+} inhibits amiloride-sensitive Na^+ channels (Chase 1984). It is clear that much additional work will be required to investigate in detail the mechanisms underlying the inhibition of amiloride-sensitive Na^+ transport in leech skin by heavy metal ions. Nevertheless, the different effects of divalent cations on amiloride-sensitive currents in vertebrate and invertebrate epithelia may serve as a basis for a new classification of amiloride-sensitive Na^+ channels.

Conclusions

The present study provided information concerning epithelial Na^+ transport and its regulation in invertebrates. Two potent blockers obviously specific for invertebrate Na^+ channels have been identified. Future studies investigating the inhibitory mechanism of the heavy metals may extend our knowledge concerning the structure-function relationship of the ancestral amiloride-sensitive Na^+ channel. Direct comparison of invertebrate Na^+ absorption with the vertebrate system may light details in the regulatory pathways controlling transepithelial Na^+ transport and even reveal new regulatory principles. In addition, our results demonstrate that leech skin is a useful model not only for the investigation of Na^+ transport regulation in invertebrates but also for analysis of toxicity mechanisms of some important environmental pollutants, such as heavy metals.

Acknowledgements. This study was supported by a grant from the Studienstiftung des Deutschen Volkes (to H. Milde). We thank Ulrike Blank for helpful discussions during the project and Harry Heidt for excellent technical assistance. Parts of this study were presented at the 1995 Spring Meeting of the German Physiological Society and were published as an abstract in *Pflügers Archiv* 429, Suppl 6:R191

References

Bloomquist, J. R. (1993) Toxicology, mode of action and target site-mediated resistance to insecticides acting on chloride channels. Comp. Biochem. Physiol.C **106**:301-304

Chase, H. S. (1984) Does calcium couple the apical and basolateral membrane permeabilities in epithelia?. Am. J. Physiol. **247**:F869-F876

Cobbold, P. H. (1971) Uptake of sodium and potassium by the leech, *Hirudo medicinalis*. J. Physiol. (Lond) **219**:10-12

Garty, H., Benos, D. J. (1988) Characteristics and regulatory mechanisms of the amiloride-blockable Na^+ channel. Physiol. Rev. **68**:309-373

Goldstein, G. W. (1993) Evidence that lead acts as a calcium substitute in second messenger metabolism. Neurotoxicology **14**:97-102

Hayes, W. J., Jr., Laws, E. R., Jr. (1991) Handbook of pesticide toxicology. Academic Press, San Diego

Milde, H., Weber, W.-M., Clauss, W. (1996) Electrogenic cation transport across leech caecal epithelium. J. Comp. Physiol. B (in press)

Ogata, N., Vogel, S. M., Narahashi, T. (1988) Lindane but not deltamethrin blocks a component of GABA-activated chloride channels. FASEB J. **2**:2895-2900

Sawyer, R. T. (1986) Leech biology and behaviour. Clarendon Press, Oxford

Smith, P. R., Benos, D. J. (1991) Epithelial Na^+-channels. Annu. Rev. Physiol. **53**:509-530

van Driessche, W., Lindemann, B. (1978) Low-noise amplification of voltage and current fluctuations arising in epithelia. Rev. Sci. Instr. **49**:52-57

Wafford, K. A., Sattelle, D. B., Gant, D. B., Eldefrawi, A. T., Eldefrawi, M. E. (1989) Noncompetitive inhibition of GABA receptors in insect and vertebrate CNS by endrin and lindane. Pestic Biochem. Physiol. **33**:213-219

Weber, W. -M., Dannenmaier, B., Clauss, W. (1993) Ion transport across leech integument. I. Electrogenic Na^+ transport and current fluctuation analysis of the apical Na^+ channel. J. Comp. Physiol. B **163**:153-159

Weber, W.-M., Blank, U., Clauss, W. (1995) Regulation of electrogenic Na^+ transport across leech skin. Am. J. Physiol. **268**:R605-R613

Zerbst-Boroffka, I., Wenning, A. (1986) Mechanisms of regulatory salt and water excretion in the leech, *Hirudo medicinalis* L.. Zool. Beitr. **30**:359-377

Exp. Biol. Online 1:9 (1996)

Respiratory movements during vocalization in the squirrel monkey

Udo Häusler

Deutsches Primatenzentrum, Abt. Neurobiologie, Kellnerweg 4, 37077 Göttingen, Germany
(e-mail: uhausl@www.dpz.gwdg.de)

Received: 26 September 1996 / Accepted: 4 December 1996
Published: December 18, 1996

Abstract. Respiratory abdominal movements during vocalization were measured in awake squirrel monkeys during spontaneous and playback-induced vocal activity. Large vocalization-correlated respiratory movements (VCRM) starting before vocalization were observed during several call types, such as peeping, trilling, cackling and err-chuck. Purring, in contrast, was accompanied by only small VCRM that started late after vocal onset. VCRM during trilling, a call with marked frequency modulation, showed a modulation in the rhythm of the frequency changes. A correlation with amplitude modulation was also present but more variable. As high frequencies need a higher lung pressure for production than low frequencies, the modulation of VCRM seems to serve to optimize the lung pressure in relation to the vocalization frequency. The modulation, furthermore, may act as a mechanism to produce different trill variants. During err-chucks and staccato peeps, which show a large amplitude modulation, a non-modulated VCRM occurred. This indicates the existence of a laryngeal amplitude-controlling mechanism that is independent of respiration.

Key words: Vocalization – Respiration – Squirrel monkey – Lung-pressure modulation

Introduction

Respiration is an essential component of vocalization. Activity of respiratory muscles is different during normal vocal activity and speech compared to quiet breathing in man (Floyd and Silver 1950; Draper et al. 1959; Hoshiko 1962;

Hoshiko and Berger 1965; McFarland and Smith 1989; Estenne et al. 1990) and primates (Jürgens and Schriever 1991). It is generally believed that "respiratory muscles act to generate and maintain lung pressures necessary for vocalization, while laryngeal and supralaryngeal structures help regulate subglottic pressure and modify the airflow to produce the appropriate sound" (West and Larson 1993). This might explain why the influence of laryngeal and supralaryngeal structures on vocal parameters like amplitude, frequency and spectral content has been extensively investigated, while the influence of respiratory temporal patterns on vocal parameters is largely neglected.

Jürgens and Schriever (1991) reported a detailed electromyographic (EMG) study of respiratory muscles during vocalization under general anesthesia in squirrel monkeys. They used electrical brain stimulation to elicit peeping or sequences of chuck calls. In contrast to quiet breathing, inspiratory and expiratory intercostal muscles are coactivated and abdominal accessory muscles are additionally recruited during vocalization. They showed that the amplitude of the uttered vocalization is most closely related to the envelope of the abdominal obliquus internus EMG activity, which therefore seems to be involved in the production and modulation of subglottic air pressure during vocalization.

Titze (1989, 1993) has developed a model based on human data that shows a dependence of fundamental frequency on lung pressure. The higher the fundamental frequency, the higher the minimal lung pressure necessary for its production. Furthermore, Hsiao et al. (1994) have shown that vocal frequency is influenced to some degree by subglottic pressure and therefore also by lung pressure.

The aim of this study was to further analyze abdominal respiratory movements related to the production of subglottic pressure during vocal activity. Two questions were addressed: (1) are respiratory movements during vocalization related only to sound intensity, or to other acoustic parameters as well? (2) Is Titze`s model also applicable to squirrel monkeys and does it help to understand how the animal can utilize respiratory modulation for the production of different calls. Clarification of these questions is regarded as an important step towards a better understanding of the central control of vocalization.

Materials and methods

Procedures

Five adult squirrel monkeys (*Saimiri sciureus*; 1 roman type, 4 gothic type) were trained to sit on a restraining chair inside a sound-attenuated chamber (Siemens + Matsushita Systems) for single sessions of about 1-2 h once a day. The chair was specially adapted to the monkey's body shape and upholstered by soft cloth. Animals were able to move their heads, arms and feet. Small pieces of banana, raisins or nuts were given as a reward during training sessions. After getting used to the situation, animals started to vocalize either as response to playback calls or spontaneously. At the end of the training period, animals were presented with trill calls of familiar individuals by a computer during the last 30 min of the day's session. Trill and kecker calls as well as isolation peeps were successfully elicited in this way.

Data recording

Respiratory movements were recorded with a magnetometric method. It is based on the effect that electromagnetic transduction between two separated coils is dependent on the square of their distance. Two coils were mounted in the midline, one over the abdomen 20 mm below the lower end of the ribcage, the other on the back in an aproximately aligned position. The coils had a diameter of 1 cm and a distance of 65 to 75 mm. The back coil had an impedance of 60 Ohm and was connected to a sine wave generator (Hameg HM8030) set to 15 kHz and used as transmitting coil. The sense coil was connected directly to the input stage of a Tectronix 5110 oscilloscope with a 5A22N type preamplifier. A resistor of 2 kOhm was connected parallel to the sense coil in order to suppress oscillations at the input. The signal was filtered with low cutoff at 10 kHz and high cutoff at 30 kHz. The amplitude of the sinewave generator and gain of the preamplifier were set to give a low noise 1.5V (P-P) signal, the amplitude modulation of which reflected abdominal diameter changes. These diameter changes are related to respiratory movements in the way that inspiration is accompanied by an increase in abdominal diameter and a decrease in the measured signal. Conversely, expiratory movements are related to a decrease in diameter and an increase in the signal. Diameter changes were in the range of 3 to 6 mm.

Vocal utterances were recorded with a Sennheiser MKH directional microphone 1 m in front of the animal. The signal was filtered with a passband from 170 Hz to 30 kHz and amplified to a level of 1.5 V. Playback vocalizations were presented with 65-70 dB SPL at the ear of the animal, depending on the actual stimulus. Calibration of the system was done with a B&K Type 2231 Sound level meter. The recording system was linear within 1.5 dB between 200 Hz and 20 kHz. Playback and recording of vocalizations, as well as recording of the movement signal, were done on line with a microprocessor system (Macintosh Quadra 950 and an NI A2100 analog i/o board that includes appropriate anti-aliasing filters). For both analog input and output, a 48 kHz rate was used. Recording was started 1.5 s before presentation of the stimulus and lasted for a total time of 5 s. In addition to online recording, the whole session was recorded on videotape (Panasonic, AG-5700), together with the vocal and movement signals.

Analysis was done offline with the program "Signalyse". Sonagrams and amplitude plots were calculated for the animal's vocalizations. The "movement" signal was analyzed by calculating an amplitude plot that represents a plot of the distance between the abdominal coils.

Vocalizations were classified according to Newman (1985) into peep calls, trilling, purring, cackling and err-chucks. The following measures were taken for each vocal utterance: (1) time between onset of expiratory movement and vocal activity; (2) cycle time of frequency and amplitude modulation as well as respiratory modulation in trill calls. An amplitude envelope was determined by calculating the moving average of the squared signal within a 15 ms time window. Cycle time of amplitude modulation was then measured if neighboring maxima and minima were different for more than 9 dB. (3) Amplitude of the vocalization-related respiratory movement relative to the movement during the non-vocal respiratory cycle. Only trails were included that were free of artifacts due to gross body movements. Measures were transferred to a statistics program (Statview) for basic statistics and ANOVA. For each animal, data for 4-6 sessions were included in the analysis.

Results

General characteristics of breathing and effects of vocal activity

Breathing movements were recorded from the abdomen, as here the movements of the body wall were largest in amplitude and gave the most consistent results. Preliminary recordings of thoracal respiratory movements showed that in the dorso-ventral direction thorac's movements were small in amplitude, and the larger ones in transverse direction were difficult to record. Breathing movements showed the typical time course illustrated in Fig 1. Inspiratory and expiratory phases are clearly differentiated. The change from expiration to inspiration is characterized by an abrupt change in movement direction, while the transition from inspiration to expiration is gradual and slow. Even though the time course differed between the animals, the abrupt change during the expiratory-inspiratory transition was always present.

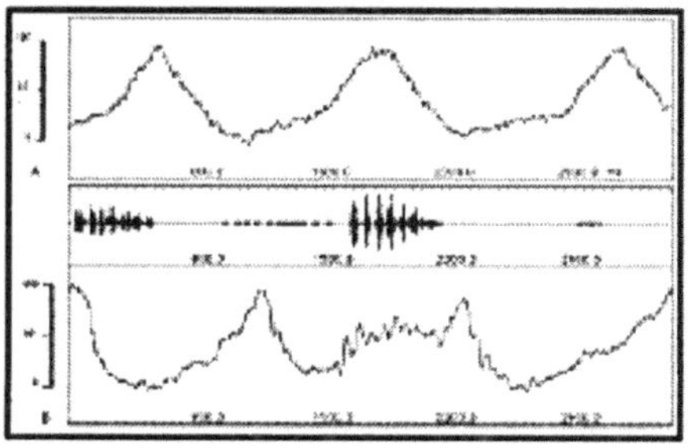

Fig. 1 A Respiratory movements without vocal activity. High values correspond to a small dorso-ventral abdominal diameter, low values to a large diameter. During normal respiration, an increase the signal indicates expiration, while a decrease corresponds to inspiration. **B** Respiratory movements with vocal activity induced by a playback stimulus. The *upper trace* shows the oscillogram of the signal picked up by the microphone, the *lower trace* the corresponding respiratory movement. The scale bars indicate the range of abdominal movements during normal respiration (in percent)

Vocalization-correlated respiratory movements (VCRM)

Respiratory movements were recorded during a total of 342 vocalizations. Compared to normal breathing, vocalization is accompanied by an additional expiratory movement, which will be called Vocalization-correlated respiratory movement (VCRM). Depending on the type of vocalization, this expiratory movement is modulated in a characteristic way. Figure 2 shows the VCRM and its modulation for different types of vocalization. The following types were recorded: isolation peeps and short peeps (n=22) (Fig. 2A,B), trills (n=286) (Fig. 2D-F), purring (n=8) (Fig. 7), cackling (n=22) (Fig. 8) and err-chucks (n= 6) (Fig. 9).

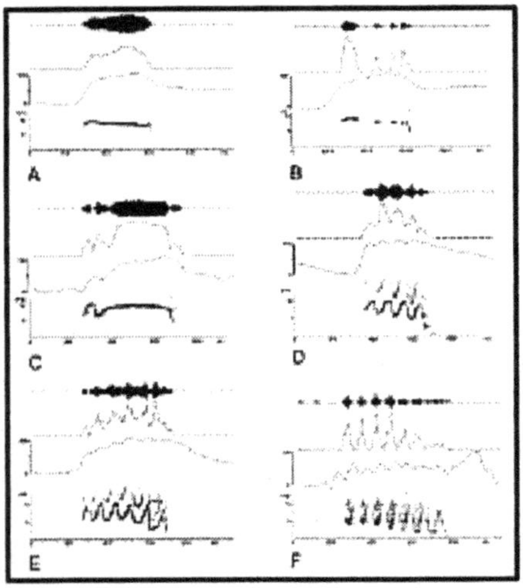

Fig. 2 A-F Respiratory movements during different peep and trill calls. In each figure the *uppermost trace* shows the oscillogram of the vocalization; the *second trace* is an amplitude plot (mean power envelope within a 15 ms time window) of the vocalization; the third displays the respiratory movement (i.e., abdominal diameter. High values correspond to a small dorso-ventral abdominal diameter, low values to a large diameter.) The *lowest trace* shows the spectrogram of the call. The *scale bars* indicate the range of abdominal movements during normal respiration in percent. **A** Isolation peep (wav-file, 35 kB); **B** series of short peeps (wav-file, 46 kB); **C** peep with preceding frequency-modulated element; **D** short trill; **E** Long trill with amplitude modulation largely in phase with respiratory movement modulation; **F** trill (wav-file, 51 kB) with amplitude modulation out of phase with VCRM

Peep calls

A very simple relationship between call structure and VCRM is found in isolation peeps; these are long-distance contact calls showing only minor frequency and amplitude modulations. The expiratory movement in this case starts with a median latency of 35 ms before vocalization and gradually increases in amplitude until the onset of vocalization, after which it remains constant (Fig. 2A). This indicates a constant expiratory effort during the course of vocalization. During 5 of the 22 cases, the VCRM showed some irregular modulations, making up about 6 % of the total respiratory movement. In a few cases, peep calls were preceded by distinct frequency-modulated elements (Fig. 2C). In these cases, the VCRM showed a strong modulation that followed the time course of the frequency modulation.

Single short peeps were either observed at the very end of the expiratory phase or they were accompanied by a single short expiratory movement of high amplitude.

Sequences of short peeps (Fig. 2B) were characterized by a continuous expiratory movement throughout the sequence that was nearly identical to the VCRM during isolation peeps.

Trill calls

Trills are tonal calls exhibiting a marked rhythmic frequency modulation with a repetition rate of about 15 Hz. This frequency will be referred to as frequency modulation rate. The frequency modulation is often accompanied by an amplitude modulation of approximately the same rate, but with a phase relationship to the frequency modulation that can vary between animals as well as between vocalizations.

A clearly rhythmic modulation of the VCRM is found in almost all trill calls. It consists of either a stepwise decrease in the abdominal diameter (Fig. 2E) or large oscillations superimposed on the expiratory movement associated with the vocal activity (Fig. 2F). The amplitude of the modulation in the latter case was above 25 % of the total respiratory movement for most calls. Out of the 286 recorded trills, only 3 lacked a rhythmic VCRM modulation.

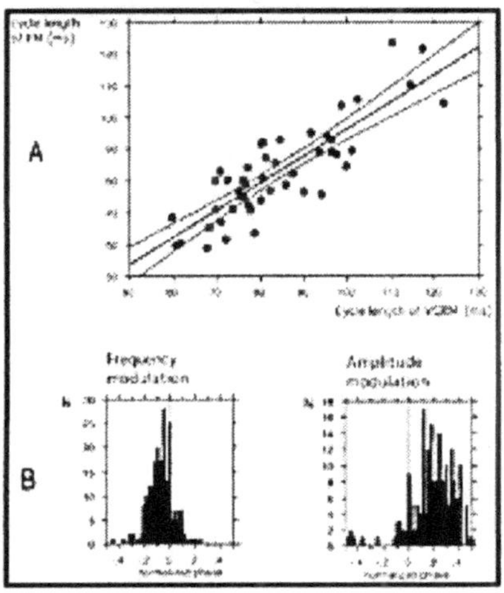

Fig. 3 **A** Relationship between cycle length of frequency modulation (*FM*) and corresponding respiratory movement (*VCRM*) modulation during trill calls; $r^2 = 0.721$. **B** Phase relationships between the VCRM modulation and frequency or amplitude modulation, respectively

The modulation of VCRM is related to both the amplitude and frequency modulation of the call. Analysis of the cycle lengths of VCRM modulation and frequency modulation shows an approximately linear relationship between the two (Fig. 3a). In addition, there is a preferred phase relationship between them. The high-frequency phase of the fundamental corresponds to the large-amplitude phase of the respiratory movement. The phase-histogram distributions (Fig. 3B) indicate that there is some variation in phase relationship. In most cases, this variation is observed between different calls; several cases, however, were also observed where the phase relationship changed within single calls. Such a phase shift correlated with changes in the cycle length of VCRM modulation or frequency modulation. These changes were always accompanied by a change in the character of the call elements (Fig. 4). Measurement of large-scale amplitude modulation during trill calls shows that these are correlated with the VCRM modulation. As can be seen from Fig. 3B, however, the correlation is not as good as that between VCRM modulation and fundamental frequency modulation. The distribution of normalized phases is significantly broader for the amplitude modulation ($p < 0.005$ *F*-test). The mean preferred position of the amplitude peak is also different ($p < 0.001$ *t*-test), with the amplitude peak preceding the peak of the VCRM.

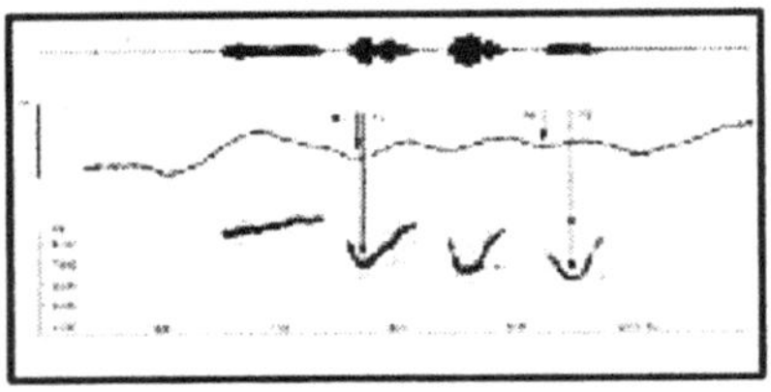

Fig. 4 Respiratoy movements during short trill call with marked change in element shape. *Upper trace:* oscillogram of the vocalization; *middle trace:* respiratory movement (i.e. abdominal diameter. High values correspond to a small dorso-ventral abdominal diameter, low values to a large diameter.); *lower trace:* spectrogram. The phase duration of frequency modulation is larger than the phase duration of VCRM modulation. Therefore, a change in phase relationship between minima of respiratory signal (R1, R2) and minima of the fundamental frequency (F1, F2) is observed during the call. The change is accompanied by a change in acoustic structure from a predominantly upward sweep to a U-shaped form

The amplitude of VCRM modulation often decreases towards the end of the vocalization (Fig. 2 D, F). This is preferentially observed for short trills that have four or five elements and show a decrease in the mean fundamental frequency, while long trills with more than seven elements that do not have a decrease in mean frequency show a constant amplitude of VCRM modulation throughout the calls.

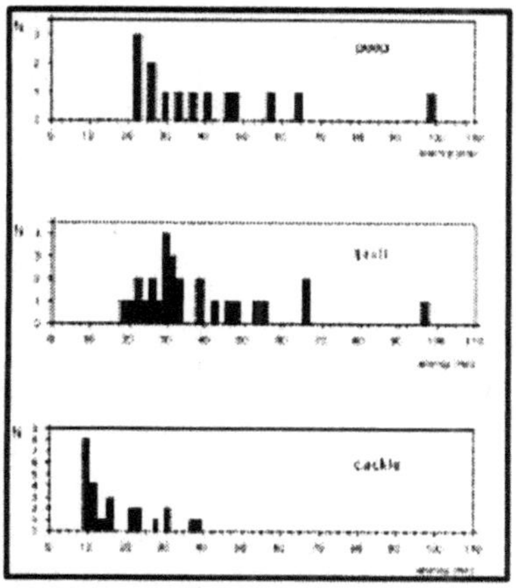

Fig. 5 Distribution of latencies for vocal onset relative to the onset of VCRM for three different call types

In trills, VCRM starts with a median of 30 ms before vocal onset (Fig. 5). In some cases, however, rhythmic modulation of VCRM was observed several hundred milliseconds before audible vocalization (Fig. 6).

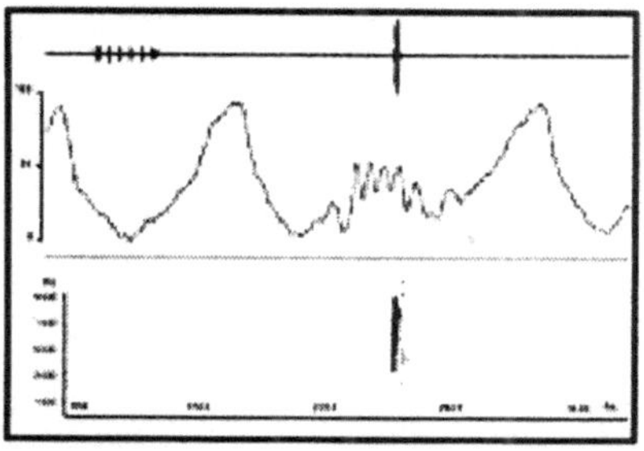

Fig. 6 Complete trill-like VCRM accompanied by a single trill element only. *Upper trace:* microphone signal; *middle trace:* respiratory movement; *lower trace:* spectrogram of vocalization

Purring

Purring is a low-intensity vocalization that consists of long pulse sequences with a rate of about 30 Hz. Animals produce these vocalizations preferably during affiliative interactions (Newman 1985). During purring, the VCRM is extremely small. Figure 7 shows a typical example. There is no extra expiratory movement detectable at the beginning of purring; only after about 250 ms, is there a small and constant expiratory movement. Mean latency of VCRM is 80 ms after vocalization onset.

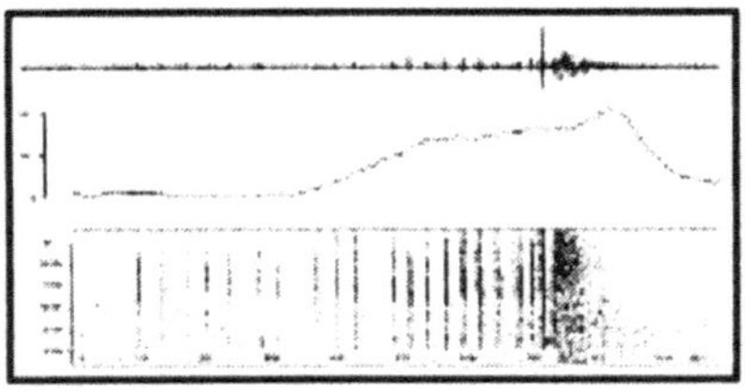

Fig. 7 Respiratory movement during purring. *Upper trace:* oscillogram of the vocalization; *middle trace:* respiratory movement (i.e. abdominal diameter. High values correspond to a small dorso-ventral abdominal diameter, low values to a large diameter.); *lower trace:* spectrogram

Cackling

Cackle calls consist of short wide-band calls with low fundamental frequency and high amplitude (Fig. 8). They may be uttered as single calls or in a sequence. These calls are accompanied by high-amplitude expiratory pulses. These pulses start with a very short latency of about 6-10 ms before vocalization onset (Fig.5). In 3 out of the 22 cases, the expiratory movement showed an oscillatory modulation consisting of two peaks that were not related to the vocal parameters.

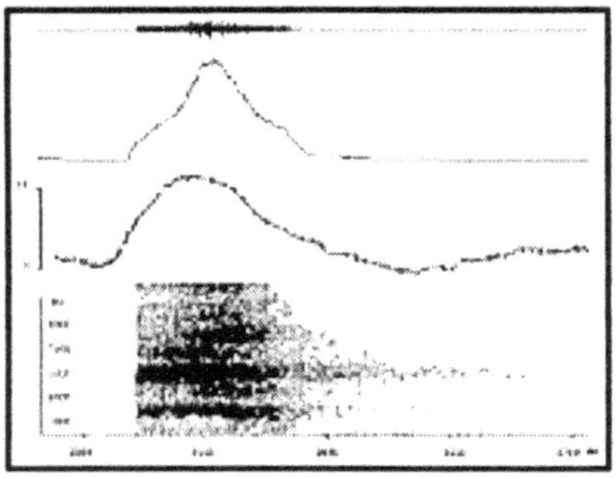

Fig. 8 Respiratory movement during cackling (wav-file, 14 kB). *Upper trace:* oscillogram of the vocalization; *middle trace:* respiratory movement; *lower trace:* spectrogram

Err-chucks

Err-chucks consist of a relatively loud growling-like element ("err") followed by an ascending and descending or only descending high, pitched element ("chuck") within 100 ms (Fig. 9). The calls show a marked amplitude modulation and a clear VCRM. VCRM starts with a median of about 35 ms before the call and continues throughout all elements without interruption. Some modulations are present, but with an irregular time course and low amplitude. The latter was below 6% of the total respiratory movement and close to the spontaneously observed fluctuations. There was no obvious correlation of the modulation with the acoustic parameters of the call.

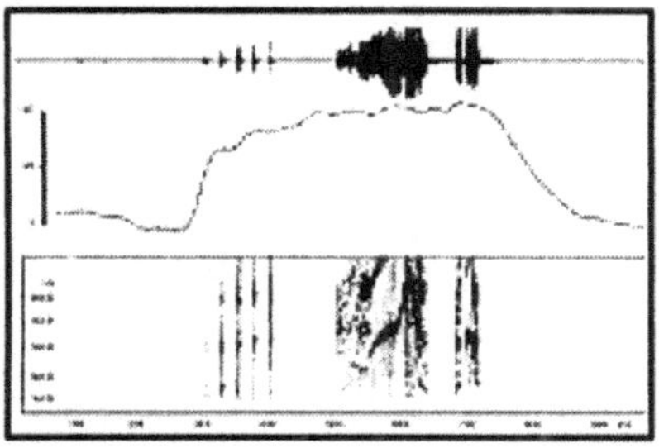

Fig. 9 Respiratory movement during err-chuck (wav-file, 51 kB). *Upper trace:* oscillogram of the vocalization; *middle trace:* respiratory movement; *lower trace:* spectrogram

Discussion

The aim of the present study was to investigate if and how respiratory movements are modulated by vocal activity. The study demonstrates that there is a modulation; its degree and character, however, depend upon the type of vocalization.

The VCRM is correlated with the intensity and duration of the call. This is evident from a comparison of purring and peeping calls. Purring is a low-intensity vocalization showing almost no additional vocalization-correlated movement if compared to spontaneous breathing. Only at the end of prolonged purring do abdominal respiratory movements come into play. Peep calls, in contrast, which are high-intensity calls, are characterized by a large additional respiratory movement, that is already present before vocal onset.

VCRM measured in this study generally consists of a decrease in abdominal diameter during vocal activity. The decrease in abdominal diameter is due to the activity of expiratory abdominal muscles that have been shown to be active during vocalization. Jürgens and Schriever (1991) demonstrated that during vocal activity in *Saimiri sciureus*, the musculi obliquus internus and externus were closely related to the amplitude of peep calls and rhythmic cackling. For echolocating bats Lancester et al. (1995) similarly have shown a close correlation of lateral abdominal wall EMG activity with high-frequency ultrasonic pulses. A close relationship of VCRM to expiratory abdominal muscles activity in *S. sciureus* is also indicated by the latency of EMG activity in the obliquus internus and externus muscles, which started between 32 and 104 ms before vocal onset (Jürgens and Schriever 1991); this corresponds to the latencies found for VCRM for peep and trill calls in the present study. Furthermore, Jürgens and Schriever (1991) showed that thoracic expiratory as well as inspiratory muscles are coactivated to stabilize the thorax against which abdominal expiratory muscles can act to produce the appropriate subglottic pressure. As the decrease in abdominal diameter measured in this study is closely related to expiratory abdominal muscle activity, it should also be related to lung pressure and subglottic pressure. It may be assumed that VCRMs during vocalization are correlated with lung pressure in such a way that a short distance between spine and abdominal surface (upward position in our plots) relates to high lung pressure, while a large distance relates to low pressure.

The results from this study differ from those characterizing respiratory movements during speech in humans (Estenne et al. 1990; Draper et al. 1959). In humans, a constant linear decrease in abdominal and thoracic diameter has been observed during speaking. This has been related to the production of a moderate, but relatively constant subglottic pressure during speaking (Estenne et al. 1990). Expiratory abdominal muscles, in this case, are only recruited after several seconds near the end of a speech period, while at the beginning, passive relaxation of the abdominal and thoracic wall is observed (Draper et al. 1959; Hoshiko and Berger 1965).

A situation partially similar to speech-related respiratory movements is observed during purring. Here, we also do not find an abrupt decrease in abdominal diameter, but a slow linear one starting sometime after vocal onset. Like speech, purring is a low-frequency vocalization that has low or moderate intensity and does not need such a high pressure for its production as high-intensity peep or trill calls. Therefore, as in human speech a relaxation of the respiratory muscles may be sufficient to support the nessessary subglottic pressure.

Sound intensity is not the only parameter with which respiratory movements are correlated. In the case of trill calls, there is a modulation of VCRM that is much better correlated with the time course of the fundamental frequency than with that of the amplitude: only during the periods where frequency modulation is present does modulation of VCRM take place.

The phase relationships indicate that a small abdominal diameter, which, as stated above, elates to high lung pressue, is normally correlated with high frequency. Surprisingly, there is not a strict relationship between the amplitude modulation present in the trill call and the breathing movement. This makes it clear that expiratory effort bears a complex relationship to frequency and amplitude, depending upon the specific call type.

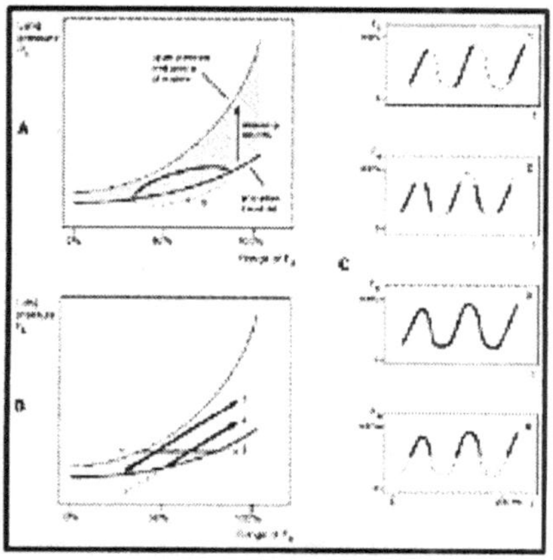

Fig. 10 A, B The *shaded area* represents the range of lung pressure (*PL*) which is compatible with the production of a specific fundamental frequency (*Fo*). Only within the shaded area is phonation possible.Within the range, intensity increases with lung pressure (*pointing up arrow*). In **B** the shading is omitted; only the upper and lower pressure limit is indicated. The *numbered lines with double arrows* represent tracks along which lung pressure and fundamental frequency can be modulated during trill calls. As phonation is only possible within the area shaded in **A**, it is only along the marked sections of the tracks that vocalization will occur: there will be no vocalization along the *thin dashed sections* of the tracks. The different tracks represent different types of modulation: (1) frequency modulation and pressure modulation are shifted in phase; (2) no pressure modulation; (3) optimal pressure modulation; (4) pressure modulation with reduced overall pressure. **C** The *figures on the right* side show schematic spectrograms that will result from three cycles along the corresponding modulation tracks of **A** and **B**. *Dashed lines* in **C** correspond to the dashed parts in **A**, **B** and indicate a fictitious frequency course during silent periods

In order to explain this relationship, a model proposed by Titze (1993) for humans may be used. According to this model, higher fundamental frequencies need higher lung pressures for their production than lower ones. According to Titze, there is not only a lower limit of lung pressure, but also an upper one. This is most relevant at low frequencies where the vocal folds have low tension; in this case high pressure can either prevent glottic closure or result in a chaotic movement so that tonality is lost. High pressure at low frequency, furthermore, will result in a high airflow, which may be incompatible with prolonged phonation. The upper limit also increases with frequency (Fig. 10 A). Data from Titze (1993) indicate that the upper limit for low frequencies is below the threshold for high frequencies. This means that in order to produce a continuous vocalization with marked frequency modulation (above 50 % of the vocal range), active modulation of lung presssure is necessary. Furthermore, a specific phase relationship between frequency modulation and breathing modulation is necessary. Figure 10 B, C shows a well-adapted (curve 3) and maladapted (curve 2, 4) lung-pressure-frequency relationship. High frequency has to be in phase with high pressure to produce a continuous trill call, the same relationship that has generally been observed during trill calls.

By varying the phase relationship between frequency (frequency is assumed to be directly related to vocal fold tension) and VCRM, the model generates either pure upward or downward sweeps (Fig. 10 C, curve 1). Additional trill types can be generated by either changing the overall lung pressure (Fig. 10 C, curves 2 and 3) or the steepness and amplitude of frequency or lung pressure modulation. In fact, all the trill variants shown in Fig. 10 C have been observed in spontaneously vocalizing animals (Winter et al. 1966; Winter 1969; Newman 1985). One might therefore speculate that the above-described mechanism is used by the animals to produce different kinds of trill calls. It is suggested that a similar mechanism underlies rhythmically frequency - modulated calls of other primates, like *Callithrix jacchus* or *Microcebus murinus* (Zimmermann et al. 1995) and perhaps mammals in general, as well.

With regard to the neuronal mechanisms underlying the above-described model, the question arises of how the respiratory and laryngeal components are appropriately synchronized. Is there a common generator for rhythmic vocal activity, or are there separate laryngeal and respiratory rhythm generators? In the case of separate generators, one would expect that under certain conditions respiratory modulation and frequency modulation can run out of phase. It also might be expected that the modulation of either frequency or respiration could disappear independently from each other. In the above-described experiments, phase shifts were observed during some calls, but the cycle length differences between frequency modulation and respiratory modulation always remained small. This indicates that the rhythm generation depends on separate, but closely coupled generators for respiratory and laryngeal components.

The large amplitude modulations observed during peep sequences and err-chucks, both of which are not accompanied by a corresponding modulation of the VCRM, cannot be explained by the mechanism described above. Here, it must be assumed that the amplitude minima are due to either a complete opening or closure of the glottis. As a complete opening of the glottis together with the high pressure during peep calls would result in a fast exhaustion of the expiratory airflow, it is likely that a transient glottic closure rather than an opening is the cause for the amplitude minima in these cases.

Acknowledgements. The author wants to thank Prof. Uwe Jürgens for his help and discussion during the preparation of the manuscript and Prof. Elke Zimmerman for her comments on an earlier version of the manuscript.

References

Draper, M.H., Ladefoged, P., Whitteridge, D. (1959) Respiratory muscles in speech. *J. Speech Hear. Res.* **2**: 16-27

Estenne, M., Zocchi, L., Ward, M., Macklem, P.T. (1990) Chest wall motion and expiratory muscle use during phonation in normal humans. *J. Appl. Physiol.* **68**: 2075-2082

Floyd, W.F., Silver, P.H.S. (1950) Electromyographic study of patterns of activity of the anterior abdominal wall muscles in man. *J. Anat.* **84**: 132-145

Hoshiko, M. (1962) Electromyographic investigation of the intercostal muscles during speech. *Arch. Physiol. Med. Rehabil.* **43:** 115-119

Hoshiko, M., Berger, K.W. (1965) Sequence of respiratory muscle activity during varied vocal attack. *Speech Monographs* **32:** 185-191

Hsiao, T.Y., Solomon, N.P., Luschei, E.S., Titze, I.R., Liu, K., Fu, T.C., Hsu, M.M. (1994) Effect of subglottic pressure on fundamental frequency of the canine larynx with active muscle tensions. *Ann. Otol. Rhinol. Laryngol.* **103:** 817-821

Jürgens, U, Schriever, S. (1991) Respiratory muscle activity during vocalization in the squirrel monkey. *Folia Primatol.* **56:** 121-132

Lancester, W.C. , Henson, O.W., Keating, A.W. (1995) Rrespiatoy muscle activity in relation to vocalization in flying bats. *J. Exp. Biol.* **198:** 175-191

McFarland, D., Smith, A. (1989) Surface recordings of respiratory muscle activity during speech: some preliminary findings. *J. Speech Hearing Res.* **32:** 657-667

Newman, J.D. (1985) Squirrel monkey communication. In: Rosenblum, L.A., Coe, C.L. (eds) *Handbook of squirrel monkey research.* Plenum Press, New York, pp. 99-126

Titze, I.R. (1989) On the relation between subglottal pressure and fundamental frequency in phonation. *J. Acoust. Soc. Am.* **85:** 901-906

Titze, I.R. (1993) Current topics in voice production mechanisms. *Acta Oto-Laryngol.* **113:** 421-427

West, R., Larson, C.R. (1993) Laryngeal and respiratory activity during vocalization in macaque monkeys. *J. Voice.* **7:** 54-68

Winter, P. (1969) The variability of peep and twit calls in captive squirrel monkeys (*Saimiri sciureus*) . *Folia Primatol.* **10:** 204-215

Winter, P., Ploog, D., Latta, J. (1966) Vocal repertoire of the squirrel monkey, its analysis and significance. *Exp. Brain Res.* **1:** 359-384

Zimmermann, E., Newman,J.D., Jürgens U. (1995) *Current topics in primate vocal communication.* Plenum Press, New York

Exp. Biol. Online 1:10 (1996)

Influence of polylysine on adhesion of fibroblasts to glass substrates visualized by total internal reflection microscopy

Jan Hornung and Günter Fuhr

Humboldt-University Berlin, Institute of Biology, Membrane Physiology, Invalidenstrasse 43, D-10115 Berlin, Germany

Correspondence to: G. Fuhr (e-mail: guenter=fuhr@rz.hu-berlin.de)

Received: 19 November 1996 / Accepted: 10 December 1996
Published: December 19, 1996

Abstract. The cell adhesion topography of mouse fibroblasts growing on glass substrates has been investigated. In order to compare cell adhesion on covered and uncovered glass, substrates were partly exposed to a solution with 0.1 mg/ml polylysine (300 kDa) for 15 min before incubation with cell suspension. After cultivation for 1, 3, 6, and 24 h their adhesion was visualised by total internal reflection microscopy. In the presence of polylysine, cells incubated for 1 h were strongly attracted to the substrate, leading to a typical cell adhesion topography characterised by round concavities under the ventral cell membrane with an approximate diameter of 1 µm. The cavity-surrounding rims were tightly bound to the glass surface. During further cell cultivation, the topography changed into a well-organised adhesion pattern with focal contact areas on the periphery of the cells. In contrast to the polylysine-mediated adhesion, cells growing on untreated surfaces did not exhibit the cavity-like topography at any stage of cultivation, but a more point spread adhesion with a dense clustering of contact-forming areas.

Key words: Total internal reflection microscopy – Cell-substratum adhesion – Polylysine – 3T3

Introduction

A widely accepted approach to study cell adhesion or cell-matrix interaction is the utilisation of artificial substrates as a supporting layer on which cells are allowed to grow. The interaction of the cells with those substrates is often investigated by optical methods, such as fluorescence microscopy, to reveal specific receptor-ligand interaction or the organisation of the cytoskeleton. However, only few optical methods allow the direct visualisation of the adhesion topography of the ventral side of the attached cells. A common technique is interference reflection microscopy (IRM), which measures the thickness of the thin liquid film between the glass and the ventral cell membrane (Curtis 1964; Izzard and Lochner 1976; Bereiter-Hahn et al. 1979; Gingell and Todd 1979). The resulting image exhibits interference fringes encoding the distance of the cell membrane to the glass. Interpretation of the images obtained is quite difficult since reflections on the ventral and dorsal cell membrane may occur, leading to overlapping interference fringes (Izzard and Lochner 1976).

Another optical method is total internal reflection microscopy (TIRM), introduced by Axelrod (1981). We used TIRM as modified by Gingell et al. (1985), so-called total internal reflection aqueous fluorescence (TIRAF). Since both methods were based on total internal reflection we use the term TIRM throughout the paper. When a light beam propagates in a glass and strikes an interface with a liquid medium with a lower index of refraction, the beam is totally internally reflected if its angle of incidence [phi] onto that interface exceeds the critical angle of total internal-reflection. Although the beam is totally internally reflected at the interface, it produces an electromagnetic disturbance in the liquid medium called the evanescent wave. This wave propagates parallel to the interface but declines exponentially in the direction perpendicular to it with a characteristic penetration depth

$$d_o = \frac{l}{2p\sqrt{n_{glass}^2 \sin(f) - n_{medium}^2}}$$

in the range of a few tens of nanometers, where is the wavelength of the incident light, n_{glass} and n_{medium} are the indices of refraction of glass and the liquid medium, respectively. Because of the restricted penetration depth, the evanescent wave can excite fluorescent dyes only within a very thin region with a thickness of about $3d_o$ without any bulk fluorescence.

TIRM was combined with different staining techniques to visualise cell compartments that are very close to the glass surface. This investigation involved the staining of the cell membrane (Axelrod 1981; Lanni et al. 1985; Truskey et al. 1992; Burmeister et al. 1994), the cytoplasm and actin (Lanni et al.

1985) or specific receptors (Axelrod 1981). In contrast, Gingell et al. (1985) used a fluorescence dye coupled to dextran of a different size to "label" the extracellular volume. Therefore, areas of the glass surfaces that are not covered by a cell appear bright, but any other area that is covered by a cell exhibits less intense fluorescence depending on the distance of the cell membrane from the glass. Thus focal contacts at a distance of about 10-15 nm (Curtis 1964; Izzard and Lochner 1976) exhibit a much less intense fluorescence, close contacts (ca. 30-50 nm) are slightly more intense, and cell contacts at a great distance (ca. 100-150 nm) appear bright. Unlike IRM, the dorsal membrane has no influence on the brightness of fluorescence unless it is separated by more than 200 nm from the glass surface. In the case of a very thin lamellipodium (ca. 100 nm) the evanescent wave can penetrate it and excite dyes on the upper side of the cell. However, by adjusting the angle of incidence the penetration depth can be reduced to less than 100 nm and therefore avoid penetration of the lamellipodia. The principle of this technique is illustrated in Fig. 1.

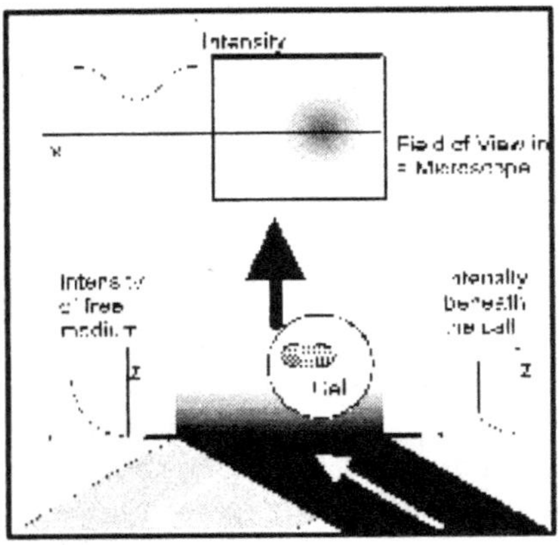

Fig. 1 Schematic illustration of the TIRM principle. The evanescent wave declines exponentially perpendicular to the glass/medium interface and excites dyes only near the interface. The integrated fluorescence intensity increases with the distance to the glass surface and reaches a maximum at about $3d_0$ (left side). If a cell approaches the surface it drives out the dye and the fluorescence intensity decreases depending on the distance between the cell membrane and the glass (right side). In the field of view of a microscope the parts of the cell at a distance of less than $3d_0$ appear grey or dark, whereas the other parts remain bright

We used TIRM to investigate the influence of polylysine on the adhesion of fibroblasts. Polylysine is known to promote cell adhesion without being significantly toxic for many mammalian cells (Freshney 1992). Polylysine is a positively charged polycation at physiological pH of mammalian cell culture media, known to mediate the electrostatic interaction of the cell membrane and the glass substrate. Recently a characteristic adhesion pattern of periodically distributed intimate contacts was found for erythrocytes flattened on a glass surface in the presence of polylysine (Thomas et al. 1996).

By covering glass and silicone substrates with polylysine Hornung et al. (1996) reported that the enhancement of cell adhesion leads to a higher survival rate of fibroblasts when they are cryopreserved in an attached state. However, this effect vanishes after cultivation times above 16 h. Here we present a broader investigation of the dynamics of polylysine-mediated adhesion and the changes in adhesion topography for fibroblasts within the first 24 h of attachment.

Materials and methods

Optical set-up

The optical set-up is illustrated in Fig. 2. In order to couple a laser beam of an argon-ion laser ([lambda]=488 nm, Laser 2000, Weßling, Germany) into an optically flat plate (BK-7, Spindler & Hoyer, Göttingen, Germany) a prism (BK-7, OWIS, Staufen i. Br., Germany) was glued (E-203, Polytech, Darmstadt, Germany) onto the plate. The glue was chosen to match the refractive index of the BK-7 glass ($n_{BK\text{-}7}$=1.52 at 488 nm). The laser beam strikes the top side of the plate under an angle of 70° to give a penetration depth d_o of about 70 nm. We used fluorescein as volume marker, since this dye is generally excluded by the cells. The detection channel consists of an upright microscope (Aristoplan, Leica, Wetzlar, Germany) equipped with differential contrast optic (DIC) and immersion objectives (40× or 100×, N.A.=1.3), a CCD camera (CS3130M, Seitner, Seefeld, Germany) which allows on-chip integration and an image processing system of a CLSM microscope (Leica). To obtain noiseless images, the gain of the CCD camera was adjusted to the minimum setting but the on-chip integration time was raised to 5 s. The video signal was directly converted into a digitised image and stored in the image processing system. Although we used an uncooled CCD chip we found an increase of thermal noise at minimum gain only at integration times exceeding 30 s.

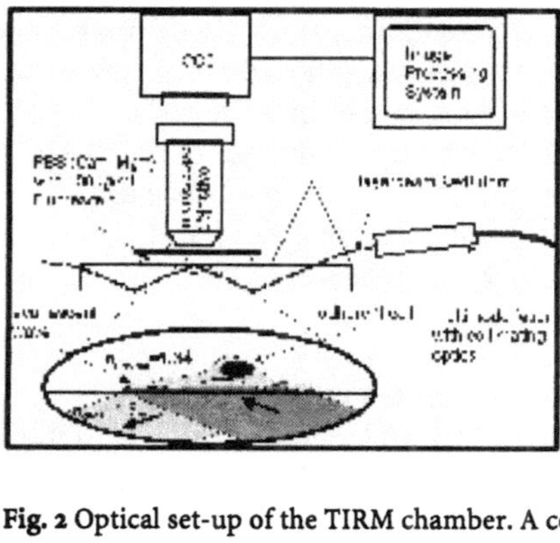

Fig. 2 Optical set-up of the TIRM chamber. A collimated laser beam enters a glass plate via a prism and is totally internally reflected within the plate under an angle of about 70°. The cells are cultivated on the top side of the plate and the culture medium is replaced by PBS supplemented with fluorescein and 10% FCS immediately before investigation. Using an upright microscope, the fluorescence is detected by a computer-controlled CCD camera connected to an image processing system

Cell cultivation and chip coating

The glass plate was partly covered with sterile 0.5 ml polylysine solution (Biochrom, Berlin, Germany) at a concentration of 0.1 mg/ml for 10-15 min in an incubator at 37°C, rinsed twice with sterile double deionised water and finally washed with 1 ml PBS containing Mg^{2+} and Ca^{2+}. NIH-3T3 (Deutsche Sammlung von Mikroorganismen und Zellkulturen, Braunschweig, Germany) were cultivated in DMEM with HEPES buffer (Sigma, Germany), supplemented with 10% FCS (Gibco, UK), 2 mM/ml glutamine and 0.1 mg/ml gentamicin (Biochrom). The cells were passaged twice a week to keep them at zhe logarithmic growing stage. Two days after passage cells were trypsinised (0.1 mg/ml trypsin/EDTA) for 5 min and resuspended in fresh culture medium. 1.5 ml of cell suspension with a final concentration of 510^4 cells/ml was inoculated onto the glass plate and allowed to grow for 1, 3, 6 or 24 h in a humidified incubator at 37°C. After incubation the attached cells were carefully washed with 100-200 μg/ml fluorescein in PBS (Mg^{2+}, Ca^{2+}) containing 10% FCS.

Results

Cells were allowed to grow for 1, 3, 6, and 24 h on the glass plate before the cell adhesion topography was visualised by TIRM. Each TIRM micrograph is accompanied by a corresponding DIC micrograph to compare the adhesion topography with cell morphology. Examples of cell adhesion and cell morphology are illustrated in Figs. 3-7. To avoid an uptake of fluorescein (Gingell et al. 1985), which we observed when cells were washed with fluorescein containing PBS, we added 10% FCS to PBS. Fluorescein is polar at a physiological pH of 7.3 and is normally excluded by the cell membrane. However, even when raising the pH up to 7.6 an uptake of fluorescein was observed, leading to a dramatic loss of fluorescence contrast within 5 min. It is known that fluorescein binds to serum albumin (Meisingset and Steen 1981), which might drastically reduce the diffusion of fluorescein into the cells. Additionally, Tsien (1989) reported that cells exhibit a permeation of the dye at a significant rate, especially at higher temperatures and greater surface to volume ratios, which is true of adherent fibroblasts. On the other hand, when cells were allowed to grow in culture medium in the presence of fluorescein, we did not observe an uptake even after 24 h.

To evaluate the diffusion limitation of fluorescein which may occur when fluorescein penetrates beneath cells that are already attached, we compared the TIRM fluorescence of cells either cultured in the presence of fluorescein, or exposed to fluorescein when already attached and spread. In the case of diffusion limitation of free or bound fluorescein a weak fluorescence is expected when fluorescein has to penetrate the thin liquid film between the glass and the cell membrane. Additionally, fine structures such as focal contacts may not be resolved. However, no differences were observed, indicating that diffusion limitation is not observed when cells are immersed in the fluorescein solution for at least 2 min, which was the mean time before we started the experiments. Similar results were observed using fluorescein-labelled dextran of molecular weight 4000 (Gingell et al. 1985).

Cells growing on uncoated glass

Figures 3 and 4 give an impression of cells growing on uncovered glass for 3 and 24 h, respectively. Within the first 3 h the cell adhesion topography is dominated by broad areas of close contacts (30-50 nm), which appear grey and cover most of the cell body (Fig. 3a). In some cases the dark grey areas are dissected by lines of higher intensity, indicating a greater distance between the cell membrane and the glass. In contrast to the close contacts the pronounced dark spots indicate focal contacts. A narrow but dense clustered area of focal contacts is visible on the lower left side of the cell very near to the periphery of the leading edge. Adjacent to this narrow band there is a wide area of close contacts, which covers about half of the cell and exhibits only few scattered focal contacts. The trailing half of the cell is again dominated by close contacts but exhibits many focal contacts arranged in a wide ring. The leading edge is rarely visible in the DIC micrograph in Fig. 3b, but a lot of visible filopodia correspond well to structures seen in Fig. 3a.

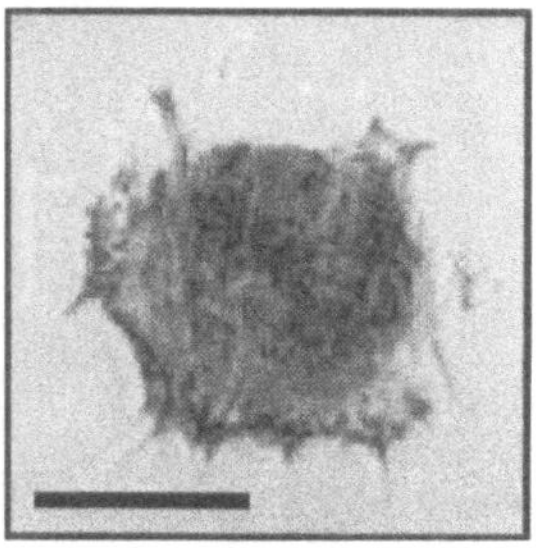

Fig. 3a 3T3 growing on uncovered glass for 3 h. The cell is partly polarised leading to intimate contact on the cell periphery at the leading edge. Many adhesion spots are scattered over the entire membrane. *Bar* 20 µm.

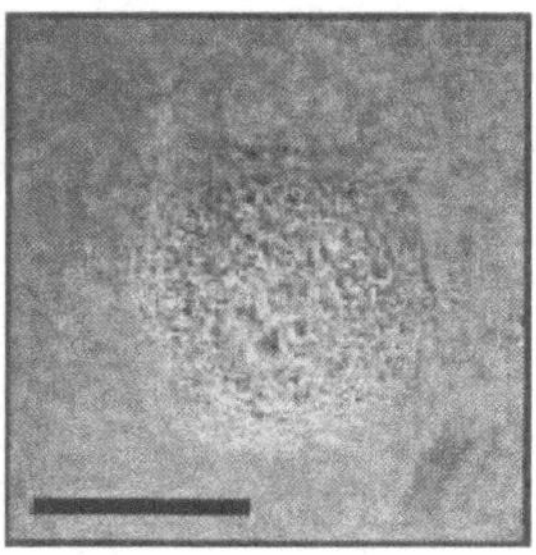

Fig. 3b DIC micrograph corresponding to a. The nucleus is clearly visible. In comparison with the TIRM micrograph only a few adhesion spots are arranged near the nucleus. *Bar* 20 µm

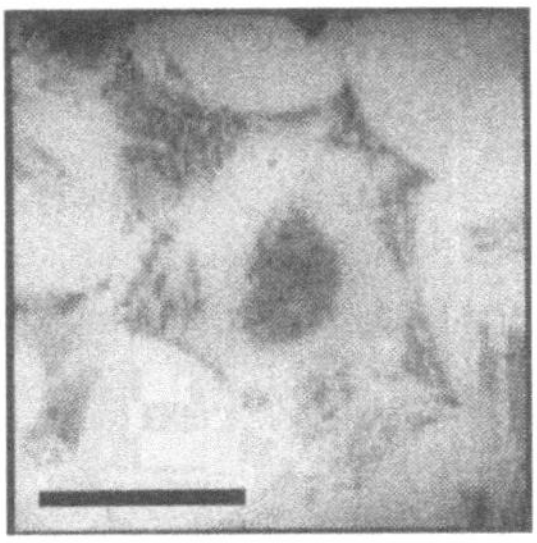

Fig. 4a Adhesion topography of cells growing for 24 h on an uncovered glass surface. The adhesion is pronounced at the periphery and in a large area full of clustered focal contacts. Since the brightness of some cell areas is similar to glass-medium interfaces not covered with cells, the distance of those areas from the cell membrane is in the range of about 150-200 nm. *Bar* 50 μm

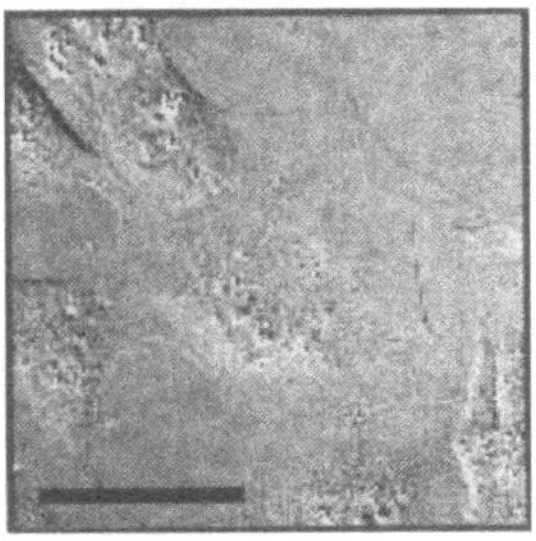

Fig. 4b The DIC micrograph clearly reveals the cell morphology. The nucleus of the widely spread cell corresponds to the bottom part of the large adhesion area seen in **a**. *Bar* 50 μm

After 24 h cultivation the cells are well spread. Only a few cell divisions have already occurred since the observed timelag is about 22-24 h. In particular, widely spread cells (Fig. 4b), which are considered to be less motile, exhibit a distinct cell adhesion topography. Pronounced contact-forming spots are visible near the cell periphery (Fig. 4a) and are mostly focal contacts, since greyish areas are rarely visible. An area with densely clustered focal contacts is visible underneath the nucleus surrounded by a wide area with a very large distance between the cell membrane to the glass. The cell topography of the ventral membrane is thus only supported on the cell periphery and in its middle.

Cells growing on coated glass

At the early stage of adhesion (3 h) the adhesion topography exhibits characteristic indentations forming a concavity-shaped pattern (Fig. 5a). The cavity-surrounding rims are tightly bound to the glass, indicated by their dark appearance. The pattern is uniformly distributed over the entire ventral membrane, including the already established leading edge. Very many cells are unpolarised at this stage and in some cases cells have already become flattened. The cell density of about 260 cells/mm^2 gives an average area of about 3800 μm^2

for each cell, which is more than a 3T3 can cover (1500 μm^2 our observation, 1800 μm^2 (Ireland et al. 1989)). Many filopodia extending from the leading edge surrounding the cells are clear visible and are in close contact with the glass (Fig. 5a). The tightly bound rims have about the same dark appearance as focal contacts but it seems unlikely that the rims correspond to the termination of actin bundles.

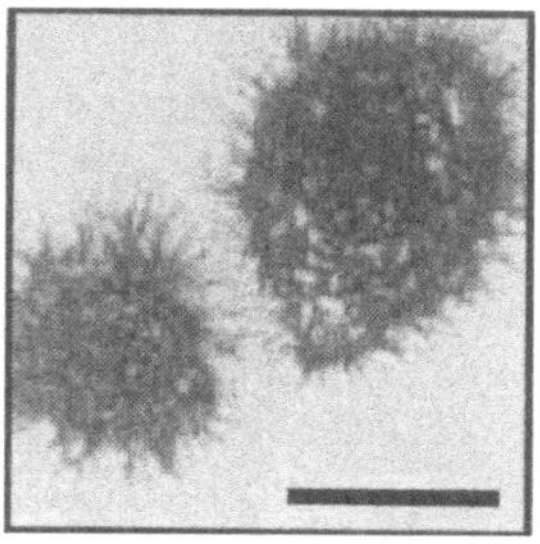

Fig. 5a Adhesion topography of 3T3 growing on glass covered with polylysine visualised by TIRM. Many cavities (bright areas) are visible, surrounded by tightly bound contact regions (dark rims). *Bar* 20 μm

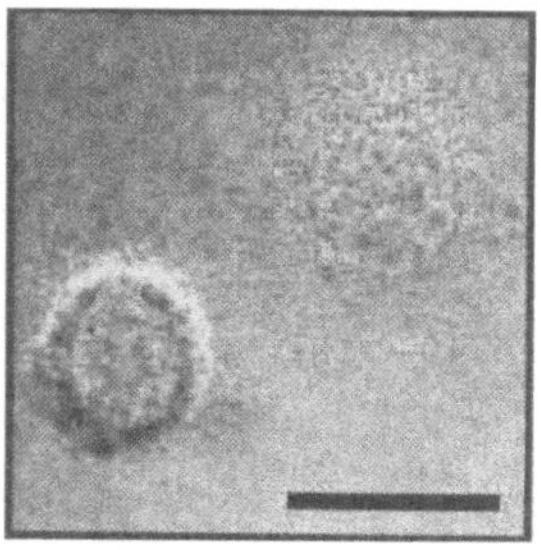

Fig. 5b DIC micrograph of the cells shown in **a**. The thin filiopodia extracted from the cell periphery correlate well with the TIRM micrograph. The broad leading edge seems to be very thin. *Bar* 20 μm

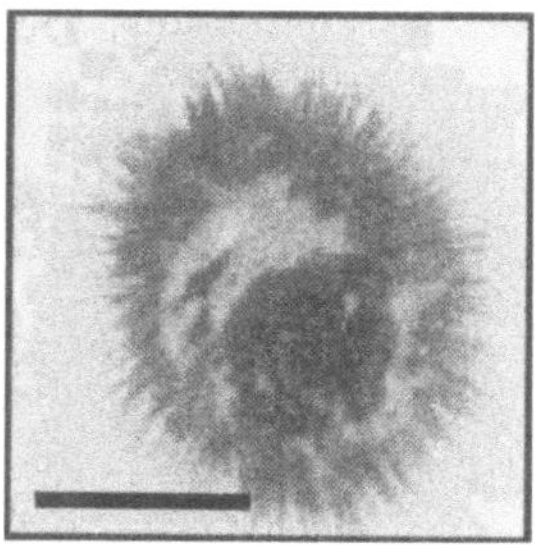

Fig. 6a TIRM micrograph of a typical cell cultured for 6 h on covered glass. A compact area with cavities is visible, but a large part of the ventral cell membrane has lost its intimate contact with the glass. The broad leading edge with many filopodia is well established and closely attached to the glass. *Bar* 20 μm

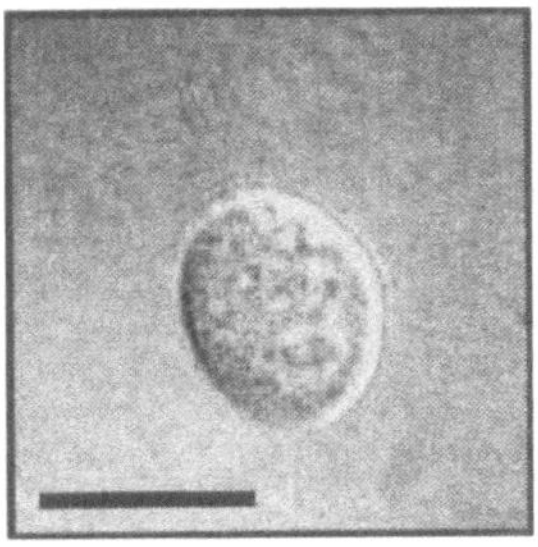

Fig. 6b The DICmicrograph corresponding to **a**. The shape resembles that of a fried egg with a very broad and thin leading edge and a compact cell body. *Bar* 20 μm

During further cell cultivation the adhesion pattern of cells growing on coated glass undergoes a remarkable change. The adhesion topography exhibits areas in close contact with the glass, but also areas at a great distance (Fig. 6a). Only near the nucleus (compare with Fig. 6b, the nucleus is on the bottom side of the round cell body) is an area resembling the concavity-shaped pattern still visible. The cavities have become smaller, but the rims are stil in close contact with the glass. On the cell periphery, the leading edge and many protruding filopodia are equidistant from the glass. Around the nucleus the cell membrane is at a much greater distance than on the periphery forming a great cavity with a few spots in close contact.

After 24 h of cultivation the cells show a variable cell morphology. Most of the cells exhibit a well spread or polarised shape, but some are also rounder (Fig. 7b). The cell adhesion topography at this cultivation stage indicates mostly close contact with the glass with a variety of pronounced focal contacts (Fig. 7a). Many long filopodia are also visible, most of which are in an intimate contact with the glass along their length, but in some cases the filipodia do not touch the glass, as explained in the legend to Fig. 7b. Figure 7c shows a magnification of the adhesion topography of cells in Fig. 7a.

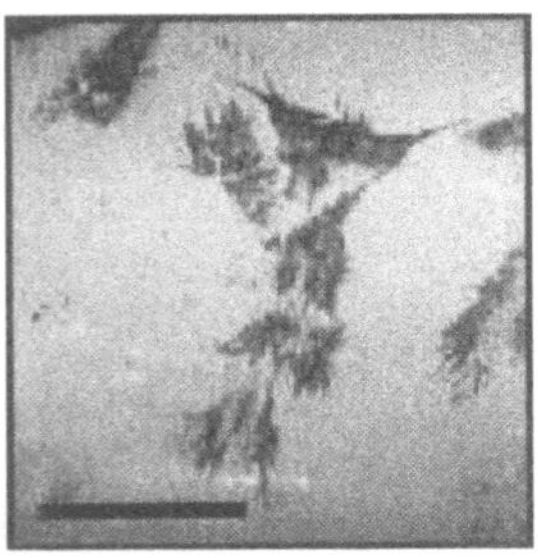

Fig. 7a Overview of the adhesion topography found after 24 h of cultivation on glass covered with polylysine. *Bar* 50 μm

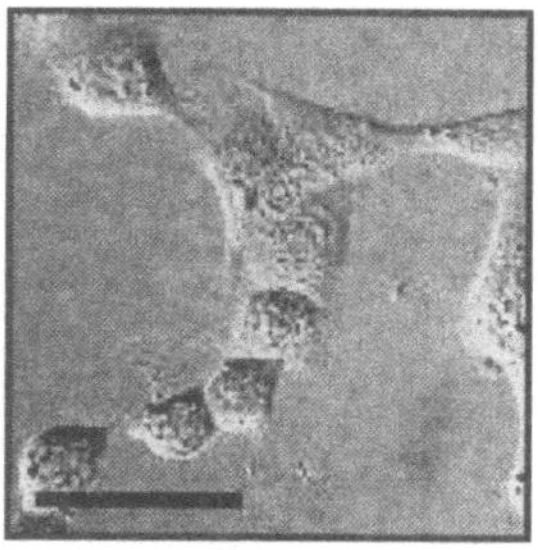

Fig. 7b DIC micrograph corresponding to **a**. Note the long filopodia extending between the cells, which in some cases have no contact along their length. The corresponding TIRM micrograph shows no dark spots underneath the filopodia. *Bar* 50 μm

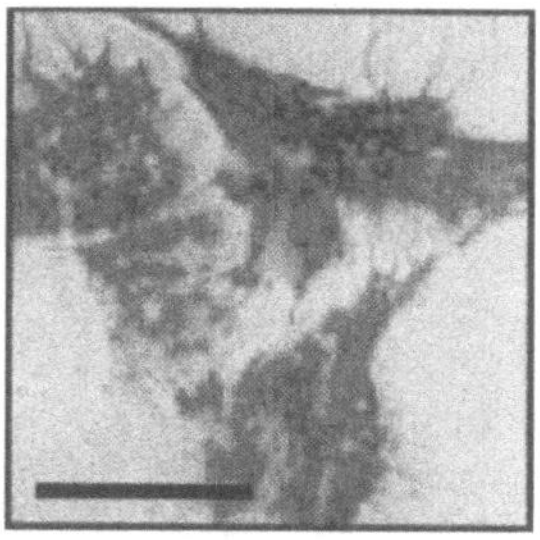

Fig. 7c Magnification of **a**. The filopodia extending underneath neighbouring cells do not touch the membrane. A bright halo is visible around each filopodium, indicating aqueous fluorescence medium. *Bar* 20 μm

Discussion

The study of the influence of the extracellular matrix on cell adhesion gives a deeper insight into cell attachment and adhesion. Artificial substrates are especially well suited since their surfaces can be modified in many ways. An easy way to do this is to cover a glass surface with a polycation like polylysine. Due to its positive charge polylysine mediates the electrostatic interaction between the cell membrane and the glass surface. This report exploits the benefits of TIRM to reveal the influence of polylysine on the adhesion of fibroblasts within the first 24 h.

By using the evanescent wave generated by total internal reflection to excite fluorescent dyes in the extracellular volume it is possible to visualise the cell adhesion topography very clearly (Gingell et al. 1985). The variation in the distance between the cell membrane and the glass leads to a fluorescence intensity modulated signal with an unambiguous correlation between membrane separation and fluorescence intensity. Any misinterpretation may arising due to very thin cellular structures such as thin lamellipodia or filopodia can be eliminated by increasing the angle of incidence and thus reducing the penetration depth (Gingell et al. 1987). Moreover, by applying an appropriate model describing the optical properties of the glass-cell system, very precise measurement of cell-substrate distance is possible (Gingell et al. 1987). By measuring the thickness of magnesium fluoride films evaporated on glass, Heavens and Gingell (1991) showned an accuracy of about 2 nm.

Although the high sensitivity of TIRM for the quantitative estimation of planar test systems has been demonstrated, difficulties may arise due to the inevitable scattering of the incident laser beam on glass surfaces that are not perfectly flat or on cellular structures in the vicinity of the surface (Gingell and Heavens 1996). Despite these difficulties, TIRM is a straightforward tool for qualitative investigation of cell adhesion topography and easy comparison with IRM, since in both methods focal contacts appear dark, close contacts grey, and areas with a great distance between cell and substrate bright when using monochromatic light in IRM and an extracellular fluorescent volume marker in TIRM. An advantage of TIRM is the high image quality we obtained combined with very good resolution of different cell-substrate separations and the high resolution of fluorescence microscopy. In addition, no cell fixation or staining is necessary when labelling the extracellular medium, which will leave the cells in a most vital state, providing a supply of sufficient fresh medium.

These benefits were used to visualise cells at different adhesion stages on uncovered glass. At the early stage cells form an intimate, but not organised, adhesion topography without any focal contact. The cells remain spherical but a thin and broad lamella is formed progressively (Bereiter-Hahn et al. 1990). During further cultivation cells begin to flatten and become well spread (Figs. 3b, 4b). At the same time, the broad close contacts seen at the beginning of the adhesion become structured to variable degrees and a clustering of focal contacts predominantly at the periphery of the leading edge can be observed (Fig. 3a). Finally, after 24 h the topography is dominated by focal contacts clearly visible at the cell periphery and in the vicinity of the nucleus separated by broad areas with a large distance between the cell membrane and the glass (Fig. 4a). These observations correlate well with investigations of cell attachment dynamics on glass visualised by TIRM discussed by Truskey et al. (1992) using cells (bovine aortic endothelial cells) fixed at different adhesion stages with subsequent labelling of the membrane.

Recently Thomas et al. (1996) investigated the contact formation of erythrocytes on glass in the presence of polylysine. The results reported were discussed in terms of the interfacial instability theory, where the aqueous layer between the glass and the membrane is assumed to be a thin film. A periodic pattern of membrane contact zones to the glass may arise if the equilibrium between the repulsive and attractive forces exerted on the membrane is disturbed by a small perturbation which increases with time. Such periodic patterns were found during the adhesion of erythrocytes if the cells were washed in a solution containing polylysine immediately before inoculation onto the glass (Thomas et al. 1992, 1996).

In contrast to erythrocytes, fibroblasts are very motile and exhibit a cortex with a dense meshwork of actin bundles. The cortex of cultured fibroblasts is about

0.2 µm thick, in contrast to the cortex of erythrocytes, which is about 0.01-0.02 µm thick (Bray et al. 1986). After the onset of adhesion the fibroblasts become polarised and begin to migrate. These processes involve a permanent rearrangement of the cytoskeleton, which leads to a variety of cell morphologies found in exponentially growing fibroblast cell cultures. The actin filaments of the cortex are intimately associated with the cell membrane, leading to a stabilised cell membrane with many protrusion-like microvilli. Due to this stabilisation, the mechanical resistance of the membrane increases many times. Therefore, the organisation of the actin filaments should dominate the adhesion topography. However, we found a similar adhesion pattern for fibroblasts (3T3) on glass surfaces covered with polylysine to that reported by Thomas et al. (1996) for erythrocytes. The attractive force exerted on the membrane due to the electrostatic interactions leads to a random modulation of the cell membrane to glass distance. The size of the formed cavities ranges from 1-2 µm and they are at a mean distance of about 2-3 µm to each other. Although the adhesion topography of 3T3 resembles that found for erythrocytes, we assume that the resulting adhesion topography is strongly influenced by the interaction between the attractive electrostatic forces and the mechanical stiffness of the membrane due to the cortex.

According to the interfacial instability theory (Thomas et al. 1996) a disturbance will grow if a strong negative (attractive) membrane-glass interaction occurs, which increases rapidly with decreasing cell membrane to glass distance, while the mechanical resistance of the membrane, expressed by its bending modulus, and the membrane tension are rather small. This prediction is supported by the periodic pattern found for erythrocytes and by the studies presented in this paper. The cell adhesion mediated by polylysine leads to an adhesion pattern within the first 3 h, which is characterised by many cavities with a concave shape indented by the intimately approached membrane. The random distribution of the cavities and their varied size probably reflects the mechanical resistance of the cytoskeleton against the strong bending of the membrane. Although we have not yet performed a quantitative study of the membrane - glass distance, the fluorescence intensity allows a first estimate. The height of the cavities seems to be dependent of their size, since larger cavities exhibit a brighter fluorescence, indicating a greater distance. In some cases the fluorescence was as bright as on pure interfaces not covered by cells. We would like to point out again that using TIRM it is only possible to measure a distance of a maximum of $3d_0$, roughly 200 nm in our experiments. Therefore, we assume a size-dependent cavity height ranging from about 50 to 200 nm.

Unlike erythrocytes the adhesion process of fibroblasts takes much longer. No flattening of the cells could be observed within 5 min. Since the adhesion topography at 1 h is not as pronounced as after 3 h we conclude that the flattening occurs within 1-3 h. Upon closer consideration the sedimentation of cells has to be taken into account.

During further cell cultivation the influence of polylysine vanishes. At 6 h the cavity-shaped topography is still partly visible, but the adhesion becomes dominated by broad areas of close contacts with focal contacts scattered over the entire membrane, whereas after 24 h no cavities are visible. However, the mean distance of the cells is much smaller than on untreated surfaces indicated by the grey colouring of the cells (Fig. 7a, c) in comparison to cells growing on uncovered surfaces (Fig. 4a). The reduced influence of polylysine is assumed to be a result of the expression of proteins and their adsorption to the surface, which leads to a loss of unsaturated surface charges. At the same time the receptor-ligand-mediated adhesion increases, characterised by the clustering of focal contacts mainly on the periphery of the cells.

In conclusion, the results presented above strongly suggest a similar instability of the membrane found on erythrocytes leading to localised intimate contact regions, which surround cavities of a mean size of about 1-2 μm. However, the process of adhesion is much slower than for erythrocytes and fibroblasts are able to establish the typical focal contact adhesion within 24 h.

Acknowledgements. J.H. would like to acknowledge a grant from the Studienstiftung des deutschen Volkes.

References

Axelrod, D. (1981) Cell-substrate contacts illuminated by Total Internal Reflection Fluorescence. *J. Cell Biol.* **89**: 141-145

Bereiter-Hahn, J., Fox, C.H., Thorell, B. (1979) Quantitative reflection contrast microscopy of living cells. *J. Cell Biol.* **82**: 767-779

Bereiter-Hahn, J., Lück, M., Miebach, T., Stelzer, H.K., Vöth, M. (1990) Spreading of trypsinized cells: cytoskeletal dynamics and energy requirements. *J. Cell Sci.* **96**: 171-188

Bray, D., Heath, J., Moss, D. (1986) The membrane-associated "cortex" of animal cells: Its structure and mechanical properties *J. Cell Sci.* (Suppl.) **4**: 71-88

Burmeister, J.S., Truskey, G.A., Reichert, W.M. (1994) Quantitative analysis of variable-angle total internal reflection fluorescence microscopy (VA-TIRFM) of cell/substrate contacts. *J. Microsc.* **173**: 39-51

Curtis, A.S.G. (1964) The mechanism of adhesion of cells to glass. A study by interference reflection microscopy. *J. Cell Biol.* **20**: 199-215

Gingell, D., Heavens, O.S. (1996) Elimination of the effects of stray light in measurements by total internal reflection aqueous fluorescence (TIRAF). *J Microsc.* **182**: 141-148

Gingell, D., Todd, I. (1979) Interference reflection microscopy. A quantitative theory for image interpretation and its application to cell-substratum separation measurement. *Biophys. J.* **26**: 507-526

Gingell, D., Todd, I., Bailey, J. (1985) Topography of cell-glass apposition revealed by Total Internal Reflection Fluorescence of Volume Markers. *J. Cell Biol.* **100**: 1334-1338

Gingell, D., Heavens, O.S., Mellor, J.S. (1987) General electromagnetic theory of total internal reflection fluorescence: the quantitative basis for mapping cell-substratum topography. *J. Cell Sci.* **87**: 677-693

Heavens, O.S., Gingell, D. (1991) Film thickness measurement by frustrated total reflection fluorescence. *Opt. Laser Technol.* **23**: 175-179

Hornung, J., Müller, T., Fuhr, G. (1996) Cryopreservation of anchorage-dependent mammalian cells fixed to structured glass and silicon substrates. *Cryobiology* **33**: 260-270

Ireland, G.W., Dopping-Hepenstal, P.J.C., Jordan, P.W., O'Neill, C.H. (1989) Limitation of substratum size alters cytoskeletal organization and behaviour of swiss 3T3 fibroblast. *Cell Biol. Int.* **13**: 781-790

Izzard, C.S., Lochner, L.R. (1976) Cell-to-substrate contacts in living fibroblasts: an interference reflection study with an evaluation of the technique. *J. Cell Sci.* **21**:129-159

Lanni, F., Waggoner, A.S., Taylor, D.L. (1985) Structural organization of interphase 3T3 fibroblasts studied by total internal reflection fluorescence microscopy. *J. Cell Biol.* **100**: 1091-1102

Maurer, H.R. (1992) Towards serum-free, chemically defined media for mammalian cell culture. In: Freshney, R.I. (ed.) *Animal cell culture: a practical approach.*, Oxford University Press, Oxford, pp. 15-46

Meisingset, K.K., Steen, H.B (1981) Intracellular bindings of fluorescein in lymphocytes. *Cytometry* **1**: 272-278

Thomas, N.E., Coakley, W.T., Galip, A. (1992) The lateral separation of contacts on erythrocytes agglutinated by polylysine. *Cell Biophys.* **20**: 125-147

Thomas, N.E., Coakley, W.T., Winters, C. (1996) Contact formation in polylysine-mediated membrane-glass interaction. *Colloids Surfaces B.* **6**: 139-147

Truskey, G.A., Burmeister, J.S., Grapa, E., Reichert, W.M. (1992) Total internal reflection fluorescence microscopy (TIRFM). II. Topographical mapping of relative cell/substratum separation distances. *J. Cell Sci.* **103**: 491-499'

Tsien, R.Y. (1989) Fluorescent indicators of ion concentrations. In: Taylor, D.L., Wang, Y. (eds.) *Fluorescence microscopy of living cells in culture: part B.*, Academic Press, San Diego, pp. 127-156

Exp. Biol. Online 2:1 (1997)

Acid-base disequilibrium in the venous blood of rainbow trout (*Oncorhynchus mykiss*)

S.F. Perry [1], C.J. Brauner [2], B. Tufts [3] and K.M. Gilmour [4]

[1] Department of Biology, University of Ottawa, 30 Marie Curie, Ottawa, Ontario K1N 6N5, Canada

[2] Department of Zoophysiology, Aarhus University, Building 131, DK-8000 Aarhus C, Denmark

[3] Department of Biology, Queen's University, Kingston, Ontario, K7L 3N6, Canada

[4] Division of Environmental and Evolutionary Biology, University of Glasgow, Graham Kerr Building, Glasgow, G12 8QQ, Scotland

Correspondence to: S.F. Perry, tel (613) 562-5800 x6005, fax (613) 562-5486, (e-mail: sfperry@oreo.uottawa.ca)

Received: 2 December 1996 / Accepted: 14 January 1997
Published: January 17, 1997

Abstract. Acid-base equilibria/disequilibria were evaluated in vivo in post-branchial arterial blood and pre-branchial venous blood of freshwater rainbow trout (*Oncorhynchus mykiss*). This was accomplished using arterial and venous extracorporeal circuits in conjunction with a stopped-flow apparatus. After the abrupt stoppage of circulating post-branchial blood within the stopped-flow apparatus, pH increased slowly ([Delta]pH = +0.032 ± 0.004 pH units; $n = 15$), thus confirming the existence of an acid-base disequilibrium state in the arterial blood of rainbow trout. The slow downstream pH changes were unaffected by prior treatment of fish with the carbonic anhydrase inhibitor benzolamide (1.2 mg kg^{-1}; [Delta]pH = +0.032 ± 0.01 pH units; $n = 5$) but were eliminated after intra-vascular injection of 10 mg kg^{-1} bovine carbonic anhydrase ([Delta]pH = -0.011 ± 0.003 pH units; $n = 8$). These results demonstrate that the acid-base disequilibrium in the arterial blood reflects a total absence of extracellular carbonic anhydrase activity. Similar stopped-flow experiments revealed the existence of a reduced, yet significant, acid-base disequilibrium in the venous blood circulating within the caudal vein ([Delta]pH = +0.004 ± 0.003 pH units; $n = 15$). Selective inhibition of extracellular carbonic anhydrase using benzolamide did not significantly influence the magnitude of the venous pH disequilibrium ([Delta]pH = +0.007 ± 0.007 pH units; $n = 8$) whereas intra-vascular injection of carbonic anhydrase eliminated the pH disequilibrium. These results demonstrate that extracellular carbonic anhydrase, although reported to be present within the skeletal muscle of rainbow trout , does not accelerate post-capillary pH changes in the venous circulation.

Key words: Carbonic anhydrase – Blood – Venous – Muscle – Carbon dioxide – Disequilibrium – Benzolamide

Introduction

Although few species have been examined, it is generally accepted that extracellular or membrane-associated carbonic anhydrase (CA) is unavailable to catalyse CO_2 - HCO_3^-- H^+ reactions in the plasma within the gills of teleost fish (Henry et al. 1988, 1993; Rahim et al. 1988; Perry and Laurent 1990). Consequently, the post-branchial arterial blood of rainbow trout (*Oncorhynchus mykiss*) exhibits an acid-base disequilibrium in which the pH slowly rises in the downstream plasma owing to the continuing uncatalysed dehydration of HCO_3^- (Gilmour et al. 1994; Gilmour and Perry 1994, 1996). Interestingly, elasmobranchs which are known to possess both plasma (Wood et al. 1994; Gilmour et al. 1996) and blood-facing gill membrane (Swenson et al. 1995) CA activity, also display an acid-base disequilibrium in the post-branchial blood (Gilmour et al. 1996). Thus, the situation in fish (elasmobranchs and teleosts) differs markedly from that in mammals (Effros et al. 1978; Klocke 1988) and turtles (Stabenau et al. 1996), where the presence of lung endothelial CA that is oriented toward the vascular compartment would suggest that post-pulmonary capillary blood is in acid-base equilibrium (Bidani and Crandall 1988; Klocke 1988).

Comparatively few studies have examined the potential contribution of systemic CA in accelerating CO_2 - HCO_3^-- H^+ reactions in the plasma of venous blood and these have focused exclusively on mammals. The results, although species- and organ-dependent, demonstrate that in mammals, CA is accessible to catalyse CO_2 - HCO_3^-- H^+ reactions within the plasma of systemic capillaries (O'Brasky et al.1979; Effros and Weissman 1979; O'Brasky and Crandall 1980; Geers*et al.*1985). In skeletal muscle, the exact location of this extracellular CA has been debated. The acceleration of plasma CO_2 - HCO_3^-- H^+ reactions in the outflowing perfusate of perfused rabbit hindlimb preparations (O'Brasky and Crandall 1980) suggests that CA is localised within the vascular compartment. However, the results of immunohistochemical studies (Sender et al. 1994) and experiments utilising conjugated CA inhibitors of variable molecular weights

(Geers et al. 1985) have yielded conflicting results. The most recent immunohistochemical evidence (Sender et al.1994) suggests that CA is associated with the capillary endothelium. In contrast, experiments incorporating conjugated CA inhibitors suggest that CA is confined to an interstitial localisation in which the enzyme is bound to the external surfaces of the sarcolemmal membranes.

Recently, Henry et al. (1996) provided evidence for the presence of extracellular CA in rainbow trout skeletal muscle, at least some of which was associated with sarcolemmal membranes. Henry et al. (1996) did not specifically examine whether any of the muscle CA activity was accessible to the vascular compartment. Thus, the goal of the present study was to determine whether the skeletal muscle of rainbow trout contains vascular CA activity capable of accelerating CO_2 - HCO_3^-- H^+ reactions in the venous circulation. This was accomplished in vivo by monitoring pH changes within a stopped-flow apparatus of blood withdrawn from the caudal vein, a vessel draining predominantly white skeletal muscle (Smith and Bell 1967).

Materials and methods

Experimental animals

Rainbow trout, *Oncorhynchus mykiss*, weighing 460 – 778 g [mean mass = 565 ± 27 (SEM) g; experimental n = 15] were obtained from Linwood Acres trout farm (Campbellcroft, Ontario) and held in large (500 l) fibreglass tanks supplied with flowing, dechlorinated City of Ottawa tap water (PO_2 = 20.0-21.3 kPa, PCO_2 = 0.033-0.047 kPa, $[Na^+]$ = 0.15 mmol l^{-1}, $[Cl^-]$ = 0.18 mmol l^{-1}, $[Ca^{2+}]$ = 0.42 mmol l^{-1}, temperature = 15 °C). All fish were maintained in the aquarium for a minimum of 4 weeks prior to experimentation and fed to satiation on alternate days with a commercial salmonid diet. They were not fed 24 h prior to experimentation. The photo-period was held at 12 h dark/12 h light.

Animal preparation

Fish were anaesthetised in a solution of ethyl-*m*-aminobenzoate (MS 222, 0.1g l^{-1}) neutralised to pH 7.0 with $NaHCO_3$. In all fish, the dorsal aorta and caudal vein were cannulated with polyethylene tubing (PE 50) according to standard surgical procedures (see Axelsson and Fritsche 1994). In two fish, the bulbus arteriosus was exposed by a ventral mid-line incision and cannulated additionally using silicone tubing.

After surgery the fish were transferred to individual opaque acrylic boxes that were supplied with aerated dechlorinated water (15°C). Fish were allowed to recover overnight prior to experimentation.

Stopped-flow experiments

pH changes in post-branchial (arterial) or post-systemic (venous) circulation were assessed using an in vivo stopped-flow procedure as developed and described in detail by Gilmour et al. (1994). Briefly, an extracorporeal circuit (for review see Thomas 1994) was established by withdrawing blood from the dorsal aorta cannula through a pH electrode chamber and then returning the blood to the fish via the caudal vein cannula. The flow of blood (1.2 ml min^{-1}) in the extracorporeal circuit was achieved using a small peristaltic pump. About 1.0 ml of blood was contained within the external loop and represented less than 4% of the total blood volume of the fish. Prior to each experiment, the circuit was rinsed with 10 ml of a concentrated ammonium heparin solution (540 units ml^{-1}) to reduce clotting.

Arterial experiments

After achieving stable baseline data (usually within 15 min of commencing the extracorporeal circuit), the flow was stopped abruptly and the pH of the blood was monitored within the stopped-flow apparatus for 6 min, after which time the flow was reinstated. Slow pH changes in the blood during stopped flow would indicate a disequilibrium state and provide evidence for CO_2 - HCO_3^-- H^+ reactions proceeding at an uncatalysed rate. However, stable pH during stopped flow would indicate an equilibrium state and provide evidence for CO_2 - HCO_3^-- H^+ reactions proceeding at a catalysed rate. Thus, the technique, when used in combination with exogenous CA or CA inhibitor treatments (see below), is capable of demonstrating whether CA is accessible to plasma CO_2 - HCO_3^-- H^+ reactions.

Venous experiments

After flow was re-initiated, the direction of the extracorporeal circuit was reversed; blood was withdrawn from the caudal vein cannula and returned to the dorsal aorta cannula. After re-stabilisation, a stopped-flow experiment was performed as described above.

Effects of CA or benzolamide

In a sub-set of fish ($n = 8$), bovine CA (10 mg kg^{-1}) was injected into the venous cannula. After 15 min, arterial and venous stopped-flow experiments were performed. In a second sub-set of fish ($n = 6$), the CA inhibitor benzolamide (1.2 mg kg^{-1}) was injected into the venous cannula and after 10 min, arterial and venous stopped-flow experiments were re-conducted. Benzolamide, a relatively impermeable CA inhibitor (Swenson and Maren 1987), has been shown previously to selectively inhibit extracellular and not red blood cell (RBC) CA, at the dose and within the time frame used in this study (Gilmour et al. 1996). However, to ensure that RBC CA activity was not being affected by benzolamide treatment, samples of blood were removed before and after benzolamide injection immediately prior to the arterial and venous stopped-flow experiments and analysed for RBC CA activity using the [Delta]pH electrometric assay (Henry 1991).

Analytical procedures and data presentation

Blood pH was monitored using a Metrohm glass combination electrode housed in a thermostatted cuvette (15°C) and connected to a Radiometer PHM73 acid-base analyser. A data acquisition system (Biopac Inc.) was used to continuously record and store the pH data.

Owing to the small size of the pH changes during stopped flow relative to the variance, the data have been normalised and are presented as changes in pH from the value immediately preceding the stoppage of flow.

The continuous pH data obtained during stopped-flow experiments were analysed by one-way analysis of variance; where statistical significance was indicated, a Bonferroni multiple comparison procedure was utilised. Differences between final venous and arterial pH disequilibria were assessed using an unpaired Student's *t*-test. In all cases, P values less than 0.05 were considered to be statistically significant.

Results and discussion

Arterial acid-base disequilibrium

The arterial (post-branchial) blood of rainbow trout displayed a marked acid-base disequilibrium as indicated by the slow rise in pH of 0.032 ± 0.004 units ($n = 15$) during the 6-min stopped-flow period (Fig. 1, Table 1). This result is in agreement with prior studies that have assessed the arterial acid-base disequilibrium in trout (Gilmour et al. 1994 1996; Gilmour and Perry 1994 1996). The basis for the disequilibrium is the rapid removal of CO_2 from the blood across the gill thereby creating a sudden imbalance in the CO_2 - HCO_3^-- H^+ equilibrium. Thus, in the absence of extracellular CA, the dehydration of plasma HCO_3^- will continue at the slow uncatalysed rate as the blood flows downstream from the gill pH will rise. Consequently, intra-vascular injection of bovine CA abolishes the disequilibrium (Table 1; Fig. 2). Interestingly, after CA injection, the pH of the blood during the 6-min stopped-flow period decreased slightly ([Delta]pH = -0.011 ± 0.003 units; $n = 8$). However, unlike the curvilinear (hyperbolic) change in pH that is characteristic of the arterial blood disequilibrium (Fig. 1), the slow decrease in pH during stopped flow after CA injection was essentially linear during the final 5 min of the 6-min stopped-flow period. Because these decreases in pH have not been observed in experiments involving buffers (Gilmour and Perry 1994), we believe that the acidification represents the RBC metabolism. Indeed, the calculated predicted pH change from metabolism during 6 min of stopped flow [using averaged metabolic rate data from Nikinmaa (1990; 6 mmol O_2/cells/h), an RQ value of 1, and buffer values from Wood *et al.* (1992; -9.7 mmol/l/pH at haematocrit = 25%)] is -0.015 units, which is similar to the measured [Delta]pH value of -0.011 units after CA injection. These metabolic effects were slightly greater when compared to previous stopped flow experiments and may reflect the higher water temperature used in the present study (15° C *versus* 10° C). Regardless, the decrease in pH during stopped-flow after CA treatment indicates that the magnitude of the acid-base disequilibrium under control conditions is being underestimated. Thus, the true extent of the acid-base disequilibrium can be calculated by subtracting the pH change during stopped flow after CA treatment from the initial control [Delta]pH value. These data are presented in Table 1 as "corrected disequilibria"; for the arterial blood the corrected disequilibrium was 0.050 ± 0.006 units ($n = 8$).

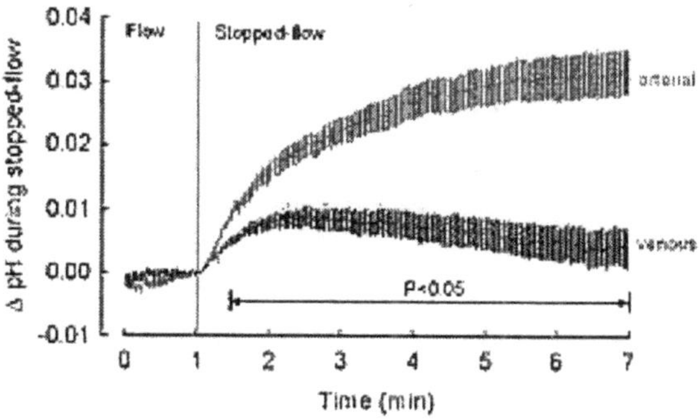

Fig. 1. Acid-base disequilibria in post-branchial dorsal aortic arterial blood (*red trace*; $n = 15$) and pre-branchial caudal vein blood (*blue trace*; $n = 15$). In each case, changes in pH ([Delta] pH) are shown during a 6-min period of stopped flow commencing at the *vertical solid line*. Arterial and venous stopped-flow experiments were performed on the same fish. The data shown are continuous traces of pH changes ± 1 SEM recorded at 1 s intervals. The *horizontal double arrow* indicates significant difference between arterial and venous [Delta] pH

Table 1. Arterial and venous blood initial pH (pH before stopped flow) and pH changes ([Delta] pH) after a 6-min stopped-flow period. + indicates a significant difference from control value; * indicates a significant difference from arterial stopped-flow value; *n* numbers are indicated in parentheses

	Arterial		Venous	
	Initial pH	[Delta] pH	Initial pH	[Delta] pH
Control	7.832 ± 0.02 (15)	0.032 ± 0.004 (15)	7.838 ± 0.02 (15)	0.004 ± 0.003* (15)
Carbonic anhydrase	7.915 ± 0.05 (8)	-0.011 ± 0.003+ (8)	7.873 ± 0.04 (8)	-0.011 ± 0.004+ (8)
Benzolamide	7.741 ± 0.04 (5)	0.038 ± 0.010 (5)	7.812 ± 0.02 (6)	0.007 ± 0.007* (6)
Corrected[a] disequilibrium		0.050 ± 0.006 (8)		0.015 ± 0.003* (8)

[a]The acid-base disequilibrium was calculated by subtracting the pH change during stopped flow after carbonic anhydrase (CA) treatment from the initial control [Delta]pH value

Benzolamide, at the dose (1.2 mg kg^{-1}) and within the time frame (30 min) used in these experiments, does not significantly inhibit RBC CA activity (Swenson and Maren 1987; Gilmour et al. 1996). In the present study, this was confirmed by assessing RBC CA activity using the electrometric [Delta]pH assay (Henry 1991). The data showed that benzolamide treatment did not significantly influence RBC activity (Table 2). The inhibitory effects of benzolamide in these experiments will therefore be exclusively restricted to extracellular CA if it is present. After benzolamide treatment, the magnitude of the arterial pH disequilibrium was unaffected (Fig. 2, Table 1). This demonstrates, for the first time, that extracellular CA is not accessible to catalyse CO_2 - HCO_3^-- H^+ reactions within the branchial vasculature. Previous studies (Gilmour et al.1994; Gilmour and Perry 1994) utilising the CA inhibitor acetazolamide were unable to unequivocally exclude a role for extracellular branchial CA because this inhibitor blocks both extracellular and RBC CA activity. Our finding that CO_2 - HCO_3^-- H^+ reactions in the branchial circulation are not catalysed by CA is consistent with the results of previous immunocytochemical (Rahim et al. 1988) and gill fractionation studies (Henry et al. 1988, 1993).

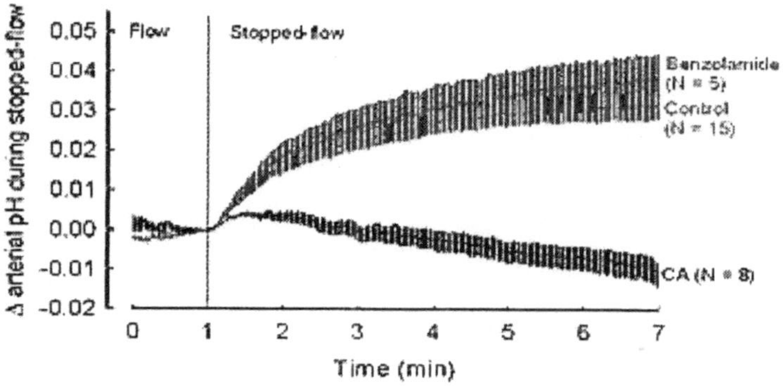

Fig. 2. Acid-base disequilibria in post-branchial dorsal aortic arterial blood under control conditions (*red trace*; $n = 15$), after injection of bovine carbonic anhydrase (*black trace*; $n = 8$) and after pre-treatment with the extracellular carbonic anhydrase (CA) inhibitor benzolamide (*green trace*; $n = 5$). In each case, changes in pH ([Delta] pH) are shown during a 6-min period of stopped flow commencing at the *vertical solid line*. The data shown are continuous traces of pH changes ± 1 SEM recorded at 1 s intervals. There were no significant differences in the [Delta] pH data between the control and benzolamide stopped-flow periods

Table 2. Red blood cell CO_2 production rates before and after injecting trout with the extracellular CA inhibitor, benzolamide. The data indicate that RBC CA activity was not significantly affected after benzolamide treatment. Data shown are means ± 1 SEM; *n* numbers are indicated in parentheses

Condition	RBC CO_2 production (mmol min^{-1} ml^{-1} RBC pellet)
Pre-benzolamide	950 ± 76 (7)
Post-benzolamide (venous stopped flow)	874 ± 91 (6)
Post-benzolamide (arterial stopped flow)	972 ± 83 (7)

Venous acid-base disequilibrium

The venous blood displayed a significant acid-base disequilibrium (Fig. 1) but its magnitude was significantly reduced when compared to the arterial blood (Table 1). As in the arterial blood, the true extent of the acid-base disequilibrium was underestimated owing to metabolic acid production; upon correction for metabolism (see above) the size of the disequilibrium after 6 min stopped flow was 0.015 ± 0.003 units (Table 1). The disequilibrium was eliminated by intra-vascular CA treatment (Figure 3; Table 1) thereby implicating slow uncatalysed HCO_3^- dehydration as the cause of the post-systemic capillary pH changes.

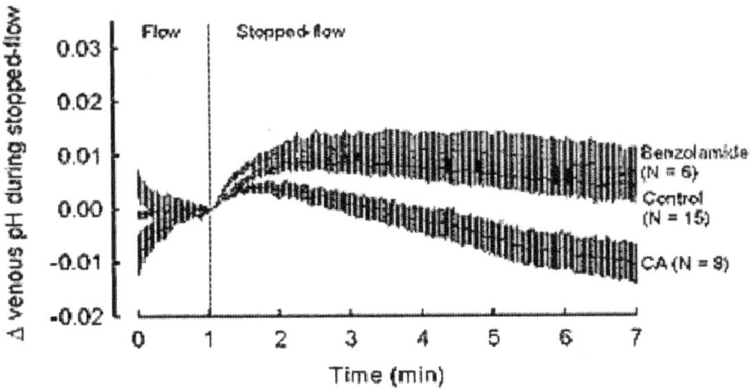

Fig. 3.. Acid-base disequilibria in pre-branchial caudal vein venous blood under control conditions (*blue trace*; $n = 15$), after injection of bovine CA (*black trace*; $n = 8$) and after pre-treatment with the extracellular CA inhibitor benzolamide (*green trace*; $n = 6$). In each case, changes in pH ([Delta] pH) are shown during a 6-min period of stopped flow commencing at the *vertical solid line*. The data shown are continuous traces of pH changes ± 1 SEM recorded at 1 s intervals. There were no significant differences in the [Delta] pH data between the control and benzolamide stopped-flow periods

The small size of the disequilibrium relative to the arterial blood suggested the potential contribution of extracellular CA catalysing CO_2 - HCO_3^-- H^+ reactions in the systemic micro-vasculature as is known to occur in some mammals (O'Brasky et al. 1979; Effros and Weissman 1979; O'Brasky and Crandall 1980; Geers et al.1985). The unchanged magnitude of the acid-base disequilibrium after benzolamide injection (Fig. 3, Table 1), however, demonstrates that extracellular CA is not accessible to plasma CO_2 - HCO_3^-- H^+ reactions in this region of the venous circulation. The extracellular CA associated with sarcolemmal membranes of trout skeletal muscle (Henry et al. 1996) , although important in facilitating CO_2 and NH_3 diffusion from tissue to blood, clearly has no role in the catalysis of CO_2 - HCO_3^-- H^+ reactions within the plasma.

The direction of the pH disequilibrium in the venous blood was the same as in the arterial blood (i.e. pH increased during stopped flow). Previous results from studies on mammals as well as theory predict a reduction in pH during stopped flow owing to rapid addition of CO_2 into the blood from tissues followed by its slow uncatalysed hydration to form HCO_3^- and H^+. Thus, the origin of the alkaline disequilibrium observed in this study is unclear. Similar alkaline disequilibria were attained using blood withdrawn from the bulbus arteriosus and the afferent branchial artery (unpublished observations) and therefore it appears to be a common feature of pre-branchial blood in trout. The in vivo stopped-flow experiments used in the present study withdrew venous blood from the caudal vein that drains predominantly skeletal muscle. Thus, although possible (in particular if shunts are present), it seems unlikely that the alkaline disequilibrium in the venous blood reflects the mixing of several blood pools containing varying PCO_2. An alternate explanation for the alkaline disequilibrium is that proton equilibrium across the RBC membrane (via the Jacob-Stewart cycle) is incomplete in the venous blood owing to the slow rate of HCO_3^- dehydration in the plasma. In this event, RBC Cl^-/HCO_3^- exchange would continue during stopped flow and raise plasma pH.

The magnitude of pH disequilibria during stopped-flow experiments is influenced by a variety of factors, including initial pH (Gilmour and Perry 1996) and blood buffering capacity. In the present experiments there were no statistically significant differences in the initial pH between arterial and venous blood or among any of the treatments (Table 1). Similarly, the buffer capacities of venous and arterial blood (assuming differences in oxygenation of 25–50%) are not likely to differ appreciably, if at all (Eddy 1974). Thus, differences observed between arterial and venous blood and among the treatments employed in the present study cannot be explained by variations in buffering or initial pH.

Regardless of the origin of the pH disequilibrium in the venous blood of trout, the data clearly demonstrate a fundamental difference from mammalian muscle vasculature. Unlike in mammals, there is no extracellular CA within the micro-circulation of the trout white muscle accessible to catalyse hydration/dehydration reactions of CO_2.

Acknowledgements. This work was supported by Natural Sciences and Engineering Research Council of Canada grants to S.F.P. and B.T. K.M.G. was the recipient of a Carnegie Trust for the Universities of Scotland travel grant. We are grateful to Dr. Ray Henry for generously providing us with the benzolamide. We also thank Drs. Chris Wood, Ray Henry and Yuxiang Wang for allowing us to cite their unpublished results.

References

Axelsson, M., Fritsche, R. (1994) Cannulation techniques. In: Hochachka, P.W., Mommsen, T.P. (eds.) Biochemistry and molecular biology of fishes, vol. 3. Analytical techniques. Elsevier, Amsterdam, pp.17-36

Bidani, A., Crandall, E.D. (1988) Velocity of CO_2 exchanges in the lungs. *Annu. Rev. Physiol.* 50: 639-652

Eddy, F.B. (1974) *In vitro* blood carbon dioxide of the rainbow trout (*Salmo gairdneri*). *Comp. Biochem. Physiol. A*, 47:129-140

Effros, R.M., Weissman, M.L. (1979) Carbonic anhydrase activity of the cat hind leg. *J. Appl. Physiol.* 47: 1090-1098

Effros, R.M., Chang, R.S.Y., Silverman, P. (1978) Acceleration of plasma bicarbonate conversion to carbon dioxide by pulmonary carbonic anhydrase.

Science **199:** 427-429

Geers, C., Gros, G., Gartner, A. (1985) Extracellular carbonic anhydrase of skeletal muscle associated with the sarcolemma. *J. Appl. Physiol.* **59:** 548-558.

Gilmour, K.G., Perry, S.F. (1994) The effects of hypoxia, hyperoxia or hypercapnia on the acid-base disequilibrium in the arterial blood of rainbow trout, *Oncorhynchus mykiss*. *J. Exp. Biol.* **192:** 269-284

Gilmour, K.M., Perry, S.F. (1996) The effects of metabolic acid-base disturbances and elevated catecholamines on the acid-base disequilibrium in the arterial blood of rainbow trout. *J. Exp. Zool.* **274:** 281-290

Gilmour, K.M., Randall, D.J., Perry, S.F. (1994). Acid-base disequilibrium in the arterial blood of rainbow trout, *Oncorhynchus mykiss*.*Respir. Physiol.* **96:** 259-272

Gilmour, K.M., Henry, R.P., Wood, C.M., Perry, S.F. (1996) Extracellular carbonic anhydrase and an acid-base disequilibrium in the blood of the dogfish, *Squalus acanthias*. *J. Exp. Biol.* (In press)

Henry, R.P. (1991) Techniques for measuring carbonic anhydrase activities *in vitro*: the electrometric delta pH and pH stat assays. In: Dodgson, S.J., Tashian, R.E., Gros, G., Carter, N.D. (eds) The carbonic anhydrases: cellular physiology and molecular genetics. Plenum, New York, pp. 119-126

Henry, R.P., Smatresk, N.J., Cameron, J.N. (1988) The distribution of branchial carbonic anhydrase and the effects of gill and erythrocyte carbonic anhydrase inhibition in the channel catfish *Ictalurus punctatus*. *J. Exp. Biol.* **134:** 201-218

Henry, R.P., Tufts, B.L., Boutilier, R.G. (1993) The distribution of carbonic anhydrase type-i and type-II isozymes in lamprey and trout - possible co-evolution with erythrocyte chloride bicarbonate exchange.*J. Comp. Physiol. B* **163:** 380-388

Henry, R.P., Wang, Y., Wood, C.M. (1996) Carbonic anhydrase facilitates CO_2 and NH_3 transport across the sarcolemma of trout white muscle. Am. J. Physiol. (In press)

Klocke, R.A. (1988) Velocity of CO_2 exchange in blood.*Annu. Rev. Physiol.* **50:** 625-637

Nikinmaa, M. (1990) Vertebrate red blood cells: adaptations of function to respiratory requirements. In: S.D. Bradshaw, W. Burggren, H.C. Heller, S. Ishii, H. Langer, G. Neuweiler, D.J. Randall (eds) Zoophysiology, vol 28. Springer Verlag, Berlin Heidelberg New York, pp 1-262

O'Brasky, J.E., Crandall, E.D. (1980) Organ and species differences in tissue vascular carbonic anhydrase activity.*Am. J. Physiol.* **49:** 211-217

O'Brasky, J.E., Mauro, T., Crandall, E.D. (1979) Postcapillary pH disequilibrium after gas exchange in isolated perfused liver.*Am. J. Physiol.* **47:** 1079-1083

Perry, S.F., Laurent, P. (1990) The role of carbonic anhydrase in carbon dioxide excretion, acid-base balance and ionic regulation in aquatic gill breathers. In: Truchot, J.P., Lahlou, B. (eds) Transport, respiration and excretion: comparative and environmental aspects. Karger, Basel, pp. 39-57

Rahim, S.M., Delaunoy, J.P., Laurent, P. (1988) Identification and immunocytochemical localization of two different carbonic anhydrase isozymes in teleostean fish erythrocytes and gill epithelia.*Histochemistry* **89:** 451-459

Sender, S., Gros, G., Waheed, A., Hageman, G.S., Sly, W.S. (1994) Immunohistochemical localization of carbonic anhydrase IV in capillaries of rat and human skeletal muscle. *J. Histochem. Cytochem.* **42:** 1229-1236

Smith, L.S., Bell, G.R. (1967). A practical guide to the anatomy and physiology of Pacific salmon. Environment Canada Fisheries and Marine Service Miscellaneous Special Publication Number 27

Stabenau, E.K., Bidani, A., Heming, T.A. (1996) Physiological characterization of pulmonary carbonic anhydrase in the turtle.*Respir. Physiol.* **104:** 187-196

Swenson, E.R., Maren, T.H. (1987) Roles of gill and red cell carbonic anhydrase in elasmobranch HCO_3^- and CO_2 excretion. *Am. J. Physiol.* **253:** R450-R458

Swenson, E.R., Lippincott, L., Maren, T.H. (1995) Effect of gill membrane-bound carbonic anhydrase inhibition on branchial bicarbonate excretion in the dogfish shark, *Squalus acanthias.Bull MDI Biol Lab* **34:** 94-95

Thomas, S. (1994) Extracorporeal circulation. In: Hochachka, P.W., Mommsen, T.P. (eds) Biochemistry and molecular biology of fishes, vol 3. Analytical techniques. Elsevier, Amsterdam, pp. 161-167

Wood, C.M., McDonald, D.G., McMahon, B.R. (1982) The influence of experimental anaemia on blood acid-base regulation *in vivo* and *in vitro* in the starry flounder (*Platichthys stellatus*) and the rainbow trout (*Salmo gairdneri*). J. Exp. Biol. **96:**:221-237

Wood, C.M., Perry, S.F., Walsh, P.J., Thomas, S. (1994) HCO_3^- dehydration by the blood of an elasmobranch in the absence of a Haldane effect.*Respir. Physiol.* **98:** 319-337

Exp. Biol. Online 2:2 (1997)

Activation of spawning in zebra mussels by algae-, cryptomonad-, and gamete-associated factors

Jörg D. Hardege[1], Jeffrey L. Ram[2], and Matthew G. Bentley[3]

[1]present location: University of Wales, Cardiff
[2]Department of Physiology, Wayne State University, Detroit, MI 48201 USA
[3]Gatty Marine Laboratory, School of Biological and Medical Sciences, University of St. Andrews, UK

Correspondence to: Jeffrey L. Ram (jeffram@med.wayne.edu)

Received: 27 October 1996 / Accepted: 17 January 1997
Published: January 22, 1997

Abstract. In zebra mussels (*Dreissena polymorpha*) simultaneous release of gametes and peaks in larval densities at particular locations suggest that spawning is triggered by synchronizing stimuli. Furthermore, spawning tends to occur only after an adequate environmental temperature is reached. To test the hypothesis that phytoplankton and gamete-associated chemicals initiate spawning in zebra mussels and that the responsiveness to such chemicals is affected by ambient temperature, the spawning response of zebra mussels to extracts from algae, a cryptomonad, and a cyanobacterium and to water associated with released gametes was assayed in animals acclimated to 12 °C and 17 °C. For animals held at 12 °C, only serotonin, a known activator of bivalve spawning used as a positive control, stimulated spawning. However, for animals acclimated to 17 °C, extracts made from a diatom (*Phaeodactylum*), a brown alga (*Fucus*), and a cryptomonad (*Rhodomonas*) stimulated spawning in both sexes; extracts from green algae (*Platymonas* and *Dunaliella*) and a cyanobacterium (*Oscillatoria*) did not cause spawning. Water associated with either released sperm or eggs elicited spawning in both females and males. Positive controls, stimulated with serotonin, spawned at a high (>90%) rate, whereas no negative control spawned. Thus, phytoplankton chemicals and gamete-associated factors may have a role in synchronizing spawning in zebra mussels once adequate ambient temperature is reached.

Key words: Algae – Cryptomonad – Diatom – *Dreissena polymorpha* – *Fucus* – Gametes – Spawning – Zebra mussel

Introduction

The zebra mussel (*Dreissena polymorpha*) is a biofouling bivalve that was accidentally introduced into North American freshwaters about 10 years ago (Hebert et al. 1989; see review by Ram et al. 1992 and The Zebra Mussel Page). Aside from the great interest in this organism engendered by efforts to control its spread and fouling activity, zebra mussels have also been amenable to studies of regulation of bivalve reproduction (Ram et al. 1996). The present study concerns itself with stimuli that may participate in triggering synchronous spawning episodes.

Spawning in zebra mussels tends to occur in synchronized bursts of gamete release by neighboring animals after an adequate environmental temperature has been reached. For example, Garton and Haag (1993) observed two distinct spawning peaks in western Lake Erie after the temperature had stabilized for nearly a month above 20 °C (Ram et al. 1996). Fraleigh et al. (1993) reported that zebra mussel veliger larvae appeared near Cleveland in several large peaks, all of which occurred when water temperature was >15 °C. The second and third peaks occurred when the temperature had stayed fairly constant in the range of 21-24 °C for several weeks. In several Swiss lakes where the annual maximum temperatures are lower than in North America, Borcherding (1991) observed synchronous decreases in zebra mussel gonad index (an indication of a spawning event) beginning at 12 °C. The occurrence of synchronous spawning in large numbers of animals at particular locales suggests that a signalling system and response pathway for coordinating spawning of nearby animals may be present. Ram et al. (1992) have suggested that once adequate temperature has been reached, spawning may be initiated by phytoplankton chemicals, and that chemicals associated with gametes provide positive feedback for amplifying and maintaining spawning activity.

Chemicals partially purified from the diatom *Phaeodactylum tricornutum* have previously been shown to activate spawning in sea urchins (Starr et al. 1992). Although crude extracts of this and other species of algae, including the macrophyte *Fucus vesiculosus*, have been shown to activate spawning in *Mytilus edulis* (Starr et al., 1990; Smith and Strehlow 1983), there has been no previous study of the partially purified algal extracts on bivalves. In the present study, algal extracts, partially effects of purified by the procedure of Starr et al. (1992), were tested for effects on zebra mussel spawning. These extracts were tested along with similar extracts from other planktonic organisms (the cryptomonad *Rhodomonas* and the cyanobacterium *Oscillatoria*). In addition, the effect of water associated with released gametes and the influence of temperature on spawning were studied in the mussels.

In this initial study on zebra mussels, we chose to test the effects of extracts of marine species instead of freshwater species because the purification procedures of Starr et al. (1992) were already well established on marine species. Also, since the dreissenids evolved relatively recently from marine ancestral species, as compared to other freshwater bivalves, it seemed possible that *Dreissena* would have a response to chemicals of marine origin that might be lacking in other freshwater bivalves, an observation that could have great significance in devising novel specific control methods.

Materials and methods

Animals

Zebra mussels were collected by H. Vanderploeg from Lake Wawasee and western Lake Erie early in the spawning season in 1995 and maintained with light feeding ("Algae Diet C" as described by Vanderploeg et al. 1996) at approximately 4 °C in L:D 12:12. Animals were gradually warmed to 12 °C 1 week prior to initiating experiments, transferred to two 70-gallon (265-l) 12 °C tanks at Wayne State University, and maintained in L:D 18:6. As described in Results, although animals spawned at high rates when tested with serotonin (5-hydroxytryptamine, 5-HT), a known bivalve spawning activator (Ram et al. 1993), virtually no spawning in response to algal extracts was obtained while animals were maintained at 12 °C. Animals in one of the tanks were then gradually warmed to 17 °C and further testing was done after maintenance at this temperature for at least 3 days (a time shown to be adequate for temperature acclimation of spawning responses; Fong et al. 1995). Results will be described for animals maintained under the two temperature regimes.

Bioassay procedures

As in previous studies (Ram et al. 1993), animals were tested individually for spawning responses in 20-ml vials, immersed in 10 ml of medium. Treatments consisted of adding either extracts or control solutions to the medium. Control solutions were 5-HT (final concentration 1 mM) as a positive control (Ram et al., 1993; Fong et al. 1994a) and water alone as a negative control. Other control groups are described below. Generally, 10-20 animals were tested in each of several experimental groups and control groups on any particular experimental day; however, since there was little day-to-day variation, results from several days of experiments were combined for statistical analysis.

Spawning responses were assessed by sampling the water surrounding the mussel. Two drops were placed on a slide, covered with a coverslip, and searched under a high-power microscope for eggs or sperm. Frequently, the response was so massive that the surrounding water became cloudy with sperm (as illustrated on the The Zebra Mussel Page) or the bottom became carpeted with oocytes, in which case the microscopic assessment was merely for corroborating the response and determining sex. Spawning was considered positive if it occurred within 4 h of addition of extracts or other solutions.

The sex and maturity of all non-spawning animals (including negative controls) was subsequently determined by addition of 5-HT (final concentration 1 mM) which, as shown in these and previous experiments, could stimulate virtually all ripe animals to spawn. Animals that did not respond to the original spawning stimulus or 5-HT were dissected whether determine if they had ripe gonads. Any animals that were not ripe or had spent gonads were not included in the statistical analysis as they were considered non-spawnable. Overall, 13.9% of the animals tested were deleted from the analysis as non-spawnable. The ratio of males to females varies somewhat from group to group because it is not possible to tell the sex of an animal in advance of an experiment.

Preparation of extracts

Extracts were made from a diatom (*Phaeodactylum tricornutum*), a brown alga (*Fucus vesiculosus*), two species of green algae (*Platymonas suecice* and *Dunaliella tertiolecta*), a cryptomonad (*Rhodomonas* sp.), and a cyanobacterium (*Oscillatoria* sp.). Extraction procedures were similar to the procedures originally used for *Phaeodactylum tricornutum* by Starr et al. (1992). Starting material, consisting of 1.7×10^9 cells (*Rhodomonas*) or 3×10^9 cells (*Phaeodactylum*, *Platymonas* , *Dunaliella*, or *Oscillatoria*), or 50 g of *Fucus*, was heated to 60 °C for 10 min in 1 M NaOH. The resultant solution was extracted in diethylether, retaining the aqueous phase. Then, after adjusting to pH 2 with HCl, the solution was extracted twice more with diethylether, this time retaining the organic phase. After adsorbing water in this phase by addition of a small amount of calcium chloride, the organic phase was removed and evaporated to dryness in a spin-vac. Residue was then dissolved in aquarium water, adjusted to pH 7.5, and bioassayed by addition of 1 ml to each test vial (containing one animal in 9 ml of aquarium water).

Gamete-associated extracts were made as follows: *Dreissena polymorpha* were treated with 1 mM 5-HT for 5-10 min to stimulate spawning. Such brief treatment with 5-HT has previously been shown to be sufficient to trigger normal levels of spawning (Fong et al. 1994b). To remove 5-HT from around the animals, the aquarium water in each vial was changed four times with aquarium water - twice rapidly, a third time for 15 min, and the fourth change of water was left in the vial until spawning occurred. After animals spawned either sperm or oocytes, the water was removed, centrifuged to remove gametes, and the supernatant tested for capability of stimulating spawning. The water removed at the end of the third rinse (termed "pre-release" water) was similarly treated and bioassayed as a control for effective washout of 5-HT prior to the release of gametes.

Results

We describe first the data from animals that were maintained at 17 °C, as these produced positive results with both algal and cryptomonad extracts (Fig. 1) and water from released gametes (Fig. 2). The important role of temperature in obtaining these results will be considered at the end. As illustrated in Fig. 1, extracts from *Fucus*, *Rhodomonas*, and *Phaeodactylum* caused spawning (depending on the extract and sex of the tested animal) in 20-90% of recipients, and 5-HT caused spawning in more than 95% of recipients. In contrast, *Platymonas* and *Dunaliella* extracts carried through exactly the same extraction procedure caused no spawning. *Oscillatoria* extracts also did not cause spawning (data not shown). Non-spawning animals included in the analysis subsequently spawned when tested for capability of spawning by exposure to 5-HT. None of more than 40 aquarium-water negative controls spawned (they were subsequently confirmed to be spawnable by post-testing with serotonin). For statistical analysis, numbers of spawning and non-spawning animals for each sex and extract were compared to the aquarium-water controls using Fisher exact tests. For males, significant differences were found for 5-HT ($p<0.0001$), *Fucus* ($p<0.0001$), *Rhodomonas* ($p<0.0002$), and *Phaeodactylum* ($p=0.023$). For females, significant differences were calculated for 5-HT ($p<0.0001$) and *Fucus* ($p<0.017$). The responsiveness of males and females to the extracts did not differ significantly ($p>0.10$). Thus, algal and cryptomonad extracts are capable of activating spawning in both male and female zebra mussels.

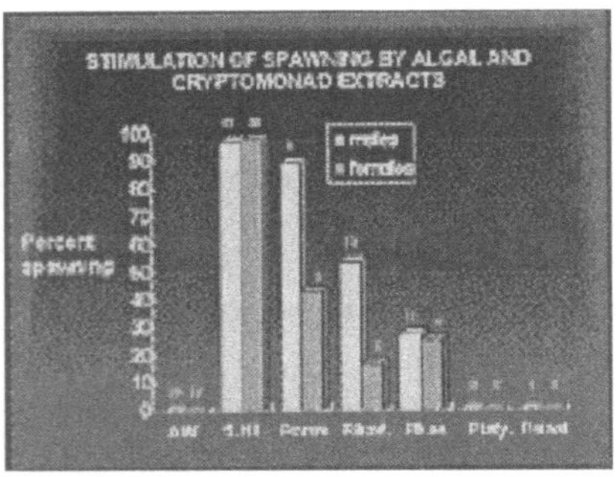

Fig. 1 Stimulation of spawning in zebra mussels by algal and cryptomonad extracts. Tested solutions were aquarium water (AW, negative control), 5-HT (positive control), and extracts of *Fucus*, *Rhodomonas* (Rhod.), *Phaeodactylum* (Phae.), *Platymonas* (Platy.), and *Dunaliella* (Dunal.). Preparation of extracts is described in the text. Animals were maintained at 17 °C prior to experiments. Animals were tested in individual vials, with the numbers of animals tested in each category indicated above each bar. Spawning was scored as positive if gametes appeared in water surrounding the mussel within 4

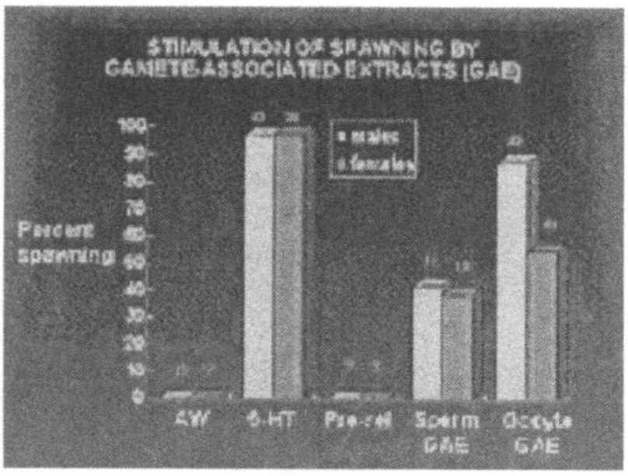

Fig. 2 Spawning induced by water associated with released gametes (GAE). Tested solutions were aquarium water (AW, negative control), 5-HT (positive control), pre-release water (Pre-rel., negative control), sperm GAE, and oocyte GAE. Preparation of GAE and pre-release water are described in the text. Animals were maintained at 17 °C prior to experiments. Animals were tested in individual vials, with the numbers of animals tested in each category indicated above each bar. Spawning was scored as positive if gametes appeared in water surrounding the mussel within 4 h

Similarly, we found that gamete-associated water was capable of activating spawning in zebra mussels (Fig. 2). All comparisons by Fisher exact test to the aquarium-water control give $p<0.01$. Furthermore, none of the recipients of the pre-release-water control spawned (significantly less than 5-HT and all gamete-associated water except for the comparison with female recipients of oocyte-associated water), indicating that spawning in response to gamete-associated water is not likely to be caused by contaminating 5-HT. Although there appears to be a tendency for male zebra mussels to respond more readily than females to oocyte-associated water, this trend did not reach statistical significance ($p=0.079$).

The temperature at which the animals were maintained appeared to have an important influence on the success of the above experiments. At the beginning of this series of experiments, extracts were tested on animals that were maintained at 12 °C, a temperature at which a high rate of spawning can usually be obtained with 1 mM 5-HT, which was also observed in these experiments (Fig. 3, illustrating a 5-HT response rate of >90% for animals at both 12 °C and 17 °C). However, only one recipient of algae extract (*Fucus*) and one of oocyte-associated water spawned in response to these stimuli. It was not until the temperature in one tank was increased to 17 °C that we began to get the results described above. Figure 3 summarizes data over the entire period of experimentation during which animals were tested at the two temperatures. Statistically significant differences between the two temperatures were present for male recipients of *Fucus* ($p=0.025$) and oocyte-associated water ($p<0.0001$), and for female recipients of oocyte-associated water ($p<0.005$). These experiments include data from one experiment in which oocyte extracts were tested on the same day on animals that had been maintained at the two different temperatures. In this experiment, spawning of 12 °C animals was 0/10 aquarium water, 29/29 5-HT, and 1/19 oocyte- associated water, whereas 17 °C animals spawned as follows: 0/7 aquarium water, 40/40 5-HT, and 13/17 oocyte-associated water recipients (significantly more frequent than at 12 °C,

$p<0.0001$). There appear to be temperature differences in spawning elicited with other extracts and recipients; however, the numbers of animals tested were too small to reach statistical significance.

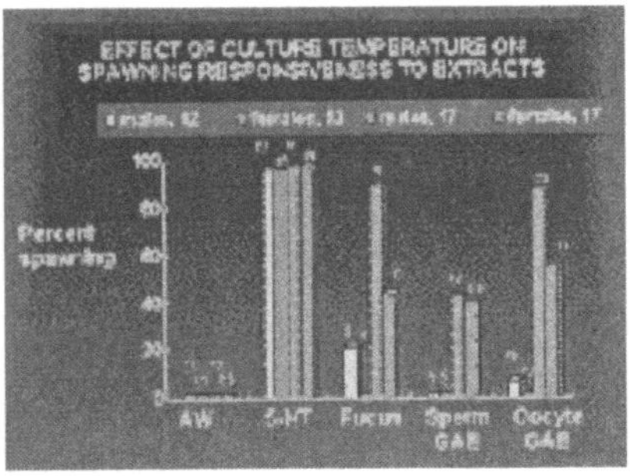

Fig. 3 Effect of temperature on spawning induced by algal extracts and water associated with released gametes (GAE). Solutions were aquarium water AW, (negative control), 5-HT (positive control), *Fucus* extract, sperm GAE and oocyte GAE. Data for 17 °C animals tested with AW, 5-HT, and *Fucus* are from the experiments shown in Fig. 1, for GAE from Fig. 2. Animals were tested in individual vials, with the numbers of animals tested in each category shown above each bar. Spawning was scored as positive if gametes appeared in water surrounding the mussel within 4 h

Discussion

This paper demonstrates that partially purified extracts from several different species of algae and water associated with released gametes can activate spawning in the zebra mussel, *Dreissena polymorpha*. The type of extract studied here is similar to that previously used by Starr et al. (1992) to induce spawning in sea urchins. Our work goes beyond Starr's and broadens its significance by eliciting spawning in a freshwater invertebrate, the zebra mussel, not only using extracts made by a multi-step purification procedure from *Phaeodactylum tricornutum* but also testing extracts made from other planktonic species (a cryptomonad, green algae, a cyanobacterium) and from a macrophyte, *Fucus*.

The occurrence of activator(s) of zebra mussel spawning in marine algae, including a macrophyte, may be indicative of the occurrence of the inducing chemicals in diverse species; however, future experiments must test extracts of freshwater algae and cryptomonads to be directly relevant to species encountered by zebra mussels. As noted in the Introduction, we started with marine species in order to follow the previous work of Starr et al. (1992) most closely and also because of the relatively recent marine origin of dreissenid bivalves. According to Mackie et al. (1989), fossils found in Poland, Bulgaria, and the former USSR indicate a Pleistocene origin for *Dreissena*. Furthermore, the reproductive behavior of *Dreissena* more closely resembles most marine bivalves in having external fertilization and a planktonic larval stage, in contrast to the internal fertilization, followed by brooding or parasitism, of most freshwater bivalves. The possibility that spawning in *Dreissena* and marine bivalves is similarly regulated by external chemicals is strengthened by the observations reported in this paper.

A curious aspect of these results is that although zebra mussels spawned in response to extracts from some planktonic organisms, there was also a strong response to a macrophyte, a type of alga that zebra mussels would never ingest. Although our hypothesis concerns the presence of the inducer chemicals in plankton, because these are used as food for zebra mussels, this does not preclude the same chemicals from being present in macrophytes. It does raise the question, however, of whether the inducing chemicals may function independently of the likely nutritive value of the source organism.

Field data provide indirect support for the hypothesis that zebra mussel spawning is triggered by phytoplankton in nature. Garton and Haag (1993) observed in western Lake Erie that synchronous spawning of zebra mussels followed the late summer rise in phytoplankton in 1989. Furthermore, in 1990, when the late summer phytoplankton recovery was much smaller than usual, the synchronous spawning peak never occurred.

In addition to chemical triggers of spawning, factors such as temperature, endocrine state, light, photoperiod, and reproductive maturity also play a role in regulating whether spawning will occur. The importance of temperature in providing a necessary precondition for spawning is supported by the results shown here; the extracts that caused spawning in zebra mussels maintained at 17 °C did not activate spawning in animals maintained at 12 °C. These results are consistent with field data indicating that spawning of North American zebra mussels usually does not occur until ambient temperature rises above 17-20 °C (see review by McMahon 1996), a higher temperature than appears to be required in many European studies (see Borcherding 1991, for example). McMahon (1996) has speculated that North American zebra mussels, having originated from near the Black Sea, the warmest region of its European distribution, may be genetically adapted to higher temperatures than mussels from colder, northern European waters on which most European studies have been conducted. Another environmental influence on zebra mussel spawning may be lunar cycle (Bosch 1996).

5-HT produced spawning in almost 100% of the animals maintained at either temperature. 5-HT is believed to work directly on zebra mussel gonads (Ram et al. 1996), to which it may have access through the zebra mussel's rather "leaky" integument (Dietz et al. 1996). In contrast, environmental chemicals (e.g. from plankton or gametes) are presumed to act via chemoreceptors whose effect on the gonad is mediated by the nervous system. Smith (1987) showed that the induction of spawning in *Mytilus* by algae required intact connections between pedal and visceral ganglia. Thus, reproductive maturation may occur first in reproductive organs, followed by maturational processes in the endocrine and nervous system that enable zebra mussels to respond to environmental stimuli which induce spawning.

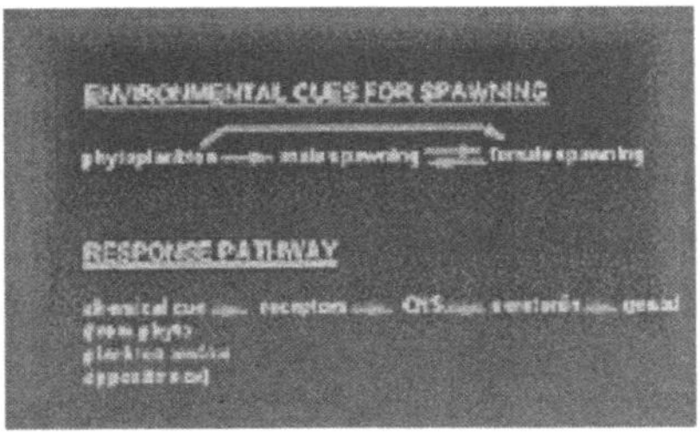

Fig. 4 Putative model of regulation of spawning in zebra mussels

These data are consistent with a model for the regulation of spawning in zebra mussels illustrated in Fig. 4. According to this model, the initial signal for spawning is released by phytoplankton, indicating that appropriate food is available. Phytoplankton chemicals trigger both males and females to spawn, with chemicals associated with released gametes acting in a positive feedback manner to synchronize and amplify the spawning event. The role of temperature is to signal that maturational processes should ensure. The process appears to take place in the reverse order of the triggering event, i.e. gonad first, peripheral inputs later.

Acknowledgements. The authors gratefully acknowledge funding in support of this research from NATO Collaborative Research Grant CRG 950825 and from NIH grant GM-08167.

References

Borcherding J (1991) The annual reproductive cycle of the freshwater mussel *Dreissena polymorpha* Pallas in lakes. Oecologia 87:208-218

Bosch I (1996) Lunar periodicity of synchronous spawnings by *Dreissena polymorpha* in Conesus Lake, New York. Sixth International Zebra Mussel and Other Aquatic Nuisance Species Conference, Abstracts, p 111

Dietz TH, Wilcox SJ, Byrne RA, Lynn JW, Silverman H (1996) Osmotic and ionic regulation of North American zebra mussel (*Dreissena polymorpha*). Am Zool 36:364-372

Fong PP, Duncan J, Ram JL (1994a) Inhibition and sex specific induction of spawning by serotonergic ligands in the zebra mussel *Dreissena polymorpha* (Pallas). Experientia 50:506-509.

Fong PP, Kyozuka K, Abdelghani H, Hardege J, Ram JL (1994b) *In vivo* and *in vitro* induction of germinal vesicle breakdown in a freshwater bivalve, the zebra mussel *Dreissena polymorpha* (Pallas). J Exp Zool 269:467-474

Fong PP, Kyozuka K, Duncan J, Rynkowski S, Mekasha D, Ram JL (1995) The effect of salinity and temperature on spawning and fertilization in the zebra mussel *Dreissena polymorpha* (Pallas) from North America. Biol Bull 189:320-329

Fraleigh PC, Klerks PL, Gubanich G, Matisoff G, Stevenson RC (1993) Abundance and settling of zebra mussel (*Dreissena polymorpha*) veligers in western and central Lake Erie. In Nalepa TF, Schloesser DW (eds) Zebra mussels: biology, impacts, and control. Ann Arbor, Lewis, pp 129-142

Garton DW, Haag WR (1993) Seasonal reproductive cycles and settlement patterns of *Dreissena polymorpha* in western Lake Erie. In Nalepa TF, Schloesser DW (eds) Zebra mussels: biology, impacts, and control. Ann Arbor, Lewis, pp 111-128

Hebert PDN, Muncaster BW, Mackie GL (1989) Ecological and genetic studies on *Dreissena polymorpha* (Pallas): a new mollusc in the Great Lakes. Can J Fish Aquatic Sci 46:1587-1591

Mackie GL, Gibbons WN, Muncaster BW, Gray IM (1989) The zebra mussel, *Dreissena polymorpha*: a synthesis of European experiences and a preview for North America. Report prepared for Water Resources Branch, Great Lakes Section. Queen's Printer for Ontario, Toronto, p 15

McMahon R (1996) The physiological ecology of the zebra mussel, *Dreissena polymorpha*, in North America and Europe. Am Zool 36:339-363

Ram JL, Fong P, Croll RP, Nichols SJ, Wall D (1992) The zebra mussel (*Dreissena polymorpha*), a new pest in North America: reproductive mechanisms as possible targets of control strategies. Invest Reprod Dev 22:77-86

Ram JL, Crawford GW, Walker JU, Mojares JJ, Patel N, Fong PP, Kyozuka K (1993) Spawning in the zebra mussel (*Dreissena polymorpha*): activation by internal or external application of serotonin. J Exp Zool 265:587-598

Ram JL, Fong PP, Garton DW (1996) Physiological aspects of zebra mussel reproduction: maturation, spawning, and fertilization. Am Zool 36:326-338

Smith JR (1987) The role of the nervous system in algae-induced gamete release by *Mytilus californianus*. Comp Biochem Physiol 86C:215-218

Smith JR, Strehlow DR (1983) Algal-induced spawning in the marine mussel *Mytilus californianus*. Int J Invertebr Reprod 6:129-133

Starr M, Himmelman JH, Therriault J-C (1990) Direct coupling of marine invertebrate spawning with phytoplankton blooms. Science 247:1071-1074

Starr M, Himmelman JH, Therriault J-C (1992) Isolation and properties of a substance from the diatom *Phaeodactylum tricornutum* which induces spawning in the sea urchin *Strongylocentrotus droebachiensis*. Mar Ecol Prog Ser 79: 275-287

Vanderploeg HA, Liebig JR, Gluck AA (1996) Evaluation of different phytoplankton for supporting development of zebra mussel larvae (*Dreissena polymorpha*): the importance of size and polyunsaturated fatty acid content. J Great Lakes Res 22:36-45

Exp. Biol. Online 2:3 (1997)

Effects of xenon on intracellular Ca^{2+} release in human endothelial cells

Christian Petzelt[1*], Juan Osés Prieto[2], Francisco Fernández Klett[3], Wolfgang Schmehl,[4] and Wolfgang J. Kox[1]

[1]Experimental Anesthesiology, Department of Anesthesiology and Intensive Therapy, University Hospital Charité, Humboldt University, Res. Bldg 31, Spandauer Damm 130, D-14050 Berlin, Germany
[2]Department of Biochemistry, University of Pamplona, E-31080 Pamplona, Spain
[3]Department of Pharmacology and Therapy, University Autonoma Madrid, E-28029 Madrid, Spain
[4]AGA Gas GmbH, D-20209 Hamburg, Germany

Correspondence to: C. Petzelt (e-mail: petzelt@ukrv.de)

Received: 6 November 1996 / Accepted: 20 January 1997
Published: January 30, 1997

Abstract. Using human endothelial cells loaded with the Ca^{2+} indicator Fura2 the effects of xenon on changes in intracellular Ca^{2+} were studied. The basal level of intracellular Ca^{2+} is not affected upon incubation of the cells in buffer saturated either with 100% xenon or with 70% xenon/30% air, a concentration which corresponds in humans to the minimum alveolar concentration necessary to induce anesthesia in 50% of patients. A defined cellular response such as the Ca^{2+} change induced by application of adenosine triphosphate (ATP) makes it possible to study the signalling chain between the stimulus and the various forms of Ca^{2+} response. ATP induces a typical Ca^{2+} fingerprint composed of an internal Ca^{2+} release consisting of several oscillations plus an additional Ca^{2+}-induced Ca^{2+} influx from the outside. The latter is absent in Ca^{2+}-free medium. When cells are incubated with xenon, only the first part of the ATP-induced Ca^{2+} response is found corresponding to the internal release of Ca^{2+}; the subsequent Ca^{2+}-induced Ca^{2+} influx does not take place. If xenon is removed, a fast recovery is observed and the cells again show both parts of the Ca^{2+} response. Such selective inhibition of Ca^{2+}-induced Ca^{2+} influx is not obtained when xenon is replaced by N_2; the ATP response of the cell remains the same as that of untreated cells. Similar effects of xenon treatment can also be observed when the cells are treated with thapsigargin, a specific inhibitor of the SERCA systems. The Ca^{2+}-induced Ca^{2+} release is almost completely suppressed in the presence of xenon. We conclude that xenon may act on the cellular level on defined sites of the mechanisms regulating the Ca^{2+}-release-activated Ca^{2+} channels of the plasma membrane and that this property may be related to its anesthetic effect.

Key words: Calcium – Endothelium – Xenon – Anesthesia

Introduction

Xenon introduced into anesthesia many years ago (Cullen and Gross 1951) but has so far not become established as a general anesthetic, mainly due to its high price and limited availability. Its many positive aspects as an anesthetic agent (Kennedy et al. 1992), the continuing progress toward perfection in the development of closed systems in anesthesia (Luttropp et al. 1991, 1994) and the requirement of abolishing and replacing halogenated anesthetic compounds in the near future have led to increased interest in its clinical application (Boomsma et al. 1990; James, 1990; Lachmann et al. 1990; Power, 1990). Consequently, the cellular and molecular effects of Xe have to be elucidated beyond our currently inadequate state of knowledge.

Xenon has shown to enhance phospholipid *N*-methylation slightly but significantly in synaptic plasma membranes of rat brain (Horn et al. 1995) and to reduce the Ca^{2+}-ATPase activity of the brain plasma membrane (Janicki et al. 1995) as well as of C6 glioma cells (Singh et al. 1995). The latter phenomenon, i.e. inhibition of synaptosomal plasma membrane Ca^{2+}-ATPase, may be common to inhalational anesthetics (Janicki et al. 1994; Franks et al. 1995 a,b).

However, clear differences between Xe and other anesthetics have been described. Nitrous oxide has been shown to inactivate methionine synthetase in mice, whereas a similar treatment with Xe remained ineffective (Koblin et al. 1981). Profound differences were also found when possible teratogenic effects were investigated. Nitrous oxide proved to be fetotoxic but Xe had no effect under the same experimental conditions (Lane et al. 1980). Studying the aggregation of human blood platelets and marine sponge cells in relation to N_2O, helium and Xe, Philp et al (1992a,b) surprisingly found that the first two agents inhibit aggregation whereas Xe potentiates it, despite the fact that Xe and N_2O are equipotent gaseous anesthetics. Likewise, the ADP-induced Ca^{2+} increase in human blood platelets is inhibited by N_2O whereas Xe has no effect (Philp et al. 1992b). Thus, a detailed study of Ca^{2+} homeostasis in single cells under the influence of Xe may provide more information on possible sites of the Ca^{2+}-modulating effect of Xe and, consequently, lead to an improved understanding of its anesthetic effects.

Materials and methods

Cell culture

An established cell line of human umbilical vein endothelial cells (HUVEC) was purchased from the European Collection of Animal Cell Cultures (Salisbury, UK), and maintained in RPMI-1640 medium supplemented with 10 % fetal calf serum (GIBCO, Grand Island, N.Y., USA) at 37°C in a 5% CO_2 atmosphere. The cells were subcultured every 3 days and the absence of mycoplasm contamination was monitored regularly. For the experiments, cells grown on delipidated coverslips were used 40 h after seeding at a confluency of ca 70%, in order to maintain reproducible measuring conditions.

Ca^{2+} measurement

The basic principle of ratiometric Ca^{2+} measurement using Fura-2 is well established (Grynkiewicz et al. 1985). The equipment and the methods used for following the Ca^{2+} fluctuations in single living cells in real time have been described (Nobiling and Buhrle 1989). Briefly, the set-up consists of a fluorescence microscope (ZEISS Axiovert with 100X-Fluar Objective) with two attachment: a fast spinning filter wheel, which carries narrow band (+ 5 nm) interference filters for 340, 360 and 380 nm, enables changes of the excitation wavelength at a rate of over 100× s^{-1}; a photon-counting multiplier, which measures emitted light, filtered at 500 + 15 nm, with high sensitivity and linearity. Wavelength control and data acquisition performed using a 68000 processor-based real-time system with a VME bus. The software used allows real-time plotting of changes in intracellular Ca^{2+} concentration on the screen. The equipment *(IONOQUANT)* is commercially available (IONOQUANT, Nussloch, Germany). R_{min} and R_{max} were determined according to Thomas and Delaville (1991). Ca^{2+} measurements were performed in the following buffer: 116 mM NaCl; 5.4 mM KCl; 0.9 mM $MgCl_2$; 0.9 mM NaH_2PO_4; 1.8 mM $CaCl_2$; 10 mM glucose; pH 7.2 adjusted with 25 mM $NaHCO_3$. ATP was added at a final concentration of 5 µM and was maintained in the solution throughout the individual measurement unless otherwise indicated.

Xe experiments

For the experiments with Xe (AGA Hamburg, Germany), a large volume of the measuring buffer was placed in a gas-tight plastic bag, taking care to avoid air entrapment. Thereafter, an equal volume of the gas, or the gas mixture in the case of the 70% Xe, was introduced into the bag and the content was shaken for

30 min at room temperature.The cells were mounted in an air-tight closed microscopic chamber in the measuring buffer described above; then the buffer was replaced by the gas-saturated buffer, taking care to maintain the air-tight conditions, and the cells were incubated for the time indicated at 37°C. At the end of the incubation period the cells were additionally exposed to 500 nM FURA-2 AM (Molecular Probes, Eugene, Ore., USA) at room temperature. Thereafter, the Fura2 was washed out, still with the same buffer, and the measurement was started. All experiments were carried out at room temperature.

For inhibition of the SERCA enzymes, thapsigargin (GIBCO), dissolved in ethanol, was added to the measuring buffer in a final concentration of 1 μM.

Results

The minimum alveolar concentration (MAC) estimates the minimum anesthetic concentration at which half the patients are anesthetized and half are not. This concept has proved to be useful for calculating the effective concentration of the anesthetic at a cellular level. For Xe it has been shown to be in the order of 70% in humans (Eger et al. 1965; de Jong and Eger 1975), whereas in rhesus monkeys a MAC of 98% has been determined (Whitehurst et al. 1994).

In order to take into account an approximately mean duration of an anesthesia we chose 2 h of incubation with the anesthetic. Thus, when the environment of the human endothelial cells studied was changed for 2 h from a normal buffer to a buffer saturated in an atmosphere of 70% Xe, 30% air, or in a 100% Xe atmosphere, no change in the basal intracellular Ca^{2+} concentration can be found (Table 1). The basal intracellular Ca^{2+} concentration was determined by monitoring an individual cell for 10 min under the measuring conditions described. Cells with an initial basal Ca^{2+} concentration of more than 250 nM were regarded as damaged cells and not counted. These usually amounted to less than 5% of the cell population studied; the percentage was found to be independent of pretreatment with Xe. In all the other cells the basal intracellular Ca^{2+} concentration was very homogeneous provided the "health" of the cells was optimized; seeding 40 h before the experiment at a final confluency of ca. 70% was especially important in this respect.

Table 1 Comparison of the basal Ca^{2+} concentrations of cells maintained in normal buffer (controls), maintained for 2 h in buffer saturated with 100% xenon (100% xenon) and maintained in buffer saturated with 70% xenon/30% air. The basal Ca^{2+} was measured after 2 h for at least 10 min. No differences were found between these three groups

	nMol Ca^{2+}	*n*
Control	118±14	18
100% Xenon	110±12	12
70% Xenon	120±11	

These results are in accordance with those of other authors showing that neither N_2O nor Xe affected the basal levels of Ca^{2+}(Philp et al. 1992b). A defined response of the cell to a distinct stimulus might yield more useful answers. ATP in micromolar concentrations is such a stimulus. Intracellular Ca^{2+} is released via activation of G proteins (reviewed in Edwards 1994), resulting in a typical ATP-induced Ca^{2+} fingerprint (Fig. 1a). Such a response can be even further dissected by changing the external Ca^{2+} concentration. If the cells are challenged in calcium-containing medium, a long-lasting Ca^{2+} elevation occurs (Fig. 1a); however, when the cells are stimulated in Ca^{2+}-free medium the first reaction of the cell to the stimulus is very much like the reaction obtained in Ca^{2+} medium but then the intracellular Ca^{2+} is reduced very quickly to its original basal level (Fig. 1b). Thus it becomes apparent that the response of the cell is a sum of the release of Ca^{2+} from intracellular stores occurring in several waves and the subsequent influx of Ca^{2+} ions from the outside.

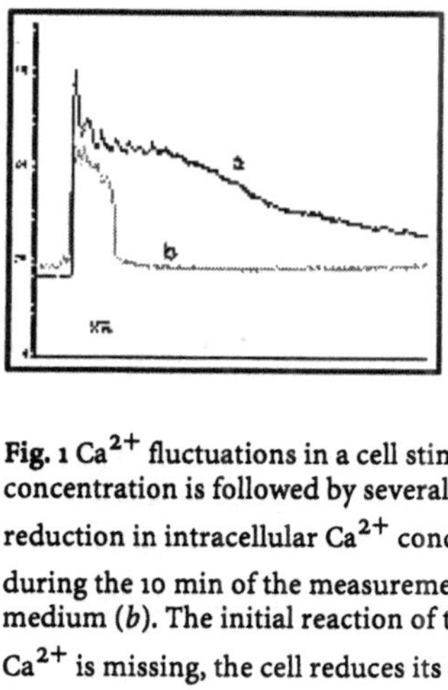

Fig. 1 Ca^{2+} fluctuations in a cell stimulated with 5 μM ATP (*a*). A steep increase in Ca^{2+} concentration is followed by several oscillations of decreasing height and a continuous but slow reduction in intracellular Ca^{2+} concentration; the original basal concentration is never reached during the 10 min of the measurement period. The ATP-induced Ca^{2+} response in Ca^{2+}-free medium (*b*). The initial reaction of the cell is very similar in both cases; if, however, the external Ca^{2+} is missing, the cell reduces its Ca^{2+} level very quickly to its original baseline value

If the same experiments are performed under the influence of Xe, the first intracellular releases of Ca^{2+} do not show any profound differences, suggesting that the internal ATP-induced release is not affected by Xe. However, the second phase of the Ca^{2+} response is missing; the intracellular Ca^{2+} concentration returns to its original level much faster, as if the cells were in a Ca^{2+}-free environment (Fig. 2). These results are obtained for a concentration of 70% Xe / 30% air (Fig. 2a), corresponding to 1 MAC, or 100% Xe (Fig. 2b). Although the original basal intracellular Ca^{2+} concentration is reached in all Xe-treated cells after the ATP stimulus, cells are consistently found that require a few minutes (up to 5 min) longer to arrive at their pretreatment level or to stabilize their baseline similar to the majority of cells. This does not appear to be an individual cell property, since if the same cell receives a second time ATP stimulus it may express a perfect fast return to its original basal Ca^{2+} concentration although it had shown a slightly slower response upon the first ATP stimulus. At present it remains unclear if such a behavior represents a transient asynchrony of the various Ca2+ sequestering systems or is based on normal biological variations.

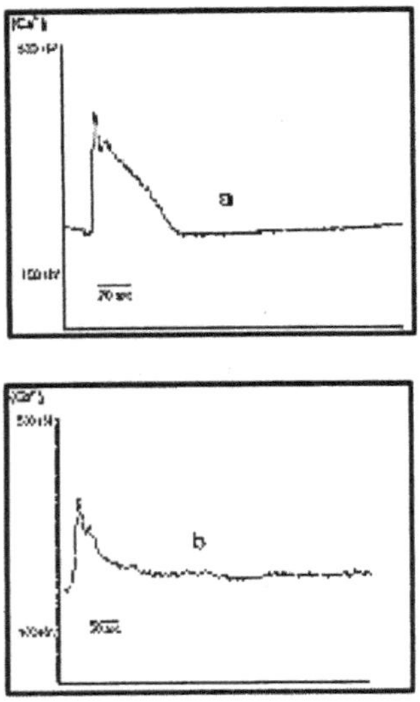

Fig. 2 Cells were incubated for 2 h in 70% xenon / 30% air-saturated buffer (**a**) or in 100% xenon-saturated buffer (**b**). A 5 μM ATP stimulus was applied still maintaining the same experimental conditions. In both cases the initial Ca^{2+} response was very similar to that of the controls. However, the second slow part of the Ca^{2+} response does not occur. No difference was observed between 70% and 100% xenon

In order to exclude the possibility that the observed effects of the fast reduction of intracellular Ca^{2+} concentration were due to the lack of oxygen, control experiments were performed in an atmosphere where Xe was replaced by N_2. In these cells the ATP-induced Ca^{2+} fingerprint was identical to that obtained from untreated cells (Fig. 3).

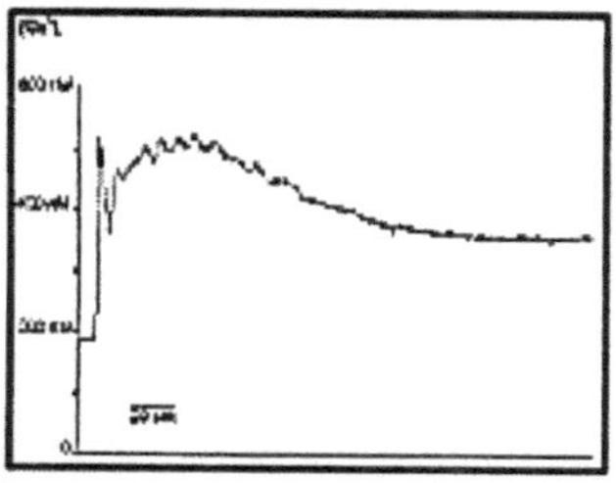

Fig. 3 Cells were incubated in a N_2-saturated buffer for 2 h and the ATP-induced Ca^{2+} changes were measured in the presence of N_2. If one compares Fig.1 with Fig. 4, it becomes evident that the cell response is very similar under these conditions; in particular, the slow decrease in Ca^{2+} concentration to the basal level is the same as in untreated cells

In order to obtain more evidence that Xe may interfere with the Ca^{2+}-induced Ca^{2+} influx, cells were treated with thapsigargin, a compound known to inhibit the intracellular Ca^{2+}-uptake systems. Such cytoplasmic membrane systems serve as storage sites where Ca^{2+} ions are sequestered by the families of SERCA systems (Koch 1990; Lytton et al. 1992; MacLennan et al. 1992) which are the specific target of thapsigargin (Thomas and Hanley 1994). Treatment of the cells with 1 μM thapsigargin causes a distinct Ca^{2+} increase in the cytoplasm, due to the inhibiting effect on the intracellular Ca^{2+} pumps, followed by an additional Ca^{2+}-induced Ca^{2+} influx (Fig. 4a). The second part of this thapsigargin-induced Ca^{2+} transient is absent in Ca^{2+}-free medium (Fig. 4b). If cells are preincubated for 2 h in Xe in Ca^{2+}-containing buffer and then treated with thapsigargin, the initial Ca^{2+} release is very similar to the reaction found in untreated cells. However, the Ca^{2+}-induced Ca^{2+} influx from outside appears to be almost completely suppressed (Fig. 4c) and a Ca^{2+} response is obtained similar to the reaction obtained in Ca^{2+}-free medium.

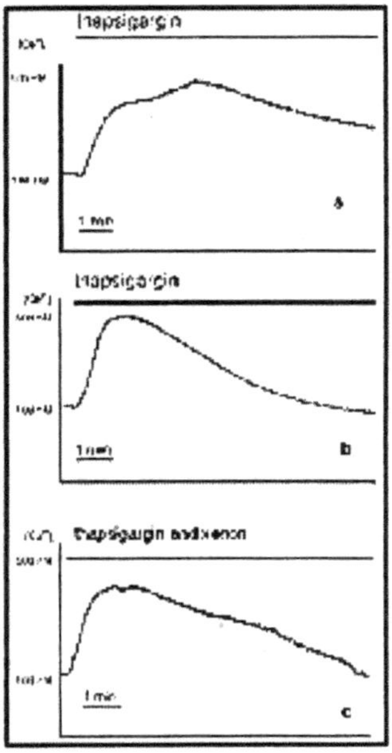

Fig. 4 Addition of 1 μM thapsigargin causes an increase in the intracellular Ca^{2+} concentration, which remains elevated in Ca^{2+}-containing buffer (**a**). In Ca^{2+}-free medium the initial Ca^{2+} rise is very similar; however, the increase is subsequently not sustained (**b**). Cells pretreated for 2 h with Xe-saturated Ca^{2+} buffer and then exposed to 1 μM thapsigargin show the initial Ca^{2+} rise but this is not sustained subsequently; there appears to be almost no Ca^{2+}-induced Ca^{2+} influx (**c**), resulting in a Ca^{2+} transient similar to the reaction in Ca^{2+}-free medium (cf. **b** with **c**)

The use of Xe as an anesthetic requires that its effects are fast and completely reversible. Cells were incubated in Xe medium and the ATP-induced Ca^{2+} fingerprint was obtained for one cell. Thereafter the buffer was changed to normal buffer, an additional ATP stimulus was applied, and the Ca^{2+} fluctuations were followed in the same cell. Figure 5 shows clearly that as soon as the Xe is removed, the internal liberation of Ca^{2+} by ATP again induces the influx of Ca^{2+} from the outside, a process which did not occur under the influence of Xe. Such complete recovery of the cells from the treatment with Xe is highly reproducible; the cells do not show any signs of damage once Xe is removed from their environment. Naturally, this requires strict adherence to the maintenance of optimal culture conditions during the measurement period. Further support for the complete recovery of cells from the Xe treatment is obtained when the growth characteristics of whole cell cultures treated for 2 h with Xe-containing buffer are compared with cultures treated with control buffer. No difference in cell proliferation was found (data not shown).

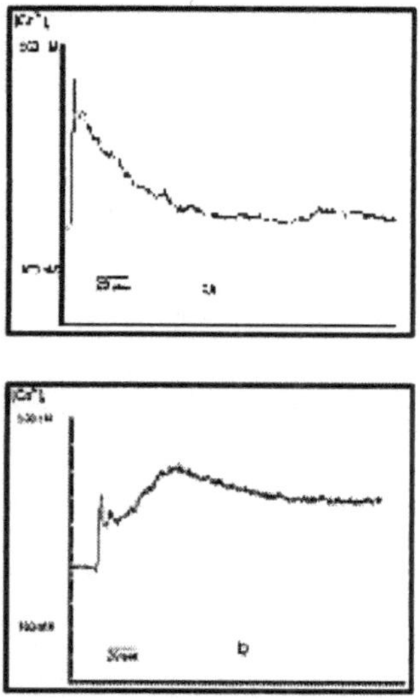

Fig. 5 Reversible effects of xenon on the Ca^{2+}-induced Ca^{2+} influx. In **a** the cell response to the ATP stimulus is shown in xenon-containing buffer. Fifteen minutes after replacing the xenon buffer with normal buffer a second ATP stimulus was applied to the same cell (**b**). It becomes evident that the xenon block of the second part of the Ca^{2+} response which is suppressed by the xenon treatment is completely removed; the cell reacts to the stimulus in the same way as an untreated cell

Ca^{2+} responses in individual cells do vary in spite of a defined stimulus. A comparison between responses of many cells will result in the establishment of a typical pattern of responses but the natural variations in the absolute values may hide effects induced experimentally. Table 2 shows the absolute values of the initial Ca^{2+} response in untreated cells and in cells pretreated with 70% and 100% xenon. It is evident that a considerable variation exists among the individual cells. No significant difference, however, is found when treated and untreated cells are compared.

Table 2 Initial increase of intracellular Ca^{2+} after addition of 5 μM ATP. The difference and its standard deviation between the basal Ca^{2+} concentration and the first Ca^{2+} maximum are indicated (nmol Ca^{2+}).

Control ($n = 12$)	Control in Ca^{2+}-free buffer ($n = 9$)	100 % xenon ($n = 13$)	70% xenon ($n = 11$)
320±74	290±90	280±86	305±75

When the absolute time course of events is not followed and compared but the kinetics are expressed to as $1/e$, a clear difference becomes apparent with respect to the time needed for the decay of the intracellular Ca. As is shown in Table 2, individual cells vary with respect to their absolute value of the first

ATP-induced Ca peak. We designed for each cell as "1" the value between the baseline and the height of the first ATP-induced Ca peak and divided this value by *e*. Then we determined the time needed for the cell to reduce the internal Ca to this concentration. As shown in Fig. 6, this treatment of the Ca^{2+} kinetics results in a clear distinction between the kinetics of the ATP-induced Ca^{2+} release in Xe-treated versus control cells. This does not imply that the Ca^{2+} decay follows simple exponential kinetics; on the contrary, it is quite possible to expect complex second- and third-order kinetics for several interrelated events. However, for the purpose of proving a clear difference with regard to the reaction of untreated and xenon-treated cells, our approach of treating the data with single-order kinetics is sufficient.

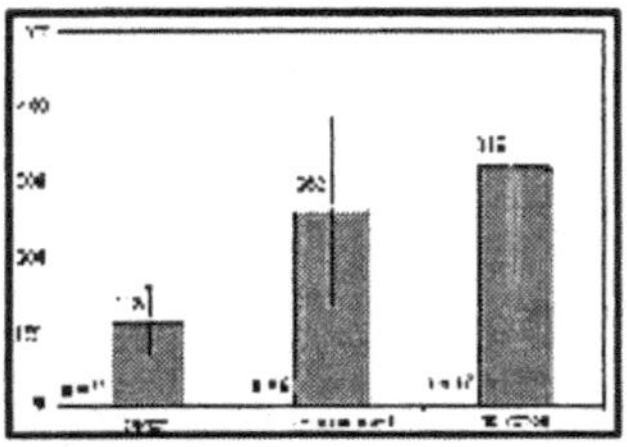

Fig. 6 Time (s) needed for the reduction of the ATP-induced increase in intracellular Ca^{2+} concentration expressed as $1/e$. The difference between the first ATP-induced Ca^{2+} maximum and the basal level was divided by *e* and the time needed for each cell to reach this Ca^{2+} level was calculated. Whereas a significant difference is found when comparing xenon-treated cells with untreated cells and cells where the Xe has been removed ($P<<0.05$), no such difference can be found when untreated cells are compared with cells incubated for 2 h from which xenon has been removed. *Ordinate*: (seconds)

Discussion

Calcium ions are such universal messengers for many types of cells that complex mechanisms have evolved to differentiate between the various inter- and intracellular signals conveyed by these ions. One major component of such a regulatory system consists of the Ca^{2+}-regulated Ca^{2+} entry, first proposed by Putney (1986); for a recent review see Berridge 1995). A specialized Ca^{2+}-**R**elease-Activated-Channel (CRAC) system resides in the plasma membrane (Hoth and Penner 1992, 1993; Luckhoff and Clapham 1994), the activity of which is controlled by the filling state of intracellular Ca^{2+} stores. When the stores reach a certain degree of emptiness, Ca^{2+} channels in the plasma membrane are activated to permit the influx of Ca^{2+} ions from the outside. At present it is still unclear how the message of the intracellular Ca^{2+} stores is transferred to the plasma membrane. Two concepts have been proposed and experimental evidence has been presented for both: those that propose the existence of a diffusible factor and those that consider information to be transferred more directly through a protein-protein interaction (Hoth and Penner 1993; Putney and Bird 1993; Fasolato et al. 1994; Gilon et al. 1995). Tsien and his co-workers have proposed the existence of diffusible factors (CIF) bridging the distance between the membraneous Ca^{2+} stores of the endoplasmic reticulum and the Ca^{2+} channels of the plasma membrane (Parekh et al. 1993; Randriamampita and Tsien 1993, 1995; Davies and Hallett 1995). G proteins have also been implicated in the signal transfer between the endoplasmic reticulum stores and the plasma membrane (Bird and Putney 1993). Recently, Innocenti et al. (1996) presented evidence that ADP may be involved in the modulation of the Ca^{2+}-release-induced Ca^{2+} influx. Protein-protein interactions, on the other hand, may obtain the signal transfer from the endoplasmic reticulum to the plasma membrane either by protein phosphorylation and dephosphorylation (Sargeant et al. 1993 a,b, 1994; Jenner et al. 1994) or by a direct conformational coupling between the two membrane systems with the IP_3 receptor functioning as the intermediary (Irvine 1990; Berridge 1991; Petersen and Berridge 1994). Such a close association between plasma membrane and endoplasmic reticulum has been described using fractionation studies (Guillemette et al. 1988; Rossier et al. 1991; Lièvremont et al. 1994, 1996) and the cytoskeleton was implicated as the scaffold keeping the endoplasmic reticulum in place close to the plasma membrane. Ankyrin, which binds to the IP_3 receptor, was suggested to be one of the regulatory links between the endoplasmic reticulum and the plasma membrane (Joseph and Samanta 1993).

Although all this evidence does not allow us yet to distinguish between these two models (and probably parts of both concepts will have to be integrated into the complex signalling structure of the "cell cortex"), any substances interfering with the capacitative calcium entry will be targetted to these components. We have presented evidence that Xe is such a substance inhibiting the Ca^{2+} influx from the outside when the intracellular stores are emptied by an ATP stimulus. In particular, the comparison of the Ca^{2+} response in normal versus Ca^{2+}-free medium and of the response in Xe-treated and Xe-washed-out cells suggests such an interpretation. Additional support has been obtained by our experiments using thapsigargin. Here the Ca^{2+}-induced Ca^{2+} increase is well documented also in other cell systems (Thomas and Hanley 1994) and we have obtained identical results in human endothelial cells. The fact that Xe suppresses the sustained Ca^{2+} elevation supports our interpretation of its interference with the Ca^{2+}-induced Ca^{2+} influx.

The kinetic features of capacitative calcium entry have been emphasized by Berridge (1995) as another of its typical characteristics. The influx component of the Ca^{2+} entry is often delayed relative to the mobilization transient (Hallam et al. 1988; Shuttleworth 1994; Martin and Shuttleworth 1994) and just such a kinetic has been found upon treatment with Xe. If one tries to identify possible sites for the observed Xe effects, some apparently unrelated reports from the literature may become helpful. Xenon has been shown to bind to proteins such as hemoglobin (Keyes and Lumry 1968; Schoenborn 1968) and probably thereby to associate preferentially with red blood cells (Goresky et al. 1988), to interfere with microtubule dynamics(Allison and Nunn 1968, Nunn and Halsey 1990) and to inhibit in vitro the Ca^{2+}-ATPase of the plasma membrane (Singh et al. 1995). Thus, one of the molecular targets of Xe is probably situated in the cell cortex, regulating a specific Ca^{2+}-channel system. Circumstantial evidence has been accumulated that anesthesia might be caused by an interference with the Ca^{2+}-channel systems of the plasma membrane (reviewed in Lynch and Pancrazio 1994; Hirota and Lambert 1996). Comparisons with other anesthetics will show if the reported effects on this Ca^{2+}-channel system are Xe-specific or shared by other substances, thus indicating a common mechanism for anesthesia. Naturally, an important part of our future work will also include experiments to see if and to what extent our results obtained with human endothelial cells can be extrapolated to the brain system, i.e. the study of Xe-induced Ca^{2+} changes in neurons, astrocytes and glia resulting in an even better understanding of the anesthetic effects of xenon.

References

Allison, A.C., Nunn, J.F. (1968) Effects of general anaesthetics on microtubules: a possible mechanism of anaesthesia. *Lancet* **II**:1326-1329

Berridge, M.J. (1991) Caffeine inhibits inositol-trisphosphate-induced membrane potential oscillations in Xenopus oocytes. *Proc. R. Soc. Lond. B. Biol. Sci.* **244**:57-62

Berridge, M.J. (1995) Capacitative calcium entry. *Biochem. J.* **312**:1-11

Bird, G.S., Putney, J.W.Jr. (1993) Inhibition of thapsigargin-induced calcium entry by microinjected guanine nucleotide analogues. Evidence for the involvement of a small G-protein in capacitative calcium entry. *J. Biol. Chem.* **268**:21486-21488

Boomsma, F., Rupreht, J., Man-in-'t-Veld, A.J., de-Jong, F.H., Dzoljic, M., Lachmann, B. (1990) Haemodynamic and neurohumoral effects of xenon anaesthesia. A comparison with nitrous oxide. *Anaesthesia* **45**:273-278

Cullen, S.C., Gross, E.G. (1951) The anesthetic properties of xenon in animals and human beings with additional observations on krypton. *Science* **113**:580

Davies, E.V., Hallett, M.B. (1995) A soluble cellular factor directly stimulates Ca^{2+} entry in neutrophils. *Biochem. Biophys. Res. Commun.* **206**:348-354

De Jong, R.H., Eger, E.I. 2nd (1975) MAC expanded: AD50 and AD95 values of common inhalation anesthetics in man. *Anesthesiology* **42**:384-389

Edwards, F.A. (1994) ATP receptors. *Cur Opin Neurobiol.***4**:347-352

Eger, E.I. 2nd, Brandstater, B., Saidman, L.J., Regan, M.J., Severinghaus, J.W., Munson, E.S. (1965) Equipotent alveolar concentrations of methoxyflurane, halothane, diethyl ether, fluroxene, cyclopropane, xenon and nitrous oxide in the dog. *Anesthesiology* **26**:771-777

Fasolato, C., Innocenti, B., Pozzan, T. (1994) Receptor-activated Ca^{2+} influx: how many mechanisms for how many channels? *Trends Pharmacol. Sci.* **15**:77-83

Franks, J.J., Horn, J.L., Janicki, P.K.,Singh, G. (1995a) Stable inhibition of brain synaptic plasma membrane calcium ATPase in rats anesthetized with halothane. *Anesthesiology* **82**:118-128

Franks, J.J., Horn, J.L., Janicki, P.K.,Singh, G. (1995b) Halothane, isoflurane, xenon and nitrous oxide inhibit calcium ATPase pump activity in rat brain synaptic plasma membranes. *Anesthesiology* **82**:108-117

Gilon, P., Bird, G.J., Bian, X., Yakel, J.L., Putney, J.W.Jr. (1995) The Ca^{2+}-mobilizing actions of a Jurkat cell extract on mammalian cells and *Xenopus laevis* oocytes. *J. Biol. Chem.* **270**:8050-8055

Goresky, C.A., Schwab, A.J., Rose, C.P. (1988) Xenon handling in the liver: red cell capacity effect. *Circ. Res.* **63**:767-778

Grynkiewicz, G., Poenie, M., Tsien, M., R.Y. (1985) A new generation of Ca^{2+} indicators with greatly improved fluorescence properties.*J. Biol. Chem.* **260**:3440-3443

Guillemette, G., Balla, T., Baukal, A.J., Catt, K.J. (1988) Characterization of inositol 1,4,5- trisphosphate receptors and calcium mobilization in a hepatic plasma membrane fraction. *J. Biol. Chem.***263**:4541-4548

Hallam, T.J., Jacob, R., Merritt, J.E. (1988) Evidence that agonists stimulate bivalent-cation influx into human endothelial cells. *Biochem. J.* **255**:179-84

Hirota, K,, Lambert, D.G. (1996) Voltage-sensitive Ca^{2+} channels and anaesthesia. *Br. J. Anaesth.* **76**:344-346

Horn, J.L., Janicki, P.K., Franks, J.J. (1995) Nitrous oxide and xenon enhance phospholipid-N- methylation in rat brain synaptic plasma membranes. *Life Sci.* **56**:455-460

Hoth, M., Penner, R. (1992) Depletion of intracellular calcium stores activates a calcium current in mast cells.*Nature* **355**:353-356

Hoth, M., Penner, R. (1993) Calcium release-activated calcium current in rat mast cells. *J. Physiol. (Lond.)* **465**:359-386

Innocenti, B., Pozza, T., Fasolato, C. (1996) Intracellular ADP modulates the Ca^{2+} release- activated Ca^{2+} current in a temperature- and Ca^{2+}-dependent way. *J. Biol. Chem.* **271**:8582- 8587

Irvine, R.F.(1990) 'Quantal' Ca^{2+} release and the control of Ca^{2+} entry by inositol phosphates - a possible mechanism. *FEBS Lett.* **263**:5-9

James, P.B. (1990) Xenon for anaesthesia. *Lancet* **336**:324

Janicki, P.K., Horn, J.L., Singh, G., Franks, W.T., Franks, J.J. (1994) Diminished brain synaptic plasma membrane Ca^{2+}-ATPase activity in rats with streptozocin-induced diabetes: association with reduced anesthetic requirements. *Life Sci.* **55**:359-364

Janicki, P.K., Horn, J.L., Singh, G., Janson, V.E., Franks, W.T., Franks, J.J. (1995) Reduced anesthetic requirements, diminished brain plasma membrane Ca^{2+}-ATPase pumping, and enhanced brain synaptic plasma membrane phospholipid methylation in diabetic rats: effects of insulin. *Life Sci.* **56**:357-363

Jenner, S., Farndale, R.W., Sage, S.O.(1994) The effect of calcium-store depletion and refilling with various bivalent cations on tyrosine phosphorylation and Mn^{2+} entry in fura-2-loaded human platelets. *Biochem. J.* **303**:337-339

Joseph, S.K., Samanta, S. (1993) Detergent solubility of the inositol trisphosphate receptor in rat brain membranes. Evidence for association of the receptor with ankyrin. *J. Biol. Chem.* **268**:477

Kennedy, R.R., Stokes, J.W., Downing, P. (1992) Anaesthesia and the 'inert' gases with special reference to xenon. *Anaesth. Intensive Care* **20**:66-70

Keyes, M., Lumry, R. (1968) Binding of anesthetics to proteins: linkage between the sixth-ligand site of heme iron ion and the nonpolar binding sites of myoglobin. *Fed. Proc.* **27**:895-897

Koblin, D.D., Watson, J.E., Deady, J.E., Stokstad, E.L., Eger, E.I. 2nd (1981) Inactivation of methionine synthetase by nitrous oxide in mice. *Anesthesiology* **54**: 318-324

Koch, G.L.E (1990) The endoplasmic reticulum and calcium storage. *BioEssays* **12**: 527-531

Lachmann, B., Armbruster, S., Schairer, W., Landstra, M., Trouwborst, A., Van-Daal, G.J., Kusuma, A., Erdmann, W. (1990) Safety and efficacy of xenon in routine use as an inhalational anaesthetic. *Lancet* **335**:1413-1415

Lane, G.A., Nahrwold, M.L., Tait, A.R., Taylor-Busch, M., Cohen, P.J., Beaudoin, A.R. (1980) Anesthetics as teratogens: nitrous oxide is fetotoxic, xenon is not. *Science* **210**:899-901

Lièvremont, J.P., Hill, A.M., Hilly, M., Mauger, J.P. (1994) The inositol 1,4,5-trisphosphate receptor is localized on specialized sub-regions of the endoplasmic reticulum in rat liver.*Biochem. J.* **300**:419-427

Lièvremont, J.P., Hill, A.M., Tran, D., Coquil, J.F., Stelly, N., Mauger, J.P.(1996) Intracellular calcium stores and inositol 1,4,5-trisphosphate receptor in rat liver cells.*Biochem. J.* **314**:189-197

Luckhoff, A., Clapham, D.E. (1994) Calcium channels activated by depletion of internal calcium stores in A431 cells. *Biophys. J.* **67**:177-182

Luttropp, H.H., Rydgren, G., Thomasson, R., Werner, O. (1991) A minimal-flow system for xenon anesthesia. *Anesthesiology* **75**:896-902

Luttropp, H.H., Thomasson, R., Dahm, S., Persson, J., Werner, O. (1994) Clinical experience with minimal flow xenon anesthesia. *Acta. Anaesthesiol. Scand.* **38**:121-125

Lynch III C, Pancrazio, J.J. (1994) Snails, spiders, and stereospecificity - is there a role for calcium channels in anesthetic mechanisms? *Anesthesiology* **81**: 1-5

Lytton, J., Westlin, M., Burk, S.E., Shull, G.E., MacLennan D.H. (1992) Functional comparisons between isoforms of the sarcoplasmic or endoplasmic reticulum family of calcium pumps. *J. Biol. Chem.* **267**: 14483-14489

MacLennan, D.H., Toyofuku, T., Lytton J. (1992) Structure-function relationships in sarcoplasmic or endoplasmic reticulum (SERCA) type Ca^{2+} pumps. In: Scarpa, A., Carafoli, E., Papa S. (eds) Ion-motive ATPases: structure, function and regulation. New York Academy of Sciences, New York, pp 1-10

Martin, S.C., Shuttleworth,T.J. (1994) Ca^{2+} influx drives agonist-activated $[Ca^{2+}]i$ oscillations in an exocrine cell. *FEBS Lett.* **352**:32-36

Nobiling, R., Bührle, C.P. (1989) A microscope fluorimeter using multiple-wavelength excitation for ultrasensitive single-cell emission spectrometry. *J. Microsc.* **156**:149-162

Nunn, J.F., Halsey, M.J. (1990) Xenon in anaesthesia. *Lancet* **336**:112-113

Parekh, A.B., Terlau, H., Stuhmer, W. (1993) Depletion of InsP3 stores activates a $Ca^{2+}$and K^{+} current by means of a phosphatase and a diffusible messenger. *Nature* **364**:814-818

Petersen, C.C., Berridge, M.J. (1994) The regulation of capacitative calcium entry by calcium and protein kinase C in *Xenopus* oocytes. *J. Biol. Chem.* **269**:32246-32253

Philp, R.B., Arora, P., Forsberg, K. McIver, D.J. (1992a) Effects of pressure and gaseous anesthetics on the aggregation of human blood platelets and marine sponge cells: similarities in responses. *Comp. Biochem. Physiol.* **101**:541-545

Philp, R.B., Arora, P., McIver, D.J. (1992b) Effects of gaseous anesthetics and ultrashort and short-acting barbiturates on human blood platelet free cytosolic calcium: relevance to their effects on platelet aggregation. *Can. J. Physiol. Pharmacol.* **70**:1161-1166

Power, K.J. (1990) Does xenon deserve a prominent place in anaesthetic practice? *Anaesthesia* **45**:783-784

Putney, J.W.Jr. (1986) A model for receptor-regulated calcium entry. *Cell Calcium* **7**:1-12

Putney, J.W.Jr., Bird, G.S. (1993) The signal for capacitative calcium entry. *Cell* **75**:199-201

Randriamampita, C., Tsien, R.Y. (1993) Emptying of intracellular Ca^{2+} stores releases a novel small messenger that stimulates Ca^{2+} influx. *Nature* **364**:809-814

Randriamampita, C., Tsien, R.Y. (1995) Degradation of a calcium influx factor (CIF) can be blocked by phosphatase inhibitors or chelation of Ca^{2+} . *J. Biol. Chem.* **270**:29-32

Rossier, M.F., Bird, G.S., Putney, J.W.Jr. (1991) Subcellular distribution of the calcium-storing inositol 1,4,5-trisphosphate-sensitive organelle in rat liver. Possible linkage to the plasma membrane through the actin microfilaments. *Biochem. J.* **274**:643-650

Sargeant, P., Farndale, R.W., Sage, S.O. (1993a) ADP- and thapsigargin-evoked Ca^{2+} entry and protein-tyrosine phosphorylation are inhibited by the tyrosine kinase inhibitors genistein and methyl-2,5-dihydroxycinnamate in fura-2-loaded human platelets. *J. Biol. Chem.* **268**:18151-18156

Sargeant, P., Farndale, R.W., Sage, S.O. (1993b) The tyrosine kinase inhibitors methyl 2,5- dihydroxycinnamate and genistein reduce thrombin-evoked tyrosine phosphorylation and Ca^{2+} entry in human platelets. *FEBS Lett.* **315**:242-246

Sargeant, P., Farndale, R.W., Sage, S.O. (1994) Calcium store depletion in dimethyl BAPTA- loaded human platelets increases protein tyrosine phosphorylation in the absence of a rise in cytosolic calcium. *Exp. Physiol.* **79**:269-272

Schoenborn, B.P. (1968) Binding of anesthetics to protein: an x-ray crystallographic investigation. *Fed. Proc.*27:888-894

Shuttleworth, T.J. (1994) Temporal relationships between Ca^{2+} store mobilization and Ca^{2+} entry in an exocrine cell. *Cell Calcium* 15:457-466

Singh, G., Janicki, P.K., Horn, J.L., Janson, V.E., Franks, J.J. (1995) Inhibition of plasma membrane Ca^{2+}-ATPase pump activity in cultured C6 glioma cells by halothane and xenon. *Life Sci.* 56:219-224

Thomas, A.P., Delaville, F. (1991) The use of fluorecent indicators for measurements of cytosolic-free calcium concentration in cell populations and single cells. In: McCormack and Cobbold (eds) Cellular calcium - a practical approach, Oxford University Press, pp 1-54

Thomas, D., Hanley, M.R. (1994) Pharmacological tools to perturb calcium storage. In: Nuccitelli, R. (ed) Methods in cell biology. A practical guide to the study of calcium in living cells, vol. 40. Academic Press, New York, pp. 65-89

Whitehurst, S.L., Nemoto, E.M., Yao, L., Yonas, H. (1994) MAC of xenon and halothane in rhesus monkeys. *J. Neurosurg. Anesthesiol.*6:275-279

Exp. Biol. Online 2:4 (1997)

A multipurpose, fast scan spectrophotometer for measuring turbid (biological) materials*

Werner Schmidt[1]

[1]Universität Konstanz, Meisenweg 7, D-78465 Konstanz, Germany
e-mail: werner.schmidt@uni-konstanz.de

*Presented at the joint meeting of the German Botanical Society and the German Zoological Society entitled: *Spectroscopy and Optical Techniques in Animal and Plant Biology*, Münster, 30th Sept-3rd October 1996

Received: 2 November 1996 / Accepted: 17 January 1997
Published: January 30, 1997

Abstract. A new optical spectrophotometer has been developed, based on a recently patented monochromator for spectral measurements of clear and, in particular, of turbid materials in the millisecond time range. The number of optical and mechanical components of the spectrophotometer has been reduced to a minimum, resulting in excellent light throughput, a low stray-light level, low cost, compactness and rigidity. The spectrophotometer has been designed for all kinds of spectral measurements such as absorption, transmission, reflection and luminescence/radiation in a single-beam mode as documented by several examples. In principle, there is no restriction of wavelengths, ranging from UV to NIR and up to the IR range. As many functions as possible are relocated from the hardware to the software part of the design, which allows for extraordinary flexibility and simplicity. An appropriate computer program providing data acquisition, control and calibration functions as well as real-time display of spectra has been developed on the basis of a compiler language; indispensable "fast routines" are written in assembler language.

Key words: Spectrophotometer, fast scan – light fiber, application – absorption of turbid materials – spectrophotometer, software – absorption measurement, reflection measurement, luminescence measurement , fluorescence measurement

Introduction

Conventional double-beam spectrophotometers require extensive scan times and typically only clear, liquid samples can be used. However, fast spectral measurements of turbid and scattering samples are often asked for in biological, (bio)chemical, medical, environmental and other studies. Unfortunately, all the fast-scan instruments developed so far are typically complex, bulky, energy consuming, expensive and not suitable for field work (e.g. the OLIS rapid scanning monochromator RSM 1996; On-Line Instrument Systems, Bogart, Georgia). As a questionable alternative in the last two decades so-called OMAs (optical multichannel analyzers) based on diode or (more recently) CCD arrays have become very popular, even if the basic idea of using a detector array and a fixed grating rather than a revolving grating with an exit slit and a single photodetector is more than 20 years old (e.g. as utilized in multiple atomic emission spectroscopy). Compared to scanning grating monochromators, OMAs allow "parallel" measurements throughout one octave of wavelengths in a reasonably short time (as little as ~10 ms) and no moving parts are involved. However, for many biological applications array spectrophotometers involve serious drawbacks. The sample is part of the optical system; thus, only completely clear samples will not disturb the light path, since dispersion takes place *after* the sample. Because of poor reproducibility of strength of the flash light and an intrinsically high stray-light level (e.g. 3% in survey mode and 0.05% in "high performance" mode) their dynamic range is moderate and their sensitivity comparably low. Nonetheless, within restrictive limits turbid materials can be measured by OMAs utilizing light fiber technology, even if it is impossible to fill the entrance cone of OMA spectrometers optimally, since light emerging from turbid samples is highly isotropic and therefore difficult to control in its path. The covered room angle is probably 1/1000 compared to the conventional light paths as utilized in the currently described single-beam set-up. In addition, irradiation with actinic broad-band white light can cause photolytic destruction of light-sensitive biological (and chemical) samples.

As required by our biophysical experiments - and by our budget - we designed an alternative spectrophotometer using a novel monochromator, a kind of modified *Czerny Turner* (Fig. 1, top) arrangement, largely composed of inexpensive and conventional parts such as a halogen-measuring lamp, entrance and exit slits, a dispersive grating, spherical mirrors and a conventional photodetector such as a photodiode or a more sensitive (*end on*) photomultiplier in the visible wavelength range (or a PbS detector for the near infrared). There is a light, vibrating surface deflection mirror inserted as a *secondary scanning component*. This alternative concept of a small, compact, low-cost and fast single-beam spectrophotometer offers unique capabilities.

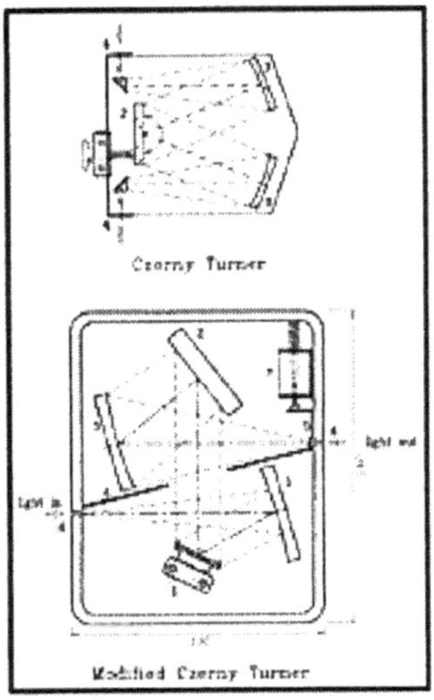

Fig. 1 Top: Conventional Czereny Turner arrangement.**Bottom:** Top view of the novel fast scan monochromator including crude dimensions [mm].
Parts:

1. scanner with first surface mirror
2. grating 50×50 mm, 1200 grooves/mm
3. spherical mirror, $f = 100$ mm
4. entrance or exit slits composed of razor blades
5. small piece of didymium filter as moved into the light path by coil 7
6. baffle
7. coil (controlled by computer)

For details see text

Monochromator

The basic design of the novel monochromator with a compartment made of black galvanized aluminum is shown in Fig. 1. Light from a 12 V/10 or 20 W halogen lamp with a reflector and a suitable lens system impinges on the entrance slit **4**, is collimated by a spherical mirror **3** and deflected by a flat surface mirror of a scanner **1** onto the flat reflection grating **2** (holographic, 50×50×12 mm). From here the dispersed light is focused via another spherical mirror **3** on the exit slit **4**. Both slits are made of razor blades and can be used either as entrance or exit. The surface mirror of 25×25 mm is part of the rugged laser scanner unit (URS-50W; Laser Scanning Products, Santa Rosa, Calif., USA) with a resonance frequency of presently 50 Hz and a maximum scan angle of 50 ° P-P. With the presently utilized grating of 1200 grooves mm^{-1} and to cover the whole visible spectrum between 400 and 800 nm, the total angle only amounts to approx. 15 ° P-P (for geometric reasons the deflection angle of the scanner mirror on the grating doubles). Other scanners with smaller scan angles and thus higher scanning frequencies are available. However, the AD converter, not the hardware, is still the speed-limiting bottleneck of the system. In order to increase the effective aperture of the grating and to optimize the light throughput of the system, a large grating of 50 x 50 mm was chosen, compared to the smaller scanner mirror of 25×25 mm. Thus, while scanning, different parts of the grating are used for dispersion.

The vibrating mirror of the scanner (Fig. 1) produces a coherent, repeating signal sequence displaying the spectrum from low to high and from high to low wavelengths. The electronic control circuit of the scanner allows the deduction of a Schmitt trigger signal. A software-controlled delay between trigger signal and data acquisition is used to select the starting wavelength precisely.

For the present demonstration we used a tiny, highly sensitive, wide dynamic range and fast time-response photomultiplier sensor module H5773-01 (300-800 nm, Hamamatsu) for the detection of visible light. For measuring absorption spectra the detector signal is converted into a logarithmic voltage signal by a log amplifier model 759P from Analog Devices (attempts to calculate the logarithm "log I_0/I" rather than to use a log amplifier failed for reasons relating to speed) and finally fed to a fast A to D converter (SAR, *successive approximation register*) with a resolution of 12 bits and a conversion time of approx. 10 µs/100 kHz (PCL-818; Advantech Co. Ltd., Taipei, Taiwan, R.O.C.).

The A to D converter card includes two additional D to A converters. These are used (1) to adjust the high voltage of the photomultiplier sensor module H5773-01 (Hamamatsu) between 0 and 1000 V and (2) to insert the didymium filter (Fig. 1, coil 7) for calibration, both software controlled. The trigger signal from the scanner electronics is fed to a separate *digital* input line on the AD card.

Specifications

The amplitude of vibration of the scanner mirror defines the spectral range. The midpoint of the spectrum is determined either by the resting position of the mirror 1 or of the (fixed) position of the grating 2 (Fig. 1). Changing the vibration amplitude and midpoint position provides a zoom function, selecting and magnifying any part of the spectrum. No attention is pais to interference from different scattering orders of the grating, since currently only approximately one octave is covered and the transmission of the present light fiber system practically drops to zero below 400 nm. Due to limitations of the present software package a spectrum is acquired only when the scanner mirror is moving in one direction, in the other direction it is simply idle. With the device in its present set up - depending on spectral resolution and range - the scan time per spectrum is approximate 5 - 10 ms.

In order to reduce the stray light level the interior of the monochromator is divided into two closed compartments (6, Fig. 1). The aperture is approximately

F/2.5. A crude estimate of the stray light level was obtained as follows. A far red line interference filter (728.5 nm) was inserted as absorber, and the excitation was set manually to the blue (450 nm). The signal measured accounts for a very low stray-light level of approx. 0.0001 % (in this visible wavelength range).

Using gratings the dispersed wavelength depends on the sine of the deflection angle [Theta] (cf. Schmidt 1994, pp. 103-132)

$$\lambda([\text{Theta}]) = c \cdot \sin\,[\text{Theta}] \quad (1)$$

with a constant *c* including the scattering order and the groove spacing. Conventionally this dependency is linearized by a costly mechanical sinusoidal drive. However, the sinusoidally vibrating scanner and the sine dependency of the grating (Eq. 1) substantially compensate for each other, yielding a linear relationship within the wavelength resolution limits (graph not shown).

The slit width is presently fixed at 0.4 mm, corresponding to approx. 6 nm wavelength resolution (Fig. 3).

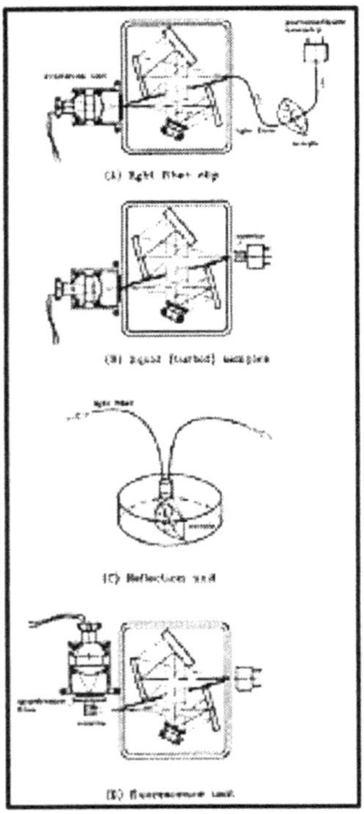

Fig. 2 A Light fiber clip for measuring absorption spectra of solid samples such as leaves or earlobes. **B** Measurement of clear and particularly *highly turbid liquid* samples (suspensions). **C** Reflection measurements using a bifurcated light fiber system with a dark compartment and **D** measurement of fluorescence (emission) spectra; excitation is determined by appropriate interference filters

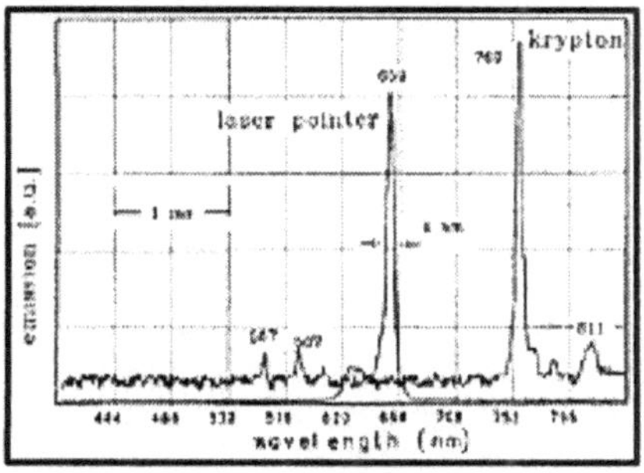

Fig. 3 Emission spectra obtained by a krypton calibration lamp and a conventional red laser pointer with well-known wavelengths. With the present arrangement a wavelength resolution of approx. 6 nm is obtained. The scan speed is indicated by a 1 ms time range as obtained by a rectangular voltage signal of a function generator set to 1 kHz

Calibration of wavelength and photometric values

Absolute photometric and wavelength calibration of commercial spectrophotometers typically vary and are always a matter of concern. We adopted a fast and reliable, even if not highly accurate, procedure. It does not depend on the typically adopted, extremely sharp lines of calibration lamps filled with Ar, Kr or Ne, but on the (relatively narrow) absorption bands of a didymium-glass filter (BG 32, d=1mm, Schott, Mainz, Germany) distributed across the whole visible spectrum (near 360, 581, 743, 800, 872 nm, mainly due to spectral properties of the f electrons of neodymium) (cf. Fig. 4). Particularly suitable are the strongest bands at 581, 743 or 800 nm. On this basis a computer calibration routine was developed and we obtained sufficient precision of +/-1.5 nm with the didymium wavelengths calibration as judged by a more precise Krypton calibration lamp (cf. Fig. 3). Photometric calibration of absorption spectra is performed simultaneously with wavelength calibration referring to the (separately determined) 581 nm peak of the individual didymium filter used (here [Delta]$A_{581/555}$) in a single scan (a simple push-button function of this PC program).

Baseline correction of single beam absorption spectra

A sequential baseline correction is performed conventionally by calculation of the difference spectrum

$$A(\lambda) = \log I_0(\lambda) - \log I(\lambda) \qquad (2)$$

where log I_0 and log I denote the logarithmic reference and sample spectra. It is very important that both curves are scanned under identical conditions, i.e. each spectrum has to be corrected with "its own" correction curve. However, there is a much more convenient procedure of a simultaneous baseline correction. Rewriting Eq. 2 yields

$$A(\lambda) = \log \frac{I_0(\lambda)}{I(\lambda)} \tag{3}$$

i.e. the baseline correction of various spectra can be performed with only one individual (linear) reference spectrum I_0 ([lambda]) which is determined just once for the specific wavelength set-up, but independent of its absolute values. For example, if the baseline I_0 ([lambda]) is multiplied by a factor of a (representing another setting of the spectrophotometer), the corrected absorption spectrum is simply shifted by log a on the absorption axis A ($\log \{a\ I_0\} = \log a + \log I_0$), but most important the spectrum itself is neither changed in shape nor in size. However, a shift does not represent a profound disadvantage because (1) absorption spectra most often exhibit regions without any absorbency which might serve as zero reference (e.g. 640 nm of the didymium filter), and (2) because at sensitive settings even conventional double-beam spectrophotometers require an individual setting of zero deflection, i.e. the offset has to be compensated for manually.

The signals are scanned and the absorption spectra are obtained by Eq. 3. Thus, with this single-beam spectrophotometer absorption spectra can be scanned and displayed in *real time* and in fully corrected form (with simultaneous elimination of the baseline the user is unaware of) on the computer monitor.

Light fibre arrangement

The monochromator is useful for various optospectral measurements such as absorption, emission, reflection or scattering spectroscopy as facilitated by a special light fiber system (Fig. 2, ST50M; Schoelly, Denzlingen, Germany; inner diameter of the fibers 3 mm, approx. T = 65% with some minor variations between 450 and 1300 nm). All kinds of samples are acceptable that are typically not adequate for conventional spectrophotometers. These include solid samples such as leaves (Fig. 5, *in situ*), earlobes, animal tissue, flowers etc. which are measured by a simple light fiber clip arrangement (Fig. 2 A). Clear liquid samples and (particularly highly) turbid suspensions such as milk, algae or vesicle systems are measured with the "end on" arrangement without light fibers included as shown in Fig. 2 B (low-cost light fibers as currently utilized often display unfavorable spectral transfer properties starting at 400 nm, including sharp "troughs" at distinct wavelengths). Reflection measurements are performed with a bifurcated light fiber (Figs. 2C, 6). Fluorescence is measured by the option shown in Fig. 2D. For this purpose the measuring lamp is replaced by a fluorescing sample which is excited by a similar halogen lamp to the one described, through an appropriate interference filter and lens system (no example shown). Figure. 3 shows the emission spectra obtained by a krypton calibration lamp and a conventional red laser pointer. With the present arrangement a wavelength resolution of approx. 6 nm is obtained. The scan speed is indicated by the 1 ms time range bar as obtained by a rectangular voltage signal of a function generator set to 1 kHz.

Absorption spectra

A demonstration of the high-speed scanning capabilities of the monochromator is shown in a 3D plot in Fig. 4: 50 subsequent absorption spectra of the didymium filter as sandwiched with white, non-fluorescent paper are monitored within approximately 1 s. Since no suitable biological sample that substantially changes its absorptivity within this time range is currently available, we simply moved the filter manually at the proper speed into the measuring light path (e.g. spectral changes of an earlobe are readily measurable but turned out to be too small and too slow for the present demonstration).

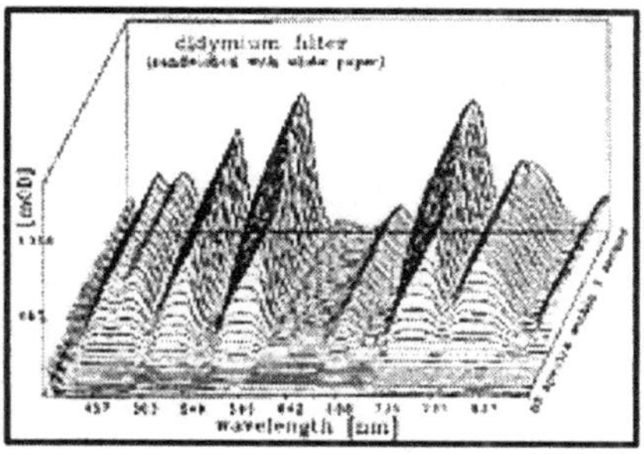

Fig. 4 Three-dimensional plot of 50 subsequent absorption spectra of a *didymium* filter (sandwiched with white, non-fluorescent paper to simulate a highly scattering, biological sample) exemplifies the fast scanning capabilities of the monochromator. For this demonstration the filter was inserted manually into the measuring light path within approx. 1 s. The well known peaks at 531 and 743 nm are utilized here for photometric and wavelength calibration by a push-button computer routine. Spectra are smoothed off-line by correlation with a 10 nm rectangular kernel

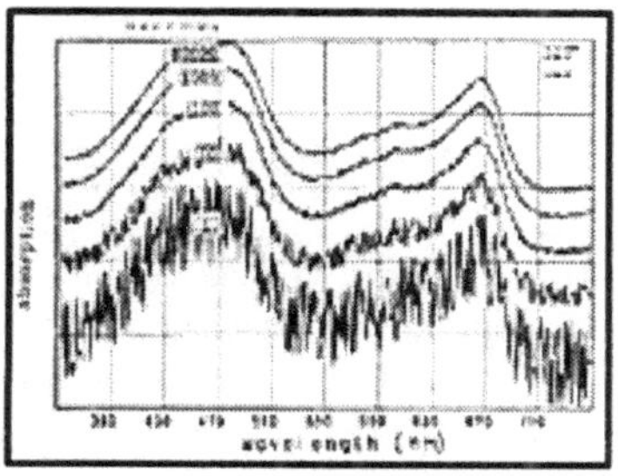

Fig. 5 *From bottom* (spectra are shifted): Averaged sequence of 1, 10, 100, 1000 and 10000 spectra of a mature green leaf with a measuring time of about 5 ms each. Clearly, the advantage of averaging decreases with the number of scans. Intentionally spectrum 1 shows a bad signal-to-noise ratio, which is *not* further improved after averaging approx. 100 scans, i.e. 2 s total scan time. The spectra in Fig. 6 are averaged accordingly

Averaging

The noise contribution of weak optical measuring signals is known to depend on the square root of the impinging number of photons: [radical]n. As a consequence the signal-to-noise ratio (SNR) is:

$$SNR = \frac{n}{\sqrt{n}} = \sqrt{n} \tag{4}$$

indicating that the advantage of measuring many quanta (i.e. spectra) decreases with the number of quanta (spectra), in other words with the measuring time and the number of spectra averaged. Extending a specific measuring time t_1 is useless when other noise sources become dominant. To test this idea we averaged a sequence of 1, 10, 100, 1000 and 10000 absorption spectra of a mature green leaf (Fig. 5; noise was intentionally enhanced by lowering the high voltage of the photomultiplier). Indeed, the 9900 spectra scanned after the first 100 did not significantly improve the SNR, i.e. after approx. 2–3 s data acquisition can be terminated without losing spectral information. One remark is cogent: the SNR is *not* limited by the AD converter, i.e. there is no readout limitation. The same result is obtained with a 16-bit AD converter and with different averaging algorithms, i.e. types of noise other than shot noise become relevant! Thus the monochromator presented here offers another feature of optical spectroscopy so far largely underrated, i.e. optimizing the scan time. This advantage really pays off if hundreds of test tubes are to be scanned routinely in a biochemical laboratory. In contrast, due to a measuring theorem, with common double-beam spectrophotometers – e.g. for a spectrum ranging from 400 to 800 nm – scan time is limited to at least 25 s; however, most spectrophotometers require even longer scan times.

Reflection measurements

Corrected reflection spectra (= R/R_0) are also quickly monitored as exemplified by Fig. 6. First a baseline (1) was recorded by pointing the branched light fiber end onto white, non-fluorescent paper and utilizing a bottomless cylindrical light tight cover box in which the sample is placed with the common light fiber end piercing the top (Fig. 2C). Reference reflection spectra from white, non-fluorescent laboratory filter paper turned out to be indistinguishable from those obtained with the more professional standards of magnesium carbonate and magnesium oxide within the scanned wavelength range in the visible part of the spectrum. Then the reflection unit was placed over white, reddened and tanned brown skin. The scale is given in arbitrary units; however the relations are valid. Note the typical contribution of blood of between 500 and 600 nm, particularly of the reddened skin. The spectra are smoothed by correlation with a rectangular kernel of 10 nm (simulating a slit width of approx. 10 nm, for simplicity, in the present software only a rectangular, less favorable smoothing function was programmed)(Schmidt 1992)

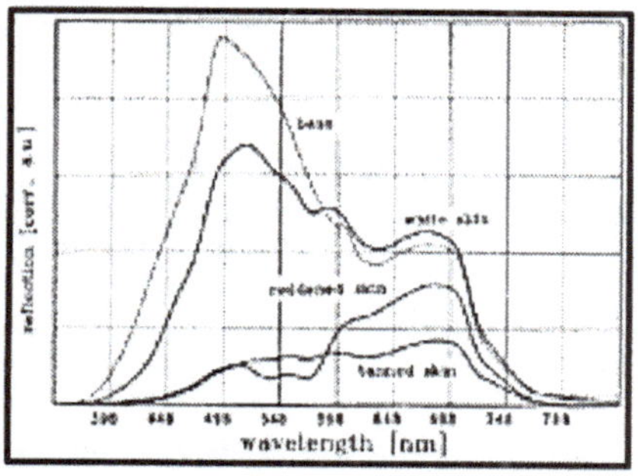

Fig. 6 Corrected reflection spectra R/R_0 of white (2), red (3) and tanned (4) human skin with arbritary reflection values. However, the relation of the different spectra is authentic

Software

In order to translocate as many functions from the hardware to the software part of the design, we developed a special computer program (particularly useful with the pentium microprocessor family). All time-limiting routines, including data acquisition itself, are written in assembler language (TASM) largely using integer numbers. A fast pulldown menu provides immediate and easy access to approximately 70 functions. It includes all control functions, *real-time* visualization, various data manipulation procedures, plotting on screen or Laser Jet (HP). Files (.HP file format) are stored with a header including various parameters such as date, file name, wavelength range, photometric values or comments. In addition to the availability of all standard procedures such as subtraction, addition, multiplication, division, normalization or peak detection, some unusual features are available. These include one-keystroke calibration of wavelength and photometric values, correlation procedures for smoothing and calculating derivatives of any order and with any "d[lambda]" (Schmidt 1992), Fourier (Steward 1987) and Kubelka-Munk (Kubelka 1948, 1954; Kortüm 1969) transformation converting reflection into absorption spectra.

Final remarks

Measurement of turbid materials is both an experimental as well as a theoretical issue (Butler 1972; Seyfried 1989; Vogelmann et al. 1991; Vogelmann 1994). The *experimental* approach is to collect as much as possible of the measuring light isotropically emanating from the sample. With conventional monochromator light pates such as the one presented here (i.e. irradiating the sample with dispersed light) this is accomplished best by placing the photodetector in close iuxtaposition to the sample (using the Ulbricht sphere is the more common but more costly and cumbersome procedure). With array detectors (dispersion after the sample!) the sample has to be placed directly at the entrance rather than at the exit slit. Only light isotropically emanating from the sample within the acceptance cone of the monochromator is processed. However, "overfilling" the monochromator with white light results in a raised stray-light level, i.e. decreased resolution of wavelengths and photometric values.

A much more challenging problem when measuring turbid samples is the *theoretical* approach. The fundamentals of the Kubelka-Munk theory were set down in the pioneering papers of Kubelka (1948, 1954). Subsequently, a lot of work has been done on this topic by various workers, particularly by Kortüm (1969) and on biological tissues by Butler (1972).

The set-up presented here allows various spectral measurements of "optically difficult" samples within the millisecond range. Fast reactions can be followed *spectrally*, whereas the conventional stopped-flow apparatus only monitors changes at one particular wavelength (assuming that the spectral shape does not change!). Since the spectra are presented as a 50 Hz signal, so-called LIACs (light-induced absorption changes; Schmidt 1994) can be monitored *spectrally*. The sample is irradiated with actinic DC light and, *simultaneously*, spectrally measured with this monochromator. This is feasible because a frequency-selective lock-in amplifier might be adopted to separate the strong DC from the weak AC measuring signals efficiently.

Stability of the system is no problem since the manual or automatic recalibration of wavelengths and photometric values as based on the fixed properties of a didymium filter (or any other suitable rare earth filter) requires less than 1 s and is performed automatically at optional time intervals. This monochromator has been patented worldwide since mid-1995 (USA: 1996; Schmidt 1996). It is flexible in terms of price/quality ratio and highly valuable for numerous analytical applications in science, medicine, technology, biotechnology, process control or ecology. The device is small, compact and suitable for field work. Depending on the individual components, it is usable from the UV through the NIR up to the IR wavelength range. Requests are welcome.

Acknowledgement. This work was financially supported by Medidata GmbH, Konstanz.

References

Butler, W.L. (1972) Absorption spectroscopy of biological materials. *Methods Enzymol* **24:** 3–25

Kortüm, G. (1969) *Reflexionsspektroskopie.* Springer, Berlin Heidelberg New York

Kubelka, P. (1948) *J. Opt. Soc. Am.* **38:** 448–457

Kubelka, P. (1954) New contributions to the optics of intensily light scattering materials. Part II J. *J. Opt. Soc. Am.* **44:** 330–335

Schmidt, W. (1992) Korrelationsmethoden in der optischen Spektroskopie. *GIT Fachz.Lab.* **1:** 45–50

Schmidt, W. (1994) *Optische Spektroskopie, eine Einführung für Naturwissenschaftler und Techniker.* Verlag Chemie, Weinheim

Schmidt, W. (1996) US patent 5,497,213: Monochromator having an oscillating mirror and diffracting element for spectrometers

Seyfried, M. (1989) Optical radiation interactions within living tissue. In: Diffey, B.L. (eds) *Radiation measurements in photobiology.* Academic Press, London, p. 191

Steward, E.G. (1987)*Fourier Optics, an Introduction*, 2nd edn. Ellis Horwood, Chichester

Vogelmann, T.C., Martin, G., Chen, G., Buttry, D. (1991) Fiber optic microprobes and measurementof the light microenvironment within plant tissues. *Adv. Bot. Res.* **18:** 256– 296

Vogelmann, T.C. (1994) Light within the plant. In: Kendrick, R.E., Kronenberg, G.H.M. (eds) *Photomorphogenesis in Plants*, 2nd edn. Kluwer, Dordrecht , pp. 491–535

Exp. Biol. Online 2:5 (1997)

Aspiration of THP1 into a micropipette. Mechanical deformation of monocytic THP-1 cells: occurrence of two sequential phases with differential sensitivity to metabolic inhibitors

Fabienne Richelme, Anne Marie Benoliel, and Pierre Bongrand

Laboratoire d'Immunologie, INSERM U 387, Hôpital de Sainte-Marguerite, BP 29, F-13274 Marseille Cedex 09, France

Correspondence to: P. Bongrand, Tel : +33 4.91.75.39.06 ; Fax +33 4.91.75.73.28; e-mail: bongrand@infobiogen.fr

Received: 2 January 1997 / Accepted: 14 February 1997
Published: February 20, 1997

Abstract. Blood leukocytes can exhibit extensive morphological changes during their passage through small capillary vessels. The human monocytic THP-1 cell line was used to explore the metabolic dependence of these changes in shape. Cells were aspirated into micropipettes for determination of the rate of protrusion formation. They were then released and the kinetics of morphological recovery was studied. Results were consistent with Evans' model (*Blood* 64:1028, 1984) of a viscous liquid droplet surrounded by a tensile membrane. The estimated values of cytoplasmic viscosity and membrane tension were 162 Pa.s and 0.0142 mN/m respectively. The influence of metabolic inhibitors on cell mechanical behavior was then studied: results strongly suggested that deformation involved two sequential phases. The cell elongation rate measured during the first 30 s following the onset of aspiration was unaffected by azide, an inhibitor of energy production, and it was about doubled by cytochalasin D, a microfilament inhibitor, and colchicine, a microtubule inhibitor. However, during the following 2 min, deformation was almost abolished in cells treated with azide and cytochalasin D, whereas the protrusion of control cells exhibited an approximately threefold increase in length. It is concluded that, although cells seemed to deform as passive objects, active metabolic processes were required to allow extensive morphological changes triggered by external forces.

Key words: Cell deformation – Monocyte – Cytoskeleton – Mechanical properties

Introduction

Leukocyte functions are heavily dependent on the cells' capacity to undergo extensive deformation following mechanical or biochemical stimulation. Thus, passage through narrow capillary vessels requires that blood leukocytes transiently acquire an elongated sausage shape (Bagge and Branemark, 1977; Worthen et al. 1989). Passage between endothelial cells during diapedesis necessitates extensive squeezing through a narrow aperture (Marchesi 1961). Migration on a surface involves acquisition of a typical hand mirror shape, with continuous emission of forward lamellipodia and rearward contraction (Zigmond 1978). Adhesion may also involve extensive deformation with concomitant spreading at the micrometer level and smoothing of microvilli on the submicrometer scale (Mège et al. 1987).

It is therefore of great interest to understand the leukocyte response to mechanical stimulation. This may yield direct information on passive cell behaviour in flowing blood. Also, this knowledge is a prerequisite for a thorough understanding of the mechanisms of shape control during active deformation, e.g. during migration through the surrounding environment. Several procedures have been devised to study cell mechanical properties, including analysis of deformation induced by centrifugal forces (Landau et al. 1954), measurement of the indentation produced by a solid rod applied on the cell surface with known force (Petersen et al. 1982), study of the displacement of intracellular magnetic particles subjected to exogenous forces (Valberg and Albertini 1985), twisting particles attached to surface receptors by means of a magnetic field (Wang et al. 1993) or aspirating cells into glass micropipettes with controlled pressure and subsequent monitoring of induced protrusion (Mitchison and Swann 1954). The last-mentioned procedure has been used extensively to study the mechanical properties of blood neutrophils. Small deformations were described with a standard viscoelastic model (Schmid-Schönbein et al. 1981). Larger changes in shape, involving the formation of a protrusion substantially greater than the pipette radius, were reported when modelling neutrophils as viscous liquid droplets surrounded by a membrane under tension (Evans and Kukan 1984; Evans and Yeung 1989; Yeung and Evans 1989). The cytoplasmic viscosity was in the order of 200 Pa.s at room temperature, and the membrane tension was about 0.035 mN/m.

It is important to know whether cells subjected to micropipette aspiration may be considered as passive objects with constant properties and to what extent material parameters are dependent on time and metabolic properties. First, micropipette aspiration may indeed trigger active cell responses. Thus, Evans and Kukan (1984) observed that neutrophils might display erratic movements a few minutes after the onset of aspiration. Further, micropipette-aspirated RBL cells (Horoyan et al. 1990) or neutrophils (Zaffran et al. 1993) often exhibited transient increases of cytosolic calcium concentration and reorganization of actin microfilaments (Horoyan et al. 1990). Second, metabolic inhibitors were reported to alter cell mechanical behavior: thus, glycolytic inhibitors such as fluoride, iodoacetate or deoxyglucose reduced the deformability of blood polymorphonuclear leukocytes (Miller and Myers 1975). A temperature decrease also reduced deformation (Sung et al. 1982). Finally, treating leukocytes with cytochalasin B, a microfilament inhibitor, increased deformability (Mège et al. 1985; Tsai et al. 1994). Similar results were obtained on fibroblasts (Erickson 1980). However, cytochalasin inhibited the formation of protrusion by micropipette-aspirated *Fundulus heteroclitus* egg cells (Tickle and Trinkaus 1977). Finally, activating neutrophils with chemotactic factors increased deformability at low concentrations, and increased rigidity at higher concentrations (Kawaoka et al. 1981).

Thus, it seemed useful to obtain more detailed information on the influence of metabolic inhibitors on leukocyte response to mechanical forces. In the present work, human monocytic THP-1 cells were subjected to micropipette aspiration under control conditions or after being treated with colchicine (a microtubule inhibitor), cytochalasin D (a microfilament inhibitor) or azide (an inhibitor of electron transport). The kinetics of protrusion formation was monitored: during the first 30 s following the onset of aspiration, protrusion length linearly increased with respect to time, and this increase was enhanced by colchicine and cytochalasin D, whereas it was not significantly affected by azide. However, during the following 2 min, both azide and cytochalasin D reduced the rate of deformation. It is concluded that mechanically induced cell deformation involved two sequential phases with differential dependence on the cell metabolic status.

Materials and methods

Cells and media

The human monocytic THP-1 cell line (Tsuchiya et al. 1980) was kindly donated by Dr. F. Birg (INSERM U 119, Marseille). Cells actively ingested opsonized erythrocytes and displayed typical antigens of the mononuclear phagocyte lineage, including CD11b, CD18, CD35, and CD64, and non-polymorphic epitopes of class I histocompatibility molecules, as checked with flow cytometry (not shown). However, they did not adhere strongly to glass surfaces.They were maintained in RPMI 1640 medium (Flow Laboratories, McLean, Va) supplemented with 20 mM HEPES, 10% fetal calf serum (Flow), 2 mM L-glutamine, 50 U/ml penicillin and 50 μg/ml streptomycin. Aspiration was performed in saline supplemented with 20 mM HEPES, 5 mM KCl, 1 mM $MgCl_2$, 1 mM $CaCl_2$, 10 mM glucose and 1 mg/ml bovine albumin.

Cell labeling

Cytoplasmic labeling was performed as described previously (André et al. 1990) by incubating cells for 15 min at 37°C in culture medium supplemented with 40 µg/ml fluorescein diacetate (Sigma). Polymerized actin was revealed (Horoyan et al. 1990) by fixing cells on the tip of the micropipette for 20 min with 3.7% paraformaldehyde, and then labeling samples by another 20-min incubation in phosphate buffer containing 10 U/ml bodipy phallacidin (Molecular Probes, Eugene, Ore.) and 0.1 mg/ml lysophosphatidylcholine (Sigma).

Metabolic inhibitors

In some experiments, cells were preincubated for 30 min at 37°C in 10 µg/ml cytochalasin D (Sigma, St. Louis, Mo.) to inhibit microfilament assembly. Alternatively, microtubules were inhibited by 30 min incubation with 0.1 mM colchicine (Sigma). Cytochrome function was inhibited by 30 min incubation with 10 mM sodium azide.

Micropipettes

Micropipettes were obtained from borosilicate capillary tubes of 0.8 mm internal diameter (Clark Electromedical Instruments, provided by Phymep, Paris), using a programmable horizontal mipropipette puller (Campden Instruments, model 773). In some cases, the aperture was enlarged by breaking the pipette tip with a deFonbrune microforge (Alcatel, Paris). The inner diameter of the pipette tip was set at 5.5 µm, i.e. about one third of the average diameter of THP-1 cells.

Microscope and aspiration stage

The micropipette was connected to a pressure generator composed of a U-tube connected to a syringe mounted on a syringe holder. The pressure was measured with a sensor (ref. 78300-04, Cole Parmer instruments) connected to an analog input card (model PC-ADC 12B8V/D, Digimetrix, Perpignan, France) mounted on an IBM-compatible desk computer. This allowed real-time pressure determination with about 1 Pa resolution.

Aspiration was performed on the stage of an Olympus IMT2 inverted microscope bearing a high sensitivity SIT videocamera (Lhesa model 4036, Cergy Pontoise, France). The output was connected to a PCVision+ digitizer (Imaging Technology, Bedford, Mass.) mounted on the computer. The digitized image was subjected to reverse digital-to-analog conversion and displayed on a monitor with continuous video recording. This procedure allowed continuous superimposition of time and pressure on video images. A representative image is shown in Fig. 1. Also, when fluorescence labeling was performed, the amount of fluorescence in the cell body and protrusion was determined as previously described (Horoyan et al. 1990).

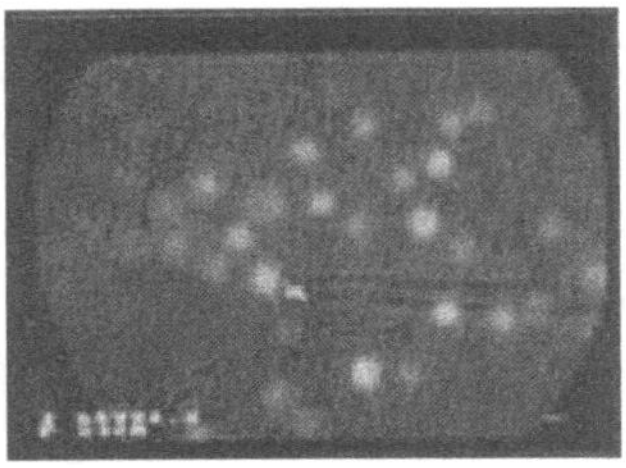

Fig. 1 Cell aspiration into a micropipette. Human monocytic THP-1 cells were labeled with Fluo-3, a fluorescent calcium probe. A cell was then aspirated into a micropipette with a pressure of about 98 Pa under continuous video recording. The microscope field was illuminated with a combination of visible and UV light. The protrusion is clearly visible (*arrow*). Time *T* is indicated (with 0.1 s accuracy) together with pressure *P* (with units of about 1.1 Pa). Bar length is 10 µm

Experimental procedure

In a typical experiment, a 300 µl aliquot of cell suspension ($4 10^5$/ml) was deposited on a glass coverslip ($20x60\ mm^2$) set on the microscope stage. After zero pressure determination, a series of typically 10-20 cells were aspirated with a negative pressure of about 100 Pa (1 cm H_2O) for 150 s each. The whole experiment was recorded for subsequent analysis. The kinetics of protrusion formation and elongation was then studied by following the motion of the protrusion with dedicated software evolved in the laboratory (Kaplanski et al. 1993). Pixel size was about 0.15 µm.

Data analysis

Model for extensive deformation

Following Evans and Kukan (1984), cells were considered as viscous liquid droplets (of viscosity µ) surrounded by an elastic membrane with constant tension *T*. Since some adhesion persisted at all protein concentrations tested (not shown), a fairly low protein concentration was used and a model for adhesion was elaborated as follows (Fig. 2):

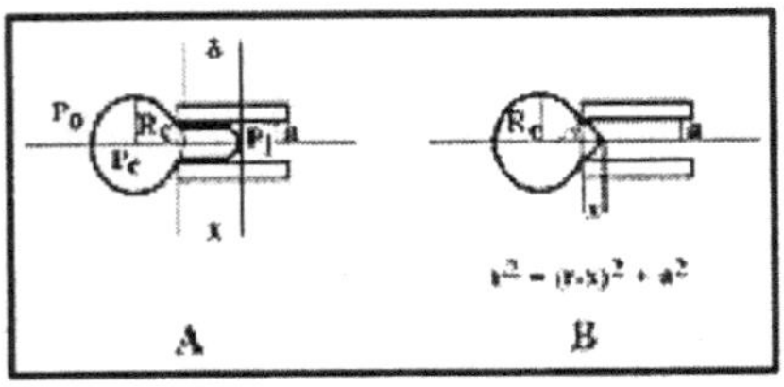

Fig. 2 Geometrical model for protrusion formation. Cells are modeled as spheres of constant radius R_c with a cylindrical protrusion with a hemispherical tip of radius *a* equal to the inner pipette radius (A). Parameter *x* is the distance between pipette entrance and protrusion tip. Parameter

[delta] is the distance between the protrusion tip and the cell boundary before deformation. The pressure is P_i within the pipette, P_c within the cell and P_o in the extracellular medium. **B** Cell response to subthreshold pressure: the protrusion is a spherical cap with radius r comprised between a and R_c

Since fairly large pipettes were used, the pressure dissipation within the cell cortex was deemed negligible, in accordance with Evans and Yeung's findings (1989). The pressure dissipation during cell deformation was thus written as the sum of two terms:

- The pressure drop upstream from the pipette mouth was approximated as the exact value calculated for a fluid flowing through a circular orifice bounded by a no-slip plane (Happel and Brenner 1991 – see Fig. 6 of Yeung and Evans, 1989), yielding:

(1) $$P_1 = 3Q\mu/a^3$$

where Q is the flow rate and a is the pipette radius.

- Then, the pressure decrease within the protrusion was calculated assuming zero velocity near the pipette wall. Thus, following Poiseuille's law, we used the following formula:

(2) $$P_2 = 8\mu LQ/[\pi]a4$$

where L is the protrusion length (Fig. 2A). By combining Eqs. (1) and (2), and replacing Q with $(dL/dt)/[\pi]a2$, The following analytical formula was obtained:

(3) $$L(t) = (-3[\pi]a/8 + (9[\pi]^2a^2/64 + a^2Pt/\mu)^{(1/2)})/2$$

Laplace`s law was used to relate the driving pressure P to the pressure difference P_0-P_i between the external medium and the pipette interior (Fig. 2A):

(4) $$P = (P_0-P_i) - 2T\,(1/a - 1/R_c)$$

where R_c is the cell radius (this is fairly constant during aspiration under our experimental conditions).The assumption of constant membrane tension T may be reconciled withassuming that there is no relative displacement between the cell surface and the pipette wall if anchoring is mediated by transmembrane proteins directly connected to the protrusion core.

Direct determination of membrane tension

We attempted to achieve direct determination of the membrane tension by subjecting aspirated cells to small sequential pressure increases (of about 20 Pa each) in order to determine the minimal pressure required to generate a fairly spherical protrusion. As shown on Fig. 2B, the displacement [delta] of the protrusion tip generated by indefinitely slow aspiration with a low pressure P_o-P_i is given by the following equations:

$$r = 1/(1/R_c + (P_o - P_i)/2T) \qquad (5)$$

(this is Laplace`s law relating the radius of curvature of the protrusion tip to sucking pressure)

$$[\text{delta}] = r - (r^2 - a^2) - R_{c+}\, x - (R_c^{\,2} - a^2)^{1/2} \qquad (6)$$

(as obtained with elementary geometrical calculations)

Thus, the membrane tension T could be obtained by numerical fit of the variations of [delta] versus P_i with a theoretical curve derived from Eqs. (5) and (6).

Use of relaxation experiments to derive the ratio between surface tension and core viscosity

A general argument supporting Evans' model is the general finding that micropipette-aspirated cells recover their spherical shape when they are released from the pipette. Further, Tran-Son-Tay et al. (1991) provided a simple numerical equation allowing easy determination of the ratio T/μ between cell tension and cytoplasmic viscosity by plotting the variations of the largest (L) and smallest (W) cell diameter versus non-dimensional time $t_{nd} = tT/\mu R_c$, where t is time. Thus, some cells were deformed into an elongated sausage shape for about 10 s, then released under video recording to determine the kinetics of shape recovery.

Results

Relaxation studies

In a first series of experiments, cells were aspirated into large micropipettes in order to induce a sausage-like shape. They were released after a 10-20 s delay, under continuous video recording. First, it was checked that cells recovered a spherical shape, in accordance with the viscous droplet model. Second, as shown in Fig. 3, the experimental plot of variation of the largest and smallest cell diameter versus time was used to derive the ratio between membrane tension (T) and interior viscosity (μ), as described by Tran-Son-Tay et al. (1991). This yielded a best fit value of 0.088 μm/s.

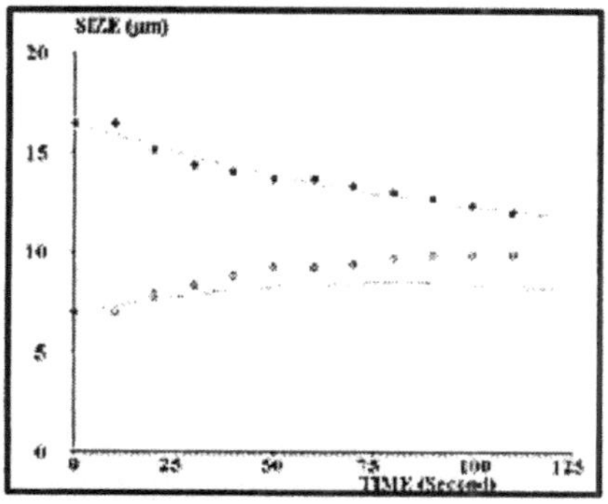

Fig. 3 Morphological recovery after deformation. A THP-1 cell was aspirated into a pipette until it acquired a sausage shape with length L and width W. After a 10-s delay, the cell was expelled and parameters L (*closed circles*) and W (*open circles*) were measured at regular intervals. The *dotted line* represents a theoretical curve calculated with a numerical formula provided by Tran-Son-Tay et al. (1991). The ratio between tension and viscosity is 8.810^{-8} m/s

Cell deformation is consistent with a model assuming constant membrane tension and cytoplasmic viscosity together with cell-to-glass adhesion

Twenty-two control cells (mean radius: 8.6 μm) were aspirated under constant pressure. The time dependence of protrusion length is shown in Fig. 4. Equations (3) and (4) were then used together with the experimental value of T/μ: the viscosity was derived by least-square error minimization. Clearly, the fitted theoretical curve displayed satisfactory agreement with experimental data. Note also that deformation remained moderate, since the protrusion did not include more than about 10% of the total cell volume, thus making it possible to disregard the variations in the radius of the spherical part of these cells.

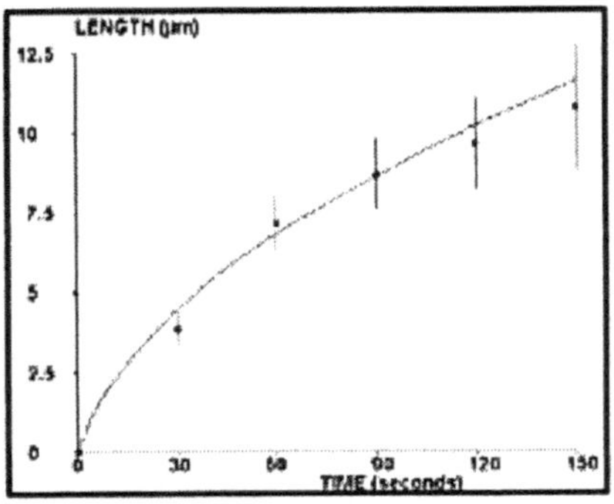

Fig. 4 Kinetics of protrusion formation in a micropipette. Twenty-two THP-1 cells were aspirated into a micropipette with a pressure of 98 Pa. Average values of protrusion length at time 30, 60, 90, 120 and 150 s are shown (length of vertical bar is twice the standard error) together with fitted theoretical curve obtained by fitting the viscosity and using the tension/viscosity ratio derived from relaxation studies

The above procedure yielded numerical estimates of 162 Pa.s and 0.0142 mN/m for the cytoplasmic viscosity and membrane tension respectively.

Direct determination of membrane tension

Seven cells were aspirated with sequential pressure increases small enough (about 30 Pa) that no extensive protrusion could appear. Also, pressure increases were separated by time intervals of several tens of s in order that balanced deformation might be achieved. Experimental data are shown in Fig. 5 together with fitted theoretical curves, yielding a membrane tension of 0.0914 mN/m.

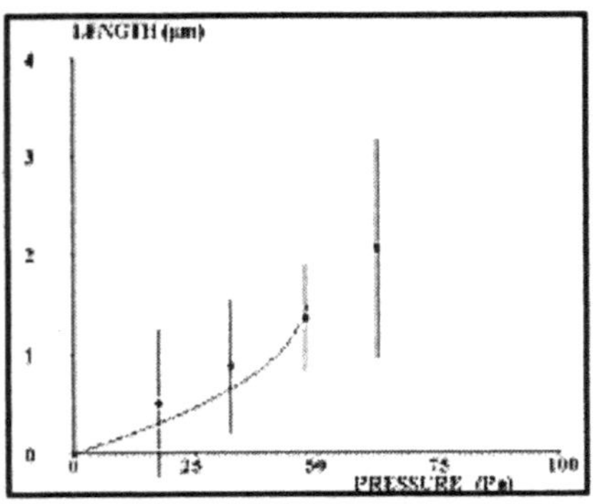

Fig. 5 Initial cell deformation induced by subthreshold pressure. Seven THP-1 cells were aspirated with increases in pressure that were too small to induce the formation of a long protrusion. The displacement of the cell boundary was determined 20-30 s after a pressure change. Mean displacements are shown (length fo vertical bar is twice the standard error) together with a theoretical curve calculated by modeling the cell surface as an elastic membrane with constant tension

Effect of metabolic inhibitors

Although a two-parameter model did not quantitatively account for all aspects of THP-1 cell behavior during aspiration, we obtained a qualitative fit between experimental sets of data and theoretical curves (Fig. 3-5), making cells appear as passive objects. It was of interest to know how metabolic inhibitors might alter experimental parameters. Therefore, aspiration experiments were repeated on cells treated with inhibitors of cytoskeletal assembly or energy production. Results are summarized in Table 1. A striking conclusion was that these inhibitors differentially affected cell behavior at different periods of time following the onset of aspiration. Indeed, during the first 30 s, protrusion elongation rate was not affected by azide but it was significantly ($P<0.001$) and substantially (about twofold) increased by colchicine or cytochalasin D. However, during the following 120 s, the deformation of azide- and cytochalasin-treated cells was nearly blocked, while the length of the protrusion in control or colchicine-treated cells increased approximately threefold, strongly suggesting that different mechanisms were involved in the early and late phase of cell deformation.

Table 1 Influence of metabolic inhibitors on cell deformation. Monocytic THP-1 cells were aspirated into micropipettes with or without being treated with metabolic inhibitors and the length of induced protrusions was measured at regular intervals. Average values obtained from a number (n) of cells (ranging between 20 and 28) are shown with the standard error of the mean (in parentheses). The significance of the effect of a given treatment was calculated using Student's t-test

	Protrusion length (μm)				
Time (s)	30	60	90	120	150
Control (n=22)	3.87 (0.51)	7.17 (0.85)	8.70(1.12)	9.67(1.41)	10.78(1.90)
Cytochalasin D (n=28)	7.26(0.47) $P<0.001$	6.96(0.33)	7.70(0.41)	7.73(0.58)	7.69(0.38)
Colchicine (n=24)	6.30(0.23) $P<0.001$	8.01(0.34)	9.49(0.51)	10.08(0.88)	10.21(0.52)
Azide (n=20)	3.28 (0.31)	3.82(0.29) $P<0.001$	3.88(0.33) $P<0.001$)	ND	ND

Cytoskeletal behavior during cell deformation

It was important to know whether cytoskeletal components remained bound to the cell membrane during deformation. This question was addressed quantitatively by comparing the distribution of cytoplasmic and cytoskeletal markers in micropipette-aspirated cells. First, cells were labeled with fluorescein diacetate and aspirated for 3 min with a pressure of 100 Pa. They were then studied with fluorescence microscopy and image analysis. The mean ratio between the total fluorescence of the cell body and the protrusion was 2.41 ± 0.10 (standard error of the mean, n = 27 cells). Polymerized actin was then labeled on individual cells. The mean ratio between the total fluorescence of the cell body and protrusion was 2.52 ± 0.22 (standard error of the mean, $n = 14$). Therefore, the density of polymerized actin was comparable in the cell body and the protrusion. However, when individual micrographs were observed, very limited fluorescence accumulation was sometimes apparent in the protrusion. A typical image is displayed on Fig. 6.

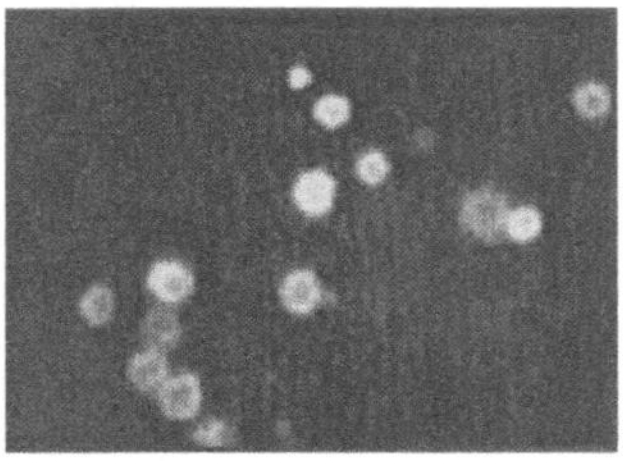

Fig. 6 Microfilament behavior during cell deformation. Fourteen THP-1 cells were aspirated for 3 min in a micropipette with a pressure of 100 Pa. They were fixed and stained with bodipy phallacidin, and then examined with fluorescence microscopy. A typical image is shown. The fluorescence is fairly homogeneous in the protrusion, but a bright fluorescent point is visible on the tip

Discussion

The aim of this work was to characterize the behavior of THP-1 cells subjected to micropipette aspiration and to analyze the effect of metabolic inhibitors on these properties.

Three main procedures were used to study the mechanical properties of cells. We studied (1) the cell relaxation properties after at least 10 s deformation in a micropipette, (2) the kinetics of extensive deformation, (3) the small deformation induced by *prolonged* exposure to weak forces. These conditions were devised to avoid the influence of short-term elastic behavior (Schmid-Schönbein et al. 1981; Tran-Son-Tay et al. 1991), thus allowing safer use of Evans' model of a viscous liquid droplet surrounded by a tensile membrane. Qualitative agreement was found between experimental data and theoretical predictions (Figs. 3-5) and the experimental estimates of 0.014 mN/m and 162 Pa.s for tension and viscosity compare well to the neutrophil values (i.e. 0.035 mN/m and 200 Pa.s, Evans and Kukan 1984; Evans and Yeung 1989; or 0.024 mN/m and 150 Pa.s, Needham and Hochmuth 1992; Tran-Son-Tay et al. 1991). Comparable values were also reported for the cytoplasmic viscosity of HL-60 promyelocytic cells (Tsai et al. 1996). The quantitative discrepancy between the tension values determined with large and small deformations might also be ascribed to either non-newtonian behavior of the cell interior (Tsai et al., 1993) or substantial cortex elasticity (Schmid-Schönbein et al. 1981). However, the main problem with our results might be the evidence of significant cell-to-glass adhesion. The qualitative agreement between theoretical and experimental data (Fig. 4) is consistent with the possibility that adhesion may be mediated by a population of cell membrane proteins likely to anchor the submembranous matrix to the glass with sufficient bilayer freedom to ensure that tension would be similar around the cell body and near the protrusion tip. This would warrant our modeling including constant surface tension combined with Poiseuille flow within the protrusion; however, more experiments are required to clarify this point. In any case, it must be emphasized that only the simultaneous use of several complementary methods of quantifying cell deformability might help to clarify the aforementioned problems.

Experiments done with metabolic inhibitors suggest somewhat unexpected conclusions: comparison between experimental and theoretical data shown in Fig. 4 was consistent with the view that cells behave as passive bodies with constant material parameters during aspiration. This interpretation was also consistent with the finding that azide, an inhibitor of energy production, did not affect protrusion elongation rate during the first 30 secondes following the onset of aspiration. The higher deformability of cytochalasin-treated cells might be ascribed to a weakening of cytoskeletal elements, in accordance with other experimental approaches (Petersen et al. 1982; Wang et al. 1993). It is not surprising that azide decreased cell deformability during the second phase of aspiration. Indeed, azide was previously reported to decrease macrophage spreading in vitro (Rabinovitch and Destefano 1974), and this inhibition was antagonized by glucose, suggesting that it might be related to a decrease of energy production. Further, the deformability of blood polymorphonuclear leukocytes was decreased by glycolytic inhibitors (Miller and Myers 1975).

However, the clearcut finding that cytochalasin also decreased cell deformability during the second phase of aspiration was somewhat unexpected. The simplest interpretation of these findings would be that some active cell participation might be required to allow cells to display a substantial level of deformation. Thus, THP-1 cells might be designed to yield to exogenous mechanical forces, which might obviously facilitate their passage through small capillary vessels. An attractive possibility would be that this yielding might be triggered by a rise in intracellular calcium, since such an increase has been described in different populations of myeloid cells (Hoyoyan et al. 1990; Zaffran et al., 1993), including THP-1 cells (not shown) subjected to micropipette aspiration. However, more experiments are needed to clarify this point, since no direct link could be demonstrated between intracellular calcium changes and mechanical responses. It is planned to address this problem by studying the effect of metabolic inhibitors and intracellular calcium manipulation on phenomenological membrane tension and cytoplasmic viscosity, as determined by aspiration and relaxation studies.

References

André, P., Capo, C., Benoliel, A. M., Buferne, M. and Bongrand, P. (1990) Analysis of the topological changes induced on cells exposed to adhesive or mechanical stimuli. Cell Biophys. 16:13-34

Bagge, U. and Branemark, P. I. (1977) White blood cell rheology - an intravital study in man. Adv. Microcirc. 7:1-17

Erickson, C. A. (1980) The deformability of BHK cells and polyoma virus-transformed BHK cells in relation to locomotory behaviour. J. Cell Sci. 44:187-200

Evans, E. A. and Kukan, B. (1984) Passive material behavior of granulocytes based on large deformation and recovery after deformation tests. Blood 64:1028-1035

Evans, E. A. and Yeung, A. (1989) Apparent viscosity and cortical tension of blood granulocytes determined by micropipet aspiration. Biophys. J. 56:151-160

Happel, J. and Brenner, H. (1991) Low Reynolds number hydrodynamics, 2nd edn, Kluwer, Dordrecht, p 153

Horoyan, M., Benoliel, A. M., Capo, C. and Bongrand, P. (1990) Localization of calcium and microfilament changes in mechanically stressed cells. Cell Biophys. 17, 243-256.

Kaplanski, G., Farnarier, C., Tissot, O., Pierres, A., Benoliel, A. M., Alessi, M. C., Kaplanski, S. and Bongrand, P. (1993) Granulocyte-endothelium initial adhesion: analysis of transient binding events mediated by E-selectin in a laminar shear flow. Biophys. J. 64:1922-1933

Kawaoka, E. J., Miller, M. E. and Cheung, A. T. W. (1981) Chemotactic factor-induced effects upon deformability of human polymorphonuclear leukocytes. J. Clin. Immunol. 1:41-44

Landau, J. V., Zimmerman, A. M. and Marsland, D. A. (1954) Temperature-pressure experiments on amoeba proteus ; plasmagel structure in relation to form and movement. J. Cell Comp. Physiol. 44:211-232

Marchesi, V. T. (1961) The site of leucocyte emigration during inflammation. Quart. J. Exp. Physiol. 46:115-133

Mège, J. L., Capo, C., Benoliel, A. M., Foa, C. and Bongrand, P. (1985) Study of cell deformability by a simple method. J. Immunol. Methods 82:3-15

Mège, J. L., Capo, C., Benoliel, A. M. and Bongrand, P. (1987) Use of cell contour analysis to evaluate the affinity between macrophages and glutaraldehyde-treated erythrocytes. Biophys. J. 52:177-186

Miller, M. E. and Myers, K. A. (1975) Cellular deformability of the human peripheral blood polymorphonuclear leukocyte: method of study, normal variation and effects of physical and chemical alterations. J. Reticuloendothelial Soc. 18:337-345

Mitchison, J. M. and Swann, M. M. (1954) The mechanical properties of the cell surface. I. The cell elastimeter. J. Exp. Biol. 31:447-459

Needham, D. and Hochmuth, R. M. (1992) A sensitive measure of surface stress in the resting neutrophil. Biophys. J. 61:1664-1670

Petersen, N. O., McConnaughey, W. B. and Elson, E. L. (1982) Dependence of locally measured cellular deformability on position of the cell, temperature and cytochalasin B. Proc. Natl. Acad. Sci. (USA) 79:5327-5331

Rabinovitch, M. and Destefano, M. J. (1974) Macrophage spreading in vitro. III. The effect of metabolic inhibitors, anesthetics and other drugs on spreading induced by subtilisin. Exp. Cell Res. 88:153-162

Schmid-Schönbein, G. W., Sung, K. L. P., Tözeren, H., Skalak, R. & Chien, S. (1981) Passive mechanical properties of human leukocytes. Biophys. J. 36:243-256

Sung, K. L. P., Schmid-Schönbein, G. W., Skalak, R., Schuessler, B. B., Usami, S. and Chien, S. (1982) Influence of physicochemical factors on rheology of human neutrophils. Biophys. J. 39:101-106

Tickle, C. and Trinkaus, J. P. (1977) Some clues as to the formation of protrusions by Fundulus deep cells. J. Cell Sci. 26:139-150

Tran-Son-Tay, R., Needham, D., Yeung, A. and Hochmuth, R. M. (1991) Time-dependent recovery of passive neutrophils after large deformation. Biophys. J. 60:856-866

Tsai, M. A., Frank, R. S. and Waugh, R. E. (1993) Pasive mechanical behavior of human neutrophils: power law fluid. Biophys. J. 65:2078-2088

Tsai, M. A., Frank, R. S. and Waugh, R. E. (1994) Passive mechanical behavior of human neutrophils: effects of cytochalasin B. Biophys. J. 66:2166-2172

Tsai, M. A., Waugh, R. E. and Keng, P. C. (1996) Cell cycle dependence of HL-60 cell deformability. Biophys. J. 70:2023-2029

Tsuchiya, S., Yamabe, M., Yamaguchi, Y., Kobayashi, Y., Konno, T. and Tada, K. (1980) Establishment and characterization of a human acute monocytic leukemia cell line (THP-1). Int. J. Cancer 26:171-176

Valberg, P. A. and Albertini, D. F. (1985) Cytoplasmic motions, rheology, and structure probed by a novel magnetic method. J. Cell Biol. 101:130-140

Wang, N., Butler, J. P. and Ingber, D. E. (1993) Mechanotransduction across the cell surface and through the cytoskeleton. Science 260:1124-1127

Worthen, G. S., Schwab III, B., Elson, E. L. and Downew, G. P. (1989) Mechanics of stimulated neutrophils : cell stiffening induces retention in capillaries. Science 245:183-186

Yeung, A. and Evans, E. A. (1989) Cortical shell-liquid core model for passive flow of liquid-like spherical cells into micropipets. Biophys. J. 56:139-149

Zaffran, Y., Lépidi, H., Benoliel, A. M., Capo, C. and Bongrand, P. (1993) Role of calcium in the shape control of human granulocytes. Blood Cells 19:115-131

Zigmond, S. H. (1978) chemotaxis by polymorphonuclear leukocytes. J. Cell Biol. 77:269-287

Exp. Biol. Online 2:6 (1997)

Three-dimensional organization of actin cytoskeleton and podosomal contact structures in neoplastic cells in vitro

Pavel Vesely[1], Luboslava Pavlikova[1], Jiri Plachy[1], Katerina Trejbalova[1], Jiri Hejnar[1], Eva Matouskova[1], Jürgen Bereiter-Hahn[2]

[1]Institute of Molecular Genetics, Academy of Sciences of the Czech Republic, Prague, Czech Republic
[2]Cinematic Cell Research Group, Biocentre, Johann Wolfgang Goethe University, Frankfurt am Main, Germany

Correspondence to: P. Vesely (email: pvy@img.cas.cz)

Received: 12 March 1996 / Accepted: 1 April 1997
Published: April 22, 1997

Abstract. In spontaneously metastasizing rat RPS sarcoma cells, a 3D structure of oblique F-actin cables was observed which was associated with active cell migration in vitro. This led us to further comparative investigations of several other neoplastic and normal cell populations in vitro for F-actin structures using confocal laser scanning microscopy (CLSM). Various forms of F-actin cytoskeleton were observed and the incidence of podosome-related contact structures appeared to be associated with malignancy, interpreted as metastatic capacity.

Key words: Actin cytoskeleton – Neoplastic cells – Podosomal contact structures

Introduction

Migration of malignant cells is considered to be one of the most important mechanisms of metastasis formation from primary solid tumours (Liotta et al., 1991; Ruenger et al., 1994). We have been studying neoplastic cell lines derived from spontaneously metastasizing RPS sarcoma in LEW/CUB inbred rats. Our focus has been the locomotory response of rat sarcoma cells in vitro to changes of extracellular pH to 6.5 designed to simulate in vivo conditions. Cells of highly metastasizing lines showed increased persistence of locomotion (Vesely et al., 1987a, 1989a) and spread in newly arising colonies in slightly acid medium (Vesely et al., 1989b; Matouskova et al., 1990; Zicha et al., 1993). The 3D confocal analysis of actin cytoskeleton in three RPS sarcoma cell lines differing in malignancy revealed a persistence of F-actin bundles in two-thirds of cells of the most malignant line. The F-actin bundles were organized into a 3D structure of oblique cables. The structure was found in highly motile cells while in poorly motile cells this structure was absent (Pokorna et al., 1994).

These observations revealed an interesting problem relating to the role of the 3D structure of F-actin in the locomotion of sarcoma cells. In the present study we used rhodamine-phalloidin staining and high-resolution confocal fluorescence microscopy for detailed, comparative 3D analysis of F-actin-related structures in several normal and neoplastic cell populations under standard and stress conditions in vitro. As the 3D organization of the actin cytoskeleton, originally revealed with stereo immunofluorescence microscopy (Osborn et al., 1978), can be seen in greater detail with confocal laser scanning microscopy (CLSM), a set of 3D configurations of F-actin cytoskeleton was examined.

The cell-to-substratum contacts imaged with staining for F-actin were dot-like structures. They are found regularly in chicken, rat and hamster Rous sarcoma cells and were called podosomes by Tarone et al. (1985), rosettes or invadopodia by Chen (1989), and point-contacts by Nermut et al. (1991). These dot-like contact structures (podosomes or invadopodia) were shown to aggregate into crescent- or ring-shaped adhesion sites called rosettes of podosomes (Gavazzi et al., 1989), thus resembling the full cycle of contact formation in resorbing osteoclasts (Vaananen and Horton, 1995). Since there is no name for the resorption organelle of the osteoclast, we decided to designate the range of forms of the cell-to-substratum contacts, as defined by the rhodamine-phalloidin-stained F-actin in normal and neoplastic cells, as podosomal contact structures (PCSs).

In MP4 cells, obtained from the metastasis of a v-*src*-induced chicken tumour (Svoboda et al, 1992), the PCSs were expressed constitutively; however, in malignant rat RPS sarcoma cells without v-*src* the PCSs were expressed very rarely, mostly only under stress conditions. All cells with the PCSs were similar in respect of an absence or alterations of the F-actin bundles.

Materials and methods

Cells

Rat normal cells. Fibroblasts were obtained as a primary culture on coverslips from fragments of skin and spleen from outbred Wistar newborn rats.

Rat sarcoma cells of the RRS family. K2: a clone of fibroblast-like cells which is isolated from neoplastic, spontaneously in-vitro-transformed rat embryo fibroblasts from the Lewis strain of inbred rats, which yields fibrosarcomas with a very low incidence of spontaneous metastasis upon subcutaneous inoculation (Vesely and Weiss, 1973; Vesely et al., 1987a).T15: a clone of spindle-shaped cells, which was selected from K2 cells, with anintermediate incidence of spontaneous metastasis (Vesely et al., 1987a). A870N: a cell line of highly metastatic spindle-shaped cells, which was selected from K2 in a different way than the T15 cells and later after their isolation (Pokorna et al., 1994). A72N: a cell line of highly metastatic (around 80% incidence of spontaneous metastasis from subcutaneous tumour to lungs, data not shown) spindle-shaped cells, which was isolated from the 10th passage of the RPS sarcoma in vivo.

Mouse cells. J774.1 mouse malignant macrophages, originally isolated from a murine reticulum cell sarcoma in a female BALB/c/NIH mouse (Ralph et al., 1975), were kindly provided by Dr. G. Griffiths, EMBL, Heidelberg, Germany.

Rous sarcoma virus (RSV) transformants. K4: a clone of rat Rous sarcoma, which was obtained after the second in vitro transformation with RSV of the spontaneously transformed fibroblasts LW13 from the Lewis rat (Vesely and Weiss, 1973). BLEF-RSV: Primary Brown Leghorn chicken fibroblasts which were freshly transformed with RSV Prague strain C. P 10, P22: chicken neoplastic cells which were isolated from transplantable tumour No. 9692 induced in chickens in vivo by transfection with LTR, v-*src*, LTR provirus (Svoboda et al., 1992). MP4: neoplastic cells which were isolated from a metastasis in the lungs shed by the tumour No. 9692.

Human cells. Human keratinocytes were cultured on a feeder layer of irradiated mouse 3T3 cells in a standard medium (Matouskova et al., 1993). Primary cells of human breast tumours in a standard medium were obtained by using the same feeder layer technique with irradiated mouse 3T3 cells as for human keratinocytes (Matouskova et al., 1993). Benign phylloides tumour of the breast and intracanalicular fibroadenoma of the breast were examined.

Culture conditions

Standard medium. Eagle's MEM with Hanks' BSS supplemented with all non-essential amino acids and 10% calf serum was used for all rat cells and mouse J774.1 cells. Brown Leghorn fibroblasts freshly transformed with RSV-P were grown on Earle's MEM supplemented with all non-essential amino acids and 10% calf serum. Chicken P10, P22 and MP4 neoplastic cells were grown on Iscove's modified Dulbecco's medium (Sigma I-7833) with 5% fetal calf serum and 5% calf serum.Human keratinocytes and breast tumour cells were grown on Eagle's MEM with Hanks' BSS supplemented with all non-essential amino acids, 120 IU1000 ml insulin, hydrocortisone, cholera toxin, epidermal growth factor (EGF) and 10% calf serum (Matouskova et al., 1993).

HBSS Stress. A 2-h incubation in Hanks BSS, set to neutral pH 7.1 with phosphate buffer, after three initial exchanges of the HBSS was used as the nutritional stress. It was followed by fixation.

Microscopy

Confocal laser scanning microscopy (CLSM). A Leica CLSM TCS 4D True Confocal Scanner (image 512x512 pixels) with Leitz upright DMRBE or inverted DMIRB microscopes (Leica, Wetzlar, Germany) were used with Planapo optics 63x, NA 1.4 (at zoom 1x picture width 161 µm) or 100x, NA 1.3 (at zoom 1x picture width 102 µm). A Bio-Rad CLSM MRC 600 Confocal Scanner (image 768x512 pixels) with Nikon Diaphot 200 with Planapo optics 60x, NA 1.4 (at zoom 1x picture width 211x141 µm) or 100x, NA 1.4 (at zoom 1x picture width 127x85 µm). In the figures presented here either a single optical section was used or the composite image, which was obtained by integrating all essential features from a specified number of sections of the 3D stack into one 2D image, thus simulating to some extent the overall image given by an ordinary incident light fluorescence microscope.

Fluorescence staining. Fixation was carried out for: 20 min at room temperature in 4% formaldehyde (Merck) in Hanks BSS pH 7.2 followed by a 5-min incubation in 100 mM glycine in PBS. Then, 20 µl of 2 µM solution of phalloidin-TRITC (Sigma P-1951) in PBS with 20% methanol was added for 20 min. After 5 min of rhodamine-phalloidin staining, 50 µl of 10 µM DiOC6 (3,3'-dihexyloxacarbocyanine iodide, Sigma D-3652) in PBS was added for the remaining 15 min for negative counterstaining of the nucleus. The specimens were mounted in Gelvatol (suggested and kindly provided by Dr. Barbara Safiejko-Mroczka, Norman, Okla., USA).

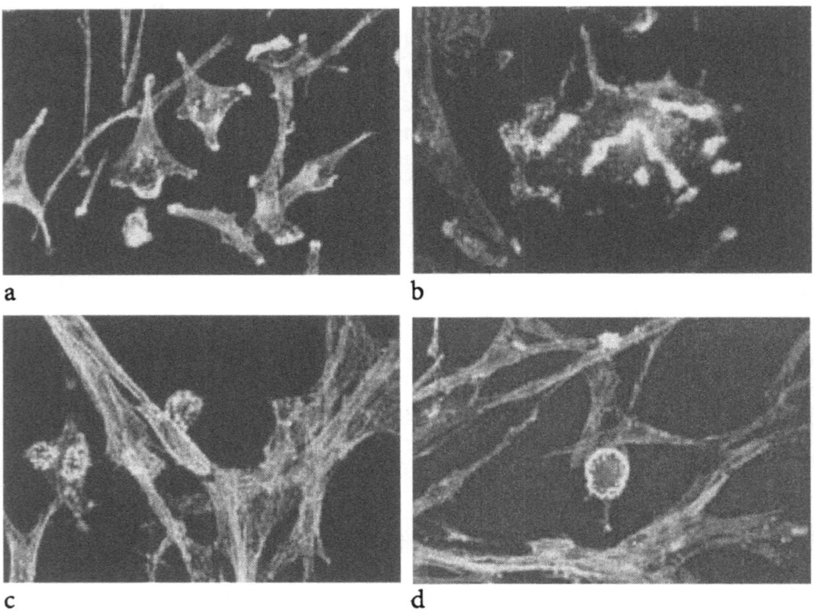

Fig. 1a–d Primary fibroblast-like cell from fragment of rat spleen in standard medium. Picture

width in a–c 102 µm, d 33 µm, Leica 63x, 1.4 NA, zoom in a–c 1.37x, d 4.91, z-step 0.34 µm. a Composite image; b ventral side at the level of cell-to-substratum contacts under nucleus; c dorsal surface 0.7 µm above the level of contacts; d detail of a prominent ridge on the top of dorsal cell surface; composite image (a) shows an ordinary picture of stress fibres and rosette-like bright dots in the leading lamella. Optical sections show ventral side (b) with circular organization of F-actin bundles, dorsal surface (c) with an apparent dehiscence in the cage of F-actin bundles opened at ventral side, in d the dehiscence is seen to be closed with an elevated ridge formed by F-actin cables

Results

Actin cytoskeleton (AC)

Spatial configuration of the AC can be observed and analysed in greater detail with the CLSM than was originally possible with stereo immunofluorescence microscopy; for example, the study of Osborn et al. (1978). Using the zoom facility, such details as a prominent ridge on the top of the dorsal surface of the cell can be identified (Fig. 1d). The targeted horizontal optical sectioning was chosen because it was considered suitable for assessing cell height up to a distance of approx. 10 µm from the coverslip and for proper selection of the number of levels, thus limiting photobleaching. Rhodamine-phalloidin staining of F-actin was combined with DiOC6 counterstaining of the mitochondria and endoplasmic reticulum. In ordinary incident light fluorescence microscopy it was useful for locating the negatively stained nucleus and for verification of the focus level inside the cell.

With this approach normal and neoplastic cells of mesenchymal and epithelial origin were compared and three basic 3D configurations of the AC were identified: (1) the classic stress fibres; (2) a cage (closed or partly opened); and (3) an absence of F-actin bundles (partial or total).

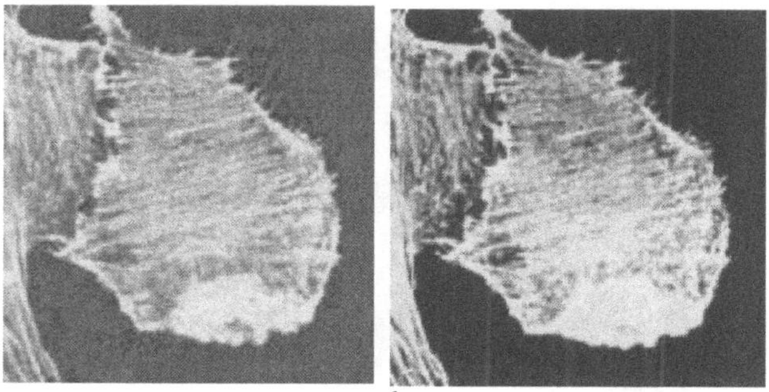

a b

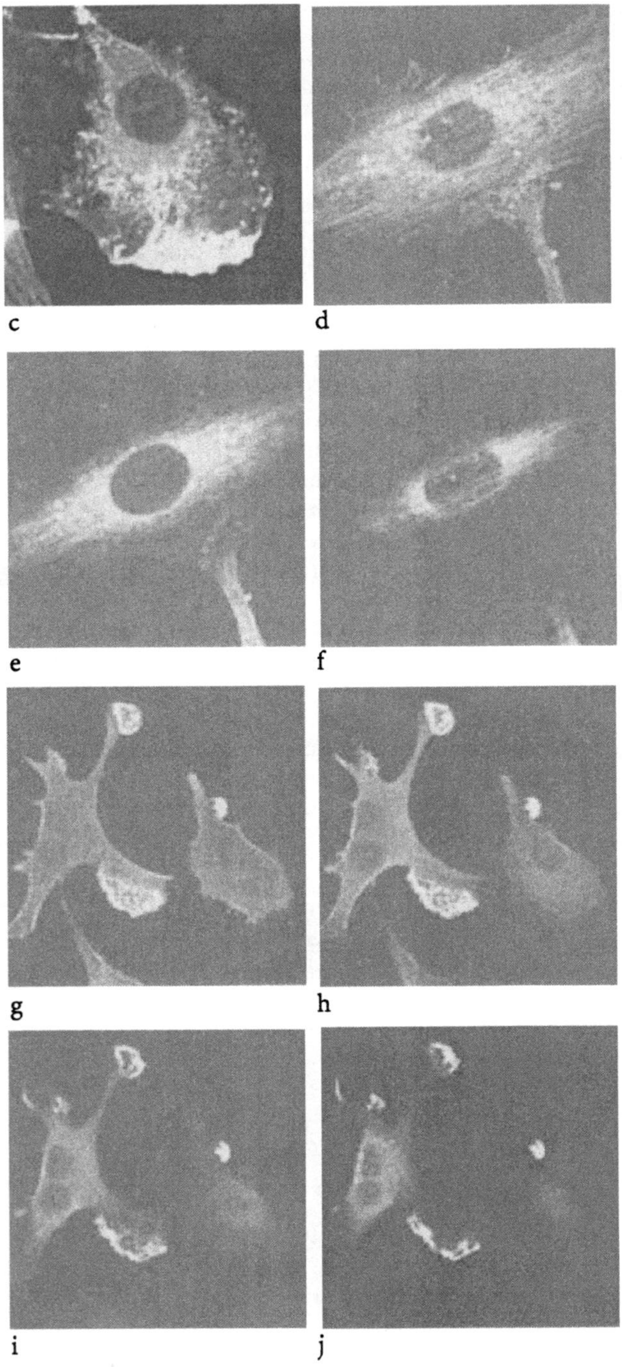

Fig. 2a–j Rat sarcoma T15 cells. **a–c** Picture width 57 μm, Leica 100x 1.3 NA, zoom 1.79, z-step 0.3 μm; **d–f** picture width 56 μm, Leica 63x 1.4 NA, zoom 2.7, z-step 0.36 μm; **g–j** picture width 120 μm,

Leica 63x 1.4 NA, zoom 1.34, z-step 0.5 μm; **a–c** T15 cell in standard medium with transverse (to the direction of cell movement) stress fibres in (**a, b**) sections under the nucleus and only F-actin in microvilli in **c** at the level of nucleus 1 μm above the ventral sides; **d–f** T15 cell in standard medium showing in three optical sections, the ventral side (**d**), nuclear level (**e**) and top dorsal surface (**f**) (0.7 μm apart from each other), the cage built of bundles of F-actin and opened under the nucleus; **g–j** T15 cells in the HBSS stress. *Left cell* which is almost devoid of recognizable bundles of F-actin shows clearly two PCSs in two lowest optical sections up to 1 μm height, their partial replacement by the image of F-actin decorated ruffles at 1.5 μm height (**i**) and image of ruffles only at 2 μm height (**j**)

Types of 3D configurations of the AC in standard medium

Classic stress fibres. Classic stress fibres in the form of a plate, made of parallel F-actin bundles on the ventral side of a cell close to cell surface under the nucleus, is not the most frequent type observed even in normal cells. They can also be found in sarcoma cells, such as locomoting T15 cells, as shown in Fig. 2. Figure 2a,b shows the transverse parallel orientation of stress fibres in respect of the leading edge of the cell while, in Fig. 2c, sectioned across the nucleus, there are only microvilli and no actin bundles on the dorsal cell surface.

Cage. The cage was found to be a common type of the AC. It is made of bundles of F-actin in the cortical layer of the cell, enveloping either fully or partly the whole cell due to dorsal or ventral, or both, "openings" in the cage. There were no apparent differences in the incidence of the various forms of the cage that could be considered to be induced simply by the culture conditions.

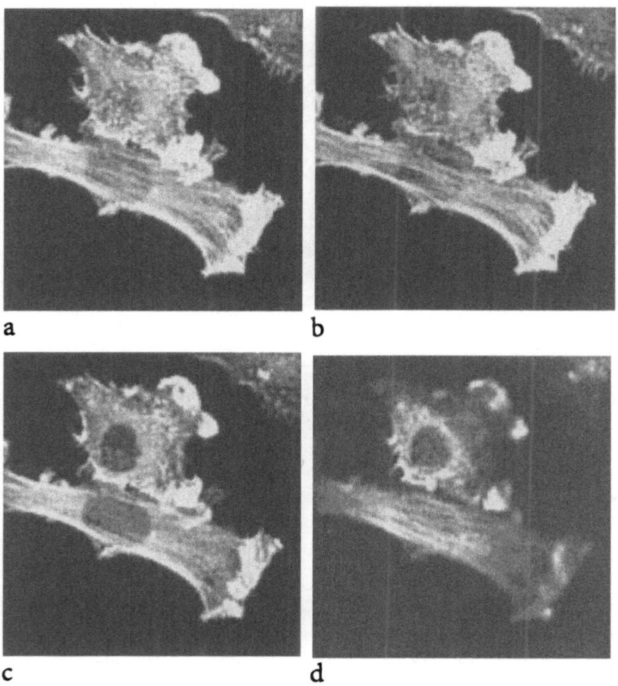

Fig. 3a–c Rat sarcoma A72N cells in standard medium. Picture width 72 m, Leica 63x, 1.4 NA, zoom

2.24x, z-step 0.71 μm. In a composite picture (**a**) the upper polygonal cell has no recognizable F-actin bundles, while the long spindle shaped cell in three optical sections (at 0.71 μm distance apart) demonstrates F-actin bundles below (**b**), at (**c**) and above (**d**) the level of nucleus. The bundles seal the closed cage structure at the sides

Closed cage. The closed cage can be found in normal fibroblast-like cells as well as in sarcoma cells of polarized shape. In Fig. 3a, the composite view shows two A72N sarcoma cells. The upper non-polarized cell shows diffuse staining for F-actin, which indicates a loose meshwork of microfilaments, while the bottom, longitudinally polarized, cell has the complete cage shown under the nucleus in Fig. 3b, at the level of the nucleus in Fig. 3c and above the nucleus in Fig. 3d.

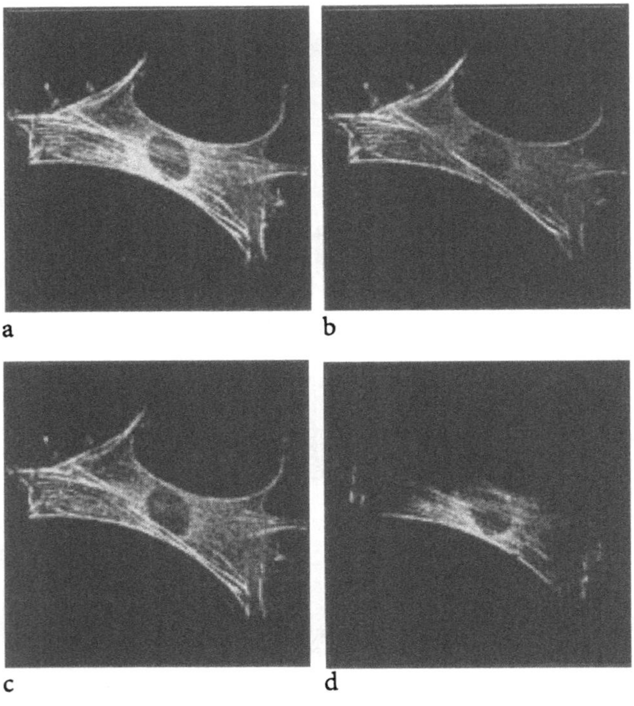

Fig. 4a–d Primary fibroblast from fragment of rat skin in standard medium. Picture width 102 μm, Leica 63x, 1.4 NA, zoom 1.53x, z-step 0.24 μm.**a** Composite image; **b** ventral side at the level of cell-to-substratum contacts under nucleus; **c** level of nucleus 0.24 μm above the level of contacts; **d** top of dorsal cell surface 1 μm above the level of contacts.Composite image (**a**) shows ordinary structure of stress fibres. Three subsequent optical sections (**b–d**) show a 3D structure of F-actin bundles forming a cage opened (i.e. actin bundles free) at the ventral side

Open cage. The open cage, particularly the one with an opening on the ventral side of the cell, appeared to be the most frequent type of AC. Figure 1 shows an example in a primary rat fibroblast. In Fig. 4a the composite view presents an ordinary view of "stress fibres", whereas three sections show the absence of F-actin bundles on the ventral side in Fig. 4b and their location under the dorsal surface in Fig. 4c,d. This type of cage is more clearly seen in T15 rat sarcoma cell in Fig. 2, in three sections under (d), across (e) and above (f) the nucleus. A similar type is found in rat cells from the primary culture of a spleen fragment (Fig. 1).

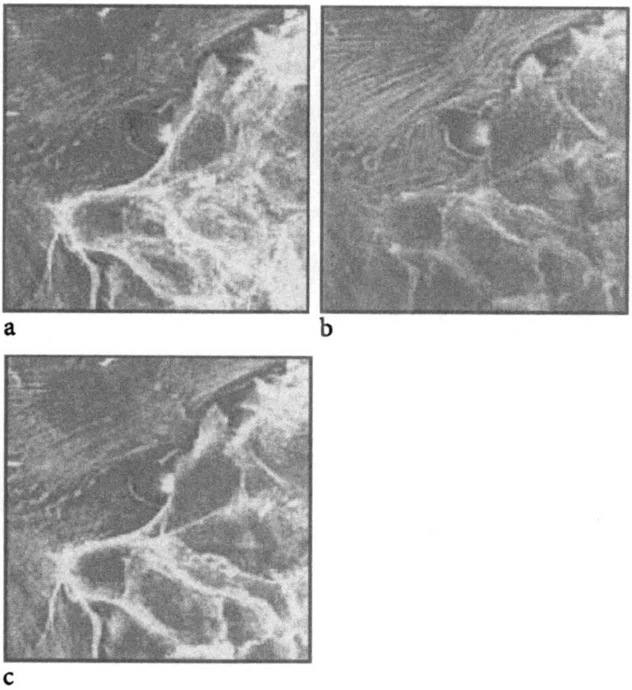

Fig. 5a–c Human keratinocytes and irradiated 3T3 cells in standard medium. Picture width 102 m, Leica 100x, 1.3 NA, zoom 1x, z-step 0.73 µm. **a** Composite image; **b** ventral side at the level of cell-to-substratum contacts; **c** a level at 0.73 µm above the level of contacts. Composite image (**a**) shows abundant stress fibres in 3T3 cells, whereas only side belts of circular F-actin bundles are seen in keratinocytes. This is further illustrated in **b** where no stress fibres are seen at the level of cell-to-substratum contacts in keratinocytes and at intermediate level in **c** with F-actin cables decorating cell-to-cell boundaries

Open cage at both ventral and dorsal cell surfaces. The open cage both at ventral and dorsal cell surfaces can be seen in epithelial cells. In normal human keratinocytes (Fig. 5) horizontal F-actin bundles are located at cell-to-cell boundaries, indicating a mostly vertical distribution of actin bundles (a,c) into side belts. This is different from the horizontal distribution seen in neighbouring 3T3 cells. No stress fibres close to the ventral side can be seen in Fig. 5b.

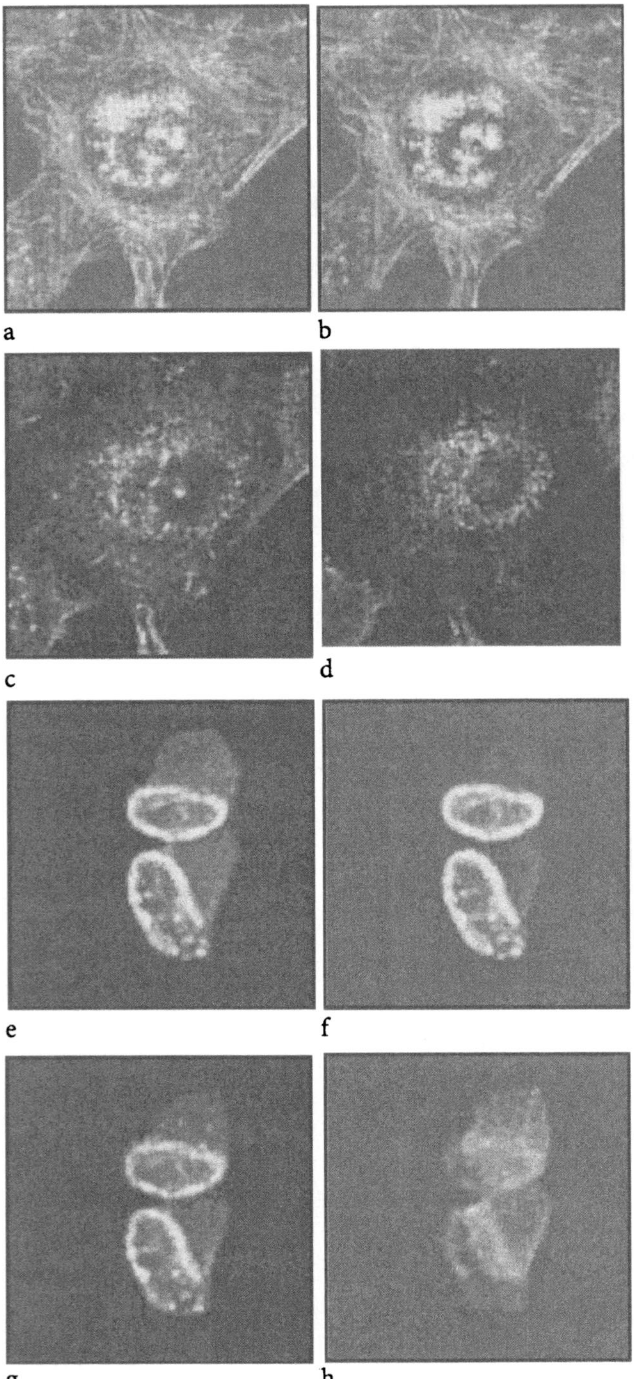
a
b
c
d
e
f
g
h

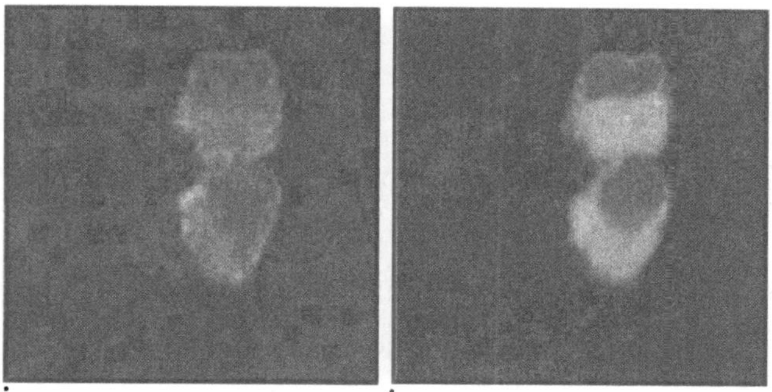

i j

Fig. 6a–j Rat Rous sarcoma K4 cells. Picture width 54 µm (**a–d**) and 46 µm (**e–j**), Leica 100x 1.3 NA (**a–d**) and 63x 1.4 NA (**e–j**), zoom 1.87x (**a–d**) and 3.57x (**e–j**), z-step 0.46 µm (**a–d**) and 0.64 µm (**e–j**). K4 cell, in standard medium (**a–d**), in a composite image (**a**) shows central ring-shaped PCS and circular basal plate of F-actin bundles. In (**b**) the PCS is localized on ventral side of the cell while in (**c**) 1 µm above the level of (**b**) it is not seen. Both (**c**) and (**d**) show absence of F-actin bundles under the entire dorsal cell surface. Two K4 cells in the HBSS stress (**e–j**) show in a composite image (**e**) an identical morphological and contact development: oval-shaped PCS over entire ventral surface with elevated bowl-shaped cell body. In half micron steps from the contact level in **f** and **g** the PVSs are seen, whereas in next micron of the cell height in **h** and **i** only diffuse subcortical layer of F-actin is detected, with no bundles. At the same level the nucleus is negatively indicated by DiOC6 staining in **j**

The absence of F-actin bundles

The partial absence of F-actin bundles is observed in the K4 cells. In Fig. 6b the circular basal plate is on the ventral side only, while the whole cortical layer on the dorsal surface is free of F-actin bundles. The total absence of F-actin bundles is clearly seen in malignant macrophages J774.1 (Fig. 7), but also in a big, central, poorly malignant K2 cell in a composite view of the layer of K2 cells, in which other cells show an abundance of actin bundles (Fig. 8a). In benign epithelial neoplasia, such as the human breast tumours in Fig. 9, F-actin bundles are absent. Cell boundaries are decorated by diffuse fluorescence of the cortical layer of actin filaments.

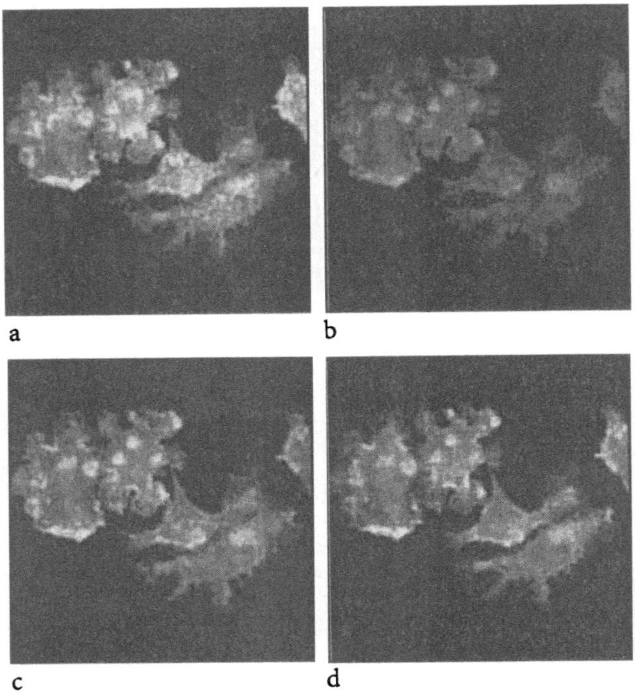

Fig. 7a–d Mouse malignant macrophages J774.1 in standard medium. Picture width 102 μm, Leica 100x 1.3 NA, zoom 1x, z-step 0.34 μm. **a** In a composite image the overall shape of the cells and more than one small PCS on one cell are shown. **c** The shape of the PCSs is shown and sections below (**b**) and above (**d**) limit the appearance ofthe PCSs into a layer adjacent to ventral side of the cell which is thinner than 1 μm while the cell height is almost 5 μm. No F-actin bundles are observed in these cells

HBSS stress and change of the 3D shape of malignant cells

Rat neoplastic and normal cells were treated by incubation in HBSS pH 7.1 for 2 h. This caused many cells to become very loosely attached and therefore washed out during fixation and staining. The most remarkable, but rarely observed, effect on sarcoma cells (T15, A72N, A8780N) was the change (transformation) of the usually elongated cell body to an almost even distribution of cell mass in space. In Fig. 10 such a case is exemplified by an A870N cell in the *x,z* section (g) and in the *x,z* composite side view (h). Together with *x,y* composite view in Fig. 10a,these images allow appreciation of the plump 3D shape of the cell. F-actin bundles are absent (e–d) and nucleus is located high in the body (f). In this case the spatial transformation is accompanied by the formation of the PCS. There were also incidences when the PCS was not formed.

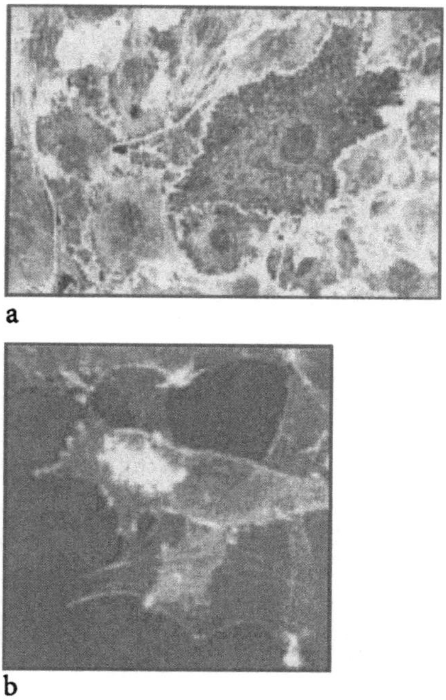

Fig. 8a,b Rat fibrosarcoma K2 cells. Picture width (**a**) 211 μm and (**b**) 65 μm, Nikon 60x, 1.4 NA (**a**), Leica 100x, 1.3 NA (**b**), zoom (**a**) 1x and (**b**) 1.57x.**a** Composite image of K2 cells in standard medium shows one cell without and many cells with ordinary stress fibres; **b** composite image of K2 cells in the HBSS stress shows one K2 cell with a podosome-like contact structure in a cell devoid of recognizable bundles of F-actin while in the surrounding cells without podosomal contacts the F-actin cables remain preserved

Podosomal contact structures (PCSs)

Fully developed PCSs in the form of ring-shaped rosettes of podosomes or even the circular continuous band resembling the sealing zone of the osteoclasts (Vaananen and Horton, 1995) were found to result from HBSS stress on all sarcoma cells derived from RPS sarcoma cells currently under investigation, i.e. T15, A72N, A870N. Lower forms of the PCSs are rarely found on these cells in standard medium. In poorly malignant K2 cells, the ancestors of the RPS sarcoma, there were no PSCs in standard medium whereas in HBSS stress lower forms were very rarely found, as is shown in Fig. 8b, where in the composite view the cell with fusing rosettes is devoid of F-actin bundles and shows a rather atypical cell shape for the fibroblast-like K2 cell.

The T15 sarcoma cell in Fig. 2 g–j is an example of location of the PCSs at the ventral side of the long cell processes (g) which change with height into ruffles (j). In Fig. 10, which shows an A870N sarcoma cell, the PCS formed by regular distribution of rosettes is located at two bottom sections, as is seen in b,c and in *x,z* sections in g,h. The morphology of the PCSs, their incidence and the related cellular changes found in the rat RPS sarcoma malignant derivatives were compared with malignant macrophages J774.1 (possibly v-*src* free), and with v-*src*-transformed rat and chicken cells.

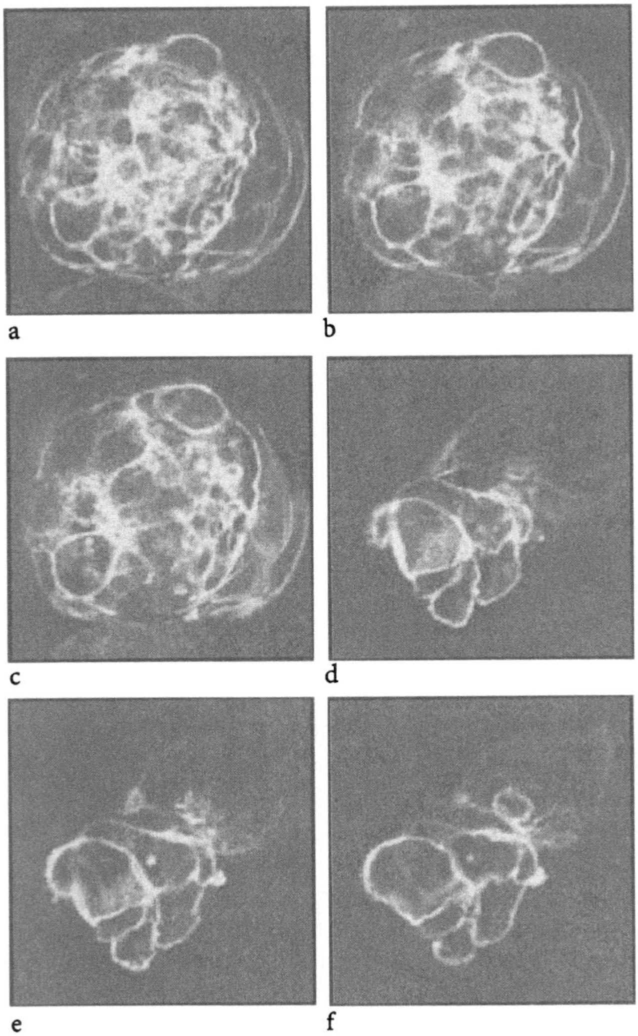

Fig. 9a–f Cells of human breast tumours in standard medium. **a–c** Benign phylloides tumour of the breast, **d–f** intracanalicular fibroadenoma of the breast.Picture width 158 µm (**a–c**) and 66 µm (**d--f**), Leica 63x 1.4 NA (**a–f**), zoom 1.55x (**a–c**) and 2.36x (**d–f**), z-step 0.77 µm (**a–c**) and 1.7 µm (**d–f**). In three subsequent sections of cellular agglomerates of these two tumours F-actin staining shows diffuse decoration of the intercellular boundaries only. No bundles of F-actin can be identified

The mouse malignant macrophages J774.1 could only be used in standard medium because of their very weak adhesion. There is a high incidence of ring-shaped PCSs on these cells in standard medium. There are frequently more than one PCS on a cell (Fig. 7). They are smaller and thinner (Fig. 7b–d), composed of small rosettes, and the contact circular band is proportionally wider with a small central clearing (Fig. 7c). There are no F-actin bundles in these cells (Fig. 7a).

Rat K4 sarcoma cells are closely related to RPS sarcoma because they are derived from the same ancestor, LW13 cells, by RSV transformation. K4 cells show the PCSs regularly in standard medium.

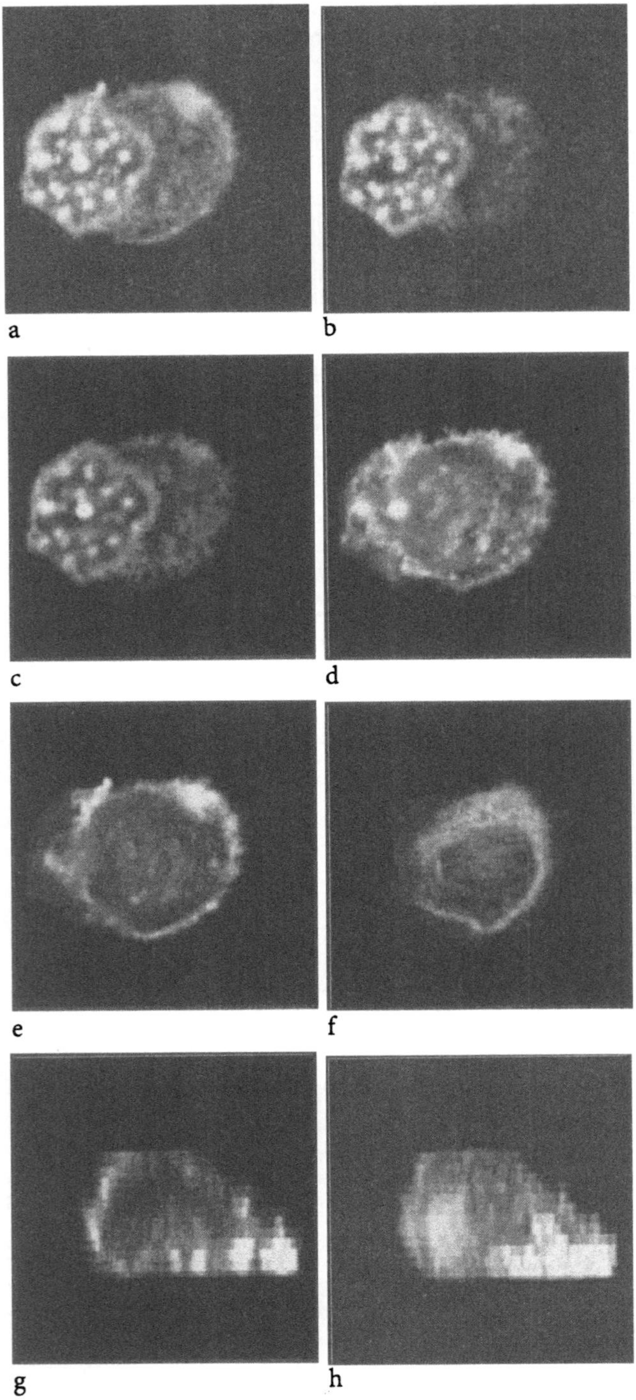

Fig. 10a–h Rat sarcoma A870N cell in HBSS stress produced a podosome and changed the 3D shape

of the cell body. Picture width 24 μm, Leica 63x 1.4 NA, zoom 6.81x, z-step 0.5 μm, inverted imaging.**a** Composite image the cell is shown with the PCS over the entire ventral side. The PCS is located only in two optical sections (**b** and **c**) through 1 μm of cell ventral side, whereas in **d** at 1.5 μm it is almost absent, and in **e** only a diffuse subcortical layer of F-actin is detected with no bundles. At the same level the nucleus is negatively indicated by DiOC6 staining in f. **g** *x,z* section shows the localization of the PCS in the height profile of the cell; **h** the *x,z* composite image indicates F-actin distribution in the side view of the cell

The PCSs are accompanied by a basal circular ring of F-actin bundles. In the cortical layer under the entire dorsal surface the F-actin bundles are absent, as is shown in Fig. 6a–d. Under HBSS stress, K4 cells react in a similar way to the A870N rat sarcoma cell, as in Fig. 10. In Fig. 6e the composite view shows the overall shape of the cell bodies of two K4 cells. In sections through the ventral side of the cells the oval-shaped PCSs are imaged only in two sections (f,g) where no circular F-actin bundles are present. Higher up only diffuse fluorescence of the cortical layer is seen (h,i), and nucleus is located at this level (j).

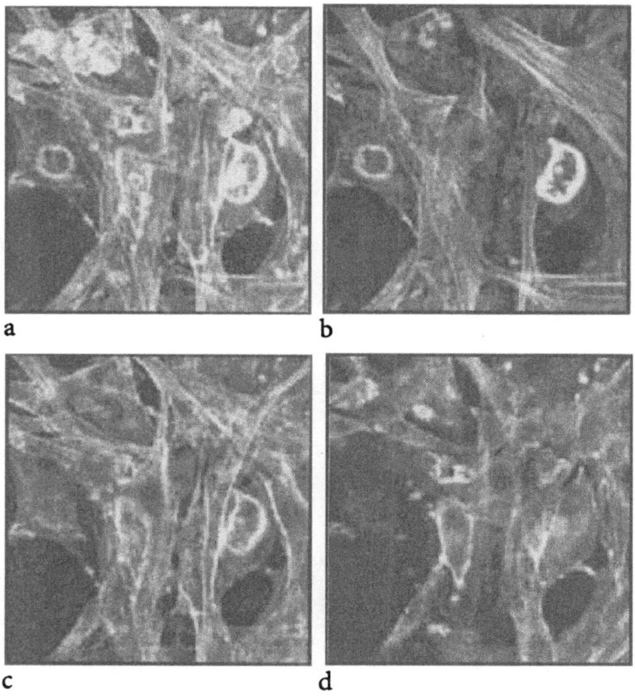

Fig. 11a–d Chicken fibroblasts transformed by Rous sarcoma virus in vitro in standard medium. Picture width 102 μm, Leica 100x, 1.3 NA, zoom 1x, z-step 0.71 μm. In a composite image (**a**) and in optical sections at cell-to-substratum contacts level (**b**) and in two steps higher (**c**) and (**d**) two cells with the PCSs typical for RSV transformants are almost devoid of F-actin bundles. Some of the neighbouring cells, in spite of belonging to the cell population infected and transformed by Rous sarcoma virus, manifest an almost normal presence of F-actin bundles

Brown Leghorn chicken fibroblasts freshly transformed in vitro with RSV Prague strain C were used as a control for podosomes/invadopodia assessment.

In Fig. 11 stratification of such cell culture is seen. PCSs are located in the cells (b,c) attached to the coverslip and there are no F-actin bundles in these cells. This is not true for all cells in culture, as is seen in Fig. 11a,c. There are cells with F-actin bundles, but a cell without PCSs and without F-actin bundles can be seen in the upper layer (d).

Two cell populations (P10, P22), derived from chicken tumour No. 9692 (induced by transfection with LTR,v-*src*,LTR provirus), and the MP4 cell population, derived from metastasis of this tumour, were examined for F-actin structures. P10 and P22 cells (Fig. 12a–d) show developing PCSs which were rather rare in the whole culture, while the MP4 cells (e,f) show fully developed PCSs which were quite numerous in the culture. It is surprising that in MP4 cells F-actin bundles are not completely lost even from cells with the PCSs.

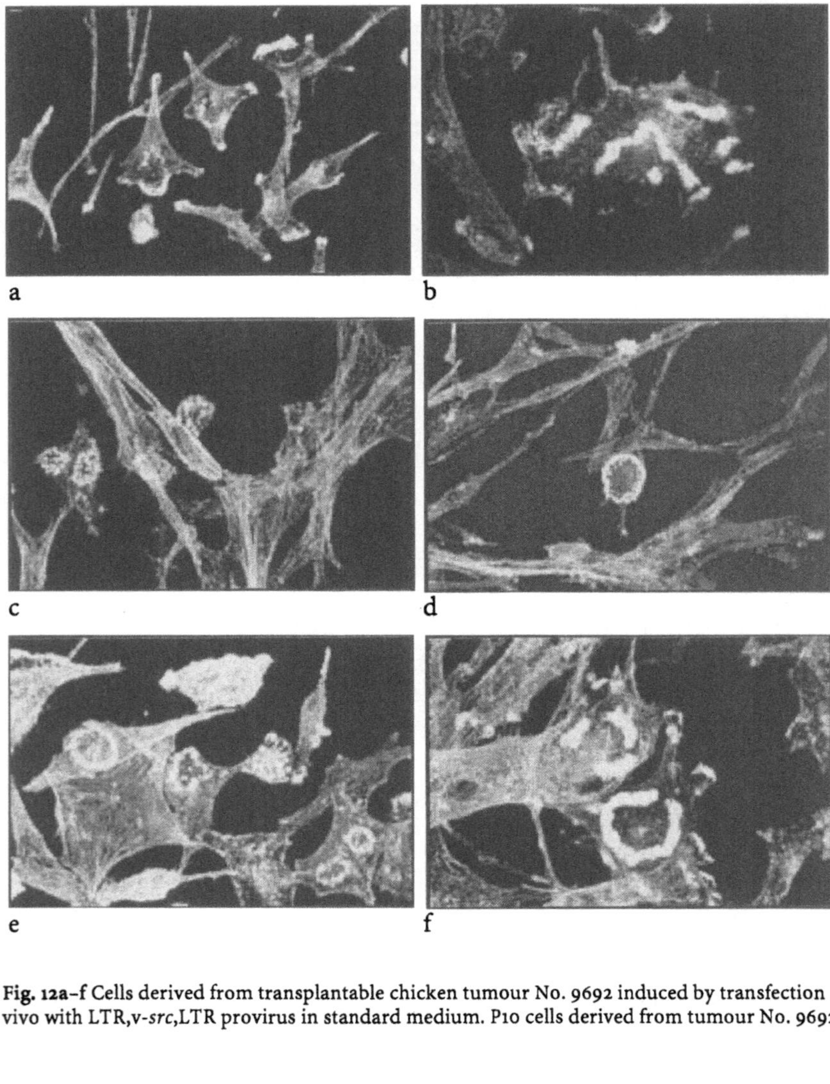

a b c d e f

Fig. 12a–f Cells derived from transplantable chicken tumour No. 9692 induced by transfection in vivo with LTR,v-*src*,LTR provirus in standard medium. P10 cells derived from tumour No. 9692 by passaging in vitro (a,b), P22 cells derived from tumour No. 9692 by passaging in vitro and in vivo (c,d), and MP4 cells isolated from metastasis in lungs of tumour No. 9692 (e,f). Picture width (a,d) 211 µm, (b) 105 µm, (c,e,f) 140 µm, Nikon 60x 1.4 NA, zoom (a,d) 1x, (b) 2x, (c,e,f) 1.4x. In a composite view of four sections through a 1.5-µm-thick basal layer of P10 cells (a) developmental stages of the PCSs such as rosettes of various densities are seen. F-actin bundles are rare but not completely lost even in cells with rosettes. In a composite view of two sections through a 1-µm-thick basal layer of P10 cell in (b) an atypical shape of the PCS is shown. In a composite view of three sections through a 1-µm-thick basal layer of P22 cells in (c) rosettes organized into roundish spots are seen in cells free of F-actin bundles.d In a composite view of three sections through a 1-µm-thick basal layer of P22 cells fully developed PCS is observed. F-actin bundles are clearly seen in the surrounding cell without the PCSs. e In a composite view of five sections through a 2-µm-thick basal layer of MP4 cells development of the PCSs is observed. F-actin bundles are clearly seen even in some cells with the PCSs. f In a composite view of a 9-µm-thick layer of MP4 cells the fully developed PCSs are observed in cells without F-actin bundles

Discussion

For the comparative study a set of neoplastic and normal cells with contrasting biological properties was selected.

A family of malignant cell populations (K2, T15, A72N, A870N) was chosen because of the different capacities of the individual types to metastasize spontaneously.

The RSV transformants (BLEF-RSV chicken fibroblasts transformed with RSV; K4 clone of rat Rous sarcoma, chicken neoplastic cells P10, P22 from transplantable tumour No. 9692; MP4 neoplastic cells isolated from metastasis in the lung shed by tumour No. 9692) were used to detect the natural incidence of the PCSs on them.

Mouse malignant macrophages J774.1 represented macrophage/monocyte type of cells.

Primary cells of a human benign phylloides tumour of the breast and intracanalicular fibroadenoma of the breast were chosen as examples of neoplastic epithelia.

Human keratinocytes and rat fibroblasts from primary fragment culture on a coverslip were chosen as examples of normal cells.

In addition to the standard conditions in full culture medium, a 2-hincubation in Hanks BSS pH 7.1 was used to challenge the cells.

In high-resolution fluorescence CLSM, a range of apparent optical thickness of the F-actin bundles from thick, stress fibres down to barely recognizable threads is observed. Even if this is not simply related to the physical size of the object observed, because it reflects the density of staining, it indicates to some extent a continuity of the organization of the actin cortex from a mesh-like layer with diffuse fluorescence to structures made of distinct bundles composed of increasing numbers of microfilaments. In this study the raw confocal sections were preferred to 3D reconstructions because they show details of resolution with all the uncertainty that primary observation provides and are less sensitive to the effects of photobleaching.

The set of 3D structures of the AC described indicates several roles of the AC in the cell. First there is the established role of the classic stress fibres in maintaining the stretchof fibroblast-like cells, which is known as anchorage dependence (Stoker et al., 1968; Slack and Higgins,). There is also an apparent need of epithelial cells for vertical support. The various forms of the cage may reflect special structural requirements imposed by static or dynamic factors. The emerging relationship between the cage structure and the enhancement of cell locomotion (Pokorna et al., 1994), its persistence (Zicha et al., 1993) and/or reappearance (Vesely et al., 1987b) will need further investigation. The absence of F-actin bundles in neoplastic cells, which is similar to the free wandering cell, can be interpreted as an increase of cell surface plasticity, which is useful in tissue penetration. It may be that even rapid changes between the presence of F-actin bundles and their partial or total absence as a response to surrounding conditions may be one of the mechanisms that endows neoplastic cells with their capacity to penetrate through surrounding tissues.

It seems that the cage is a primary organization of the AC. Thus, it only depends on the actual thickness of the actin bundles as to whether we see it as diffuse cortical fluorescence when stained with, for example, rhodamine-phalloidin, or actin bundles of various thickness. There can be rapid changes in the observed shape of the cage in order to satisfy the biomechanical needs of the situation. These changes, as well as the role of the various configurations of the cage in locomotion, can only be investigated by direct visualization of the AC using living cells with microinjected fluorescent analogues of actin and video-enhanced microscopy, as was done for the dynamics of stress fibres (Wang, 1984; Guiliano and Taylor, 1990).

In this context it is of interest that MP4 cells show some presence of F-actin bundles even in cells with the PCSs. This could be of importance for their metastatic potential, by influencing their locomotory behaviour. If this proves to be so then it is similar to the anticipated role of F-actin bundles in the A870N cells, which was the finding that initiated this study (Pokorna et al., 1994). It will also be in accordance with the finding that, in the V2 rabbit carcinoma cells in vitro, there is an absence of large microfilament bundles (stress fibres) in locomoting cells while small microfilament bundles were contained in focal contacts of cells detaching from others (Jockusch et al., 1983).In the current study strong stress treatment was used. Exposure to HBSS pH 7.1 at 37 C for 2 h was used mainly as a nutritional shock with the aim of analysing changes of F-actin-related structures. Only rat sarcoma cells demonstrated the same adaptive behaviour, which is presented for A870N cell. The change of the 3D shape of the cell accompanied by loss of F-actin bundles and appearance of the PCS can be understood in terms of a redistribution of the F-actin between bundles and the PCS. This is similar to the unchanged quantity of polymerized actin found in the Rat-1 cell line transformed by LA29, a temperature-sensitive mutant of RSV, on changing from permissive to restrictive temperatures (Felice at al., 1990). The induced PCSs could also prevent the cell from being washed out, by allowing them to stick better to the coverslip. The result of the transformation is reminiscent of the morphotype of activated free wandering cells of the macrophage/monocyte lineage with the PCSs and the AC without F-actin bundles, as maximum plasticity of the cell surface is required.

The PCSs were not previously observed in the family of RPS sarcoma cell lines in either conditions, i.e. standard or stressing, possibly because the stresses applied were not sufficient to elicit such a transformation of the cells. It is not clear whether the spatial transformation of the cell has to precede the formation of the PCSs. PCSs are also called invadopodia and are considered to be expressed on neoplastic cells in order to facilitate invasive penetration into the surrounding tissue (Chen, 1990, 1996; Monsky et al., 1994). Then it can be assumed that the kind and number of the PCSs on neoplastic cells can reflect their malignancy potential. We found only exceptional expression of the PCSs under the HBSS stress, even on a very poorly malignant K2 cell, without very pronounced spatial transformation of the cell body. More frequent cases of the changes in 3D cell shape, accompanied by formation of the PCSs, were seen in malignant T15, A72N and A870N sarcoma cells after the same challenge, which seems to confirm this assumption. Moreover, the fact that RPS sarcomas are not locally invasive can be considered a consequence of the very low incidence of the PCSs under normal conditions while their capacity to spontaneously metastasize from a subcutaneous tumour can be related to the inducible expression of the PCSs by an appropriate stress. In an attempt to localize and follow the PCSs in living cells with high-resolution reflection interference microscopy (Bereiter-Hahn and Vesely, 1994), a full adhesion plaque, not only the circular band, was seen. This means that follow up of the PCSs in living cells can be accomplished only when this approach is used with simultaneous fluorescence labelling of F-actin in living cells or with subsequent fluorescence identification of the F-actin ring in the adhesion plaque studied.

There is an increased incidence and maturation of the PCSs on MP4 cells derived from metastasis of the chicken tumour No. 9692 induced by transfection with LTR,v-*src*,LTR provirus (Svoboda et al., 1992) compared to two tumour cell populations (i.e. P10, P22) derived from the same tumour, i.e. No. 9692. It is necessary to point out that chicken tumour No. 9692 (Svoboda et al., 1992) resembles Rous sarcomas in mammals in that there is v-*src* integrated but no RSV produced and released. This is different from the chicken cells transformed by RSV that were studied by Chen (1989). The MP4 cells from the metastasis that show numerous PCSs can be considered to be an indication of the role of these contacts in local as well as in distant malignant spread. There are other pieces of evidence for the correlation between the rate of expression of the PCSs and the state of malignancy of the cells. The pp110 tyrosine-phosphorylated protein associated with pp60src colocalizes with actin stress filaments in normal cell while in *src*-transformed cells it is found in podosomes (Flynn et al., 1993). The product of the human *ems1* gene, the 80- to 85-kDa protein (85% homology with chicken protein which is a substrate for the *src* oncogene), is immunocytochemically localized in the cytoplasm of human epithelial cells, whereas in human carcinoma cells (breast and squamous carcinomas) it accumulates in the podosome-like adherens junctions associated with the cell-substratum contact sites (Schuuring et al., 1993).

All these findings seem to support our interpretation that the HBSS stress induction of the PCSs in the rat RPS sarcoma cells is not related to the RSV. Hence, the induction of developed PCSs in rat sarcoma cells by the HBSS stress, which is usually accompanied by spatial transformation of these cells, is considered to be a promising marker of their malignancy potential. Preliminary results of the polymerase chain reaction (PCR) detection of v-*src* in K2 and T15 cell were negative, which means that the PCSs in these cells are not caused by v-*src* expression. The question of whether they are related to c-*src* overexpression (Neet and Hunter, 1995; Hall et al., 1996) or can be controlled by Rho GTPase CDC42Ds (Dutartre et al., 1996) similarly as focal complexes associated with stress fibres are regulated by Rho, Rac and CDC42 GTPases (Nobes and Hall, 1995) requires further study. The functional similarity of fully developed PCSs with the resorbing organelle of the osteoclasts (Vaananen and Horton, 1995) promptsfurther investigation, in spite of the evidence that human calcitonin-receptor negative macrophage polykaryons only rarely have the capacity to excavate bone slices (Flanagan et al., 1992).

Our findings indicate that, if appropriate stress or shock is used to challenge cells in vitro, malignant cells may show their otherwise hidden ability to adopt actively and survive. Such an approach is currently difficult to apply to neoplastic cells from biopsied material, as a degree of basic adaptation to in vitro conditions seems to be necessary. However, for assessing the functional significance of the phenotypic traits of malignant cells in model situations in vitro and thenanalysing the genetic background of these cellular properties, such an approach shows promise.

Acknowledgements. The work was supported by grants No. 312/94/1481, 312/97/1222 and 523/96/0670 of the Grant Agency of the Czech Republic. Financial support from the J. W. Goethe University for visiting the Biocentre and performing research in Frankfurt AM is gratefully acknowledged.

References

Bereiter-Hahn, J., Vesely, P. (1994) Reflection interference microscopy. In: Cellis, J.E. (ed) *Cell biology: a laboratory handbook*, part 2, vol. 7A. Academic Press, San Diego, pp. 15–24

Chen, W.T. (1989) Proteolytic activity of specialized surface protrusions formed at rosette contact sites of transformed cells.*J. Exp. Zool.* **251**: 167–185.

Chen, W.T. (1990) Transmembrane interactions at cell adhesion and invasion sites. *Cell Differ. Dev.* **32**: 329–335.

Chen, W.T. (1996) Proteases associated with invadopodia, and their role in degradation of extracellular matrix. *Enzyme Prot.* **49**: 59–71.

Dutartre, H., Davoust, J., Gorvel, J.P., Chavrier, P. (1996) Cytokinesis arrest and redistribution of actin-cytoskeleton regulatory components in cells expressing the Rho GTPase CDC42Hs. *J. Cell Sci.* **109**: 367–377.

Felice, G.R., Eason, O., Nermut, M.V., Kellie, S. (1990) pp60v-src association with the cytoskeleton induces actin reorganization without affecting polymerization status. *Eur. J. Cell Biol.* **52**: 47–59.

Flanagan, A.M., Horton, M.A., Dorey, E.L., Collins, D.A., Evely, R.S., Moseley, J.M., Firkin, F.C., Chambers, T.J., Helfrich, M.H., Martin, T.D.J. (1992) An assessment of the ability of human bone marrow cultures to generate osteoclasts. *Int. J. Pathol.* **73**: 387–401.

Flynn, D. C., Leu, T.H., Reynolds, A.B., Parsons, J.T. (1993) Identification and sequence analysis of cDNAs encoding a 110-kilodalton actin filament-associated pp60src substrate. *Mol. Cell. Biol.* **13**: 7892–7900.

Gavazzi, I., Nermut, M.V., Marchisio, P.C. (1989) Ultrastructure and gold-immunolabelling of cell-**substratum** adhesions (podosomes) in RSV-transformed BHK cells. *J. Cell Sci.* **94**: 85–99.

Guiliano, K.A., Taylor, D.L. (1990) Formation, transport, contraction, and disassembly of stress fibres in fibroblasts. *Cell Motil. Cytoskeleton* **16**: 14–21.

Hall, C.L., Lange, L.A, Prober, D.A., Zhang, S., Turley, E.A. (1996) pp60 c-src is required for cell locomotion regulated by the hyaluronan receptor RHAMM. *Oncogene* **13**: 2213–2224.

Jockusch, B.M., Haemmerli, G., In Albon, A. (1983) Cytoskeletal organization in locomoting cells of the V2 rabbit carcinoma. *Exp. Cell. Res.* **144**: 251–263.

Liotta, L.A., Starcke, M.L., Aznavoorian, S.A., Beckner, M.E., Schiffmann, E. (1991) Tumour cell motility. *Semin. Cancer. Biol.* **2**: 111–114.

Matouskova, E., Zicha, D., Urbanec, P., Vesely, P. (1990) Quasi-dynamic test of in vitro cell migration developed to characterize dividing cells of spontaneously metastasizing rat sarcomas. *Folia Biol. (Praha)* **36**: 117–129.

Matouskova, E., Vogtova, D., Koenigova, R. (1993) A recombined skin composed of human keratinocytes cultured on cell-free pig dermis. *Burns* **19**: 118–123.

Monsky, W.L., Chen-Yong, L., Aoyama, A., Kelly, T., Akiyama, S.K., Mueller, S.C., Chen, W.T. (1994) A potential marker protease of invasiveness, seprase, is localized on invadopodia of human malignant melanoma cells. Cancer Res. **54**: 5702–5710.

Neet, K., Hunter, T. (1995) The nonreceptor protein-tyrosine kinase CSK complexes directly with the GTPase-activating protein-associated p62 protein in cells expressing v-Src or activating c-Src. *Mol. Cell. Biol.* **15**: 4908–4920.

Nobes, C.D., Hall, A. (1995) Rho, Rac, and Cdc42 GTPases regulate the assembly of multimolecular focal complexes associated with actin stress fibers, lamellipodia, and filopodia. Cell 81: 53-62.

Nermut, M.V., Eason, P., Hirst, E.M., Kellie, S. (1991) Cell/substratum adhesions in RSV-transformed rat fibroblasts. *Exp. Cell Res.* **193**: 382–397.

Osborn, M., Born, T., Koitsch, H.J., Weber, K. (1978) Stereo immunofluorescence microscopy: I. Three-dimensional arrangement of microfilaments, microtubules and tonofilaments. *Cell* **14**: 477–488.

Pokorna, E., Jordan, P.W., O'Neill, C.H., Zicha, D., Gilbert, C.S., Vesely, P. (1994) Actin cytoskeleton and motility in rat sarcoma cell populations with different metastatic potential. *Cell Motil. Cytoskeleton* **28**: 25–33.

Ralph, P., Prichard, J., Cohn, M. (1975) Reticulum cell sarcoma: an effector cell in antibody-dependent cell-mediated immunity. *J. Immunol.* **114**: 898–905.

Ruenger, T.M., Klein, C.E., Becker, J.C., Broecker, E.B. (1994) The role of genetic instability, adhesion, cell motility, and immune escape mechanisms in melanoma progression. *Curr OpinOncol* **6**: 188–196.

Schuuring, E., Verhoeven, E., Litvinov, S., Michalides, R.J. (1993) The product of the EMS1 gene, amplified and overexpressed in human carcinomas, is homologous to a v-src substrate and is located in cell-substratum contact sites. *Mol. Cell. Biol.* **13**: 2891–2898.

Slack, J.K., Higgins, P.J. (1996) Cytoarchitecture and cell growth control. *Cell Motil. Cytoskeleton* **33**: 83–87.

Stoker, M.G.P., O'Neill, C., Berryman, S., Waxman, V. (1968) Anchorage and growth regulation in normal and virus-transformed cells. *Int. J. Cancer* **3**: 683–693.

Svoboda, J., Plachy, J., Hejnar, J., Karakoz, I., Guntaka, R.V., Geryk, J. (1992) Tumor induction by the LTR, v-src, LTR DNA in four B (MHC) congenic lines of chickens. *Immunogenetics* **35**: 309–315.

Tarone, G., Cirillo, D., Giancotti, F.G., Comoglio, P.M., Marchisio, P.C. (1985) Rous sarcoma virus-transformed fibroblasts adhere primarily at discrete protrusions of the ventral membrane called podosomes. *Exp. Cell Res.* **159**: 141–157.

Vaananen, H.K., Horton, M. (1995) The osteoclast clear zone is a specialized cell-extracellular matrix adhesion structure. *J. Cell Sci.* **108**: 2729–2732.

Vesely, P., Weiss, R.A. (1973) Cell locomotion and contact inhibition of normal and neoplastic rat cells. *Int. J. Cancer* **11**: 64–76.

Vesely, P., Chaloupkova, A., Urbanec, P., Urbancova, H., Bohac, L., Krchnakova, E., Franc, F., Sprincl, L., Vousden, K., Moss, R., Heaysmann, J., Dilly, P.N. (1987a) Patterns of in vitro behaviour characterizing cells of spontaneously metastasizing K2M rat sarcoma. *Folia Biol. (Praha)* **33**: 307–324.

Vesely, P., Bershadsky, A.D., Kren, V., Wyke J. (1987b) Partial reappearance of actin cables in rat tumour cells supertransformed by avian retrovirus B77 in vitro. *Wiss. Ztschr. Friedrich-Schiller-Univ. Jena, Naturwiss. R,* **36**: 192–197.

Vesely, P., Urbanec, P., Zicha, D., Chaloupkova, A., Matouskova, E., Urbancova, H., Krchnakova, E. (1989a) Patterns of in vitro locomotory behaviour characterizing cells of spontaneously metastasizing rat sarcomas. In: Kotyk, A., Skoda, J., Paces, V., Kostka, V. (eds) *Highlights of modern biochemistry* VSP International Publishers, pp. 973–981.

Vesely, P., Melezinkova, H., Krenova, D., Kren, V., Matouskova, E., Chaloupkova, A. (1989b) Spontaneously metastasizing rat sarcomas LW13K2 and RPS: assessment by the immunogenetic test of malignancy, in vitro behaviour and karyology. *Folia Biol. (Praha)* **35**: 1–12.

Wang, Y.L. (1984) Reorganization of actin filament bundles in living fibroblasts. *J. Cell Biol.* **99**: 1478–1485.

Zicha, D., Pokorna, E., Urbanec, P., Chaloupkova, A., Vesely, P. (1993) Acid pH induced persistent motility in vitro of metastasizing sarcoma cells. *J. Comput Assist Microsc* **5**: 273–279.

Exp. Biol. Online 2:7 (1997)

Relationship of Human Neutrophil Morphology and Actin Distribution to Dispersive Locomotion caused by a steroid induced factor

J.D. Young[1], A.G. MacLean[1], A.J. Lawrence[1], R.D. Stevenson,[2] and S. Chettibi[1]

[1]Institute of Biomedical and Life Sciences, Division of Infection and Immunity, Joseph Black Building, University of Glasgow, Glasgow G12 8QQ, UK
[2]Department of Respiratory Medicine, Royal Infirmary, Alexandra Parade, Glasgow G31 2ES, UK

Correspondence to: J.D.Young (email: gbaa19@udcf.gla.ac.uk)

Received: 7 March 1996 / Accepted: 1 April 1997
Published: April 22, 1997

Abstract. A monocyte-derived steroid-induced factor has been shown previously to induce dispersive locomotion in human neutrophils and to lower adhesion to an albumin-coated glass surface. In this paper we show that this factor inhibits adhesion of neutrophils to bovine aorta and human endothelial cells by an undetermined mechanism. It induces unique changes in neutrophil shape with a characteristic monopolar pattern of F-actin distribution, which may correlate with the dispersive locomotion observed in the absence of a concentration gradient. This factor also inhibits *N*-formyl-methionyl-leucyl-phenylalanine-induced chemotaxis of neutrophils in a modified Boyden chamber assay. The reduction of adhesion and the inhibition of chemotaxis by the factor in vitro indicate a possible in vivo anti-inflammatory role.

Key words: Neutrophils – Monocyte – Actin – Locomotion – Adhesion – Glucocorticoids

Introduction

Neutrophils possess sensitive machinery that responds to inflammatory signals, directing the cells to the sites of infection. Upon stimulation these cells undergo dramatic morphological changes associated with the activation of motile function (Stossel, 1988). Migration to the inflammatory site appears to involve weak, reversible, "rolling" adhesion mediated by selectins (Bevilacqua and Nelson, 1993), followed by margination involving strong adhesion mediated through the $[beta]_2$ integrins Mac-1 (CD11b/CD18) and LFA-1 (CD11a/CD18) which interact with ICAM-1 on the endothelial cells (Arnaout, 1993; Hogg and Berlin, 1995). Transmigration through the endothelium involves active, directed locomotion and the activation of the cells' motility machinery. Anti-inflammatory agents could interfere with any of these steps. For example, glucocorticoids are known to inhibit the expression of ICAM-1 (Cronstein et al., 1992) and to down-regulate transcription of several pro-inflammatory cytokines including interleukin 1 (IL-1) (Lew et al., 1988), tumour necrosis factor-[alpha] (TNF-[alpha]), granulocyte-macrophage colony stimulating factor (GM-CSF), IL-3, IL-4, IL-5, IL-6 and IL-8 (Kwon et al., 1993). Although the anti-inflammatory effects of steroids are usually attributed to the down-regulation of these pro-inflammatory mediators, there has been evidence that anti-inflammatory mediators of steroid action may exist (Goulding et al., 1990). Early work on the putative role of lipocortins as anti-inflammatory mediators has been extremely productive and has led to the characterisation of the family of annexins whose role as modulators of inflammation is still unclear (Flower, 1990; Wong et al., 1991).

Stevenson (1974) used culture conditions that closely resembled those in which lipocortin was generated and described a factor secreted by human monocytes in the presence of glucocorticoids. This factor enhanced the migratory behaviour of human neutrophils and the data suggested that the locomotor activity observed was not attributable to the absence of steroid-suppressed inhibitors. Since this work the range of known cytokines has increased greatly, but studies of Chettibi et al. (1993) have supported the original conclusion that a neutrophil pro-migratory and anti-adhesive factor is released by monocytes under the influence of glucocorticoids. The response of neutrophils to this factor (termed STMS for steroid-treated monocyte supernatant), characterised in a novel locomotion assay, clearly distinguished it from the activities of some of the known neutrophil activating molecules such as fMLP, IL-8 and LPA (Chettibi et al., 1994) and leukotriene B_4, TNF[alpha], platelet activating factor (PAF) and transforming growth factor (TGF[beta]) (unpublished results). STMS alone induced persistent and highly dispersive locomotion in the absence of a concentration gradient. A more dramatic demonstration of the motility characteristics was obtained using a tilted-stage assay where STMS alone of these agonists caused highly polarised locomotion down-slope that was quite distinct from free rolling or floating (Chettibi et al., 1994).

In this paper we examine the effect of STMS on neutrophil morphology using scanning electron and confocal microscopy to try to correlate structural changes with the characteristics of induced persistent locomotion. We also investigate other aspects of neutrophil behaviour comparing the effects of STMS with pro-inflammatory mediators. These include examination of the chemotactic response and adhesion to endothelial cells to see whether the phenomenon of persistent locomotion could be assigned a specific role in the anti-inflammatory process.

Materials and methods

Reagents

TNF[alpha] was purchased from Genzyme, dissolved in phosphate-buffered saline (PBS) and stored at 1µg per ml at -70°C. Dexamethasone and fMLP were purchased from Sigma. Dexamethasone was prepared as a 10^{-2} M stock solution in ethanol and fMLP as a 10^{-2} M stock solution in dimethyl sulphoxide (DMSO).

Neutrophil purification

Neutrophils were obtained as described by Chettibi et al.(1993). Briefly, whole blood was mixed with 1:10 v/v of 5% dextran and allowed to sediment at 37°C for 1 h. The leucocyte-rich plasma was layered over Nycoprep 1.077 and centrifuged at 750 g for 15 min. Erythrocytes in the resulting pellet were removed by hypotonic lysis and neutrophils were washed twice with balanced salt solution (BSS). The cells were checked for viability by trypan blue exclusion (generally greater than 96% viable).

Cell-tracking assay

Automated cell-tracking migration chambers were made as previously described (Chettibi et al., 1993) and placed on the stage of an inverted phase-contrast microscope within a temperature-controlled (37°C) P>C) transparent box and locomotion observed by means of a video camera connected to a monochrome monitor and also to an Acorn A5000 computer with a Watford video digitiser programmed to capture and analyse one frame every 5 s. Data were obtained from a maximum of 80 selected cells and used to calculate instantaneous speed, the two-dimensional diffusion coefficient and the locomotion persistence time, using the described procedure to eliminate the contribution of systematic drift (Chettibi et al., 1994).

STMS preparation

STMS was obtained essentially as described by Chettibi et al. (1993) by the culture of human monocytes in Hams F-10 medium with 10% fetal calf serum (FCS) at a concentration of approximately 5×10^7 cells per ml for 60 min, rinsed with PBS, and then cultured in Hams F-10 medium without FCS for 24 h in the presence of 10^{-6}M dexamethasone. Parallel cultures in which dexamethasone was omitted were used to prepare control monocyte supernatants (CMS). Purification of STMS was carried out using the two-dimensional diffusion coefficient to identify active fractions. Partial purification was achieved using gel filtration and ion-exchange chromatography on Mono-Q resin. Highly purified material was obtained by the additional use of reverse-phase HPLC.

Adhesion assay

Bovine aorta endothelial cells were cultured on 13-mm-diameter glass coverslips in a multi-well dish in Dulbecco's modified Eagles medium with 10% fetal calf and 10% horse serum and grown to confluence. Human endothelial cells were grown in Endothelial growth medium (TCS Biologicals). Human neutrophils suspended in 0.1% bovine serum albumin (BSA) in BSS (w/v) were labelled by incubating them at 1×10^6 cells/ml for 1 h with 20 µCi/ml (740 kBq/ml) [^{51}Cr]sodium chromate. Free ^{51}Cr was removed by three washes with BSS 0.1% BSA. Then, 200 µl of neutrophils was mixed with 800 µl of test substance, added to the wells and incubated for 30 min at 37°C. Non-adherent cells were washed gently three times with BSS containing 0.1% BSA and coverslips were placed in a Wilj gamma counter.

Electron microscopy

Neutrophils were stimulated with the various agonists for 20 min before fixing in 2% buffered gluteraldehyde for 1 h and washed twice in PBS. Post-fixation in osmium tetroxide was followed by washing in distilled water. Uranyl acetate was then added to the samples and left in the dark for at least 1 h before washing. The cells were passed through a graded series of acetone (or alcohols) ranging from 30% to dried absolute, before critical point drying and mounting. The specimens were then coated in gold and viewed in the scanning electron microscopy (SEM).

Actin staining and confocal microscopy

Purified neutrophils were placed on albumin-coated glass coverslips before treatment with various stimuli and incubated at 37°C for 30 min. The cells were fixed in 1% paraformaldehyde solution for 1 h, washed with BSS and permeabilised with 1% Triton X-100 for 15 min at room temperature. Cells were washed three times with BSS and treated with 0.1 µg/ml TRITC-labelled phalloidin for 20 min at room temperature. Cells were washed three times in BSS at 5-min intervals and mounted on glass slides with 50% glycerol. Results were analysed using a Noran Odyssey confocal microscope.

Chemotaxis assays

Filters (Sartorious membranfilter 3 µm pore) were cut and glued to a modified 1-ml syringe barrel. Then 5 x 10^5 neutrophils were added to antagonist and 200 µl of the suspension was added to the upper chamber (syringe barrel). The lower chamber (a 5-ml beaker) contained 3.6 ml of agonist. Following incubation at 37°C for 45 min, the cells were fixed in 70% ethanol (v/v) for 5 min. This procedure also removes the filter from the syringe barrel by dissolving the glue. The filters were placed in a multi-well dish and treated as follows:

Distilled water for 2 min, Harris haemotoxylin for 1 min, distilled water for 1 min, Scotts Tap water [equal volumes of 0.7% sodium bicarbonate (w/v) and 4% magnesium sulphate (w/v)] for 5 min, 70% ethanol (v/v) for 3 min, 95% ethanol (v/v) for 3 min and 80%:20% ethanol : butanol (v/v) for 5 min. The filters were cleared in xylene for 5 min, mounted in DEPEX and examined under bright-field illumination with a 40x objective. Five randomly selected fields were counted for each filter. The leading front was determined as the focal plane that contained the two cells that had migrated furthest into the filter.

Results

Characteristics of neutrophil locomotion

Agonists were used at concentrations which caused similar, sub-maximal stimulation of basal motility. Previous studies of crude and partially purified STMS showed that it stimulated neutrophils to undergo dispersive locomotion in a uniform concentration. This was in marked contrast to responses to other agents and in particular to fMLP where the locomotory characteristics suggested that the cells, though highly motile, could not readily break their initial adhesions to the substrate. In addition to the determination of quantitative locomotory parameters, observation of the cells during the assay showed very clear and characteristic patterns of behaviour when the cells were treated with different stimulating agents. Neutrophils exposed to STMS rapidly become phase-dark, corresponding to flattening and adhesion, but then regain the phase-bright state and become motile. In contrast IL-8 induces characterised cyclic behaviour in which the cells darken and brighten reversibly, whilst cells treated with fMLP remain phase-bright. STMS-treated cells also showed a very characteristic appearance, giving the subjective impression that the cell is attached to the substrate at a single site while the cell body is dynamically active above it. All other stimulants tested appeared to cause the neutrophils to form several attachment points to the substrate. This observation is consistent with previous adhesion studies which showed that STMS-treated neutrophils were very readily washed off a protein-coated glass surface. These observation suggested that treatment of neutrophils with STMS induced a highly unusual mode of cytoskeletal organisation, but did not cast any direct light on the underlying basis for persistent locomotion. The apparent correlation of behaviour with adhesion under the light microscope was therefore extended by SEM and confocal microscopy studies.

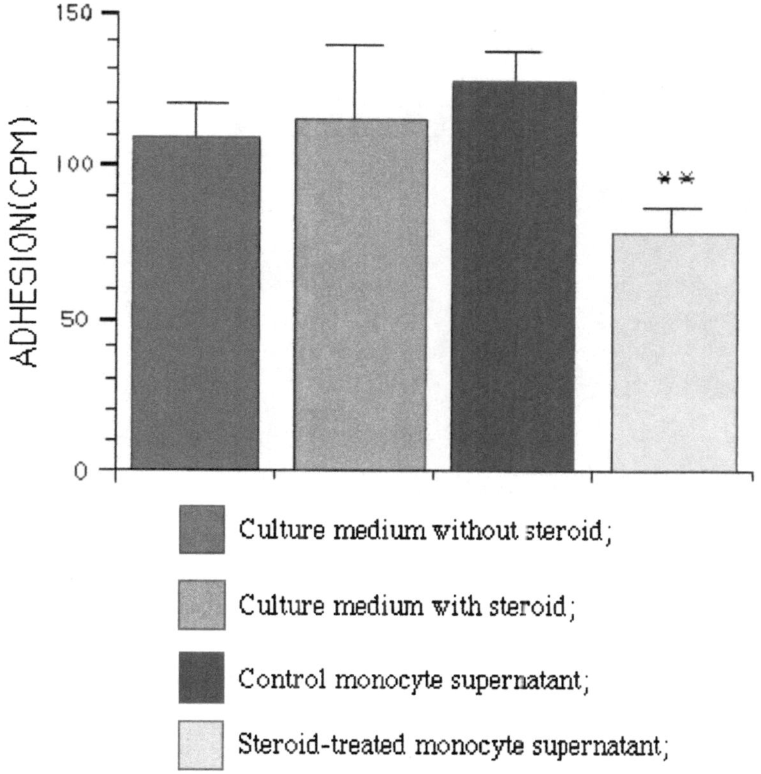

Fig. 1 Effect of supernatants from human monocytes cultured for 24 hours at 37°C, with and without 10^{-6} M dexamethasone, on the adhesion of human neutrophils to bovine aorta endothelial cell monolayers. **A** culture medium without dexamethasone; **B** culture medium with 10^{-6} M dexamethasone; **C** control monocyte supernatant (CMS); **D** steroid-treated monocyte supernatant (STMS).

Mean ± SEM (*vertical bars*) $n = 3$. $P < 0.05$ given by **

Neutrophil adhesion to bovine aorta endothelial cells

Our previous data showing that STMS diminished neutrophil adhesion to protein-coated glass (Chettibi et al., 1993) were extended using the more physiological substrate of bovine aorta and human endothelial cells (Fig. 1).

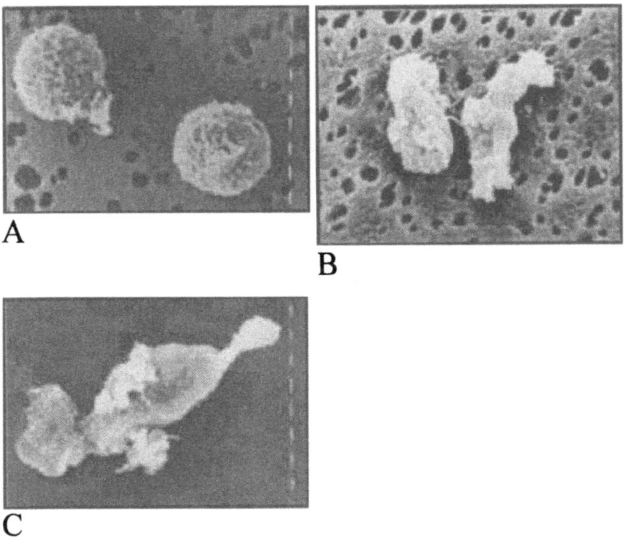

Fig. 2A–C Comparison of the morphology of human neutrophils treated with STMS and various neutrophil locomotion stimulators, using scanning electron microscopy. **A** Control (culture medium); **B** fMLP 10^{-8}M and **C** STMS. *Bar* 2 μm

Scanning electron microscopy

The results of this study demonstrated that control cells were round with no visible activation, whereas cells treated with fMLP showed classic polarisation with abundant areas of ruffled membrane and no clear leading edge (Fig. 2B). In contrast, the morphology induced by STMS was unique and readily described (Fig. 2C). The cells were of extended bipolar shape but the two ends that appeared to be involved in adhesive contact with the substratum were not identical, one showing an apparent focal adhesion, whilst the other was more rounded with no obvious adhesive focus. The membrane was relatively smooth with numerous small protrusions and there was no very clear difference in surface appearance between the cell body and the apparent pseudopodia.

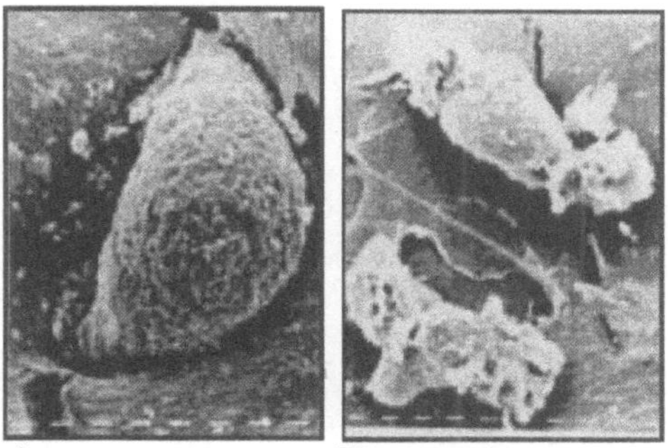

A B

Fig. 3A,B Comparison of the morphology of human neutrophils treated with STMS and various neutrophil locomotion stimulators and seeded on bovine aorta endothelial cell monolayers, using scanning electron microscopy. **A** fMLP (10^{-8} M); **B** STMS. *Bar* 2 μm

Neutrophils were seeded on bovine aorta endothelial monolayers for 30 min and prepared for SEM. This treatment made little difference to the appearance of the surface membrane, which was ruffled in response to fMLP, but relatively smooth in response to STMS. The most significant difference was that whereas fMLP-treated neutrophils (Fig. 3A) were almost exclusively found at endothelial cell junctions, STMS-treated cells (Fig. 3B) tended to be found on the body of the endothelial cell.

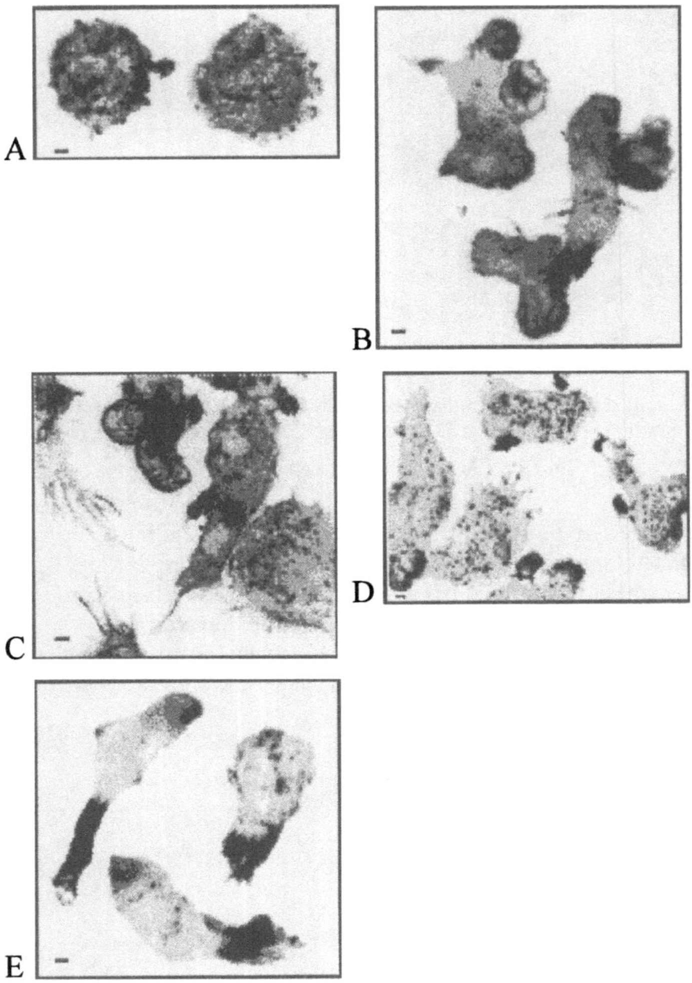

Fig. 4A–E Comparison of F-actin distribution in human neutrophils treated with STMS and various neutrophil locomotion stimulators. *Darker colours* indicate higher concentration of F-actin:

Black	blue	red	yellow	white
High »		»	low »	none

A Culture medium with and without 10^{-6} M dexamethasone; **B** 10^{-8} M fMLP; **C** CMS; **D** TNF[alpha]; **E** STMS. *Bar* 2 μm

Confocal microscopy

To investigate the underlying basis of the shape differences and to understand the nature of dispersive locomotion, we examined the distribution of polymerised actin in the cells. Control, non-activated neutrophils showed a weak, rather punctate distribution of fluorescence, but with no sign of a major polarised focus corresponding to a relatively even distribution of cortical actin (Fig. 4A). The results obtained for fMLP were similar to those presented by other workers (Coates et al., 1992) and are consistent with the interpretation that actin polymerisation is much more intense than in control cells and is associated with points of adhesive contact that are foci of active locomotion (Fig. 4B). Cells that were highly spread in response to CMS or to TNF[alpha] showed an extremely punctate distribution of F-actin (Fig. 4C,D). The pattern of actin staining in STMS-treated cells was highly unusual and distinct. Staining was only present in the extremes of the bipolar cells and of these two ends, which appear to be points of adhesive contact, one was invariably more intensely stained than the other (Fig. 4E).

Chemotaxis

Preliminary observations of neutrophil chemotaxis using the modified Boyden chamber revealed a striking contrast between the response to STMS and fMLP. Neutrophils showed massive invasion of the filter when fMLP was present in the lower chamber, but gave no such response to STMS. However, when the filters were examined by the leading front method it became clear that a few STMS-treated cells were able to invade successfully, in keeping with the predictions for persistent locomotion. These observations suggested that a more suitable analysis would be to measure total invasion, or to determine the number of cells present midway between the leading front and the surface of the filter. Figure 5 presents the analysis of chemotactic assays using both the leading front and the average invasiveness methods. The results showed that STMS was not itself chemotactic, but, when present in the upper chamber, dramatically inhibited the response to fMLP. In a uniform concentration of fMLP (i.e. present in both chambers) was a marked reduction in invasiveness.

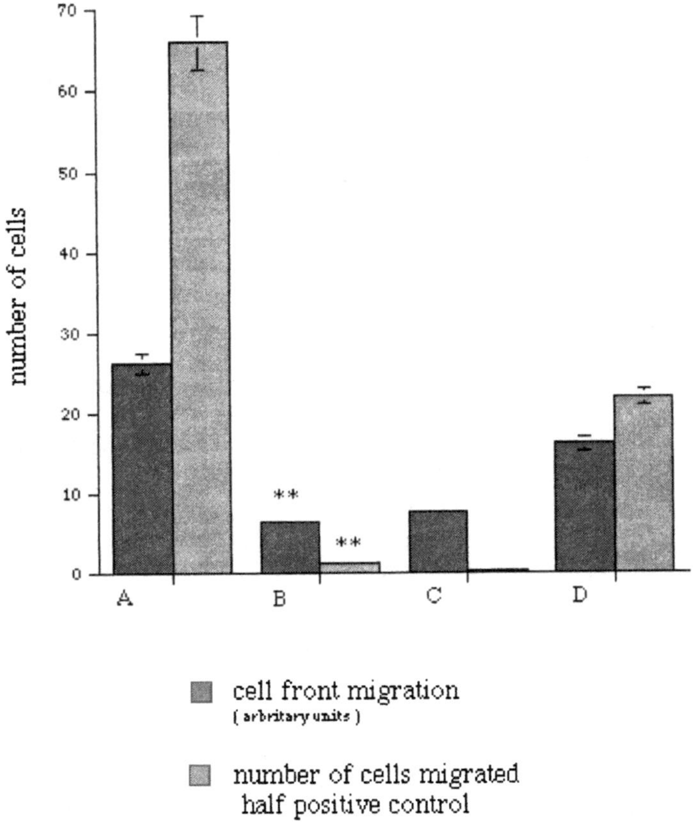

Fig. 5 The inhibition of fMLP-induced chemotaxis of human neutrophils by STMS measured using a modified Boyden chamber assay.

A *Upper chamber* – culture medium with dexamethasone; *lower chamber* – 10^{-9} M fMLP; **B** *upper chamber* – STMS; *lower chamber* –

10^{-9} M fMLP; **C** *upper chamber* – 10^{-9} M fMLP; *lower chamber* –10^{-9} M fMLP; **D** *upper chamber* – culture medium with dexamethasone; *lower chamber* – STMS. Number of cells which have migrated half of the mean migration distance of positive control (**A**). Cell-front migration distance. Five randomly selected fields were counted for each filter. Values shown are mean ± SEM (*vertical bars*) $P < 0.05$ given by ** against positive control (**A**)

Discussion

STMS activity was first observed as an agent or group of agents that enhanced neutrophil locomotory activity (Stevenson, 1974). This is not one of the expected features of an anti-inflammatory meditator and therefore the activity was not widely considered to be relevant to the possible mode of action of steroids as global anti-inflammatory agents. However, the original assay, as noted at the time, did not distinguish between the consequences of reduced adhesiveness and increased motility and it is generally accepted that the steps involved in adhesion of neutrophils to endothelium are a key target for anti-inflammatory agents (Gahmberg et al., 1992;Makgoba et al., 1992; Cronstein, 1994). STMS released into the blood stream by monocytes in response to glucocorticoids would present itself to neutrophils as a uniform concentration and could therefore modify neutrophil behaviour in very much the way that we observe in our assays. Thus, the cells should undergo some of the classic signs of activation, including polarisation and increased motility (Coates et al., 1992; Fernandez et al., 1995; Ehrengruber et al., 1996); however, they would also undergo a dramatic decrease in adhesiveness with the expected consequence that they would not be able to adhere under conditions of even modest shear stress.

The evidence presented here shows that the response to STMS contrasts sharply to the effects of the pro-inflammatory, chemotactic species exemplified by IL-8, fMLP and lipid derived pro-inflammatory mediators. The mechanisms of neutrophil chemotaxis are still not clearly understood, but the quantitative features have been well described. Chemotactic locomotion of human neutrophils in a gradient of fMLP, for example, is characterised by a persistent locomotion up-gradient in which the cells adopt a largely bipolar morphology (Tranquillo and Moghe, 1994). This bears superficial resemblance to the behaviour induced by a uniform concentration of STMS (Chettibi et al., 1994). However, the activity of the cells in a uniform concentration of chemotactic agent is characterised by very active exploratory motion, with pseudopodia forming at apparently random positions on the cell periphery and acting in conflict with one another. Each pseudopodium has high locomotory activity, but also high apparent adhesiveness to the substrate and the consequence for each cell is seen as high apparent locomotory activity as measured by the speed, with very low persistence (Chettibi et al., 1994). A simple interpretation of this behaviour is that it underlies the mechanisms of gradient sensing and that pseudopodia that reach up-gradient will be reinforced by the effects of the gradient and dominate the down-gradient pseudopodia. The observations suggest that adhesive effects may be more important than pure locomotory effects in determining the outcome of the conflict.

The role of actin polymerisation in activated neutrophils has been the subject of intensive study (Coates et al., 1992). Our results indicate that pseudopodia are the sites of high concentrations of F-actin and the observations of motile activity indicate that this actin will be in a state of constant reorganisation. The accepted mechanisms for pseudopodium-driven locomotion require some form of actin cycling between the leading edge and a juxtanuclear location behind it (Stossel, 1988). For a highly directed chemotactic response, virtually all of the actin could be concentrated into this one active area. In our experiments we have never observed such extreme bipolarity in response to uniform concentrations of chemotactic agents on protein-coated glass and neither do we see this morphology in cells treated, and then fixed, in suspension. However, STMS-treated neutrophils are highly bipolar and, furthermore, appear to have a major concentration of F-actin at one pole and a secondary, minor, concentration at the opposite pole. These cells do not have the overall degree of activity of cells responding to a chemotactic agent, but appear to be more highly motile because the locomotion is more persistent. The mechanism of locomotion is obscure because the cells appear to be weakly tethered to the glass by one "tail" whilst the other moves freely until the cell appears to attach to, and move to, a distant site. The obscurity lies in the fact that because the cell appears to have a single site of adhesion, and is freely motile, there is no obvious reason why movement should continue in the previously determined direction. It is also not clear what role actin may play in this type of locomotion. There are two extreme models: the looping model and the tumbling model. In the looping model the tail of the cell would be the major point of adhesion whilst the head would make exploratory movements and then adhere, causing the tail to detach and move close to the head. In the tumbling model each end would act in turn as an adhesion point and the corollary is that the polymerised actin would have to pass from one pole of the cell to the other. Here the speed of locomotion would be governed by the rate at which loss of actin from the trailing pole allowed the leading adhesion to dominate over the trailing adhesion. In this model the directed flow of actin to the distant pole would underly directional persistence.

The effect of STMS on chemotactic responses also sheds considerable light on its possible mode of action. Naturally, in the Boyden chamber type of assay, chemotactic agents in the upper well inhibit chemotactic responses to agents in the lower well, as seen here for fMLP. However STMS is not a chemotactic agent, but is a powerful inhibitor of the chemotactic response to fMLP. It appears to short-circuit chemotactic responses by inducing a highly polarised structure without any aspect of gradient sensing, but, unlike other polarising stimuli, it causes a dramatic decrease in adhesiveness (Chettibi et al., 1993). Observations of the morphology of the cells and of the internal organisation give some indication why this should be. The surface of STMS-treated neutrophils, as seen in SEM is readily distinguished from that induced by chemotactic agents by its extreme smoothness. There is evidence that some forms of adhesion, for example that mediated by L-selectin is greatly increased by the clustering of the lectin on microvilli (vonAndrian et al., 1995). Following this example we propose that neutrophil adhesion may depend on the presence of surface projections, the lack of which in STMS-treated cells could explain the lowered levels of adhesion observed.

Acknowledgments. The authors would like to thank Professor A.S.G. Curtis for his helpful discussion. This work was supported by the Sylvia Aitken Trust, the Fraser Foundation and a grant from the Wellcome Trust.

References

von Andrian, U.H., Hasslen, S.R., Nelson, R.D., Erlandsen, S.L., Butcher, E.C. (1995) A central role for microvillous receptor presentation in leukocyte adhesion under flow. *Cell* **82**: 989–999.

Arnaout, M.A. (1993) Dynamics and regulation of leukocyte-endothelial cell interactions. *Curr. Opin. Haematol.* 1:113–122.

Bevilacqua, M.P., Nelson, R.M. (1993) Selectins. *J. Clin. Invest.* **91**: 379–387.

Chettibi, S., Lawrence, A.J., Stevenson, R.D., Young, J.D. (1993) A factor released by monocytes in the presence of dexamethasone stimulates neutrophil locomotion. *Br. J. Pharmacol.* **108**: 248–254.

Chettibi, S., Lawrence, A.J., Young, J.D., Lawrence, P.D., Stevenson, R.D. (1994) Dispersive locomotion of human neutrophils in response to a steroid-induced factor from monocytes. *J. Cell Sci.* **107**: 3173–3181.

Coates, T.D., Watts, R.G., Hartman, H., Howard, T.H. (1992) Relationship of F-actin distribution to development of polar shape in human polymorphonuclear neutrophils. *J. Cell Biol.* **117**: 765–774.

Cronstein, B.N., Kimmel, S.C., Levin, R.I., Martinuik, F., Weissmann, G. (1992) A mechanism for the anti-inflammatory effects of corticosteroids. The glucocorticoid receptor regulates leukocyte adhesion to endothelial cells and expression of endothelial leukocyte adhesion molecule-1 and intercellular adhesion molecule-1. *Proc. Natl. Acad. Sci. USA* **89**: 9991–9995.

Cronstein, B.N. (1994) Adhesion molecules in inflammation: current research and new therapeutic targets. *Clin. Immunother.*1:323–326.

Ehrengruber, M.U., Deranleau, D.A., Coates, T.D. (1996) Shape oscillations of Human Neutrophil Leukocytes: characterisation and relationship to cell motility. *J. Exp. Biol.* **199**: 741–747.

Fernandez, S., Garcia, J.M., Santos, J.L., Campos, A. (1995) Shape, F-actin, and surface-morphology changes during chemotactic peptide-induced polarity in human neutrophils. *Anat Rec* **241**:519–528.

Flower, R.J. (1990) Lipocortin. In:*Cytokines and lipocortins in inflammation and differentiation.* Wiley-Liss, New York, pp. 11–25.

Gahmberg, C.G., Nortamo, P., Li, R., Valmu, L. (1992) Leukocyte cell_adhesion proteins. From molecular dissection to clinical applications. *Ann Med* **24**: 329–335.

Goulding, N.J., Godolphin, J.L., Sharland, P.R., Peers, S.H., Sampson, M., Maddison, P.J., Flower, R.J. (1990) Anti-inflammatory lipocortin 1 production by peripheral blood leukocytes in response to hydrocortisone. *Lancet* **335**: 1416–1418.

Hogg, N., Berlin, C. (1995) Structure and function of adhesion receptors in leukocyte trafficking. *Immunol Today* **16**: 327–330.

Kwon, O.J., Collins, P.D., Au, B., Adcock, I.M., Yacoub, M., Chung, K.F., Barnes, P.J. (1993) Glucocorticosteroid inhibition of TNF-alpha-induced IL-8 gene expression in human primary cultured airway epithelial cells (abstract). *Am Rev. Respir. Dis.* **147**: A752.

Lew, W., Oppenheim, J.J., Matsushima, K. (1988) Analysis of the suppression of IL-1[alpha] and IL-1[beta]fl production in human peripheral blood mononuclear adherent cells by a glucocorticoid hormone. *J. Immunol.* **140**: 1895–1902.

Makgoba, M.W., Bernard, A., Sanders, M.E. (1992) Cell adhesion/signalling: biology and clinical applications. *Eur. J. Clin. Invest.* **22**: 443–453.

Stevenson, R.D. (1974) Polymorph migration stimulator, a new factor produced by hydrocortisone-treated monocytes. *Clin. Exp. Immunol.* **17**: 601–606.

Stossel, T.P. (1988) The mechanical responses of white blood cells. In: Gallin, J.I., Goldstein, I.M., Snyderman, R. (eds) *Inflammation: basic principles and clinical correlates.* Raven Press, New York, pp. 325–342.

Tranquillo, R.T., Moghe, P.V. (1994) Stochastic model of chemoattractant receptor dynamics in Leukocyte chemosensory movement. *Bull. Mathemat. Biol.* **56**: 1041–1093.

Wong, W.T., Frost, S.C., Nick, H.S. (1991) Protein-synthesis dependent induction of annexin 1 by glucocorticoid. *Biochem. J.* **275**: 313–319.

Exp. Biol. Online 2:8 (1997)

Structural and physiological properties of leech giant glial cells as studied by confocal microscopy

C. Lohr and J.W. Deitmer

Abteilung für Allgemeine Zoologie, FB Biologie, Universität Kaiserslautern, Postfach 3049, D-67653 Kaiserslautern, Germany

Correspondence to: C. Lohr (email: clohr@rhrk.uni-kl.de)

Received: 2 April 1997 / Accepted: 12 May 1997
Published: June 3, 1997

Abstract. We have studied the architecture of giant neuropile glial cells of the medicinal leech *Hirudo medicinalis* L. using confocal laser scanning microscopy. We also measured changes in the intracellular Ca^{2+} concentration ($[Ca^{2+}]_i$) induced by activation of glutamate receptors or voltage-gated Ca^{2+} channels in different glial cell compartments. Glial cells of isolated segmental ganglia were filled iontophoretically with the Ca^{2+} indicator dye Fluo-3. The three-dimensional structure, calculated from serial sections, showed that numerous fine glial branches extend within the whole neuropile, where most of the synapses between neurones are established. Activation of glial glutamate receptors by glutamate or kainate, or depolarizing the cell membrane by elevating the external K^+ concentration resulted in a transient increase in $[Ca^{2+}]_i$, as measured by Fluo-3 fluorescence. The comparison of $[Ca^{2+}]_i$ changes in glial cell branches with changes in the cell body demonstrated that transients in the branches were 2–3 times larger than those in the cell body. The results suggest that glutamate receptors and voltage-gated Ca^{2+} channels are located in the membrane not only of the glial cell body but also of the cellular branches, which may extend close to synaptic domains.

Key words: Fluo-3 – Glutamate receptors – *Hirudo medicinalis* – Intracellular calcium – Three-dimensional images – Voltage-gated calcium channels

Introduction

Pioneering studies of the leech central nervous system by Kuffler and Potter (1964) and Coggeshall (1974) have addressed morphological and physiological properties of the glial cells. In recent years, neuron-glia interactions in this preparation have been studied (reviewed by Deitmer and Munsch 1994), and it was also shown that the glial cell membrane potential and intracellular Ca^{2+} concentration ($[Ca^{2+}]_i$) are affected by glutamate-mediated neuronal excitation in the leech central nervous system (Rose et al. 1995). The synaptic regions are within the central neuropile, where two large glial cells are located (Coggeshall and Fawcett 1964). We have re-investigated some morphological and physiological relationships of glial cells with respect to neuron-glia interactions in the leech ganglia using confocal laser scanning microscopy.

We studied the three-dimensional structure of the giant neuropile glial cells of the leech and changes in Ca^{2+}-dependent Fluo-3 fluorescence induced by voltage- and ligand-gated ion channel activation in different glial cell compartments.

Materials and Methods

Preparation

Leeches (*Hirudo medicinalis* L.) were purchased from a local supplier and held in freshwater tanks at room temperature. Before an experiment, individual segmental ganglia were dissected and pinned, ventral side upwards, into a Sylgard-lined experimental chamber, as described previously (Munsch and Deitmer 1992). The ventral part of the outer capsule was then removed, and the neurones overlying the neuropile glial cells were sucked off with a pipette after treatment with 2 mg/ml collagenase/dispase (Boehringer-Mannheim).

Solutions

The normal superfusion saline contained (in mM): NaCl, 85; KCl, 4; $CaCl_2$, 2; $MgCl_2$, 1; 4-(2-hydroxyethyl)-1-piperazineethanesulphonic acid (HEPES), 10; pH adjusted to 7.4 with NaOH. In solutions with a raised K^+ concentration, equimolar amounts of NaCl were replaced by KCl. L-Glutamate and kainate (Sigma) were kept as stock solutions and added to the saline shortly before an experiment.

Confocal fluorescence microscopy

Microelectrodes were filled with a solution of 10 mM of the Ca^{2+} indicator dye Fluo-3 (Molecular Probes, Eugene, Ore., USA) and bevelled to a resistance of 150–180 MOhm. A microelectrode was inserted into the glial cell body, and the dye was injected iontophoretically by a constant negative current of 1.5–2 nA. For about 45 min, Fluo-3 fluorescence in the cell body and proximal branches increased and then reached a constant value, indicating that dye diffusion into distal cell branches and out of the cell was compensated for by continuous dye injection. Measurements were carried out when Fluo-3 fluorescence was near constant in the compartments of interest. Fluo-3 fluorescence was monitored by a Leica TCS 4D confocal laser scanning microscope (Leica, Bensheim, Germany) equipped with a 10×/0.2 objective (Leica Plan) and a long-distance 40×/0.75 water immersion objective (Zeiss Achroplan). The lateral resolution of the latter at a wavelength of 525 nm, the emission maximum of the dye Fluo-3, was approximately 430 nm, referring to the numerical aperture of 0.75. The axial resolution has been determined experimentally and varied between about 1.5 µm at a pinhole diameter of 20 µm, and 12 µm at a pinhole diameter of 600 µm.

The dye was excited by the 488 nm line of an Kr/Ar laser, and fluorescence emission was collected by a photomultiplier using a 535 ± 15 nm bandpass filter. Images (8 bit dynamic range) were sampled and evaluated by the use of PC software (Scanware, Leica). Stereo images were calculated from a stack of optical serial sections and can be viewed through red/green spectacles (red: right eye).

In order to calculate changes in $[Ca^{2+}]_i$, a sequence of images in a constant focal plane was sampled at a rate of 0.125–0.67 Hz. Regions of interest were defined at the first image, and the mean fluorescence intensities of these regions were measured at each image. Since Fluo-3, in contrast to Fura-2, does not allow dual excitation ratiometric measurements, the fluorescence is given as the ratio F/F_0, were F is the fluorescence intensity at any time and F_0 is the basic fluorescence intensity at the beginning of an experiment (see also Eilers et al. 1995).

Ratiometric $[Ca^{2+}]_i$ measurements were performed, as described previously (Munsch and Deitmer 1995), with a conventional fluorescence microscope, using the Ca^{2+}-sensitive dye Fura-2. In brief, glial cells were filled iontophoretically with Fura-2. The dye was excited alternately at 350 and 380 nm, and the fluorescence intensity was measured at 515 ± 15 nm. $[Ca^{2+}]_i$ was calculated referring to the method of Grynkiewicz et al. (1985).

Statistics

Measurements are given as mean values ± standard error of mean (SEM) with n indicating the number of experiments. Statistical differences were checked using the paired t-test.

Results

Structure of the segmental ganglion

Stereo images of the three-dimensional structure of an intact midbody segmental ganglion, sampled from autofluorescence, are shown in Fig. 1A, B. Figure 1A shows the dorsal aspect, where the lateral cell packets are separated by the median axonal pathways which enter the ganglion from the connectives. Each packet contained the somata of 50–70 neurones, enveloped by a packet glial cell. The diameter of the neurones varied between 10 and 65 µm. A pair of nerve roots left the ganglion at both sides. The ventral aspect (Fig. 1B) shows, besides the lateral packets, the ventro-medial cell packets.

A schematic cross-section of an intact ganglion is given in Figure 1C. The ganglion consists of the peripheral cell packets and the central neuropile. The neuropile comprises neuronal processes, synaptic regions, and two giant neuropile glial cells. The neuropile glial cells were exposed after removing the ventro-medial cell packets (Fig. 1D).

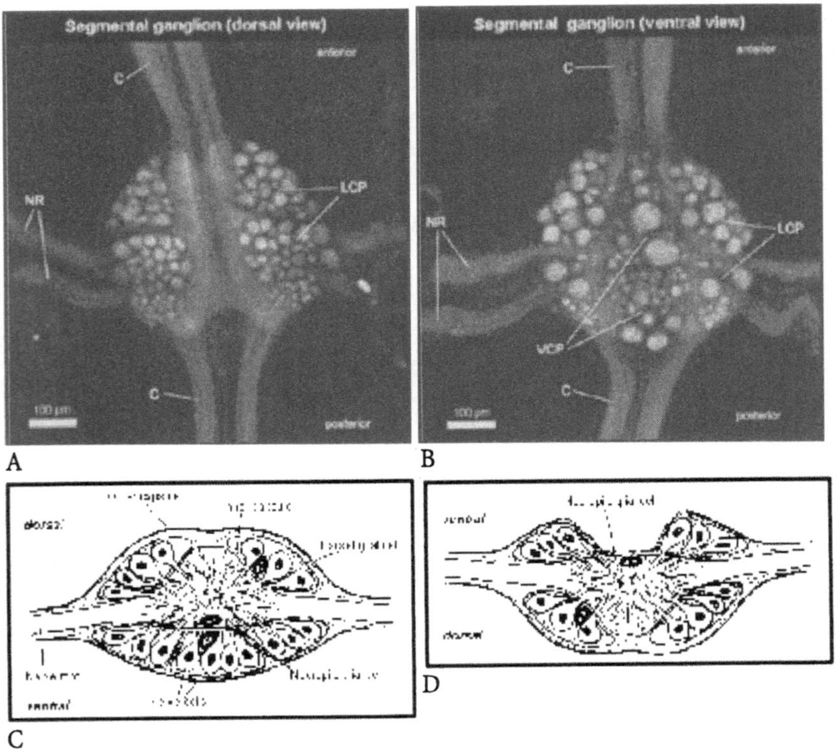

Fig. 1A Red/green stereo image of the dorsal aspect of a midbody segmental ganglion. **B** Ventral aspect of a midbody segmental ganglion. **C** Schematic cross-section of a segmental midbody ganglion (modified from Kuffler and Potter, 1964). **D** Schematic cross-section of the *in situ*

preparation of the neuropile glial cell. (*C* Connective, *LCP* lateral cell packets, *VCP* ventro-medial cell packets, *NR* nerve root)

The fine structure of the giant neuropile glial cell

At higher magnification, the fine structure of the giant neuropile glial cell could be examined. Figure 2 compares a confocal image of a posterior glial cell (Fig. 2A) with an image taken by a conventional microscope (Fig. 2B). The glial cells consist of a cell body of up to 100 µm in diameter and enormously elaborate arborizations. These cell branches extend within the whole neuropile, which is the central part of the ganglion and which is filled with neurites and synaptic contacts. Compared to the confocal image, the conventional image displayed less contrast, and the fine branches appeared blurred due to out-of-focus fluorescence.

As shown in Fig. 2A, some of the glial processes passed to the anterior part of the neuropile and made contacts with the anterior glial cell via gap junctions. A small amount of the dye had diffused into the anterior glial cell (lower right) as well as into the neighbouring packet glial cell (upper right).

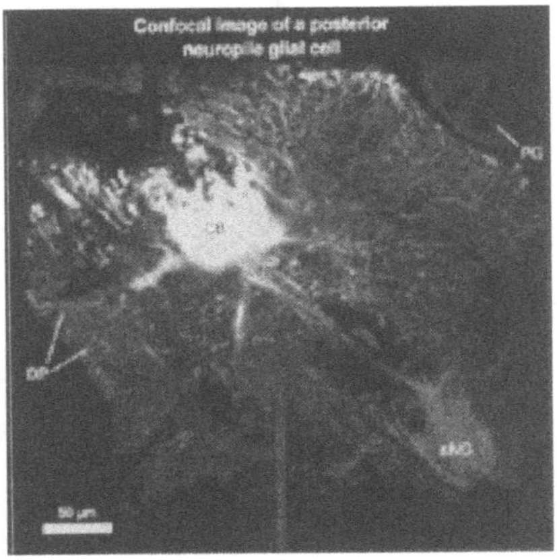

A

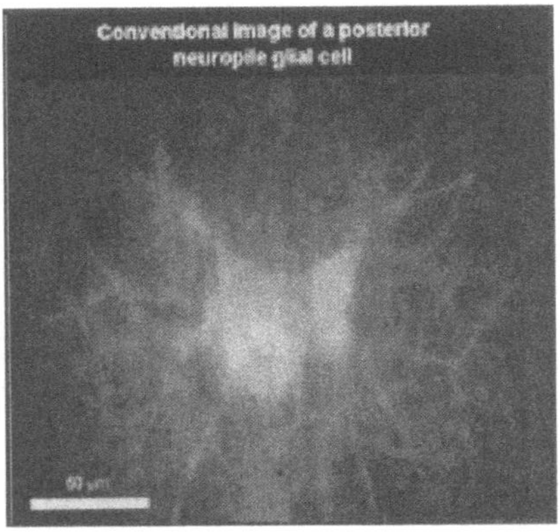

B

Fig. 2A Confocal fluorescence image of a posterior neuropile glial cell, filled with Fluo-3. The fine dendritic processes of the cell extend within the whole neuropile. Some dye passed through gap junctions into the anterior neuropile glial cell (*lower right*), and into a packet glial cell (*upper right*). **B** Conventional fluorescence image of a posterior neuropile glial cell (modified from Pfeiffer 1992). [*CB* Cell body (of the posterior glial cell), *DP* dendritic processes, *aNG* anterior neuropile glial cell, *PG* packet glial cell]

The giant glial cell arborization could be estimated by combining images from different spatial depth, as shown in Fig. 3. The maximum depth of tissue at which images of appropriate quality could be sampled was found to be approximately 100 µm. Figure 3A is a projection from 40 confocal serial sections within a depth of approximately 40–80 µm. It demonstrates that the fine branches arborized in all spheres of the neuropile. In most cases, fine individual branches with a diameter of down to 1 µm were only a few micrometres apart from one another. This is further illustrated by the stereo image in Fig. 3B. This image was calculated from a stack of 25 confocal images, each one at an axial distance of 1 µm from the next. It gives a three-dimensional impression of glial branching (use red/green spectacles).

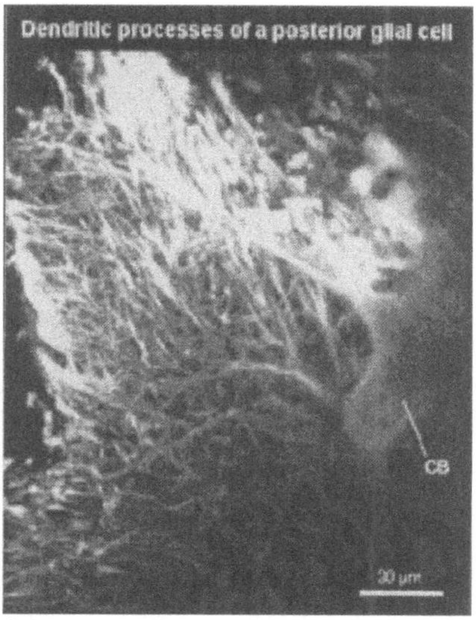

A

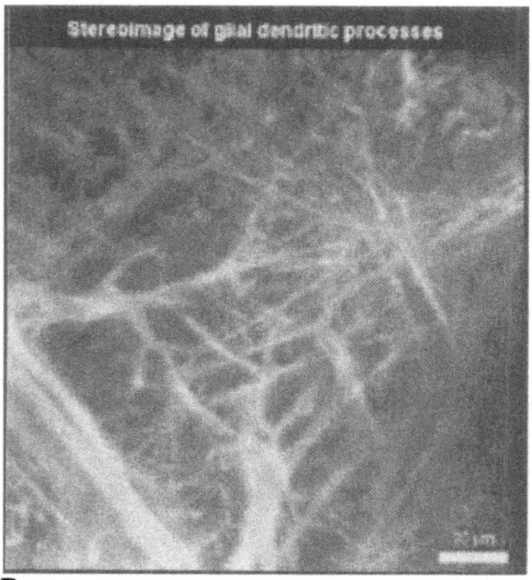

B

Fig. 3A A part of the arborization of a posterior neuropile glial cell, filled with Fluo-3. The image was projected from a stack of 40 serial sections, each at an axial distance of 1 µm from the next. (*CB* Part of the glial cell body) **B** Stereo image of glial processes, calculated from a stack of 25 confocal serial sections (use red/green spectacles)

The dye coupling between the giant neuropile glial cells and the packet glial cells is further demonstrated in Fig. 4. Several neuronal cell bodies were wrapped by a thin layer of a packet glial cell. The glial cell accompanied the axons of the unipolar neurones to the inner ganglion capsule where the axons entered the neuropile. The neurones and the packet glial cell were embedded in tissue which only weakly emitted fluorescence.

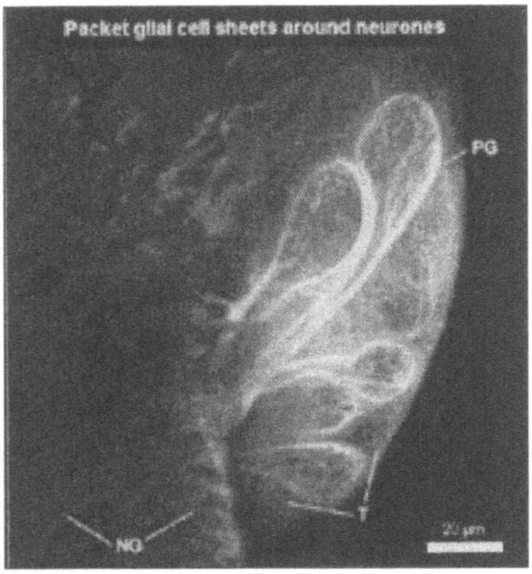

Fig. 4 Thin packet glial cell sheets, filled with Fluo-3, enveloping neuronal cell bodies. The dye was injected into a neuropile glial cell, and then passed through gap junctions into the packet glial cell. (*PG* Packet glia, *NG* branches of a neuropile glial cell, *T* tissue)

Confocal Ca^{2+} imaging

In previous reports, $[Ca^{2+}]_i$ transients evoked by agonist application or membrane depolarization were measured by ratiometric Fura-2 fluorescence microscopy in the neuropile glial cell, where fluorescence was collected only from the large cell body (Deitmer and Munsch 1992; Munsch and Deitmer 1997). We have now compared increases in $[Ca^{2+}]_i$, as indicated by Fluo-3 fluorescence changes, in different compartments of the glial cells. Fluorescence changes in glial cell bodies were measured by conventional and confocal fluorescence microscopy (Fig. 5). $[Ca^{2+}]_i$ transients were evoked by elevating the external K^+ concentration from 4 to 20 mM to depolarize the cell, which opens voltage-gated Ca^{2+} channels, and by application of the excitatory amino acid glutamate (500 μM) and the ionotropic glutamate receptor agonist kainate (10 μM).

Compared to the measurements made by the conventional technique, the fluorescence changes measured by confocal imaging showed approximately similar time courses and relative amplitudes (Fig. 5). The mean $[Ca^{2+}]_i$ increase evoked by high K^+ was 397.4 ± 86.9 nM (n=4), as measured by conventional microscopy, and was reflected by a Fluo-3 fluorescence increase of 0.5 ± 0.17 ratio units (n=4) in the confocal microscopy image. Application of 500 μM glutamate elicited an increase in $[Ca^{2+}]_i$ of 30.2 ± 3.6 nM (n=11) and 0.14 ± 0.02 ratio units (n=11), respectively, whereas 10 μM kainate caused $[Ca^{2+}]_i$ to increase by 58.6 ± 7.4 nM (n=11) and the Fluo-3 fluorescence by 0.22 ± 0.02 (n=4) ratio units.

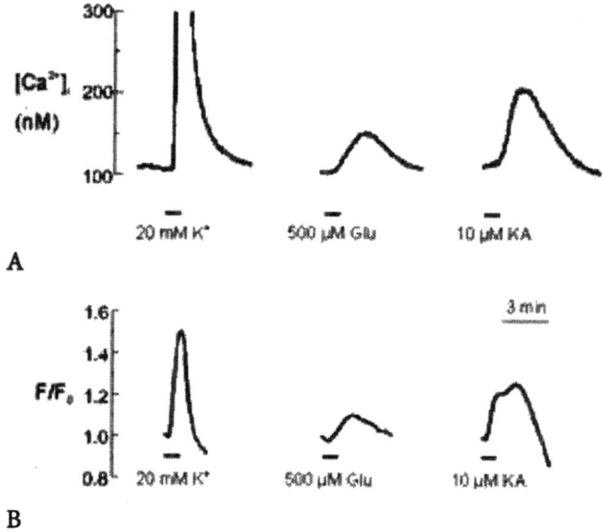

Fig. 5 Comparison of $[Ca^{2+}]_i$ transients measured in the glial cell body by **A** conventional (Fura-2) and **B** confocal microscopy (Fluo-3). $[Ca^{2+}]_i$ transients induced by an elevation of the external K^+ concentration from 4 to 20 mM (20 mM K^+), or by application of glutamate (*500 μM Glu*) or kainate (*10 μM KA*) showed similar time courses and relative amplitudes with both methods

An example of a rise in Fluo-3 fluorescence induced by the elevation of the external K^+ concentration from 4 to 20 mM for 1 min is given in Fig. 6 (MOV file, 330kB). It shows a sequence of 36 confocal images (1 frame/8 s). The fluorescence increased in all parts of the glial cell, indicating that $[Ca^{2+}]_i$ not only rose in the cell body, but also in the cellular branches.

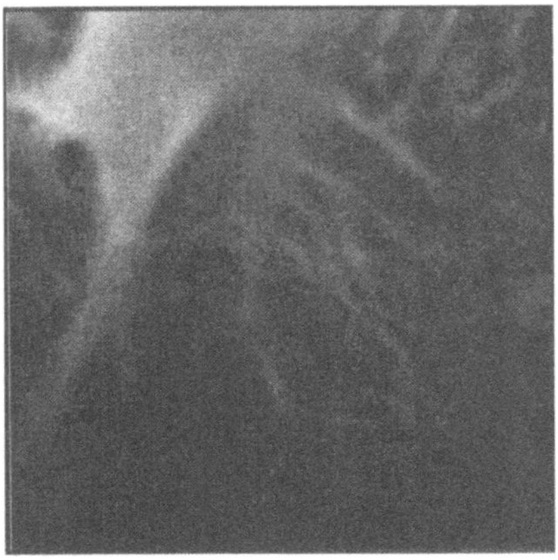

Fig. 6 Changes in the Ca^{2+}-dependent Fluo-3 fluorescence in a posterior glial cell induced by elevating the external K^+ concentration from 4 to 20 mM (1 min), displayed in an image sequence (sample rate: 1 frame/8 s)[MOV file, 330kB (click on picture to start)]

Spatial distribution of agonist-induced $[Ca^{2+}]_i$ changes

The spatial distribution of $[Ca^{2+}]_i$ changes induced by glutamate receptor agonists or high K^+ in the glial cells was studied by taking Fluo-3 fluorescence images. The mean fluorescence from the different cell compartments was measured throughout a sequence of confocal images, which were sampled in one focal plane over a certain period (3–7 min).

Transient $[Ca^{2+}]_i$ changes evoked by application of 500 μM glutamate for 1 min, measured at the indicated locations, are shown in Fig. 7. $[Ca^{2+}]_i$ transients measured in the glial cell branches rose 2–3 times above those measured in the cell body. On average, 500 μM glutamate evoked fluorescence rises of 0.14 ± 0.02 ratio units (n=11) in the cell body, and of 0.31 ± 0.04 ratio units (n=18) in the branches (Fig. 8). Kainate (50 μM, 30 s) elicited changes of 0.14 ± 0.02 ratio units (n=10) and 0.46 ± 0.05 ratio units (n=18), in the cell body and branches, respectively, while changes induced by high K^+ reached 0.5 ± 0.17 ratio units (n=4) and 1.15 ± 0.11 ratio units. Thus, in all three cases fluorescence changes in the cell branches were significantly larger than those in the cell body (P<0.01).

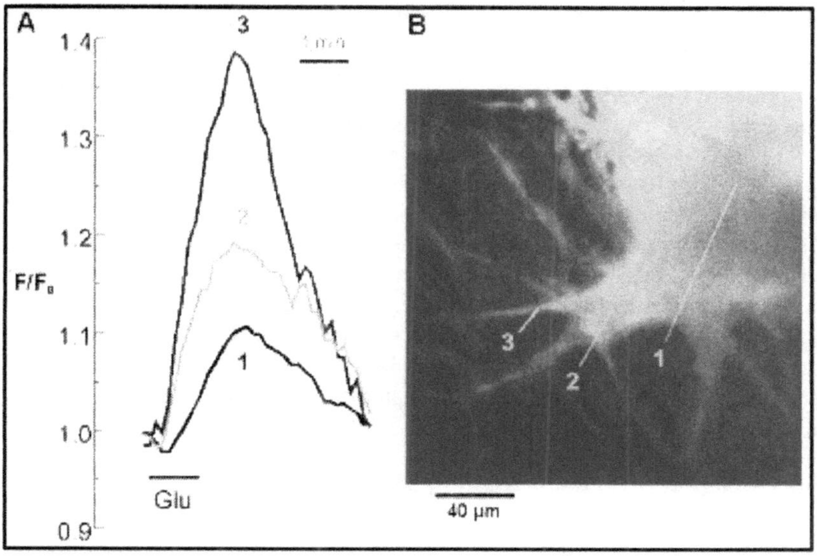

Fig. 7 Glutamate-induced $[Ca^{2+}]_i$ changes **A** in different compartments of a neuropile glial cell, as indicated in the original image in **B**: (*1* Cell body, *2,3* cell branches)

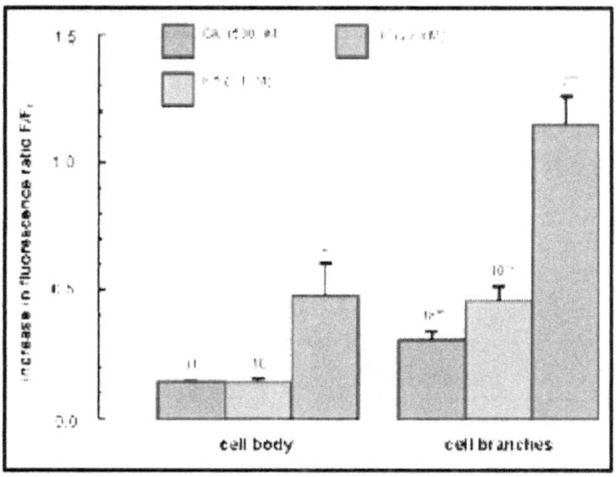

Fig. 8 Summary of the Ca^{2+}-dependent Fluo-3 fluorescence changes induced by 500 µM glutamate (*Glu*, 1 min), 50 µM kainate (*KA*, 30 s), or 20 mM external potassium (*K+*, 1 min) in different cellular compartments. Fluorescence changes in glial cell branches were significantly larger than those in the cell body ($P<0.01$)

Discussion

The structure of the central nervous system of the medicinal leech has been examined using light and electron microscopes (Retzius 1891; Coggeshall and Fawcett 1964; Payton 1981). Transmission electron microscopy, although indispensable for ultrastructural studies, has the disadvantages that only fixed material can be examined and that the three-dimensional reconstruction from serial sections spans only a few micrometres in depth or less. On the other hand, three-dimensional images from confocal microscopy have to be interpreted carefully, since they may imply artefacts due to geometric distortions and photometric errors, such as chromatic aberrations (White et al. 1996). In the present study, we have shown data from serial sections of living cells *in situ* within a range of up to 50 µm in depth revealed by optical sectioning with a confocal laser scanning microscope.

The morphology of the giant neuropile glial cell

Compared to images taken by conventional fluorescence microscopy, confocal images are distinguished by an enhanced contrast and a more detailed reproduction of fine glial branches, even within a depth of almost 100 µm in the tissue. In addition, confocal microscopy enables the recording of optical sections of a thickness of 1–2 µm. These optical sections can be used to reconstruct the cell branches and give detailed information about the structure in three dimensions. The three-dimensional architecture of the neuropile glial cell indicates that glial branches pass through the whole neuropile. It has been demonstrated by electron microscopy studies that most of the synapses are established within the neuropile (Coggeshall and Fawcett 1964), and the present data confirm that glial branching and synaptic regions may build a close morphological unit.

The two giant neuropile glial cells of a ganglion, as well as the neuropile glial cells and the packet glial cells, have been shown to be dye-coupled (Figs. 2A, 4). Dye-coupling between leech glial cells has also been reported by Pfeiffer (1992). Presumably, the fluorescence dye passes through gap junctions which build low-resistance junctions between the cells (Kuffler and Potter 1964); this had been also inferred from electron microscopy studies (Coggeshall 1974). The packet glial cells wrap the neuronal cell bodies in a thin layer, whereas the major part of the interstices of the neurones is filled with tissue that is not, or only poorly, dye-coupled to the neuropile glial cell, and thus is unlikely to be part of the packet glia. This is in contrast to earlier studies, in which the packet glial cells have been described to fill *all* the interstices of the neurones (Coggeshall and Fawcett 1964; Kuffler and Potter 1964). However, this was not further examined in the present study.

$[Ca^{2+}]_i$ transients

In the leech giant glial cell, activation of glutamate receptors may lead to Ca^{2+} influx from the extracellular space and Ca^{2+} release from internal stores (Deitmer and Munsch 1992; Lohr et al. 1996a; Munsch and Deitmer 1997). Application of 20 mM K^+ depolarizes the glial cell membrane sufficiently to open voltage-gated Ca^{2+} channels (Munsch and Deitmer 1992, 1997). Ca^{2+} influx into the cells, induced by transmitter application or nervous activity, is reflected by an decrease in the extracellular Ca^{2+} concentration, which can be measured with Ca^{2+}-selective microelectrodes in the extracellular space of the synaptic regions of the neuropile (Rose et al. 1995; Lohr et al. 1996b).

Since synaptic regions are expected to be located closer to glial cell branches than to the large glial cell body, differences in $[Ca^{2+}]_i$ transients induced by ligand- or voltage-gated ion channel activation in these distinct cell compartments were of particular interest. Application of glutamate, or of the ionotropic glutamate receptor agonist kainate, as well as depolarizing the cell by an elevated external K^+ concentration elicited Ca^{2+}-dependent fluorescence changes in the cell bodies.

The most striking result is that the $[Ca^{2+}]_i$ changes in the glial branches exceed the changes in the somata by 2 to 3 times. This may be explained by the higher surface/volume ratio of fine structures compared to that of the cell body. On the other hand, an unequal distribution of ion channels or receptors in the glial membrane cannot be ruled out.

When comparing $[Ca^{2+}]_i$ transients in different cell compartments it should be kept in mind that the concentration of applied glutamate or K^+ may be higher at the exposed glial cell body than at cell branches, since diffusion within the tissue of the ganglion may be hindered by the narrow extracellular spaces. In addition, these glial cells have been shown to take up glutamate (Deitmer and Schneider 1997) or K^+ ions (Wuttke 1990). These uptake processes may operate more efficiently at the larger surface/volume ratio of the branches compared to the cell body, which might reduce the effective concentration of glutamate or K^+ near the fine branches in the neuropile. Both mechanisms would tend to *decrease* the $[Ca^{2+}]_i$ responses in the cell branches; however, we observed a relative increase in the $[Ca^{2+}]_i$ responses to glutamate and high K^+ in the cell branches as compared to the cell body.

Transmitter-induced $[Ca^{2+}]_i$ transients measured in the glial cell body by conventional fluorescence microscopy usually reach amplitudes of several tens or hundreds of nM (Deitmer and Munsch 1992; Munsch and Deitmer 1992, 1997; this study). Therefore, $[Ca^{2+}]_i$ transients in glial branches are likely to exceed 1 μM, and activity-induced decreases in the extracellular Ca^{2+} concentration may partly be due to a significant Ca^{2+} influx into glial cell branches. This large intracellular Ca^{2+} rise may have profound effects on cell metabolism and/or different ion channel activity in glial cell processes, including those in the vicinity of synaptic domains.

Acknowledgments. We thank Drs. Jens Eilers and Arthur Konnerth (I. Physiologisches Institut der Universität des Saarlandes, Homburg, Germany) for introducing us to confocal Ca^{2+} imaging, and W. Nett (Allgemeine Zoologie, Kaiserslautern) for comments on an earlier version of the manuscript. C. Lohr is a recipient of a stipend from the Graduierten-Förderungsprogramm des Landes Rheinland-Pfalz. This study was supported by a grant from the Deutsche Forschungsgemeinschaft (De 231/9-3).

References

Coggeshall, R.E. (1974) Gap junctions between identified glial cells in the leech. *J. Neurobiol.* **5**:463–467

Coggeshall, R.E., Fawcett, D.W. (1964) The fine structure of the central nervous system of the leech, *Hirudo medicinalis. J. Neurophysiol.* **27**:229–289

Deitmer, J.W., Munsch, T. (1992) Kainate/glutamate-induced changes in intracellular calcium and pH in leech glial cells. *NeuroReport* **3**:693–696

Deitmer, J.W., Munsch, T. (1994) Neurone-glia dialogue in the leech central nervous system: glial cell responses to glutamate and kainate. *Verh. Dtsch. Zool. Ges.* **87**:185–194

Deitmer, J.W., Schneider, H.-P. (1997) Intracellular acidification of the leech giant glial cell evoked by glutamate and aspartate. *Glia* **19**:111–122

Eilers, J., Augustine, G.J., Konnerth, A. (1995) Subthreshold synaptic Ca2+ signalling in fine dendrites and spines of cerebellar Purkinje neurones. *Nature* **373**:155–158

Grynkiewicz, G., Poenie, M., Tsien, R.Y. (1985) A new generation of Ca2+ indicators with greatly improved fluorescence properties. *J. Biol. Chem.* **260**:3440–3450

Kuffler, S.W., Potter, D.D. (1964) Glia in the leech central nervous system: physiological properties and neurone-glia relationship. *J. Neurophysiol.* **27**:290–320

Lohr, C., Nett, W., Deitmer, J.W. (1996a) Glutamate-mediated signalling in the giant leech glial cell. *Verh. Dtsch. Zool. Ges.* **89**:67

Lohr, C., Rose, C.R., Deitmer, J.W. (1996b) Extracellular Ca2+ changes during transmitter application in the leech central nervous system. *Neurosci. Lett.* **205**:57–60

Munsch, T., Deitmer, J.W. (1992) Calcium transients in identified leech glial cells in situ evoked by high potassium concentrations and 5-hydroxytryptamine. *J. Exp. Biol.* **167**:251–265

Munsch, T., Deitmer, J.W. (1995) Maintenance of fura-2 fluorescence in glial cells and neurones of the leech central nervous system. *J. Neurosci. Methods* **57**:195–204

Munsch, T., Deitmer, J.W. (1997) Intracellular Ca2+, Na+ and H+ transients evoked by kainate in the leech giant glial cells in situ. *Neurosci. Res.* **27**:45–56

Payton, B. (1981) Structure of the leech nervous system. In: Muller, K.J., Nicholls, J. G., Stent, G.S. (eds) *Neurobiology of the leech.* Cold Spring Harbor Laboratory, New York, pp. 35–50

Pfeiffer, F. (1992) Farbstoffmarkierung zur Identifizierung von Nerven- und Gliazellen in Ganglien des Blutegels *Hirudo medicinalis.* Diplomarbeit der Universität Kaiserslautern.

Retzius, G. (1891) Zur Kenntniss des centralen Nervensystems der Würmer. *Biol Unters. Neue Folge* II:1–28

Rose, C.R., Lohr, C., Deitmer, J.W. (1995) Activity-induced Ca2+ transients in nerve and glial cells in the leech CNS. *NeuroReport* **6**:642–644

White, N.S., Errington, R.J., Fricker, M.D., Wood, J.L. (1996) Aberration control in quantitative imaging of botanical specimens by multidimensional fluorescence microscopy. *J. Microsc.* **181**:99–116

Wuttke, W.A. (1990) Mechanism of potassium uptake in neuropile glial cells in the central nervous system of the leech. *J. Neurophysiol.* **63**:1089–1097

Exp. Biol. Online 2:9 (1997)

Nuclear and cytosolic calcium levels in NIH 3T3 fibroblasts

S. Giovannardi and A. Peres

Laboratorio di Fisiologia cellulare e molecolare, DBSF - Dipartimento di Biologia Strutturale e Funzionale, Università di Milano, Via Ravasi, 2, I-21100 Varese, Italy

Correspondence to: A. Peres (email: peres@imiucca.csi.unimi.it)

Received: 11 June 1997 / Accepted: 8 July 1997
Published: July 16, 1997

Abstract. The spatial distribution of intracellular calcium in resting NIH 3T3 fibroblasts loaded with Fura-2 has been studied by digital image analysis. Calibration parameters were determined separately for the nucleus and the cytosol to take into account possible differences in the physico-chemical properties of the two compartments and were found not to differ significantly. The apparent resting calcium concentration in these cells was found to be significantly lower in the nucleus than in the cytoplasm; however, this difference appears to be an artefact arising from the presence in the cytoplasm of regions with higher calcium levels. Application of thapsigargin, to block active uptake of calcium into these compartments, substantially eliminated the differences between nuclear and cytosolic calcium concentrations. These observations indicate that nuclear and cytosolic calcium are in equilibrium in the resting fibroblasts and argue against the existence of diffusional barriers between these two compartments.

Key words: Cytosolic calcium - Image analysis - NIH3T3 fibroblasts - Nuclear calcium

Introduction

The issue of whether there is independent control of nuclear and cytosolic Ca^{2+} levels is particularly difficult to address, as complicated technical aspects add to real cell-type-related differences. The great relevance of this problem has been recently boosted by reports which show that differences in spatial and temporal patterns of Ca^{2+} elevation can specifically regulate gene expression (Bito *et al.* 1996; Deisseroth *et al.* 1996; Dolmetsch *et al.* 1997; Ginty 1997; Hardingham *et al.* 1997). These effects appear to occur through both direct and indirect actions of Ca^{2+} in the nucleus. By blocking nuclear Ca^{2+} rises by injection of a Ca^{2+} chelator in a pituitary cell line, Hardingham *et al.* (1997) found differential effects of cytosolic and nuclear Ca^{2+} elevation on the cAMP- and serum-response elements, (CRE and SRE, respectively). Furthermore, Ca^{2+}-dependent phosphorylation of CREB, the transcription factor which activates CRE, has been shown to be induced by particular patterns of synaptic activity in hippocampal neurons (Bito *et al.* 1996; Deisseroth *et al.* 1996).

The hypothesis that nuclear Ca^{2+} does not passively follow the cytosolic changes but is instead independently regulated, as first suggested by Bachs *et al.* (1992), remains therefore very interesting, and it is strengthened by the presence in the nuclear envelope and in the nucleoplasm of elements essential for such independent control, i.e. an active Ca^{2+} transport system (Lanini *et al.* 1992; Malviya *et al.* 1990; Nicotera *et al.* 1989), receptors for the Ca^{2+} releasing agonists inositol 1,4,5- trisphosphate (InsP_3) (Malviya *et al.* 1990; Nicotera *et al.* 1990), and Ca^{2+}-binding proteins [reviewed in Bachs *et al.* (1992)]. In addition, the existence of the polyphosphoinositide cycle in the cell nucleus has also been reported (Divecha *et al.* 1991; Payrastre *et al.* 1992).

Determinations of nuclear versus cytosolic Ca^{2+} levels encounter several technical difficulties and have often given variable results (Chandra *et al.* 1989; Himpens *et al.* 1992; Tucker and Fay 1990; Waybill *et al.* 1991). Many reports have indicated that in resting cells the nuclear Ca^{2+} concentration is lower than that in the cytosol [see Bachs *et al.* (1992) for review], although on some occasions the opposite has been observed (Tucker and Fay 1990). In other cases no differences between the Ca^{2+} concentrations in the nucleus and cytosol have been found (Brini *et al.* 1994) or have been attributed to artefacts (Connor 1993).

In this paper we have addressed the question of whether there are true nucleo-cytosolic Ca^{2+} differences at rest in NIH 3T3 cells, by performing accurate evaluations of the fluorescent indicator properties in the various cellular compartments.

Materials and Methods

Cells culture and loading

NIH 3T3 fibroblasts were cultured in Dulbecco modified Eagle's medium (DMEM, Irvine Scientific) supplemented with 10% fetal calf serum (FCS, Irvine Scientific) and grown at 37°C in a 95% air 5% CO_2 atmosphere. Cells were plated on round glass coverslip and induced to reach quiescence by substituting the culture medium with DMEM at 0.5% FCS, 36 h before the experiments.

Cells were loaded with Fura-2 acetoxymethyl ester (Fura-2/ AM, Sigma) by incubation in a solution of Fura-2 (3 µM) containing (in mM): 125 NaCl, 5 KCl, 1.2 $MgCl_2$, 2 $CaCl_2$, 25 HEPES-NaOH and 6 glucose (pH 7.3), for 30 min at room temperature.

Ca^{2+} imaging

Coverslips containing the cells loaded with Fura-2 were placed in a Zeiss Axiovert microscope stage equipped for epifluorescence; observation and light measurements were done using a Zeiss Achrostigmat 40x objective (NA 1.3). Light from a Xenon arc lamp passed through a software-controlled filter changer on which a 340-nm and a 380-nm interference filter were mounted. Light emitted by the cells was collected by an Extended-Isis intensified charge-coupled device (CCD) camera (Photonic Science) connected to a DIS 3000 Image Processor (Digital Imaging Systems). Each image (512 × 512 pixels) was obtained by averaging 8 or 16 frames depending on the acquisition frequency. Pairs of images at the two wavelengths were collected in rapid succession and transferred to the RAM of the host computer. The fastest acquisition frequency of the system was 0.5 Hz (for a couple of averaged images). Background fluorescent images at the two wavelengths were also acquired by placing the cells out of focus or by dissolving the cells with Triton X-100 and quenching the Fura-2 with Mn^{2+}.

Images of Ca^{2+} concentration were constructed off-line by dividing the short-wavelength by the long-wavelength image after subtraction of the respective backgrounds; the resulting pixel values were then scaled according to Grynkiewicz *et al.* (1985). The nucleus is clearly identifiable in these cells under bright-field conditions, due to their flatness (Tucker and Fay 1990); one such image was acquired at the beginning of each experiment and used to delimit the nuclear and cytosolic regions.

Numerical Ca^{2+} values are expressed as R-R_{min}. The statistical significance of differences between means was assessed with the paired Student's t test , n indicates the number of cells analysed.

Results

Intracellular calibration

The determination of the absolute Ca^{2+} concentration in the cell interior is subject to various problems: it is known that R_{min} and R_{max} values determined using a droplet of solution of known composition are generally lower than those in the cytoplasm (Poenie 1990; Poenie *et al.* 1986), due to differences in viscosity and ionic strength; these properties may also be different in the cytosol and in the nucleoplasm, the two compartments in which we are interested. Finally the K_d for the Ca^{2+}-Fura-2 complex may also differ in the two compartments.

We believe that it is necessary to have separate calibration parameters for the nucleus and cytosol. In order to do so, we followed the procedure outlined by Williams and Fay (1990), with some modifications. The cells, loaded with Fura-2 and prepared as for the experiments, were bathed in one of the three solutions listed in Table 1.

Table 1 Ionic composition of solutions used for intracellular calibration. All solution contained 10 μM ionomycin, 10 μM monensin and 10 μM nigericin. The pH of each solution was adjusted to pH 7.2 with KOH

Component	Solution		
	A	B	C
Free calcium	0	379 nM	~ 2 mM
KCl (mM)	130	120	115
$MgCl_2$ (mM)	2	2	2
HEPES (mM)	20	20	20
EGTA (mM)	10	10	-
$CaCl_2$ (mM)	-	7	2

All these solutions contained ionomycin (10 μM) to equilibrate Ca^{2+}, monensin (10 μM) to equilibrate Na^+ and consequently avoid influences from Na^+/Ca^{2+} exchange activity, and nigericin (10 μM) to minimize the functioning of the Na^+/K^+ pump which could restore the Na^+ gradient. In solution B the free Ca^{2+} was calculated to be 379 nM. We assumed that after 10 min under these conditions a sufficient equilibration of Ca^{2+} between exterior and all the internal compartments could occur (Williams and Fay 1990). Ratio images of groups of cells in these three solutions are shown in Fig. 1. The value of the ratio (R) appears very uniform throughout each cell at free Ca^{2+} concentrations of 0 and 379 nM, while R_{max} in the nuclear region is somehow higher than in the cytoplasm.

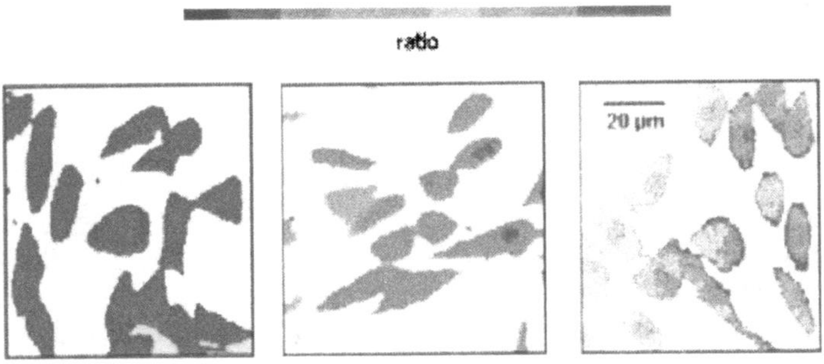

Fig. 1 Intracellular calibration ratio images obtained by exposing the cells to three solutions with different Ca^{2+} concentrations added with a mixture of ionophores; from left to right respectively 0 Ca^{2+}, 379 nM Ca^{2+} and 2 mM Ca^{2+}. As shown in Table 2, under each condition, the cytoplasmic and nuclear ratio values are not significantly different

Average values of R_{min} and R_{max} can be found in Table 2. The ratio value corresponding to the solution containing 379 free Ca^{2+} (R_{379}) was used to determine the value of the effective dissociation constant $K_d{}^* = K_d(S_{f2}/S_{b2})$ from the well known equation (Grynkiewicz *et al.* 1985):

$$K_d{}^* = 379 \frac{R_{max} - R_{379}}{R_{379} - R_{min}}$$

Nuclear and cytosolic $K_d{}^*$ values are shown in Table 2, which indicates also, for comparison, the calibration parameters obtained fromin *vitro* determinations.

Table 2 Intracellular calibration parameters. Mean values ± SD ($n = 8$)

	R_{min}	R_{379}	R_{max}	$K_d{}^*$ (µM)
Ccytosol	0.276 ± 0.007	0.514 ± 0.087	2.69 ± 0.97	3.465
Nnucleus	0.271 ± 0.011	0.481 ± 0.082	2.75 ± 0.78	4.095
In vitro droplet	0.274	0.94	6.60	3.22

It is interesting to note that for all three parameters there is no significant difference between nuclear and cytosolic estimates; also R_{min} and $K_d{}^*$ from the *in vitro* calibration are very similar to the cellular values; only R_{max} from the *in vitro* condition is grossly different.

These results are in agreement with those of Poenie (1990) who reported a large decrease in the value of R_{max} due to increased viscosity of the medium, while R_{min} was much less affected. In addition, no significant difference appears to exist in the properties of Fura-2 between the nucleus and the cytoplasm in our cells. A similar result was previously obtained by Himpens *et al.* (1992) in a study of cultured smooth muscle cells. In the following, Ca^{2+} concentrations were therefore calculated using as calibration parameters the mean values between nucleus and cytosol, i.e. $R_{min} = 0.274$, $R_{max} = 2.72$, $K_d^* = 3.760$.

Resting cytosolic and nuclear calcium levels

Quiescent NIH 3T3 fibroblasts plated on a glass coverslip are very flat cells with a thickness of few micrometrers at most, in which the nucleus is clearly visible. Tucker and Fay (1990) have shown that the layers of cytoplasm below and above the nucleus are very thin so that the fluorescence collected from that area may be considered as arising mainly from the nucleus. After Fura-2 loading, a weaker nuclear fluorescence intensity is frequently detectable upon excitation at 340 nm, as shown in Fig. 2A. The corresponding image at 380 nm excitation shows in contrast an enhanced nuclear fluorescence compared to the cytosol (Fig. 2B). Consequently the ratio image, shown in Fig. 2C, exhibits a clear difference in Ca^{2+} calcium concentration between nucleus and cytosol.

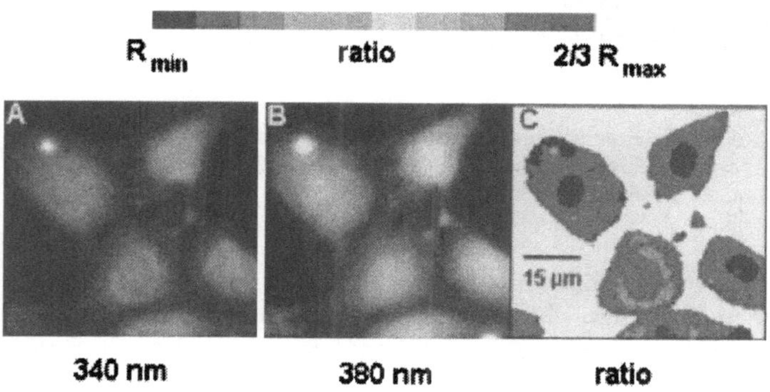

Fig. 2 Original fluorescent images of a group of quiescent NIH 3T3 fibroblasts excited at 340 nm (**A**) and at 380 nm (**B**). A lower emission in a central area corresponding to the nuclear region can be clearly seen in several cells at 340 nm excitation. The same area is strongly fluorescent when excited at 380 nm. This results in a lower calcium Ca^{2+} concentration in the nuclear region (**C**)

Average values, expressed as $R - R_{min}$, give a nuclear Ca^{2+} concentration of 0.149 ± 0.041 and a cytosolic value of 0.193 ± 0.055 (mean ± SD, $n = 88$); the difference, 0.044 ± 0.036, is significantly different from zero ($P < 0.001$).

Previous determinations of nuclear and cytosolic free Ca^{2+} levels in mouse fibroblasts have given contrasting results; for instance, Tucker and Fay (1990) reported higher nuclear versus cytosolic Ca^{2+} levels, while Chandra *et al.* (1989) observed the opposite. In order to check whether our observations depended upon compartmentalization in cytoplasmic organelles with higher Ca^{2+} levels at rest (Connor 1993), we performed measurements before and after permeabilization of the cells with saponin. As can be seen in Fig. 3, this treatment causes a complete loss of fluorescence from the nucleus but not from the cytoplasm, suggesting that some Fura-2 remains trapped in cytoplasmic organelles.

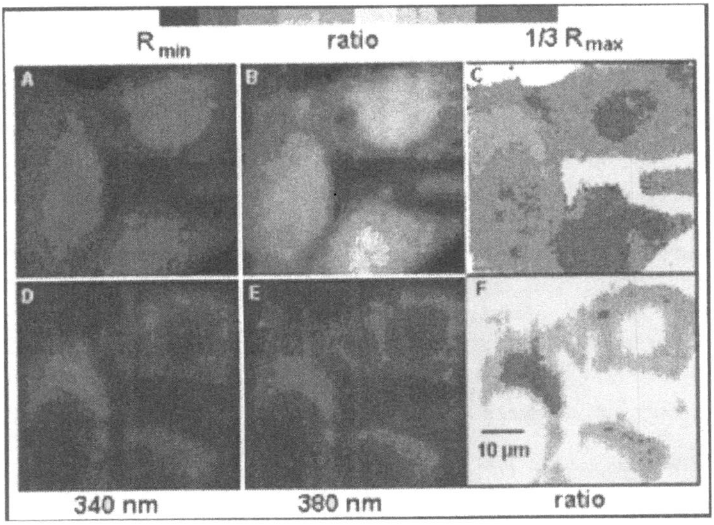

Fig. 3A-F Permeabilization with saponin reveals compartmentalization of Fura-2 into cytosolic organelles. **A, B** Fura-2 loaded cells excited at 340 and 380 nm respectively, before permeabilization; **C** ratio image from **A** and **B**; **D, E** images of the same cells after saponin (60 µg/ml) treatment; **F** ratio image from **D** and **E**

Panel F of Fig. 3 shows that indeed this occurs, and the Ca^{2+} level in the organelles appears to be higher, with an average value of 0.493 ± 0.258 (mean ± SD, $n = 21$), than in the intact cytosol (0.193 ± 0.056); for comparison the nuclear level in this group of cells was 0.159 ± 0.056.

To gain an idea of the real free cytosolic Ca^{2+} level, we used panels D and E of Fig. 3 as background images to be subtracted respectively from panels A and B in order to construct a ratio image, shown in Fig. 4, depurated from the contribution of the compartmentalized Fura-2. Quantitative analysis from this kind of procedure gives a nuclear Ca^{2+} value 0.051 ± 0.052 compared with a cytosolic value of 0.075 ± 0.066 (mean ± SD, $n = 21$). Although the absolute values of these determinations appear particularly low, and may reflect the complex subtraction procedure rather than a reliable estimate, the important result is that they are not significantly different from each other ($P > 0.05$), confirming the suggestion made by Connor (1993).

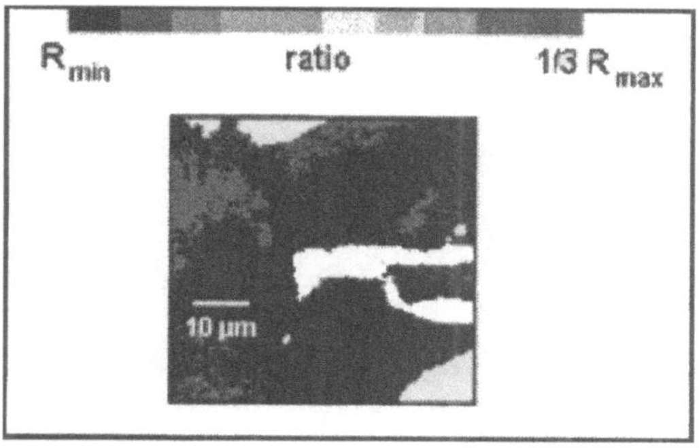

Fig. 4 Ca^{2+} map obtained from the experiment illustrated in Fig. 3, using the images after saponin treatment as backgrounds, i.e. performing the image operation: (A–D)/(B–E)

Effect of thapsigargin

The results of the saponin permeabilization experiments imply that the Ca^{2+} level in the cytoplasmic organelles must be higher than that in the cytosol. To maintain this difference, the continuous operation of an active Ca^{2+} transport system pumping Ca^{2+} from the cytosol to the organelles is required. To test this hypothesis we have applied thapsigargin, an agent which is reputed to selectively inhibit the Ca^{2+} ATPases of the internal membranes (Inesi and Sagara 1992; Thastrup *et al.* 1990). The experiment depicted in Fig. 5 illustrates the results of such.

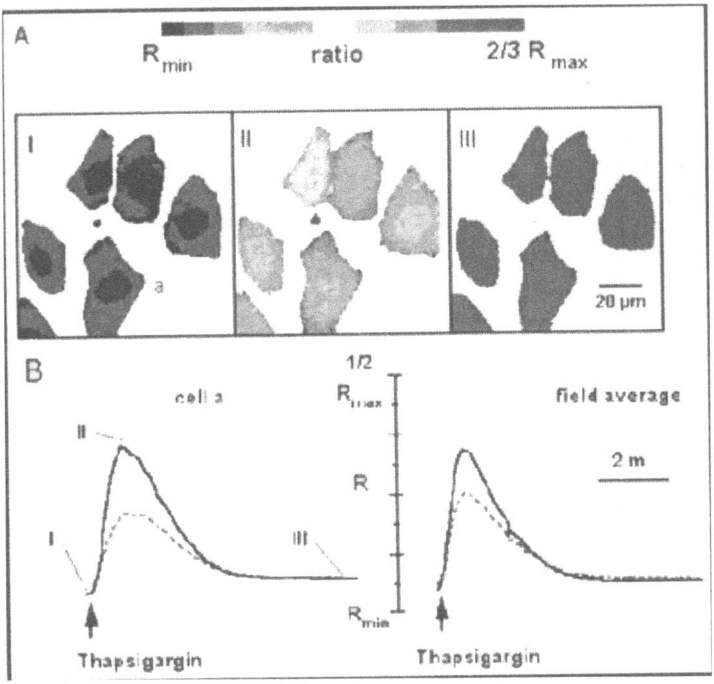

Fig. 5A, B Ca^{2+} changes upon treatment with thapsigargin (200 nM). **A** images taken at the time points indicated in **B** by the corresponding roman numerals. **B** time course of nuclear (*solid black lines*) and cytosolic (*red dashed lines*) Ca^{2+} in the indicated cell (*left*) and from all the cells in the field

Shortly after the addition of thapsigargin (200 nM), Ca^{2+} rises in both the nucleus and the cytosol, presumably because of an imbalance between uptake and release from the internal stores due to the block of the Ca^{2+} pump. The level attained by the nuclear Ca^{2+} concentration apparently exceeds that in the cytosol. However, this effect is an artefact due to the contribution of Fura-2-containing organelles observed above, as demonstrated by Connor (1993). The Ca^{2+} elevation is transient, because plasma membrane Ca^{2+} transport systems are still normally functioning and extrude the ions out of the cell. The Ca^{2+} level returns close to the resting one and a tendency for the nuclear and cytosolic concentrations to equalize can be noted (panel III of Fig. 5A). The average values of Ca^{2+} after 8 min of thapsigargin application are 0.228 ± 0.061 in the nucleus and 0.238 ± 0.051 in the cytosol (R-R_{min}, mean ± SD, n = 55). The average difference, 0.010 ± 0.009, although still significant at the 99.9% confidence level, is much reduced compared to the initial difference in this group of cells which was 0.057 ± 0.014 (mean ± SD).

Discussion

Quantitative determination of intracellular Ca^{2+} using fluorescent dyes is subject to several possible artefacts which have been extensively discussed by various authors (Connor 1993; Moore *et al.* 1990; Tucker and Fay 1990). We have thus evaluated the Fura-2 calibration parameters intracellularly, using the procedure suggested by Williams and Fay (1990), and estimated two sets of parameters, cytosolic and nuclear. Our results show that no significant differences exist in the fluorescence properties of Fura-2 in the two intracellular compartments, while the parameters obtained by extracellular calibration are much higher, according to previous determinations (Poenie 1990). Fura-2 is certainly present in sufficient quantities in the nucleoplasm, as shown in Fig. 2B, and it is active, as demonstrated by the fact that the nuclear signal can change faster than the cytosolic one (Fig. 5). Indeed, the fact that the nucleo-cytosolic Ca^{2+} gradient is able to apparently reverse its direction indicates that Fura-2 is not excessively buffering the Ca^{2+} concentration and it is in a Ca^{2+}-sensitive form in both compartments.

Given the thinness of our quiescent fibroblasts (Tucker and Fay 1990), we are confident that only a small fraction of the fluorescence signal may arise from cytosolic spaces below or above the nucleus.

The main results of our experiments show that the nucleo-cytoplasmic Ca^{2+} gradient observed in resting fibroblasts does not correspond to real differences between the free cytosolic and nuclear Ca^{2+} levels, but arises instead from the presence of organelles containing a higher Ca^{2+} level. When active transport of Ca^{2+} inside these organelles is inhibited by thapsigargin, the nucleo-cytosolic gradient disappears.

Our observations therefore are in agreement with previous findings (Connor 1993) and do not favour an independent control of Ca^{2+} by the nucleus in fibroblasts. However, it must be stressed that our results do not imply that nuclear Ca^{2+} changes are necessarily dominated by passive influx from the cytosol. Indeed the presence in the nuclear envelope of all the molecular components for Ca^{2+} release and uptake (Divecha *et al.* 1991; Lanini *et al.* 1992; Malviya *et al.* 1990) suggests an active role of this compartment in nuclear Ca^{2+} signaling. Furthermore, the existence of a Ca^{2+}-induced Ca^{2+}-release mechanism located in the nuclear envelope cannot be excluded (Hernández-Cruz *et al.* 1991).

Acknowledgements. This work was partially supported by a grant of the Associazione Italiana per la Ricerca sul Cancro (AIRC) to Prof. L. Alberghina.

References

Bachs, O., Agell, N., Carafoli, E. (1992) Calcium and calmodulin function in the cell nucleus. *Biochim Biophys Acta* **1113**: 259–270

Bito, H., Deisseroth, K., Tsien, R.W. (1996) CREB phosphorylation and dephosphorylation: a Ca^{2+} - and stimulus duration-dependent switch for hippocampal gene expression. *Cell* **87**: 1203–1214

Brini, M., Marsault, R., Bastianutto, C., Pozzan, T., Rizzuto, R. (1994) Nuclear targeting of aequorin. A new approach for measuring nuclear Ca^{2+} concentration in intact cells. *Cell Calcium* **16**: 259–268

Chandra, S., Gross, D., Ling, Y.-C., Morrison, G.H. (1989) Quantitative imaging of free and total intracellular calcium in cultured cells. *Proc Natl Acad Sci USA* **86**: 1870–1874

Connor, J.A. (1993) Intracellular calcium mobilization by inositol 1,4,5-trisphosphate: Intracellular movements and compartmentalization. *Cell Calcium* **14**: 185–200

Deisseroth, K., Bito, H., Tsien, R.W. (1996) Signaling from synapse to nucleus: postsynaptic CREB phosphorylation during multiple forms of hippocampal synaptic plasticity. *Neuron* **16**: 89–101

Divecha, N., Banfic, H., Irvine, R.F. (1991) The polyphosphoinositide cycle exists in the nuclei of Swiss 3T3 cells under the control of a receptor (for IGF-I) in the plasma membrane, and stimulation of the cycle increases nuclear diacylglycerol and apparently induces translocation of protein kinase C to the nucleus. *EMBO J* **10**: 3207–3214

Dolmetsch, R.E., Lewis, R.S., Goodnow, C.C., Healy, J.I. (1997) Differential activation of transcription factors induced by Ca^{2+} response amplitude and duration. *Nature* **386**: 855–858

Ginty, D.D. (1997) Calcium regulation of gene expression: isn't that spatial? *Neuron* **18**: 183–186

Grynkiewicz, G., Poenie, M., Tsien, R.Y. (1985) A new generation of Ca^{2+} indicators with greatly improved fluorescence properties. *J Biol Chem* **260**: 3440–3450

Hardingham, G.E., Chawia, S., Johnson, C.M., Bading, H. (1997) Distinct functions of nuclear and cytoplasmic calcium in the control of gene expression. *Nature* **385**: 260–265

Hernández-Cruz, A., Sala, F., Connor, J.A. (1991) Stimulus-induced nuclear Ca^{2+} signals in fura-2-loaded amphibian neurons. *Ann NY Acad Sci* **635**: 416–420

Himpens, B., De Smedt, H., Droogmans, G., Casteels, R. (1992) Differences in regulation between nuclear and cytoplasmic Ca^{2+} in cultured smooth muscle cells. *Am J Physiol* **263**: C95–C105

Inesi, G., Sagara, Y. (1992) Thapsigargin, a high affinity and global inhibitor of intracellular Ca^{2+} transport ATPases. *Arch Biochem Biophys* **298**: 313–317

Lanini, L., Bachs, O., Carafoli, E. (1992) The calcium pump of the liver nuclear membrane is identical to that of endoplasmic reticulum. *J Biol Chem* **267**: 11548–1552

Malviya, A.N., Rogue, P., Vincendon, G. (1990) Stereospecific inositol 1,4,5-[^{32}P]trisphosphate binding to isolated rat liver nuclei: Evidence for inositol trisphosphate receptor-mediated calcium release from the nucleus. *Proc Natl Acad Sci USA* **87**: 9270–9274

Moore, E.D.W., Becker, P.L., Fogarty, K.E., Williams, D.A., Fay, F.S. (1990) Ca^{2-} imaging in single living cells: theoretical and practical issues. *Cell Calcium* **11**: 157–179

Nicotera, P., McConkey, D.J., Jones, D.P., Orrenius, S. (1989) ATP stimulates Ca^{2+} uptake and increases the free Ca^{2+} concentration in isolated rat liver nuclei. *Proc Natl Acad Sci USA* **86**: 453–457

Nicotera, P., Orrenius, S., Nilsson, T., Berggren, P.-O. (1990) An inositol 1,4,5-trisphosphate-sensitive Ca^{2+} pool in liver nuclei. *Proc Natl Acad Sci USA* **87**: 6858–6862

Payrastre, B., Nievers, M., Boonstra, Breton, M., Verkleij, A.J., Van Bergen en Henegouwen, P.M.P. (1992) A differential location of phosphoinositide kinases, diacylglycerol kinase, and phospholipase C in the nuclear matrix. *J Biol Chem* **267**: 5078–5084

Poenie, M. (1990) Alteration of intracellular Fura-2 fluorescence by viscosity: a simple correction. *Cell Calcium* **11**: 85–91

Poenie, M., Alderton, J., Steinhardt, R.A., Tsien, R.Y. (1986) Calcium rises abruptly and briefly throughout the cell at the onset of anaphase. *Science* **233**: 886–869

Thastrup, O., Cullen, P.J., Drobak, B.K., Hanley, M.R., Dawson, A.P. (1990) Thapsigargin, a tumor promoter, discharges intracellular Ca^{2+} stores by specific inhibition of the endoplasmic reticulum Ca^{2+}-ATPase. *Proc Natl Acad Sci USA* **87**: 2466–2470

Tucker, R.W., Fay, F.S. (1990) Distribution of intracellular free calcium in quiescent BALB/c 3T3 cells stimulated by platelet-derived growth factor. *Eur J Cell Biol* **51**: 120–127

Waybill, M.M., Yelamarty, R.V., Zhang, Y., Scaduto, R.C., Jr., LaNoue, K.F., Hsu, C.-J., Smith, B.C., Tillotson, D.L., Yu, F.T.S., Cheung, J.Y. (1991) Nuclear calcium gradients in cultured rat hepatocytes. *Am J Physiol Endocrinol Metab* **261**: E49–E57

Williams, D.A., Fay, F.S. (1990) Intracellular calibration of the fluorescent calcium indicator Fura-2. *Cell Calcium* **11**: 75–83

Exp. Biol. Online 2:10 (1997)

Second-messenger-induced signalling events in pollen tubes of *Papaver rhoeas**

B.K. Drøbak[1], F.C.H. Franklin[2], P.J. Shaw[1], G.M. Calder[1], A.J. Trewavas[3], A.C. Allan[4] and V.E. Franklin-Tong[2]

[1]Department of Cell Biology, John Innes Centre, Norwich, NR4 7UH, UK
[2]School of Biological Sciences, University of Birmingham, Birmingham, B15 2TT, UK
[3]University of Edinburgh, Edinburgh, EH9 3JH, UK
[4]Department of Genetics, Weizmann Institute, Rehovot, Israel

* Presented at the 1997 SEB Annual Meeting: *Interactive MultiMedia Biology - Experimental Biology Online Symposium*, Canterbury, 7-11 April

---

Correspondence to: V.E. Franklin-Tong (e-mail: V.E.Franklin-Tong@bham.ac.uk)

Received: 8 March 1997 / Accepted: 12 May 1997
Published: July 22, 1997

Abstract. A role for cytosolic free Ca^{2+} ($Ca^{2+}{}_i$) in the regulation of growth of *Papaver rhoeas* pollen tubes during the self-incompatibility response has recently been demonstrated [Franklin-Tong et al. *Plant J.* 4:163–177 (1993); Franklin-Tong et al. *Plant J.* 8:299–307 (1995); Franklin-Tong et al. submitted to *Plant J.*]. We have investigated the possibility that $Ca^{2+}{}_i$ is more generally involved in the regulation of pollen tube growth using confocal laser scanning microscopy (CLSM). Data obtained using Ca^{2+} imaging, in conjunction with photolytic release of caged inositol 1,4,5-trisphosphate [Ins(1,4,5)P_3], point to a central role of the phosphoinositide signal transduction pathway in the control of Ca^{2+} fluxes and control of pollen tube growth. These experiments further revealed that increases in cytosolic levels of Ins(1,4,5)P_3 resulted in the formation of distinct Ca^{2+} waves. Experiments using the pharmacological agents heparin, neomycin and mastoparan further indicated that Ca^{2+} waves are propagated, at least in part, by Ins(1,4,5)P_3-induced Ca^{2+} release rather than by simple diffusion or by "classic" Ca^{2+}-induced Ca^{2+} release mechanisms. We also have data which suggest that Ca^{2+} waves and oscillations may be induced by photolytic release of caged Ca^{2+}. Ratio-imaging has enabled us to identify an apical oscillating Ca^{2+} gradient in growing pollen tubes, which may regulate normal pollen tube growth. We also present evidence for the involvement of Ca^{2+} waves in mediating the self-incompatibility response. Our data suggest that changes in Ca^{2+}_i and alterations in growth rate/patterns are likely to be closely correlated and may be causally linked to events such as Ca^{2+}-induced, or Ins(1,4,5)P_3-induced wave formation and apical Ca^{2+} oscillations.

Key words: Ca^{2+} – Inositol 1,4,5-trisphosphate – *Papaver rhoeas* L. – Pollen – Second messengers – Signal transduction

Introduction

Pollination is a crucial step in fertilization in flowering plants and the study of pollination is important for the understanding of the control of fertility and reproduction. The growing pollen tube is also an attractive model system for the investigation of polarized growth. One of the first stages in the process of pollination in flowering plants is the alighting of pollen grains on the stigma of the recipient plant. Not all pollinations will produce seed. There are several mechanisms, including inter-and intra-species breeding barriers, with which many plants regulate the acceptance or rejection of pollen, in order to prevent fertilization by unwanted pollen. One of the most important outbreeding devices found in angiosperms is self-incompatibility (SI), which is used to prevent self-fertilization within species. This genetically controlled mechanism is, in many species, controlled by a single, multi-allelic (*S*-) gene. Self-fertilization is prevented when the growth of pollen carrying an *S*-allele is inhibited when it is genetically identical to that carried by the pistil on which it lands, while pollens carrying other *S*-alleles are not inhibited. SI is an example of an event where very precise cell–cell communication and signalling are essential.

We are interested in the signalling pathways and mechanisms involved both in normal pollen tube growth and in the SI response. One of the most common second messengers employed by higher plants is cytosolic free calcium (Ca^{2+}_i) (Trewavas and Gilroy 1991; Gilroy et al. 1993). Much evidence also suggests that it is likely to be involved in the regulation of pollen tube growth. It has been noted for many years that pollen tube growth requires exogenous Ca^{2+} (Brewbaker and Kwack 1963; Jaffe et al. 1975; Picton and Steer 1983), and it is now well established that Ca^{2+}_i plays a key role in the regulation of the pollen tube growth (Derksen et al. 1995; Feijó et al. 1995; Taylor and Hepler 1997). A number of studies have demonstrated that growing pollen tubes have a tip-focused gradient of Ca^{2+}_i (Obermeyer and Weisenseel 1991; Rathore et al.

1991; Miller et al. 1992; Malhó et al. 1994; Pierson et al. 1994, 1996), which results from an apical concentration of open Ca^{2+} channels (Malhó et al. 1995; Pierson et al. 1996), but the precise physiological role of the Ca^{2+} gradient is not yet established (Pierson et al. 1994; Malhó et al. 1995).

We are using pollen of *Papaver rhoeas* as a model system. *P. rhoeas* has a well-established SI system (Lawrence et al. 1978) and the development of an experimental system which permits the SI reaction of *Papaver rhoeas* to be reproduced in vitro (Franklin-Tong et al. 1988) has allowed us to begin to dissect the molecular and signalling events involved in SI in this species. We have identified proteins secreted by the stigma which are encoded by the *S*-gene (Franklin-Tong et al. 1989; Foote et al. 1994) and have cloned the stigmatic component of several *S*-alleles of the *S*-gene from this species (Foote et al. 1994; Walker et al. 1996). We describe here some of our studies which have investigated a role for Ca^{2+}-mediated signal transduction pathways in pollen of *P. rhoeas*, using Ca^{2+}-imaging techniques.

Imaging of Ca^{2+}_i using Calcium Green-1

In many of our studies we have used pollen tubes microinjected with the Ca^{2+}-sensitive fluorescent dye Calcium Green-1 (CG-1), which has high affinity and selectivity for free Ca^{2+} (Haughland 1992). Although CG-1 has the disadvantage that it is a single-wavelength dye, we chose to use it for these studies since it may be visualized using a confocal scanning laser microscope (CSLM) equipped with an Ar laser. It also enables us to perform experiments with caged probes, which is not possible with ratiometric dyes such as fura-2 and Indo-1, since they are usually excited at 340–350 nm and photolysis of caged probes requires a flash of UV light. There are, however, disadvantages with the use of CG-1. One is that although relative changes in Ca^{2+}_i may be visualized, it is virtually impossible to obtain an accurate measurement of the intracellular Ca^{2+} concentration ($[Ca^{2+}]_i$). Another drawback with this dye is that no apical gradients of Ca^{2+}_i were detectable using CG-1 in these pollen tubes, although they are known to be present. This is thought to be due to differences in the dye concentration and path length within the cell. Another problem was that dye fluorescence decreased with time, due to sequestration of dye away from the tip. This problem has been partly overcome by the use of dyes conjugated to 10-kDa dextran (Franklin-Tong et al. submitted; Calder et al. 1997).

A role for Ca^{2+}_i in the SI response

We have begun a detailed analysis of the signalling pathway(s) which may be involved in the *Papaver rhoeas* SI response in pollen and have recently established that the SI response is mediated by a signal transduction pathway involving modulation of $[Ca^{2+}]_i$ (Franklin-Tong et al. 1993, 1995). Figure 1 illustrates the spatial distribution of the alterations in $Ca^{2+}{}_i$ stimulated in these pollen tubes (Fig. 1a,b), together with images which show the general morphology and organelle distribution (Fig. 1c–i). Figure 1a shows that partially purified stigmatic proteins which contain S-glycoproteins encoded by the self-incompatibility (S-) gene specifically induce transient increases in $[Ca^{2+}]_i$ in incompatible pollen tubes (Franklin-Tong et al. 1993). The increases in $[Ca^{2+}]_i$ were initiated within 60 s of the addition of S-proteins, and lasted for 6–10 min. In a number of cases, we observed a second increase in $Ca^{2+}{}_i$, which may be due to a Ca^{2+}-induced Ca^{2+} release (CICR). Treatment of pollen tubes with compatible stigmatic extracts or with heat-denatured-incompatible glycoproteins did not result in increases in $[Ca^{2+}]_i$, and pollen tube growth continued normally (Franklin-Tong et al. 1993). Use of CSLM generally allows good spatial resolution (Fricker et al. 1991) and it enabled us to spatially visualize, to some extent, $[Ca^{2+}]_i$ dynamics. The increases in $[Ca^{2+}]_i$ as a consequence of the SI response appeared, at least superficially, to be generated from internal Ca^{2+} stores. We have previously suggested that the increases may correlate with the pollen nuclear complex, comprising the generative cell and vegetative nucleus, as shown in Fig. 1c and g, rather than the vacuole (see Fig. 1h,i) (Franklin-Tong et al. 1995). The pollen nuclear complex is often associated with endoplasmic reticulum (Lancelle and Hepler 1992), as indicated in Fig. 1d,e. A role for the involvement of mitochondria appears not to be implicated (see Fig. 1f). Our data also showed that pollen tube growth was linked to the $Ca^{2+}{}_i$ response, which provided important evidence for a connection between the rise in $Ca^{2+}{}_i$ and the inhibition of pollen tube growth. Thus, an increase in $[Ca^{2+}]_i$ induced by incompatible S-glycoproteins, if large enough, results in the inhibition of growth of incompatible pollen.

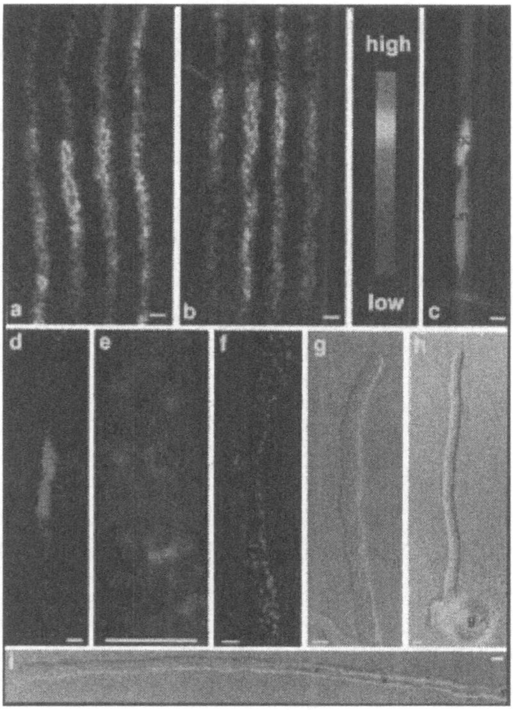

Fig 1a–i Spatial distribution of the $Ca^{2+}{}_i$ transient generated by an incompatible SI response. **a, b** Laser-scanning confocal microscopy of growing pollen tubes microinjected with Calcium Green-1 showing changes in $Ca^{2+}{}_i$ following treatment with incompatible S-glycoproteins and recombinant S-protein. Each block represents a 100 µm region of a single pollen tube starting in (**a**) 75 µm and in (**b**) 40 µm behind the tip, before and at the peak of the response after challenge with (**a**) S-glycoproteins and (**b**) with recombinant S-proteins. The pollen tubes are presented as pseudocoloured images, where "warm" colours indicate high levels of Ca^{2+} and "cold" colours indicate low levels of Ca^{2+} (see colour scale). A transient increase in $Ca^{2+}{}_i$ is clearly observed and pollen tube growth was inhibited after the $[Ca^{2+}]_i$ transient peaked. Increases in $Ca^{2+}{}_i$ appear to be spatially localized in or immediately around the "nuclear complex". **c** A pollen tube stained with DAPI to show the generative cell (*gc*) and vegetative nucleus (*vn*). This is similar in size, shape and position to the $Ca^{2+}{}_i$ transients illustrated in **a** and **b**, and may implicate intracellular Ca^{2+} release. **d** Pollen tube stained with $DiOC_6$ to show the general distribution of a concentrated region of ER in the area around the nuclear region. **e** Part of this region "zoomed" to show some detail of the polygonal ER structures. **f** Pollen tube stained *in vivo* with rhodamine 123 to show the distribution of the mitochondria. This optical section cuts through the nuclear region, which has no mitochondria associated with it. Images which do not cut through this region show an even distribution of mitochondria throughout. All of the images in **a–f** are of the region of the pollen tube where the "nuclear complex" is localized; in all cases the pollen tube is orientated with the tip pointing towards the top of the page. Note the similarity of the size and position of the cell components and compare with the $Ca^{2+}{}_i$ transients in **a** and **b**. **g–i** The cellular structures visible in typical *P. rhoeas* pollen tubes grown *in vitro* and visualized using DIC microscopy. **g** shows the general region imaged in **a–f**, but also includes the tip. Note the nuclear region (*n*) and numerous small organelles, which circulate due to ccytoplasmic streaming. **h** A representative pollen tube

used for the imaging experiments, showing the pollen grain (*g*). Note there is no marked vacuolation at this stage (compare with i). i Pollen tube grown for several hours. Note the marked vacuolated zone, which starts at ~135 μm behind the tip in this particular tube, which is typical of pollen tubes grown for much longer time periods than those used for the imaging experiments. Figure from Franklin-Tong et al., 1995 © *Plant Journal*

Caged Ca^{2+} [and caged Ins(1,4,5)P_3] have been used in other systems to demonstrate a direct link between an increase in $[Ca^{2+}]_i$ and biological responses (Gilroy et al. 1990; Shacklock et al. 1992). We have also used caged Ca^{2+} in this way to demonstrate that the intracellular release of $Ca^{2+}{}_i$ can also induce inhibition of pollen tube growth. These experiments provided the first direct evidence that $[Ca^{2+}]_i$ plays a mediatory role in the regulation of *P. rhoeas* pollen tube growth during SI reactions.

We have also used recombinant S-proteins expressed in *Escherichia coli* from the cloned S_1 and S_3 alleles of the S-gene (Foote et al. 1994; Walker et al. 1996). Since this recombinant protein has S-specific biological activity, we tested if they would elicit the $Ca^{2+}{}_i$ signalling response. Data revealed that the recombinant S-protein elicits a Ca^{2+} response in pollen tubes which is indistinguishable from that elicited by the relatively crude glycoprotein fraction obtained from stigmatic tissue, as shown in Fig. 1b (Franklin-Tong et al. 1995). Several important conclusions about the nature of the components involved in the SI response can be made on the basis of these results. First, it enabled us to demonstrate that it is the S-protein itself that elicits the $[Ca^{2+}]_i$ transient detected in the pollen tubes, and that the SI response is elicited by a single protein, which presumably acts as a signal molecule. These studies also established that although recombinant S-proteins differed from the stigmatic S-(glyco)proteins in a number of ways due to the lack of post-translational processing in *E. coli* (see Foote et al. 1994), they still elicited the SI Ca^{2+} signal.

These data all point to alterations in $[Ca^{2+}]_i$ as being central to modulation of pollen tube growth, and suggest that the S-proteins exert their effect by stimulating changes in $[Ca^{2+}]_i$. More recently, we have identified Ca^{2+}-dependent increases in protein phosphorylation specifically triggered by the SI response (Rudd et al. 1996, 1997). The increased phosphorylation of a 26-kDa cytosolic protein, p26.1, is specifically induced by the SI response (Rudd et al. 1996). Hyper-phosphorylation of p26.1 appears to be dependent on Ca^{2+} and calmodulin (CaM), which suggests that a Ca^{2+}-dependent protein kinase requiring CaM-like domains is activated during the SI response, and acts as an intracellular signal mediating the SI response in *P. rhoeas* pollen.

A role for Ins(1,4,5)P_3 in regulation of $Ca^{2+}{}_i$ in pollen tubes

As the increases in $[Ca^{2+}]_i$ stimulated by the SI response appeared (at least superficially) to be generated intracellularly, this led us to investigate the possibility that the phosphoinositide (PI) signalling pathway could be involved in the generation of $[Ca^{2+}]_i$ transients in pollen tubes. A primary mechanism for stimulus-induced elevation of $[Ca^{2+}]_i$ in mammalian cells is signal transduction via the PI system (Berridge 1993). A consensus is emerging that a plant PI system exists (for a review see Drøbak et al. 1996), and it has been demonstrated that Ins(1,4,5)P_3 can mobilize Ca^{2+} from intracellular stores both in vitro (Drøbak and Ferguson 1985) and in vivo (Gilroy et al. 1990). Furthermore, a putative Ins(1,4,5)P_3 receptor has been identified and partially purified from plant cells (Brosnan and Sanders 1990; Biswas et al. 1995). We have investigated the action of Ins(1,4,5)P_3 in pollen tubes.

Photolysis of caged Ins(1,4,5)P_3 can induce release of $[Ca^{2+}]_i$ in the form of a Ca^{2+} wave

We began our investigations by examining the effect of release of caged Ins(1,4,5)P_3 in pollen tubes of *P. rhoeas* growing in vitro using Ca^{2+} imaging. The pollen tubes were microinjected with the Ca^{2+}-sensitive dye CG-1 and caged Ins(1,4,5)P_3, which was subsequently released by UV photolysis, and $Ca^{2+}{}_i$ was imaged using CSLM. Figures 2 and 3a show typical examples of the increases $[Ca^{2+}]_i$ stimulated by increases in Ins(1,4,5)P_3 in *P. rhoeas* pollen tubes. A slow increase in $[Ca^{2+}]_i$, which continued for 300 s and was sustained at high levels for a further 400 s, was observed (Franklin-Tong et al. 1996). The kinetics suggest that only a small proportion of the Ca^{2+} visualized was directly mobilized by the photolysed Ins(1,4,5)P_3 and that further mobilization is dependent on other reactions. Imaging also revealed that a Ca^{2+} "wave" was initiated, in this case approximately 150 μm behind the tip. This Ca^{2+} wave traversed the pollen tube reaching the tip within ~2 min. Thus, the initial wave travelled at a speed of ~0.8 μm s^{-1}. Additional Ca^{2+} waves continued to move forward after the initial wave had reached the pollen tube tip, and $[Ca^{2+}]_i$ continued to increase. After a few minutes the increases in $[Ca^{2+}]_i$ declined (i.e. it is a transient increase, typically lasting 6–10 min). These increases in $[Ca^{2+}]_i$ were not detected in pollen tubes in which Ins(1,4,5)P_3 had not been released.

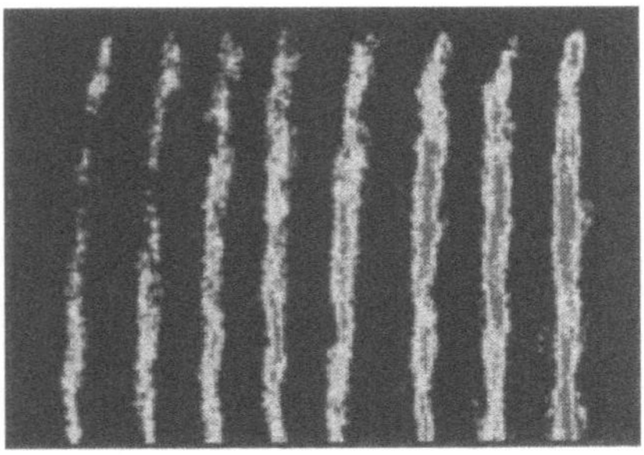

Fig. 2 Ca^{2+} wave initiated by release of caged $InsP_3$. Laser scanning confocal microscopy imaging of a time series of a growing pollen tube of *Papaver rhoeas* before (first image) and after (20-s intervals between images) photolytic release of caged Ins(1,4,5)P_3. The pollen tube was microinjected with Calcium Green-1 and caged Ins(1,4,5)P_3. Ca^{2+}-imaging revealed a gradual mobilization of $Ca^{2+}{}_i$ which slowly increased and moved as a "Ca^{2+} wave" towards the pollen tube tip. Pollen tube inhibition occurred as a result of this treatment. (Figure adapted from Franklin-Tong et al., 1996 © *Plant Cell*)

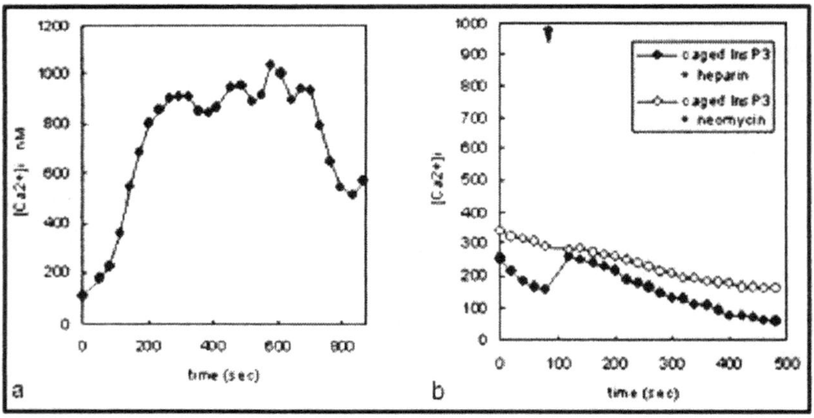

Fig. 3 Evidence that Ca^{2+} waves are $InsP_3$-induced. **a** shows profile of Ca^{2+}-release from a pollen tube obtained as a result of photolysis of caged $InsP_3$. **b** shows a profile of the same reactions in the presence of neomycin and heparin. Growing pollen tubes were microinjected with the calcium-sensitive fluorescent dye Calcium Green-1 and caged Ins(1,4,5)P_3. The pollen tube was exposed to 20 s of UV photolysis at time zero. Plots were obtained from a series of confocal images of a single pollen tube and relative Ca^{2+} levels recorded as pixel values. (Figures from Franklin-Tong et al., 1996 © *Plant Cell*)

The spatial and temporal pattern of the $[Ca^{2+}]_i$ release observed was unexpected. Our data establish that Ca^{2+} waves can also be generated in growing *P. rhoeas* pollen tubes. Ca^{2+} waves have not previously been reported in plant cells, although Felle (1988) and more recently McAinsh et al. (1995) have described oscillations of $[Ca^{2+}]_i$ in coleoptiles and guard cells. Ca^{2+} waves have been described extensively in the animal literature, especially in egg cells, where Ins(1,4,5)P_3-induced Ca^{2+}release (IICR) leads to increases in Ca^{2+} which propagate from the point of initial fertilization across the cell (see Jouaville et al. 1995 for a recent discussion). Jaffe (1991) has provided a "unifying hypothesis" for Ca^{2+} waves in animal cells, based on a collection of data which suggests that the velocities of these waves are remarkably conserved across species, at ~10 µm s^{-1} in activating egg cells and ~25 µm s^{-1} in other cells. However, there is considerable variation within these samples, with velocities varying from 5 to 14 µm s^{-1} for egg cells and from 15 to 40 µm s^{-1} for other cells. Furthermore, there are exceptions which fall outside this general grouping, having much faster waves (~100 µm s^{-1} for heart myocytes) and much slower waves (~3 µm s^{-1} for some echinoderm eggs). Although we prefer, at this stage, not to make a "comparison" of our Ca^{2+} waves with animal Ca^{2+} waves, it is clear that our "slow Ca^{2+} wave" at 0.8 µm s^{-1} is much slower than any of these Ca^{2+} waves described in animal cells. However, we have also described faster Ca^{2+} waves stimulated in pollen tubes by mastoparan (Franklin-Tong et al. 1996), which have a several-fold higher velocity, and which would fall within the lower range of those in animal cells. As we do not know the mechanisms underlying the plant waves yet, it is inappropriate at this stage to speculate on the differences/similarities further.

Demonstration of Ca^{2+}-dependent PIC activity in pollen

As our experiments indicated that Ins(1,4,5)P_3 has the potential to be a physiological regulator of $Ca^{2+}{}_i$ fluxes in *Papaver* pollen, we investigated this more directly by testing whether a phosphatidylinositol 4,5-bisphosphate-[PtdIns(4,5)P_2-] specific phospholipase C (PIC), which is responsible for the in vivo production of Ins(1,4,5)P_3, was present in *Papaver* pollen microsomes. Our studies provided the first direct evidence for the presence of a Ca^{2+}-dependent PIC activity which is capable of mediating the diesteratic cleavage of PtdIns(4,5)P_2 leading to the production of Ins(1,4,5)P_3 in pollen (Franklin-Tong et al. 1996). This pollen PIC is membrane associated and its overall activity is comparable to that observed in a wide range of other plant tissues (see Drøbak 1992). We also established that neomycin, which has been shown to inhibit PIC activation and inhibit IICR in both plant and animal systems, caused virtually complete inhibition of PIC activity (at and above 150 µM) in *P. rhoeas* pollen microsomes. Relatively few investigations of PI metabolism in pollen tubes have been carried out, but our data and those of Helsper et al. (1987) suggest that there is indeed a functional PI signalling pathway present in growing pollen tubes.

Mastoparan induces Ins(1,4,5)P_3 production and increases in $[Ca^{2+}]_i$

Having established that *P. rhoeas* pollen has Ca^{2+}-dependent PIC activity, we investigated whether measurable levels of Ins(1,4,5)P_3 could be detected in growing pollen tubes. Mastoparan has been shown to induce rapid increases in the cytosolic levels of Ins(1,4,5)P_3 in animal and plant cells, although its precise mode of action is currently open to debate (see Franklin-Tong et al. 1996 for a discussion of this). We treated pollen tubes with mastoparan and measured Ins(1,4,5)P_3 using an Ins(1,4,5)P_3 receptor assay. Substantial rises in Ins(1,4,5)P_3 levels were measured; these were comparable to levels observed in stimulated carrot cells (Drøbak and Watkins 1994). Increases were biphasic and, initially, a very rapid increase in the level of Ins(1,4,5)P_3 was observed within 40–60 s after stimulation. Ins(1,4,5)P_3 levels returned to basal within 120 s and was followed by a more sustained increase after approximately 5–6 min. The speed with which Ins(1,4,5)P_3 can be degraded in pollen tubes suggests the presence of active Ins(1,4,5)P_3-degrading enzymes. However, this does not preclude very rapid Ins(1,4,5)P_3 production, provided that the turnover rate of PtdIns(4,5)P_2 is upregulated in synchrony with PIC activation (B.K. Drøbak and V. Stavar, unpublished calculations). Thus, there is scope for sustained increases in $[Ca^{2+}]_i$. As mastoparan induces a rise in Ins(1,4,5)P_3 levels, we also investigated whether it could affect $[Ca^{2+}]_i$ in pollen tubes. Ca^{2+} imaging of pollen tubes subjected to the addition of mastoparan revealed that there was a very rapid increase in $Ca^{2+}{}_i$ which traversed the pollen tube within 20 s. This was followed by a longer term increase in $[Ca^{2+}]_i$. Although this might resemble a fast-moving Ca^{2+} wave, the changes are much faster than those stimulated by photolysis of caged Ins(1,4,5)P_3. Thus, it remains unclear whether the same mechanisms are operating.

Are the $Ca^{2+}{}_i$ increases propagated by IICR?

In order to further investigate the possibility of the involvement of IICR in the stimulation or propagation of the Ca^{2+} waves, the effects of the two inhibitors neomycin and heparin were tested. Heparin is a widely used competitive inhibitor of Ins(1,4,5)P_3 receptor binding, and neomycin, which inhibits IICR in both plant and animal systems, has also been shown to inhibit PIC activation in *P. rhoeas* pollen tubes. There are limitations to this approach and interpretation of data obtained using pharmacological agents such as mastoparan, neomycin and heparin should be approached with caution, as there are a number of problems associated with their use due to a lack of information about their specificity. Nevertheless, these agents remain useful tools for the study of Ca^{2+}/PI-related effects (see Franklin-Tong et al. 1996 for discussion).

Growing *P. rhoeas* pollen tubes were either microinjected with CG-1 and caged $Ins(1,4,5)P_3$ or CG-1, caged $Ins(1,4,5)P_3$ and heparin. In the absence of heparin, slow increases in $Ca^{2+}{}_i$ were observed, as shown in Fig. 3a, whereas, in the presence of heparin, $Ins(1,4,5)P_3$-induced increases in $Ca^{2+}{}_i$ were blocked either completely or partially, as shown in Fig. 3b. Typically, any increases in $Ca^{2+}{}_i$ in the presence of heparin were small, highly localized and short-lived, returning to resting levels within approximately 100 s, whereas, in the absence of heparin, they continued to increase for more than 600 s. Pre-treatment of pollen tubes with neomycin resulted in the complete inhibition of $Ins(1,4,5)P_3$-induced $Ca^{2+}{}_i$ increases, as also shown in Fig. 3b. We further confirmed that the increases in $Ca^{2+}{}_i$ induced by mastoparan were mediated by $Ins(1,4,5)P_3$ since neomycin-treated pollen tubes treated with mastoparan revealed no detectable increases in $Ca^{2+}{}_i$. These data implicate PIC activity in generating the mastoparan-induced Ca^{2+} increases and provide good evidence for the detected increases in $Ca^{2+}{}_i$ being induced by $Ins(1,4,5)P_3$. These data further suggest the involvement of an IICR mechanism in the generation of the increases in $Ca^{2+}{}_i$. As both heparin and neomycin potently inhibited the increases in $Ca^{2+}{}_i$ generated by $Ins(1,4,5)P_3$ and mastoparan, it indicates that PIC activity and/or heparin-sensitive $Ins(1,4,5)P_3$ receptors ($InsP_3R$) are likely to be involved in the observed increases in Ca^{2+}.

A model for IICR in P. rhoeas pollen tubes

We have proposed a possible model for IICR in pollen tubes which is illustrated in Fig. 4. There are caveats which must be considered in the interpretation of our data, especially as we cannot rule out that CICR may occur in addition to, or in combination with, IICR. We also cannot, at present, rule out the idea that Ca^{2+} influx across the plasma membrane may play some part in this increase. Our data, however, indicate that the slow increases in $Ca^{2+}{}_i$ involved in the generation of the Ca^{2+} waves observed in pollen tubes are most likely to be a result of a series of $Ins(1,4,5)P_3$-generating and Ca^{2+}-mobilizing reactions. A scenario can be envisaged whereby a small initial rise in $Ins(1,4,5)P_3$ levels leads to a limited, and localized, release of Ca^{2+} from $Ins(1,4,5)P_3$-sensitive Ca^{2+} stores. This increase in Ca^{2+} may lead to further activation of other/additional Ca^{2+}-dependent PICs, renewed synthesis of $Ins(1,4,5)P_3$ and additional Ca^{2+} release. A similar cascade, or "domino-effect" could be initiated by a rise in Ca^{2+} (from either internal or external sources) which would stimulate PIC activity, increase the rate of $Ins(1,4,5)P_3$ formation and lead to Ca^{2+} release. This model should provide an exciting starting point for further detailed investigation of PI-mediated Ca^{2+} signalling mechanisms in pollen tubes.

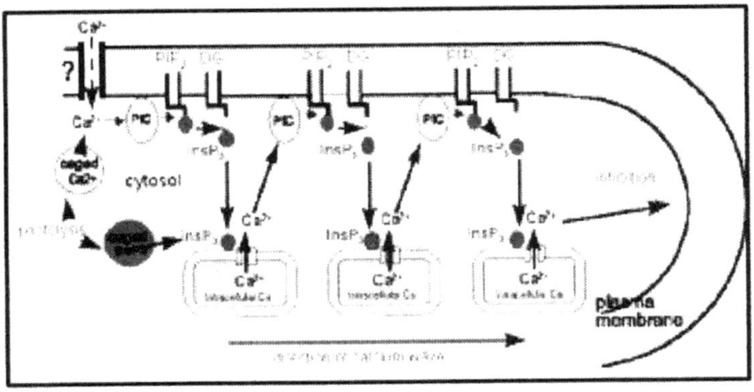

Fig. 4 Model for the propagation of the Ca^{2+}. The Ca^{2+} wave may be initiated by a small rise either in Ca^{2+} and/or $Ins(1,4,5)P_3$. Increases in $[Ca^{2+}]_i$ will activate Ca^{2+}-sensitive phosphoinositide-specific phospholipases C (PICs) which will then hydrolyse the membrane lipid $PtdIns(4,5)P_2$ (PIP_2) to diacylgycerol (DG) and $Ins(1,4,5)P_3$. Increased cytosolic levels of $Ins(1,4,5)P_3$ will stimulate the release of additional Ca^{2+} from $Ins(1,4,5)P_3$-sensitive intracellular stores. Cytoplasmic streaming in the forward direction occurs in the cytoplasm immediately adjacent to the plasma membrane and may contribute to the wave direction. The continued Ca^{2+} increase may occur as a result of a cascade or "domino" effect of activation of PICs resulting in continued generation of Ca^{2+} waves from the initiation site. When the wave reaches the pollen tube tip, inhibition of growth results, presumably due to the disruption of the Ca^{2+} gradient at the tip. (From Franklin-Tong et al., 1996. © *Plant Cell*)

Increases in Ca^{2+} can induce Ca^{2+} waves and oscillations

We have also investigated whether Ca^{2+} waves can be initiated by increases in Ca^{2+} itself, using the Ca^{2+} ionophore Br A23187. Ca^{2+} imaging revealed a slow increase in $Ca^{2+}{}_i$ which moved towards the pollen tube tip and continued to rise for many minutes. This suggests that Ca^{2+} influx can stimulate slow increases in Ca^{2+}which take the form of a wave. Although A23187 can equilibrate Ca^{2+} across membranes with some specificity, it is likely also to affect membrane permeability. However, we have also observed that photolysis of NITR-5 (caged Ca^{2+}) can also induce prolonged increases in $Ca^{2+}{}_i$ (Franklin-Tong et al. 1993; Malho and Trewavas 1996; G. Calder, B.K. Drøbak, P.J. Shaw, V.E. Franklin-Tong, unpublished observations). Thus, an elevation in $[Ca^{2+}]_i$ itself, not necessarily involving Ca^{2+} influx, can be sufficient to initiate Ca^{2+} waves. Use of heparin and caged Ca^{2+} in pollen tubes revealed that although there was an immediate, but small, increase in $[Ca^{2+}]_i$, Ca^{2+} levels declined almost immediately (Franklin-Tong et al. 1996). These data imply that these increases in $Ca^{2+}{}_i$ may be initiated by Ca^{2+} and but that the subsequent propagation of the

Ca^{2+} signal is likely to be dependent on Ins(1,4,5)P_3.

We have also obtained data which demonstrate that Ca^{2+} oscillations may be initiated in response to release of caged Ca^{2+} (G. Calder, B.K. Drøbak, P.J. Shaw, V.E. Franklin-Tong, unpublished observations). Photolysis of microinjected NITR-5 (caged Ca^{2+}) in pollen results in an initial and rapid elevation in $[Ca^{2+}]_i$ with $Ca^{2+}{}_i$ levels returning to basal values after approximately 15 s. Following a period of approximately 75–90 s after the initial NITR-5 release and the first $Ca^{2+}{}_i$ spike, a second major $Ca^{2+}{}_i$ spike occurs. This second $[Ca^{2+}]_i$ increase is also relatively short-lived and again returns to near-basal levels after approximately 15 s. Following these two initial $Ca^{2+}{}_i$ spikes, a regular oscillatory pattern in $Ca^{2+}{}_i$ is set up, with a periodicity of approximately 30–45 s between peaks, as shown in Fig. 5. We have found that tip swelling and resumption of tip growth coincide with the peak of the second $Ca^{2+}{}_i$ spike (i.e. approximately 90 s after NITR-5 photolysis). In some experiments we also observed that the oscillations stopped after long-term incubations (>8–10 min) and a sustained increase in $[Ca^{2+}]_i$ ensued. Thus, there appears to be (at least in certain circumstances) a key role for rapid increases in $[Ca^{2+}]_i$ in the control of gradient formation and triggering of tip-based oscillations.

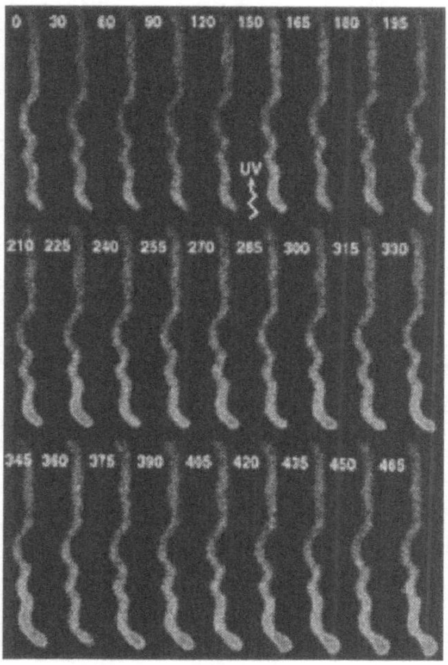

Fig. 5 Release of caged Ca^{2+} in pollen tubes can stimulate initiation of oscillations of $Ca^{2+}{}_i$. This shows a single pollen tube, microinjected with Calcium-Green dextran and Nitr-5 (caged Ca^{2+}),

imaged over time (at the intervals indicated) using confocal laser scanning microscopy. $[Ca^{2+}]_i$ is indicated in pseudocolour, with green indicating low , and red indicating high $[Ca^{2+}]_i$. At 150 s, the caged Ca^{2+} was released by photolysis, using the UV laser. Note the immediate increase in $[Ca^{2+}]_i$ after this treatment, after which $[Ca^{2+}]_i$ fell back to basal levels by 165 s. However, at 180 s $[Ca^{2+}]_i$ increased again, and continued to rise for 45 s, whereupon a set of oscillations of $[Ca^{2+}]_i$ at the pollen tube tip was set up. these were visualized at approximately 30- to 40-s intervals. At around 270 s after treatment, the $[Ca^{2+}]_i$ became more constant at the tip, although it remained high. This pollen tube was not growing at the time of photolytic release of Ca^{2+}; however, the setting up of the $Ca^{2+}{}_i$ oscillations at the tip appeared to stimulate tip extension, as growth had resumed by the end of this sequence.(© *Biochem. Biophys. Res. Comm.* 1997)

A role for Ca^{2+} waves and oscillations in mediating pollen tube growth

IICR and modulation of pollen tube growth

A key question in current plant biological research is how the perception of extracellular signals leads to modulation of growth patterns and directionality of plant cells. It is becoming increasingly clear that many plant cells are likely to utilize the "phosphoinositide-Ca^{2+} signalling cassette" for signalresponse coupling. We have experimental data which strongly suggest that a functional PI signalling pathway controls the modulation of pollen tube growth (Franklin-Tong et al. 1996).

The photolysis of caged Ins(1,4,5)P_3 resulted in increases in $[Ca^{2+}]_i$ and Ca^{2+} waves which correlated well with changes in *P. rhoeas* pollen tube tip growth. This was observed as a variety of responses, including inhibition of growth in approximately 70% of all cases. In 45% of the treated pollen tubes, altered tip morphology was observed as tip swelling and flattening, accompanied by growth cessation. In other pollen tubes, elongation stopped completely when Ca^{2+} in the tip was elevated. Other pollen tubes showed a change in tip morphology and a temporary cessation of growth subsequent to Ca^{2+} increases, after which growth recommenced, often with a change of tip orientation. A large proportion of pollen tubes with increases in $[Ca^{2+}]_i$ generated a Ca^{2+} wave, but at present it is unclear why not all of them respond in the same way. The variation may be due to differences in amounts of caged molecules injected or released by photolysis, or cells may differ in their state of responsiveness.

The pharmacological modulators of Ins(1,4,5)P_3 metabolism, i.e. mastoparan, neomycin and heparin, were found to all have effects on pollen tube growth. Mastoparan, at concentrations below those required to stimulate $Ca^{2+}{}_i$ release, did not significantly inhibit pollen tube growth, whereas concentrations sufficient to stimulate $Ca^{2+}{}_i$ release (25 μM) rapidly inhibited pollen tube growth. This provides indirect evidence supporting the view that increases in Ins(1,4,5)P_3 stimulated by mastoparan may be responsible for the inhibition of pollen tube growth. Low-molecular-weight heparin (3 kDa), applied externally, resulted in virtually complete inhibition of pollen tube growth. Microinjection of heparin also resulted in immediate inhibition of pollen tube growth. This suggests a possible requirement for Ins(1,4,5)P_3 binding to its receptor for normal pollen tube growth. The heparin experiments were complemented with experiments using neomycin and it was found that pollen tube growth was severely inhibited when concentrations of neomycin of 100 μM or greater were used. At 150 μM neomycin, pollen tube growth and pollen tube elongation were inhibited virtually immediately, although cytoplasmic streaming continued for some time (Franklin-Tong et al. 1996). As this is the concentration at which neomycin is effective in inhibiting PIC activity and $Ca^{2+}{}_i$ release, it suggested a possible requirement for PI turnover for normal pollen tube growth.

Our data indicate that a two-tier level of control involving PI signalling may act to modulate pollen tube growth. Inhibition of PI turnover and inhibition of Ins(1,4,5)P_3 binding to its receptor results in inhibition of pollen tube growth, suggesting that this pathway is required for normal pollen tube growth, and increases in Ins(1,4,5)P_3 and the consequent elevation of $[Ca^{2+}]_i$ also inhibit pollen tube growth. Although the idea that Ca^{2+} plays a central role in the control of pollen tube growth is not new (see Introduction), these data are the first direct demonstration of IICR in pollen tubes. The control of pollen tube tip growth is clearly a complex process and we envisage that signals controlled by Ca^{2+} influx, IICR and perhaps CICR are capable of giving complex control of $[Ca^{2+}]_i$ levels with the possibility of multi-level feedback mechanisms.

Alterations in $[Ca^{2+}]_i$ control modulation of pollen tube growth

We have also obtained experimental data which provide further evidence that alterations in $[Ca^{2+}]_i$ are involved in the modulation of normal pollen tube growth. Whether these events are the result of IICR, CICR or Ca^{2+} influx is not, at present, known. However, these recent data clearly implicate a major role for Ca^{2+} waves and oscillations in pollen tube growth.

Growing pollen tubes of *P. rhoeas*, when imaged using ratiometric Ca^{2+} dyes, exhibit a steep tip-focused $Ca^{2+}{}_i$ gradient (V.E. Franklin-Tong, G. Hackett, P.K. Hepler, submitted; G. Calder, B.K. Drøbak, P.J. Shaw, V.E. Franklin-Tong, unpublished observations) which was not evident in the studies with CG-1 (Franklin-Tong et al. 1993, 1995, 1996). This apical $Ca^{2+}{}_i$ gradient appears to be quite characteristic of growing pollen tubes in several species (Obermeyer and Weisenseel 1991; Rathore et al. 1991; Miller et al. 1992; Malho et al. 1994; Pierson et al. 1996), thus supporting the contention that they are a typical feature of all growing pollen tubes. The $Ca^{2+}{}_i$ gradient in *P. rhoeas* pollen tubes was not constant, but oscillated over time; the oscillations were observed in pollen tubes imaged at 10-s intervals, and they varied between ~1 and 2 μM at their peak and between 200 and 250 nM at their trough. Apical $Ca^{2+}{}_i$ oscillations in pollen tubes have been previously reported to occur in *Lilium*, where there is good evidence that the oscillations are due to large fluctuations in Ca^{2+} entry, which is restricted to a small region at the extreme apex of the pollen tube (Pierson et al. 1996). Detailed measurements have established that the oscillations in $[Ca^{2+}]_i$ correlate positively with pulses in growth rate, suggesting that there is a direct causal link between these two processes (Pierson et al. 1996).

Evidence for Ca^{2+} waves generated by the SI response

Early imaging studies of pollen tubes challenged with the S-proteins suggested that there was a movement of the Ca^{2+} transient, which originated around the "nuclear complex" region in the "shank" of the pollen tube, and that the changes were then propagated further along the pollen tube. Detailed analysis of a series of confocal images of $Ca^{2+}{}_i$ provides evidence suggesting that the SI response leads to the production of a Ca^{2+} wave, as illustrated in Fig. 6. While Ca^{2+} waves have been reported in many stimulated animal cells, the experimental evidence for this phenomenon in higher plants remains very limited. We have recently extended our studies of the SI response in *P. rhoeas* pollen tubes using ratiometric Ca^{2+} imaging with the dual-wavelength dye fura-2 dextran, which has provided further preliminary evidence for the propagation of Ca^{2+} waves being elicited in the pollen by an incompatible SI reaction (V.E. Franklin-Tong, G. Hackett, P.K. Hepler, submitted). Using recombinant S-proteins to elicit the SI response, we observed rapid and dramatic changes in $[Ca^{2+}]_i$ in both the shank and apical region of the pollen tube, with the elevations in $[Ca^{2+}]_i$ in the shank preceding the decreases in $[Ca^{2+}]_i$ at the apex. The rapidity of the Ca^{2+} response is in line with the idea that the elevation of $[Ca^{2+}]_i$ is likely to be one of the first events in the signalling pathway. The dramatic changes in $[Ca^{2+}]_i$ observed involve increases and a redistribution of $[Ca^{2+}]_i$, which appears to propagate both backwards and forwards from a region behind the sub-apical region of the pollen tube. The mechanism for this is, at present, unknown. On the basis of previous studies (Franklin-Tong et al. 1996) a model of IICR might be proposed also to account for the Ca^{2+} waves seen in the SI response. However, a role for CICR and Ca^{2+} influx in the SI response cannot be ruled out at this stage.

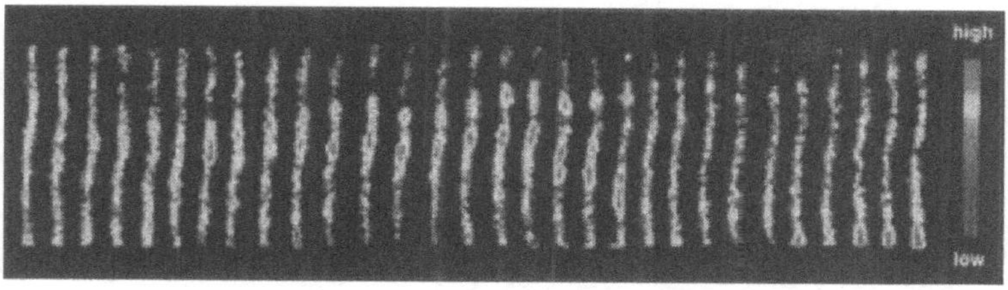

Fig. 6 $Ca^{2+}{}_i$ imaging of pollen tubes challenged with incompatible recombinant S proteins reveals a Ca^{2+} wave. Laser-scanning confocal microscopy of growing pollen tubes microinjected with Calcium Green-1 showing changes in $Ca^{2+}{}_i$ following treatment with incompatible recombinant S-proteins at 20-s intervals after challenge. A transient increase in $Ca^{2+}{}_i$ in the form of a Ca^{2+} wave is clearly observed. The images are pseudocoloured images, where "warm" colours indicate high levels of Ca^{2+} and "cold" colours indicate low levels of Ca^{2+} (see colour scale). The animated series shows first a pollen tube before challenge, then challenged halfway through the series with incompatible S-glycoproteins. The Ca^{2+} wave can be seen moving forward in the pollen tube from an unidentified source. Each image is 20-s intervals apart.

Clearly, there is now good evidence for the presence of Ca^{2+} waves and oscillations in pollen tubes and our observations indicate that they may act to modulate a variety of events related to pollen tube growth. There appear to be several ways in which these waves can be elicited, including localized increases in Ca^{2+} and Ins(1,4,5)P_3. These observations, together with the evidence for Ca^{2+} waves in the SI response, provide an exciting starting point for further investigation of the Ca^{2+}-mediated signalling pathways controlling these processes.

Acknowledgments. This work was carried out through a series of collaborations between V.E. F.-T. and F.C.H.F. (University of Birmingham), with A.J.T. (University of Edinburgh), B.K.D. and P.J.S. [John Innes Centre (JIC), Norwich]. V.E. F.-T. is funded by a Biotechnology and Biological Sciences Research Council (BBSRC) Advanced Fellowship. Thanks are due to Sue Clarke, Antony Jones (Birmingham), Peter Watkins (JIC, Norwich) and Grant Hackett (University of Massachusetts) for technical assistance and to John Martin and gardening staff (Birmingham) for expert horticultural services. Thanks are also due to Jon Ride and David Marshall for their help with the imaging files. Thanks also to Kinetic Imaging for their help in producing the animated figures for use at the Society for Experimental Biology EBO symposium. Work in the laboratory of B.K.D. and P.J.S. is supported by the BBSRC Intracellular Signalling Initiative; V.E. F.-T., F.C.H.F. and A.J.T. are funded by BBSRC. Part of this work was funded by a NSF/DOE/USDA Interagency Research Coordination Grant via the Cytonet Group.

References

Berridge, M.J. (1993) Inositol trisphosphate and calcium signalling. *Nature* **361**:315–325

Biswas, S., Dalal, B., Sen, M., Biswas, B.B. (1995) Receptor for *myo*-inositol trisphosphate from the microsomal fraction of *Vigna radiata*. *Biochem. J.* **306**:631–636

Brewbaker, J.L., Kwack, B.H. (1963) The essential role of calcium ions in pollen germination and pollen tube growth. *Am. J. Bot.* **50**: 859–863

Brosnan, J.M., Sanders, D. (1990) Inositol trisphosphate-mediated Ca^{2+} release in beet microsomes is inhibited by heparin. *FEBS Lett.* **260**:70–72

Calder, G., Franklin-Tong, V.E., Shaw, P.J., Drøbak, B.K. (1997) Ca^{2+} oscillations in plant cells: initiation by rapid elevation in cytosolic free Ca^{2+} concentrations. *Biochem. Biophys. Res. Commun.* (in press)

Derksen, J., Rutten, T., van Amstel, T., Doris, F., Steer, M. (1995) Regulation of pollen tube growth. *Acta Bot. Neerl.* **44**:93–119

Drøbak, B.K. (1992) The plant phosphoinositide system. *Biochem. J.* **288**:697–712

Drøbak, B.K., Ferguson, I.B. (1985) Release of Ca^{2+} from plant hypocotyl microsomes by inositol 1,4,5-trisphosphate. *Biochem. Biophys. Res. Commun.* **130**:1241–1246

Drøbak, B.K., Watkins, P.A.C. (1994) Inositol (1,4,5)-trisphosphate production in plant cells: stimulation by the venom peptides, melittin and mastoparan. *Biochem. Biophys. Res. Commun.* **205**:739–745

Drøbak, B.K., Dove, S.K., Staiger, C.J. (1996) Inositides and plant signalling. In: Moller, I.M., Brodelius, P. (eds.) *Plant membrane biology*. Clarendon, Oxford, pp. 29–50

Feijó, J.A., Malhó, R., Obermeyer, G. (1995) Ion dynamics and its possible role during *in vitro* pollen germination and tube growth. *Protoplasma* **187**:155–167

Felle, H. (1988) Auxin causes oscillations of cytosolic free calcium and pH in *Zea mays* coleoptiles. *Planta* **174**:495–499

Foote, H.C.C., Ride, J.P., Franklin-Tong, V.E., Walker, E.A., Lawrence, M.J., Franklin, F.C.H. (1994) Cloning and expression of a distinctive class of self-incompatibility (*S*) gene from *Papaver rhoeas* L. *Proc. Natl. Acad. Sci. USA.* **91**:2265–2269

Franklin-Tong, V.E., Lawrence, M.J., Franklin, F.C.H. (1988) An *in vitro* bioassay for the stigmatic product of the self-incompatibility gene in *P. rhoeas* L. *New Phytol.* 110:109–118

Franklin-Tong, V.E., Lawrence, M.J., Franklin, F.C.H. (1989) Characterization of a stigmatic component from *Papaver rhoeas* L. which exhibits the specific activity of a self-incompatibility (S-) gene product. *New Phytol.* 112:307–315

Franklin-Tong, V.E., Ride, J.P., Read, N.D., Trewavas, A.J., Franklin, F.C.H. (1993) The self-incompatibility response in *P. rhoeas* is mediated by cytosolic free calcium. *Plant J.* 4:163–177

Franklin-Tong, V.E., Ride, J.P., Franklin, F.C.H. (1995) Recombinant stigmatic self-incompatibility (S-) protein elicits a Ca^{2+} transient in pollen of *P. rhoeas*. *Plant J.* 8:299–307

Franklin-Tong, V.E., Drøbak, B.K., Allen, A.C., Watkins, P.A.C., Trewavas, A.J. (1996) Growth of pollen tubes of *Papaver rhoeas* is regulated by a slow-moving calcium wave propagated by inositol 1,4,5-trisphosphate. *Plant Cell* 8:1305–1321

Franklin-Tong, V.E., Hackett, G., Hepler, P.K. (1997) Ratio-imaging of $Ca^{2+}{}_i$ in the self-incompatibility response in pollen tubes of *Papaver rhoeas. Plant J.* (submitted)

Fricker, M.D., Gilroy, S., Read, N.D., Trewavas, A.J. (1991) Visualization and measurement of the calcium message in guard cells. In Jenkins, G.I., Schuch, W. (eds) *Molecular biology of plant development.* The Company of Biologists, Cambridge, pp.177–190

Gilroy, S., Read, N.D., Trewavas, A.J. (1990) Elevation of cytoplasmic calcium by caged calcium or caged inositol trisphosphate initiates stomatal closure. *Nature* 346:769–771

Gilroy, S., Bethke, P.C., Jones, R.L. (1993) Calcium homeostasis in plants. *J. Cell Sci.* 106:453–462

Haugland, R.P. (1992) *Molecular probes: handbook of fluorescent probes and research chemicals.* Molecular Probes, Eugene, Ore.

Helsper, J.P.F.G., Heemskerk, J.W.M., Veerkamp, J.H. (1987) Cytosolic and particulate phosphatidylinositol phospholipase C activities in pollen tubes of *Lilium longiflorum. Plant Physiol.* 71:120–126

Jaffe, L.F. (1991) The path of calcium in cytosolic calcium oscillations: a unifying hypothesis. *Proc. Natl. Acad. Sci. USA.* 88:9883–9887

Jaffe, L.A., Weisenseel, M.H., Jaffe, L.F. (1975) Calcium accumulation within the growing tips of pollen tubes. *J. Cell Biol.* **67**:488–492

Jouaville, L.S., Ichas, F., Holmuhamedov, E.L., Camacho, P., Lechleiter, J.D. (1995) Synchronization of calcium waves by mitochondrial substrates in *Xenopus laevis* oocytes. *Nature* **377**:438–441

Lancelle, S.A., Hepler, P. K. (1992) Ultrastructure of freeze-substituted pollen tubes of *Lilium longiflorum*. *Protoplasma* **167**:215–230

Lawrence, M.J., Afzal, M., Kenrick, J. (1978) The genetical control of self-incompatibility in *Papaver rhoeas* L. *Heredity* **40**:239–285

Malhó, R., Trewavas, A.J. (1996) Localized apical increases of cytosolic free calcium control pollen tube orientation. *Plant Cell* **8**:1935–1949

Malhó, R., Read, N.D., Pais, M.S., Trewavas, A.J. (1994) Role of cytosolic free calcium in the reorientation of pollen tube growth. *Plant J.* **5**:331–341

Malhó, R., Read, N.D., Trewavas, A.J., Pais, M.S. (1995) Calcium channel activity during pollen tube growth and reorientation. *Plant Cell* **7**:1173–1184

McAinsh, M.R., Webb, A.A.R., Taylor, J.E., Hetherington, A.M. (1995) Stimulus-induced oscillations in guard cell cytosolic free Ca^{2+}. *Plant Cell* **7**:1207–1219

Miller, D.D., Callaham, D.A., Gross, D.J., Hepler, P.K. (1992) Free Ca^{2+} gradient in growing pollen tubes of *Lilium*. *J. Cell Sci.* **101**:7–12

Obermeyer, G., Weisenseel, M.H. (1991) Calcium channel blocker and calmodulin antagonists affect the gradient of free calcium ions in lily pollen tubes. *Eur. J. Cell Biol.* **56**:319–327

Picton, J.M., Steer, M.W. (1983) Evidence for the role of calcium ions in tip extension in pollen tubes. *Protoplasma* **115**:11–17

Pierson, E.S., Miller, D.D., Callaham, D.A., Shipley, A.M., Rivers, B.A., Cresti, M., Hepler, P.K. (1994) Pollen tube growth is coupled to the extracellular calcium ion flux and the intracellular calcium gradient: effect of BAPTA-type buffers and hypotonic media. *Plant Cell* **6**:1815–1828

Pierson, E.S., Miller, D.D., Callaham, D.A., van Aken, J., Hackett, G., Hepler, P.K. (1996) Tip-localized calcium entry fluctuates during pollen tube growth. *Dev. Biol.* **174**:160–173

Rathore, K.S., Cork, R.J., Robinson, K.R. (1991) A cytoplasmic gradient of Ca^{2+} is correlated with the growth of lily pollen tubes. *Dev. Biol.* **148**:612–619

Rudd, J.J., Lord, J.M., Franklin, F.C.H., Franklin-Tong, V.E. (1996) Increased phosphorylation of a 26 kD pollen protein is induced by the self-incompatibility response in *Papaver rhoeas. Plant Cell* 8:713–724

Rudd, J.J., Franklin, F.C.H., Franklin-Tong, V.E. (1997) Ca^{2+}-independent phosphorylation of a 68kD pollen protein is stimulated by the self-incompatibility response in *Papaver rhoeas. Plant J.* (in press)

Shacklock, P.S., Read, N.D., Trewavas, A.J. (1992) Cytosolic free calcium mediates red light-induced photomorphogenesis. *Nature* 358:753–755

Taylor, L.P., Hepler, P.K. (1997) Pollen germination and tube growth. *Annu. Rev. Plant Physiol. Plant Mol. Biol.* 48 (in press)

Trewavas, A.J., Gilroy, S. (1991) Signal transduction in plant cells. *Trend. Genet.* 7:356–361

Walker, E.A., Ride, J.P., Kurup, S., Franklin-Tong, V.E., Lawrence, M.J., Franklin, F.C.H. (1996) Molecular analysis of two functionally identical S3 homologues of the self-incompatibility gene of *Papaver rhoeas* from two different populations. *Plant Mol. Biol.* 30:983–994

Exp. Biol. Online 2:11 (1997)

Effects of salinity on solute clearance from the freshwater bivalve, *Dreissena polymorpha* Pallas

T.H. Dietz[1] and R.A. Byrne[2]

[1] Department of Biological Sciences, Louisiana State University, Baton Rouge, LA 70803, USA
[2] Department of Biology, State University of New York, College at Fredonia, NY 14063, USA

Correspondence to: T.H. Dietz (email: zothom@lsuvm.sncc.lsu.edu)

Received: 14 July 1997 / Accepted: 11 August 1997
Published: August 19, 1997

Abstract. Clearance of polyethylene glycol (PEG), inulin, or dextran that had been injected into the hemolymph of the mussel, *Dreissena polymorpha*, was measured in animals acclimated to pondwater (PW) or 10% seawater (SW). In addition, we measured the clearance of PEG from mussels acutely transferred into 10% SW and following return to PW after acclimation to 10% SW. Clearance values calculated for PW-acclimated mussels ranged from 2.0 to 3.3 ml (g dry tissue · h)$^{-1}$ and declined to 0.28 ml (g dry tissue · h)$^{-1}$ in 10% SW-acclimated animals. Transferring mussels into 10% SW resulted in a reduction in PEG clearance from the blood, coincident with the reduction of osmotic gradient. When 10% SW-acclimated mussels were returned to PW the clearance of PEG increased to rates observed in PW-acclimated animals within 1 h. The PEG clearance remained constant during the re-acclimation to PW even though the osmotic gradient declined from about 100 to 30 mosmol kg^{-1}. Clearance of the solutes used in this study was likely to be a measurement of renal filtration rate. The clearance values appeared to be maximal when the animals were in PW. The limited capacity to increase clearance in the face of an osmotic challenge may be a critical factor in restricting *D. polymorpha* to freshwater or lower salinity environments with small ranges in salinity.

Key words: Dextran clearance – Freshwater bivalve – Inulin clearance – PEG clearance – Renal clearance – Salinity tolerance – Zebra mussel

Introduction

Freshwater bivalves are osmoregulators and the maintenance of water balance is an essential component of their hydromineral regulation. Kidney function and especially filtration rate are important variables, but are not readily studied in freshwater mussels (Picken 1937; Florey and Cahill 1977; Hevert 1984), as the renal pore is located in an anterior location within the suprabranchial chamber. However, an indirect estimate of kidney function can be obtained by measurement of clearance of specific "marker" solutes from the hemolymph of an animal (Kirschner 1991). The marker solute injected into the hemolymph of bivalves will be cleared from the hemolymph and appear in the bathing medium if the marker is subject to filtration by the kidney (Potts 1954b; Murphy and Dietz 1976). This technique allows an estimate of filtration rate and urine formation, but does not provide information on excretion rate, as there could be some water reabsorption.

The freshwater zebra mussel, *Dreissena polymorpha* Pallas, maintains a low hemolymph solute concentration (Dietz *et al.* 1996), displays high epithelial permeability to solutes (Dietz *et al.* 1995; Dietz *et al.* 1997), and has a low transepithelial electrical potential (Wilcox and Dietz 1995). These physiological traits are characteristic of an animal having "leaky" epithelia (Frömter and Diamond 1972). A high passive permeability of the epithelium requires that *D. polymorpha* has a high rate of epithelial ion transport in order to maintain water and ion balance (Horohov *et al.* 1992; Dietz *et al.* 1995).

Dreissena polymorpha also has several aspects of its life history that are unique among North American freshwater bivalve species, including the production of free swimming veliger larvae, and the use of a byssus for attachment to the substratum. The introduction and spread of *D. polymorpha* in North America has been dramatic and there is some evidence of its possible invasion of brackish coastal estuaries. Indeed, populations of *D. polymorpha* are present in the tidal reaches of the St. Lawrence and in portions of the Hudson River in New York (Mellina and Rasmussen 1994; Caraco *et al.* 1997).

In this study we examined the rates of clearance of several radioactively labeled solutes injected into the hemolymph space of *D. polymorpha*. We compared clearance rates of pondwater-acclimated (PW-acclimated) mussels to those of animals incubated in dilute seawater (SW). We also determined the changes in clearance when animals were acutely transferred into 10% SW and when SW-acclimated mussels were returned to PW. The similar values obtained with the different marker solutes suggest that they are freely filtered by the renal organs and provide a useful index of hemolymph clearance. We demonstrate that *D. polymorpha* has a limited ability to regulate water balance in brackish-water habitats that experience periodic changes in salinity.

Materials and Methods

Acclimation conditions

Dreissena polymorpha (zebra mussels) were collected from the Raisin and Huron Rivers in Michigan and from the Mississippi River near Baton Rouge, Louisiana. Mussels were stored, unfed, in aerated artificial PW at room temperature 22 ± 2°C. The composition of PW (in mmol l^{-1}) is 0.5 NaCl, 0.4 $CaCl_2$, 0.2 $MgSO_4$, 0.2 $NaHCO_3$, 0.05 KCl, pH 7.6 and approximately 4 mosmol kg^{-1} (Dietz *et al.* 1994). Larger animals (1.5–3.0 g total live mass) attached by byssal threads were selected for experiments.

Some mussels were acclimated to an artificial SW diluted with PW to nominally 10% SW (~105 mosmol kg^{-1}) for >20 days (Dietz *et al.* 1997). The stock SW ion concentration (in mmol l^{-1}) is 449.1 NaCl, 27.5 $MgSO_4$, 24.4 $MgCl_2$, 9.9 $CaCl_2$, 6.6 KCl, 2.4 $KHCO_3$, 0.8 KBr, 0.4 H_3BO_3, 1076 mosmol kg^{-1} total solute (35‰) (Chambers and De Armendi 1979). No buffer was added to the SW and the pH of diluted SW was 7–8. All acclimation media were changed twice a week and all water and containers treated with 1% chlorine bleach for 24 h before being discarded.

Solute analyses

Hemolymph samples (150–200 µl) were obtained from mussels by pericardial puncture and centrifuged at 24,000 *g* · min before analysis (Fyhn and Costlow 1975). The total osmolality of hemolymph or bathing medium was determined using undiluted samples by freezing-point depression. Sodium and potassium concentrations were determined by flame emission photometry using samples diluted with distilled water. Calcium and magnesium concentrations were determined by atomic absorption spectroscopy from samples diluted with $LaCl_3$/HCl. The chloride concentration was determined by electrometric titration. The difference between the osmolality of the hemolymph and the measured ion concentrations was identified as "other" and in PW-acclimated mussels it is mostly bicarbonate (Horohov *et al.* 1992; Byrne and Dietz 1997).

Renal clearance

Radioactive tracer ([^{3}H]polyethylene glycol, PEG, 4 kDa; [^{3}H]inulin 5.2 kDa; [^{3}H]- or [^{14}C]-dextran 70 kDa) dissolved in deionized water was injected (10 µl, 1 µCi or 37 kBq) into the body mass of each mussel and the animal was returned to the appropriate acclimation medium to allow equilibration of the isotope. The clearance of radioisotope from the body fluids of *D. polymorpha* was determined using procedures adapted from Murphy and Dietz (1976). Because of the high clearance of solutes from the blood, the equilibration time was reduced to 3 h.

The animals were rinsed 3 times to remove adsorbed isotope and transferred to 25 ml of appropriate experimental medium. When the mussels resumed siphoning (usually 10–15 min) a bath sample was collected initially and after 1 h, and the radioactivity determined by liquid scintillation procedures (Weigman *et al.* 1975). After collecting the final bath sample, the mussel was removed from the bath, blotted dry, and a hemolymph sample collected. The soft tissue was removed, the byssus dissected, and the remaining tissue dried overnight at 90°C.

The clearance of isotope was calculated from the amount of radioactivity that accumulated in the bathing medium during the 1-h interval ($dpm \cdot h^{-1}$) divided by the isotope specific activity in the hemolymph ($dpm \cdot ml^{-1}$) at the end of the clearance study. The clearance was expressed as ml $(g\ dry\ tissue \cdot h)^{-1}$. Because of the small size of the animals and the potential damage due to pericardial sampling, we assumed, for our calculations, that the hemolymph specific activity remained constant during the 1-h clearance measurement, although it would have been declining at an exponential rate. However, we have estimated that the change in hemolymph concentration of inulin would be less than 22.0±0.4% (n=33) during the 1-h clearance and the change would be relatively linear. Previous studies have indicated that the assumption of a constant hemolymph marker concentration would result in an underestimate of clearance, by 20% or less (Murphy and Dietz 1976).

Urine samples cannot be collected so the urine/hemolymph ratio of radioactive tracer was not determined and the urine volume was not measured. The clearance value represents removal of the marker from the animal by all routes, but the results indicate that most of the loss is renal (see below). Previous studies indicate that contamination of the hemolymph sample with bathing medium is rare (Dietz *et al.* 1997). However, a maximal estimate of bathing medium contamination of hemolymph is 6.4%. A recalculation of the inulin clearance applying a correction for the maximal bathing medium contamination would reduce the clearance value by 5.5±0.0% (n=36). Isotope purity was determined by thin layer chromatography and no low-molecular-weight fragments were present in the PEG or dextran solutions (see Dietz *et al.* 1997). There were approximately 20% high-molecular-weight fragments of inulin in the stock solution.

Clearance studies were performed using mussels acclimated (>20 days) to PW and 10% SW. Clearance values were obtained 1, 4, 12, or 24 h after an acute transfer of mussels from PW into 10% SW. In addition, clearance measurements were obtained from animals acclimated to 10% SW and acutely transferred to PW for 1, 4, 24, or 72 h.

Net water uptake in mussels was examined by measuring weight gain after acute transfer of 10% SW-acclimated mussels to PW. The 10% SW-acclimated mussels were removed from the bath and the mantle cavity fluid allowed to drain on absorbent towels for 10 ± 2 min. Mussels were weighed (± 1 mg) and transferred to PW for 2 h. The mussels were removed from the bath, the mantle cavity allowed to drain as before, and reweighed. The soft tissues were removed, the byssus dissected and the tissue dried overnight at 90°C. Net water uptake, ml (g dry tissue · h)$^{-1}$, was calculated as the difference in total animal weight divided by dry tissue mass.

Statistical analyses

All data are expressed as mean ± 1 standard error. An animal was used only once and "*n*" indicates the number of animals used in a treatment group. Overall differences between treatment groups were determined by one-way analysis of variance (ANOVA) followed by the Tukey-Kramer post-hoc test to determine differences between specific means ($P<0.05$). Regression analysis was used to determine relationships between animal size and clearance. Analyses were performed by Statview and SuperANOVA programs (Abacus Concepts). Data were tested for homogeneity of variance with some data transformed by natural logarithm before performing the ANOVA.

Results and Discussion

Clearance determination

Clearance of radioactive markers from the hemolymph of *D. polymorpha* acclimated to PW ranged from 2.0 to 3.3 ml (g dry tissue · h)$^{-1}$ (Table 1). Recently, we demonstrated that smaller molecular weight molecules (<5 kDa) rapidly moved across the epithelia of *D. polymorpha* by paracellular routes, but that the 70-kDa dextran penetrated very slowly over a 6-h exposure (Dietz *et al.* 1995; Dietz *et al.* 1997). Because the clearance of the high-molecular-weight dextran molecule (70 kDa) was not different from that of PEG or inulin we conclude that the clearance measurements were largely renal, and that non-renal contributions were minor. The clearance of inulin was significantly higher than the values for PEG (Table 1). It is possible that some of the radioactive fragments of inulin were sufficiently small to penetrate the epithelia and thus were "cleared" by both renal and extra-renal routes. We have not explored the possible extra-renal routes of clearance in these animals beyond the paracellular pathway. Because inulin decomposes even when stored frozen, we discontinued its use after the initial studies, relying on dextran and PEG for the majority of the study.

Table 1 Clearance of radioactive solutes from the hemolymph of pondwater-acclimated *D. polymorpha*. Values are mean ± SE. Values with different letters are significantly different using the Tukey–Kramer test of the mean values ($P<0.05$).

Solute	*n*	Clearance ml (g dry tissue · h)$^{-1}$
Polyethylene glycol	38	1.99±0.13a
Dextran	15	2.69±0.31a,b
Inulin	32	3.26±0.24b

These clearance measurements provide information on the maximal volume of fluid filtered by the bivalve kidney tubules. In the absence of urine samples, it is not possible to determine the amount of water reabsorbed and the urine flow. Previous studies have shown that water reabsorption is variable in freshwater invertebrates and vertebrates (see Kirschner 1991). Although some animals may reabsorb half of the filtered water, many animals reabsorb little water, thus, clearance values represent 50–100% of the true excretion rate (Kirschner 1991). These clearance studies allow an estimate of the maximal excretion rate and provide an estimate of the osmotic water uptake for an animal maintaining its water balance.

The measured clearance values observed in *D. polymorpha* are 4–8 times higher than values of 0.41–0.59 ml (g dry tissue · h)$^{-1}$ observed for the unionid bivalves *Ligumia subrostrata* and *Toxolasma (Carunculina) texasensis* (Murphy and Dietz 1976; Scheide and Dietz 1986). However, *D. polymorpha* used in the study were less than 150 mg dry mass compared to about 1 g of dry mass for the unionids used in previous studies. There was a significant relationship between logarithm of dry mass and clearance (Fig. 1). The extrapolated clearance of inulin for a standard 1-g *D. polymorpha* using the regression analysis was 1.4 ml (g dry tissue · h)$^{-1}$, or about 40% of the value reported in Table 1, but still over twice the value of similarly sized unionid bivalves, and other freshwater animals that have been studied (Kirschner 1967; Kirschner 1991).

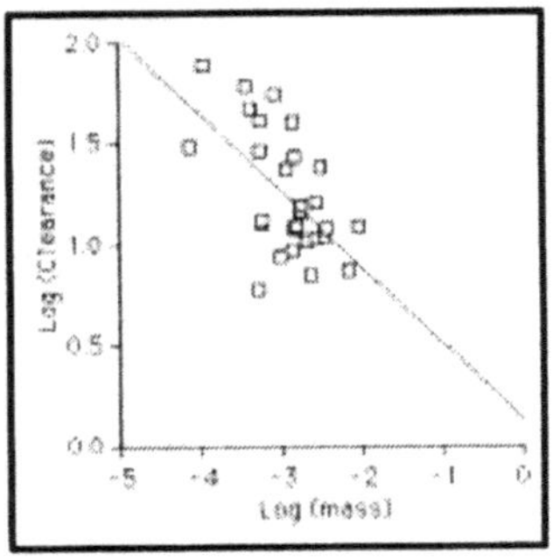

Fig. 1 Relationship between logarithm of clearance, ml (g dry tissue · h)$^{-1}$, of inulin and dry tissue mass (g) in *D. polymorpha*. The data were fit to the equation: $Y = -0.379X + 0.135$ (r=0.58, P<0.01, n=29). The line extends beyond the data points to demonstrate the extrapolated clearance of a 1.0 g (log mass = 0) animal

The clearance of radioactive PEG was reduced when PW-acclimated mussels were acutely transferred to 10% SW (Fig. 2). Clearance of PEG decreased 73–87% between 4 and 24 h in 10% SW, relative to the PW controls. During the first hour in 10% SW the reduction in clearance was evident, but was variable because some animals had low clearance while others had a PEG clearance which was about the same as that of mussels acclimated to PW. In a separate study, we measured the clearance of dextran from *D. polymorpha* 12 h after being transferred from PW into 10% SW. The clearance of dextran was 0.32±0.15 (n=10) ml (g dry tissue · h)$^{-1}$, a value that does not differ from that of mussels exposed to 10% SW for 24 h shown in Fig. 2. These data indicate that even when subjected to hyperosmotic challenge the clearance values represent renal not extra-renal clearance because the rates measured with dextran (70 kDa) and PEG (4 kDa) were identical.

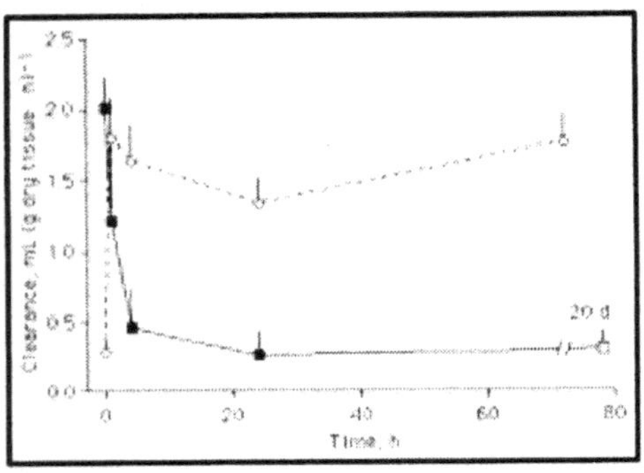

Fig. 2 Clearance of PEG from *D. polymorpha* transferred from pondwater (PW) into 10% seawater (*SW*, filled squares) or from 10% SW-acclimated animals transferred into PW (*circles*). The clearance was significantly elevated by 1 h when 10% SW-acclimated mussels were placed in PW and was significantly reduced 4 h after animals were transferred into 10% SW. The *open square* is a replicate of the clearance of >20 days 10% SW-acclimated mussels. Each point represents data from 6–18 animals and the *vertical lines* represent 1 SE

Clearance of radioactive PEG was measured from mussels acclimated to 10% SW for >20 days and during the transition when mussels were returned to PW (Fig. 2). The clearance of PEG was least in the animals acclimated to 10% SW with values less than 15% of those observed in PW-acclimated mussels. When mussels were transferred from 10% SW to PW the clearance of PEG returned within the first hour to clearance values of PW-acclimated mussels.

Osmotic regulation

Dreissena polymorpha acutely transferred into 10% SW became isosmotic with the bathing medium (103 mosmol kg^{-1}) between 4 and 24 h (Table 2). The mussels rapidly gained Na, Cl and Mg while losing Ca during the 24-h acclimation period. The transient reduction in hemolymph K at 1 h was likely due to shifting of K into the intracellular compartment to maintain cellular volume (Dietz *et al.* 1997). PW-acclimated mussels were hyperosmotic to their bathing medium, with Na and Cl as the major cation and anion, respectively. In previous studies we observed similar changes in hemolymph ion concentrations when mussels were acclimated to dilute SW (Dietz *et al.* 1997; Wilcox and Dietz 1997). *D. polymorpha* transferred into hyperosmotic 10% SW did not completely acclimate within 24 h (compare Tables 2 and 3).

Table 2 Concentration of solutes in the hemolymph of *D. polymorpha* acclimated to pondwater (PW), or transferred to 10% seawater (SW) for various intervals. Values are mean ± SE. Values in a column with different letters are significantly different using the Tukey-Kramer test to compare the means ($P<0.05$). Composition of the 10% SW bathing medium: mosmol kg^{-1} total solute 103±1, other -1.9±1.0; mmol l^{-1} Na 46.4±1.3, K 1.1±0.1, Ca 1.4±0.2, Mg 3.8±0.1, Cl 51.6±1.5 (n=3).

		Ion concentration						
		mosmol kg^{-1}		mmol l^{-1}				
Acclimation	*n*	Total	Other	Na	K	Ca	Mg	Cl
PW	10	37±1a	3.1±0.8a	12.1±0.7a	0.5±0.0b	4.6±0.2c	1.1±0.1a	15.4±0.6a
1 h in SW	10	57±2b	5.5±0.6a,b	23.1±1.0b	0.3±0.0a	4.0±0.2b,c	2.1±0.1b	22.3±1.0b
4 h in SW	9	82±2c	5.0±1.6a,b	36.5±1.0c	0.6±0.1b	3.5±0.3a,b	3.6±0.1c	33.1±1.9c
24 h in SW	10	111±1d	7.3±0.7b	50.0±1.1d	0.8±0.0c	2.9±0.3a	5.2±0.1d	44.3±0.9d

Table 3 Concentration of solutes in the hemolymph of *D. polymorpha* acclimated to PW, 10% SW for >20 days, or returned to PW for various intervals. Values are mean ± SE. Values in a column with different letters are significantly different using the Tukey-Kramer test of the means ($P<0.05$). *Values transformed by natural logarithm before performing ANOVA.

		Ion concentration						
		mosmol kg^{-1}		mmol l^{-1}				
Acclimation	*n*	Total	Other	Na*	K	Ca	Mg*	Cl*
SW	9	110±1d	-0.3±1.0a	51.6±0.3e	1.5±0.0c	3.4±0.2a,b	4.4±0.1d	49.2±0.5d
1 h in PW	10	73±1c	3.5±1.1b,c	32.0±0.7d	1.0±0.5b	3.1±0.2a	3.2±0.1c	29.7±0.9c
4 h in PW	10	51±2b	5.2±1.0c,d	20.2±1.1c	0.9±0.0b	4.2±0.2b,c	2.2±0.1b	18.4±1.5b
24 h in PW	10	33±1a	7.7±0.9d	9.6±0.7a	0.5±0.0a	5.7±0.3d	0.9±0.0a	9.0±0.7a
72 h in PW	10	34±1a	5.4±0.5c,d	13.2±0.7b	0.4±0.0a	3.5±0.2a,b	0.9±0.1a	10.6±0.8a
PW	19	36±1a	1.8±0.5a,b	12.8±0.5b	0.5±0.0a	4.6±0.1c	1.0±0.0a	15.7±0.5b

When *D. polymorpha* were transferred into 10% SW the reduction in clearance of PEG or dextran was coincident with the decline in osmotic gradient achieved within 12–24 h and no further change in clearance was observed after 20 days of acclimation to 10% SW (see Fig. 2). Although the osmotic gradient was less than 10 mosmol kg^{-1} when mussels were in 10% SW for longer than 12 h, the measured clearance of PEG or dextran by *D. polymorpha* was approximately the same as that observed in the freshwater unionids acclimated to PW (Murphy and Dietz 1976; Scheide and Dietz 1986). Thus, even with a reduced osmotic gradient, *D. polymorpha* has the highest clearance values of any freshwater vertebrate or invertebrate that has been studied (see Kirschner 1991).

The hemolymph solute concentrations for *D. polymorpha* acclimated to PW, 10% SW, or during the transition of re-acclimating to PW are shown in Table 3. The mussel hemolymph solute concentrations for PW- and 10% SW-acclimated mussels are similar to those reported previously (Potts 1954a; Dietz *et al.* 1996; Dietz *et al.* 1997; Wilcox and Dietz 1997). Mussels acclimated to 10% SW were isosmotic to the medium, and approximately isoionic also. During the return to PW the solute concentration in the hemolymph of the animals was rapidly reduced and the osmolality was restored to values found in PW-acclimated animals within 24 h. Na, K, and Mg concentrations were the same as those found in PW-acclimated animals following 3 days of re-acclimation to PW. However, Ca and Cl concentrations remained significantly less than those found in PW-acclimated animals and the "other" solute remained elevated even after 3 days in PW. Some of the unidentified solute present in animals that had been exposed to 10% SW is likely to be sulfate. We have observed that radioactive sulfate placed in the bath will accumulate slowly in the hemolymph of *D. polymorpha* (unpublished observations). Sulfate may remain elevated in the hemolymph for some time after mussels are returned to PW, as we have found that the sulfate clearance of 0.45 ml (g dry tissue · h)$^{-1}$ was significantly less than PEG clearance (unpublished observations). The uptake and loss of sulfate in freshwater bivalves has not been studied. The negative "other" reported for the 10% SW-acclimated animals was due to the complete ionization during specific ion analysis, over-estimating their contribution to osmolality.

When 10% SW-acclimated animals are returned to PW, they gain water and appear to swell (Dietz *et al.* 1997; Wilcox and Dietz 1997). An acute transfer to PW resulted in an initial osmotic gradient of nearly 100 mosmol kg^{-1} compared to the 30–40 mosmol kg^{-1} gradient experienced by PW-acclimated animals. The higher osmotic gradient contributes to the net uptake of water at the rate of 1.51 ± 0.12 (n=7) ml (g dry tissue · $h)^{-1}$, resulting in an apparent osmotic water uptake of approximately 3.3 ml (g dry tissue · $h)^{-1}$ (net water uptake + clearance) in the first 2 h after transfer from 10% SW to PW. It is noteworthy that the clearance of PEG was not elevated over that of PW-acclimated mussels at any time during the study when 10% SW-acclimated mussels were returned to PW (Fig. 2). Apparently, *D. polymorpha* clearance of PEG is near the maximal rate at all times when the animals are in PW. We have observed that most *D. polymorpha* acclimated to <260 mosmol kg^{-1} SW will survive an acute transfer back to PW (Wilcox and Dietz 1997); however, when mussels are acutely transferred from 377 mosmol kg^{-1} SW to PW, less than half of the animals survive beyond 2 days. It is likely that the high mortality is due, in part, to the inability of the kidney tubules to filter the excess water gained by the mussels subjected to a 377 mosmol kg^{-1} osmotic challenge.

Studies have indicated that *D. polymorpha* has exceptionally high epithelial permeability to water and solutes (Dietz *et al.* 1995; Dietz *et al.* 1997; Wilcox and Dietz 1997; this study). When subjected to osmotic challenge, these mussels will survive an acute transfer provided that the gradient is less than 150 mosmol kg^{-1} when they are transferred into dilute SW, or less than 270 mosmol kg^{-1} when transferred from dilute SW to PW (Wilcox and Dietz 1997). The kidney of *D. polymorpha* is able to eliminate the water gained osmotically, but appears to function at maximal clearance when this species lives in freshwater. *D. polymorpha* was able to reduce solute and water clearance when the chemical gradient was lower, but does not appear to have the capacity to increase the clearance when challenged by markedly elevated osmotic gradients.

Dreissena polymorpha has evolved in freshwater connected to inland seas, but this species can tolerate dilute SW (Kilgour *et al.* 1994; Wright *et al.* 1996; Dietz *et al.* 1997; Wilcox and Dietz 1997). Their limited ability to control epithelial permeability (Dietz *et al.* 1995; Wilcox and Dietz 1995) and the limited capacity of the kidney to increase clearance or filtration rate when challenged by elevated osmotic gradients may be critical factors that limit *D. polymorpha* to lower salinity environments with small changes in salinity.

Acknowledgements. We thank S. J. Nichols for providing animals and for many suggestions and comments. We thank Harold Silverman and the anonymous reviewers for many helpful suggestions. Julie Cherry and Paul Bruce provided technical assistance. This work was supported by Louisiana Sea Grant NOAA 46RG00960 Project R/ZMM-1.

References

Byrne, R.A., Dietz, T.H. (1997) Ion transport and acid-base balance in freshwater bivalves. *J. Exp. Biol.* **200:** 457–465

Caraco, N.F., Cole, J.J., Raymond, P.A., Strayer, D.L., Pace, M.L., Findlay, S.E.G., Fischer, D.T. (1997) Zebra mussel invasion in a large, turbid river: phytoplankton response to increased grazing. *Ecology* **78:** 588–602

Chambers, E.L., De Armendi, J. (1979) Membrane potential, action potential and activation potential of eggs of the sea urchin *Lytechinus variegatus. Exp. Cell Res.* **122:** 203–218

Dietz, T.H., Lessard, D., Silverman, H., Lynn J.W. (1994) Osmoregulation in *Dreissena polymorpha*: the importance of Na, Cl, K, and particularly Mg. *Biol. Bull.* **187:** 76–83

Dietz, T.H., Byrne, R.A., Lynn, J.W., Silverman, H. (1995) Paracellular solute uptake by the zebra mussel, *Dreissena polymorpha. Am. J. Physiol.* **269:** R300–R307

Dietz, T.H., Wilcox, S.J., Byrne, R.A, Lynn, J.W., Silverman, H. (1996) Osmotic and ionic regulation of North American zebra mussels (*Dreissena polymorpha*). *Am. Zool.* **36:** 364–372

Dietz, T.H., Wilcox, S.J., Byrne, R.A., Silverman, H. (1997) Effects of hyperosmotic challenge on the freshwater bivalve, *Dreissena polymorpha*: Importance of K^+. *Can. J. Zool.* **75:** 697–705

Florey, E., Cahill, M.A. (1977) Hemodynamics in lamellibranch molluscs: confirmation of constant-volume mechanism of auricular and ventricular filling. Remarks on the heart as site of ultrafiltration. *Comp. Biochem. Physiol.* **57A:** 47–52

Frömter, E., Diamond, J. (1972) Route of passive ion permeation in epithelia. *Nature N.B.* **235:** 9–13

Fyhn, H.J., Costlow, J.D. (1975) Anaerobic sampling of body fluids in bivalve molluscs. *Comp. Biochem. Physiol.* **52:** 266–268

Hevert, F. (1984) Urine formation in the lamellibranchs: evidence for ultrafiltration and quantitative description. *J. Exp. Biol.* **111:** 1–12

Horohov, J., Silverman, H., Lynn, J.W., Dietz, T.H. (1992) Ion transport in the freshwater zebra mussel, *Dreissena polymorpha. Biol. Bull.* **183:** 297–303

Kilgour, B.W., Mackie, G.L., Baker, M.A., Keppel, R. (1994) Effects of salinity on the condition and survival of zebra mussels (*Dreissena polymorpha*). *Estuaries* **17:** 385–393

Kirschner, L.B. (1967) Comparative physiology: Invertebrate excretory organs. *Annu. Rev. Physiol.* **29:** 169–196

Kirschner, L.B. (1991) Water and ions. In: Prosser, C.L. (ed) *Environmental and metabolic animal physiology*, 4th edn.. Wiley-Liss, New York, pp. 13–107

Mellina, E., Rasmussen, J.B. (1994) Occurrence of zebra mussel (*Dreissena polymorpha*) in the intertidal region of the St. Lawrence estuary. *J. Freshw. Ecol.* **9:** 81–84

Murphy, W.A., Dietz, T.H. (1976) The effects of salt depletion on blood and tissue ion concentrations in the freshwater mussel, *Ligumia subrostrata* (Say). *J. Comp. Physiol.* **108:** 233–242

Picken, L.E.R. (1937) The mechanism of urine formation in invertebrates. II. The excretory mechanism in certain mollusca. *J. Exp. Biol.* **14:** 20–34

Potts, W.T.W. (1954a) The inorganic composition of the blood of *Mytilus edulis* and *Anodonta cygnea. J. Exp. Biol.* **31:** 376–385

Potts, W.T.W. (1954b) The rate of urine production of *Anodonta cygnea. J. Exp. Biol.* **31:** 614–617

Scheide, J.I., Dietz, T.H. (1986) Serotonin regulation of gill cAMP production, Na transport and water uptake in freshwater mussels. *J. Exp. Zool.* **240:** 309–314

Weigman, T., Woldring, M.G., Pratt, J.J. (1975) A new cocktail for liquid scintillation counting of aqueous radioimmunoassay-samples. *Clin. Chim. Acta* **59:** 347–356

Wilcox, S.J., Dietz, T.H. (1995) Potassium transport in the freshwater bivalve *Dreissena polymorpha. J. Exp. Biol.* **198:** 861–868

Wilcox, S.J., Dietz., T.H. (1997) Salinity tolerance of the freshwater bivalve, *Dreissena polymorpha. Nautilus* **110:** (in press)

Wright, D.A., Setzler-Hamilton, E.M., Magee, J.A., Kennedy, V.S., McIninch, S.P. (1996) Effect of salinity and temperature on survival and development of young zebra (*Dreissena polymorpha*) and Quagga (*Dreissena bugensis*) mussels. *Estuaries* **19:** 619–628

Exp. Biol. Online 2:12 (1997)

Nucleotide pool changes in coelomic cells (eleocytes) of the polychaete *Nereis virens* during sexual maturation

U. Hoeger and C. Märker

Institut für Zoologie, Universität Mainz, Saarstr. 21, D-55099 Mainz, Germany

Correspondence to: U. Hoeger (email: hoeger@mzdmza.zdv.uni-mainz.de)

Received: 23 June 1997 / Accepted: 19 August 1997
Published: August 22, 1997

Abstract. Eleocytes (a type of coelomic cell) of the polychaete *Nereis virens* can store large amounts of adenine nucleotides at certain times. Since eleocytes have specific functions related to gametogenesis, we tested whether the presence of these large nucleotide stores in eleocytes is specific to gender or related to specific events during gametogenesis. Nucleotide pools in eleocytes isolated at different stages of sexual maturation from *N. virens* were analysed using high-performance liquid chromatography. Eleocytes from immature and male animals had extremely high concentrations of both AMP and ADP (each > 10 µmol/ml of packed cell volume). In eleocytes from male animals, the high nucleotide stores were maintained throughout the maturation phase and decreased at a late stage, while in female animals the nucleotides were degraded at an early stage of maturation. In male eleocytes, the decrease in the adenine nucleotide pool may be the result of its conversion to inosine which is then released by the eleocytes and reutilized by male germ cells for nucleic acid biosynthesis, as has been suggested previously. Our study shows that the time of degradation of the adenine nucleotide pool coincides with the period of spermatogonia proliferation which involves intense nucleic acid synthesis. ATP levels (0.4–1.5 µmol/ml packed cell volume) and the guanine nucleotide pool (GTP+GDP+GMP; 0.08–0.18 µmol/ml packed cell volume) were similar in both sexes, did not change during germ cell development and were decreased only in eleocytes from prespawning females. The GTP/GDP ratios were initially higher (up to 14) in eleocytes from females compared to ratios in eleocytes from immature (4–9) and male animals (up to 8), and decreased during the maturation phase of the animals. GTP levels were correlated with those of ATP; this correlation was much closer in eleocytes from females than from males. The results further support the hypothesis that the adenine nucleotide stores in the eleocytes are maintained as a supply of purine precursors for the growing germ cells.

Key words: Coelomic cells – Eleocytes – *Nereis* – Nucleotides – Polychaetes – Sexual maturation

Introduction

Eleocytes of polychaetous annelids represent a type of coelomic cell which has specialized functions related to gametogenesis (for recent reviews, see Fischer and Hoeger 1993; Fischer *et al.* 1996). Our studies of the physiological changes in eleocytes related to the sexual maturation of *Nereis virens* (Hoeger *et al.* 1995) revealed that there are extremely high concentrations of both ADP and AMP in these cells, each ranging between 10 and 20 μmol/ml of packed cell volume (pcv). ^{31}P nuclear magnetic resonance studies of living eleocytes (Hoeger *et al.* 1995) suggested that these high nucleotide concentrations are located in the large acidic vacuole which is characteristic of eleocytes, thus preventing these highly concentrated nucleotides from interfering with cellular metabolism. These nucleotide stores were found only in eleocytes isolated from animals before or at the onset of germ cell development and in eleocytes from male but not female animals. In light of the specific functions of eleocytes in gametogenesis, these high nucleotide concentrations raise the question of whether their presence is related to specific events during the growth and development of the germ cells. In this study, we compared the concentrations of these nucleotides within the rich stores of eleocytes from *N. virens* of both sexes, isolated at different stages of germ cell development. In addition, we compared adenine nucleotide pools in eleocytes with the levels of other cellular nucleotides to detect additional stage-specific differences in eleocytes during the course of the sexual maturation of *N. virens*.

Materials and Methods

Animals

N. virens were collected between 1989 and 1995 from the mudflats of the Oosterscheldt Bay at Yerseke, The Netherlands. The animals were transferred on ice to the laboratory within 8 h of capture and kept at 8–12°C under a natural light regime in 40-l trays with artificial seawater and a sandy bottom. Coelomic cells were isolated (see below) from immature (i.e. with no gamete cells) male and female animals at different stages of sexual maturation. In the Yerseke population, *N. virens* has a 3-year life cycle (A. Fischer and M. Lehmann, unpublished observation) and *N. virens* reaches sexual maturity during the last 16 months of the life cycle. Thus, animals at different stages of germ cell development co-exist. However, in a population of maturing animals, the growth and development of the germ cells is highly synchronized between individuals (see Fischer and Hoeger 1993), thus allowing precise identification of the different stages of germ cell development (see below). Eleocytes isolated at six different stages of germ cell development were analysed in this study (Table 1). The first set of samples represents immature individuals (collected in January), the next four sets comprise animals between the onset of sexual maturation and the prespawning stage of development. The last set represents animals which are ready to spawn. One set of samples may contain eleocyte samples from different years of collection.

Table 1. Characterization of the eleocyte samples isolated from *Nereis virens* at different stages of germ cell development. Data for female animals are taken from Hoeger (1991). Data for developmental stages of male germ cells are taken from Geier and Hoeger (1997). Number in parentheses indicate number of samples analysed

Sample Nr.	Time of eleocyte preparation	Stage of sexual maturation	
1	Jan. 1991	**immature animals** (7) (no germ cells present)	
		female animals (mean oocyte volume [10^6 µm^3])	**male** animals (germ cell stage)
2	June 1989	0.11 (9)	Spg stage I [a] (12)
3	Jul./Aug. 1989	0.19 (3)	Spg stage I (5)
4	Nov. 1989	0.94 (5)	Spermatocytes (6)
5	Jan/Feb. 1990,1991	2.55 (7)	Spz (immotile) [b] (6)
6	Apr. 1990,1991 (prespawning animals)	3.55 (6)	Spz (motile) (3)

[a] spg: spermatogonia
[b] spz: spermatozoa

Determination of maturation stage

The gender and the stage of maturation were identified by the presence and the developmental stage of the germ cells in samples of coelomic fluid obtained by puncturing the coelomic cavity with a fine glass capillary. In females, the stage of maturation was characterized as the mean oocyte volume calculated from the diameters of 25–30 oocytes. Males at different stages of sexual maturation were identified by the presence of spermatogonia clusters, spermatocytes and spermatozoa, respectively (Geier and Hoeger 1997).

Isolation of coelomic cells

Coelomic cells were isolated as described in detail previously (Hoeger 1991). In brief, the body cavity of the anaesthetized animals was opened and the coelomic cells were flushed out with Ca/Mg-free seawater. Eleocytes were separated from other coelomic cells by passing the suspension through nylon sieves with mesh sizes of between 40 and 150 μm. The eleocyte suspensions were further purified by layering onto a linear Ficoll gradient, and from the finally purified suspensions 50- or 100-μl aliquots of packed cells were frozen in liquid nitrogen until analysis.

Preparation of metabolite extracts

To the frozen pellets, a ninefold volume of ice-cold perchloric acid (0.4 M) was added and the cells were broken up by three 5-s bursts (50 W) of an ultrasonic disintegrator. The homogenates were centrifuged for 3 min at 15,000 *g* in a refrigerated centrifuge and the extracts were recovered by puncturing the microcentrifuge tube near the bottom to separate the floating lipid layer. The perchloric acid extracts were neutralized by adding a twofold volume of 0.5 M *N*-trioctylamine in chloroform (1:3.55 v/v; Dobson *et al.* 1991). After vigorous mixing, the phases were separated by brief centrifugation and the aqueous (top) phase was removed. The aqueous phase was further purified by a brief centrifugation (1500 *g*, 1 min) through a 1-ml solid phase extraction column (Adsorbex RP 18, Merck) previously equilibrated with water and precentrifuged under the same conditions. The purified extracts were stored at -20°C for no longer than 2 weeks. The recovery of the purification procedure was determined using standard solutions for each of the compounds of interest. Depending on the compound, recoveries were between 77 and 97% (SD 4.6–6.5%, n=5).

Determination of the nucleotides

Nucleotides were separated by anion exchange HPLC at 55°C on a Nucleosil 5SB column (250 X 4.6 mm; Macherey-Nagel) using a phosphate gradient elution (buffer A: 25mM K-phosphate, pH 2.8, 10% methanol; buffer B: 600 mM K-phosphate, pH 2.8, 10% methanol; modified after Abe and Okuma 1991) and a flow rate of 1 ml/min. Elution of the nucleotides was monitored by simultaneous UV detection at 254 and 280 nm. Compounds were identified by comparison with the retention times of standard solutions chromatographed under identical conditions and their characteristic absorbance ratios at 254 and 280 nm. The concentration was calculated from the area of the peak and corrected for recovery of the purification procedure (see above).

Statistical treatment

Mean nucleotide concentrations and GTP/GDP ratios were compared using Student's *t*-test after log transformation of the data. ATP and GTP concentrations were correlated using the Spearman's rank sum test.

Results and Discussion

Adenine nucleotides

Our previous work (Hoeger *et al.* 1995) suggested that the majority of total cellular AMP and ADP is stored in the large acidic vacuole that is characteristic of eleocytes. For this reason, the sum of AMP and ADP given in Fig. 1a reflects the large intra-vacuolar nucleotide store. The sum of AMP and ADP concentrations reached an extremely high value of up to 50 µmol/ml pcv in eleocytes of immature and male animals. Considering their location in the vacuolar compartment of the eleocyte, their intra-vacuolar concentrations can be assumed to be almost twice as high taking a vacuole volume of 40–50% of the cell volume into account (Hoeger *et al.* 1995). Such high concentrations of AMP and ADP have not been reported to occur in any other invertebrate tissues, while the ATP levels are comparable to those of other non-muscular tissues of both vertebrates and invertebrates (Beis and Newsholme 1975; Traut 1994).

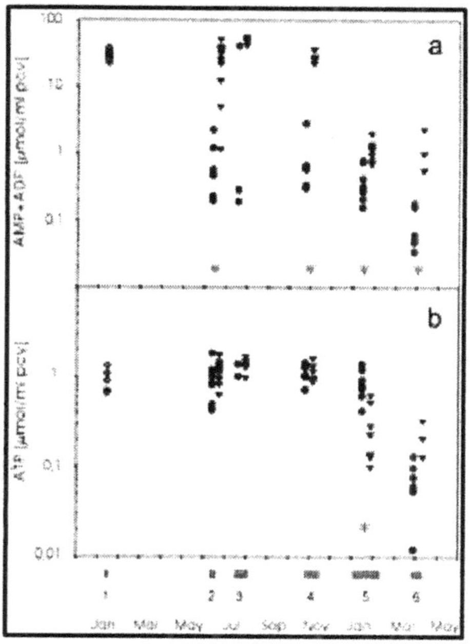

Fig. 1 Seasonal changes in the concentrations (expressed as µmol/ml of packed cell volume) of ADP+AMP (**a**) and of ATP (**b**) in eleocytes from immature (*green symbols*), female (*red symbols*) and male (*blue symbols*) *Nereis virens* isolated at different stages of sexual maturation. For explanation of maturation stages see Table 1. *Asterisks* indicate significant differences between male and female eleocyte samples (unpaired *t*-test; $P < 0.002$). *Hatched squares* below the *abscissa* designate the time at which eleocyte samples were prepared. *Numbers 1–6* correspond to the samples as characterized in Table 1. (*pcv* Packed cell volume)

After germ cells have formed and developed in the coelomic cavity (samples 2–5, see Table 1), the concentration of both AMP and ADP decreased in female eleocytes to the much lower levels (< 0.5 µmol/ml pcv; separate values for AMP and ADP not shown) commonly observed in animal tissues. Male eleocytes on the other hand maintained the high adenylate concentrations for a prolonged period (Fig. 1a). In contrast, ATP levels (0.4–1.5 μmol/ml pcv; Fig. 1b) remained unchanged in eleocytes of both sexes until the later phases of sexual maturation during which they decreased to levels below 0.2 μmol/ml pcv.

The decrease in the adenylate stores observed between November and January in male eleocytes (samples 4 and 5 see Fig. 1a) may be related to the increase in the intracellular levels of inosine, which is a main breakdown product of adenine nucleotides, found during this period (Hoeger *et al.* 1996). The concentrations of inosine monophosphate (IMP), which is a potential intermediate in the breakdown of adenine nucleotides, were not found to be increased when the AMP+ADP concentration was decreased (data not shown); however, this could be the result of rapid further conversion to inosine. Inosine is released by eleocytes in culture, at a much higher rate by male eleocytes compared with female eleocytes (Hoeger *et al.* 1996). Labelled inosine, in turn, has been shown to be taken up by both male (Geier and Hoeger 1997) and female (G. Geier and U. Hoeger, in preparation) germ cells, where it is utilized for anabolic pathways such as nucleic acid biosynthesis. Interestingly, the increase in the inosine concentration in male eleocytes coincides with the period of spermatogonia proliferation in male animals. This is accompanied by a sharp increase in the rate of inosine uptake by the spermatogonia cells *in vitro*, reflecting increasing utilization of exogenous inosine for DNA synthesis (Geier and Hoeger 1997). Although our data indicate that this increase in intracellular inosine concentration takes place between August and November, and thus precedes the observed degradation of adenylate nucleotides between November and January (see Fig. 1a), the conversion of adenine nucleotides to inosine may not be initially visible as a decrease in their concentrations due to the much larger size of the adenine nucleotide pool (20–50 μmol/ml pcv) compared to that of inosine (< 2 μmol/ml pcv). In female eleocytes, such a correlation between adenine nucleotide breakdown (observed between January and May; see Fig. 1a) and the time of highest inosine uptake rates by the female germ cells (measured in August; Geier and Hoeger 1997) was not found. Furthermore, our (unpublished) data show only a transient increase in the intracellular inosine levels in female eleocytes during this time. Further studies are needed to determine the fate of the degraded adenine nucleotides in female eleocytes.

Guanine nucleotides

Considering the intra-sample variability, guanine nucleotides (GMP+GDP+GTP; Table 2) did not change significantly over the time course of maturation, with pool sizes between 0.08 and 0.3 μmol/ml pcv, and there were no gender-specific differences, except for prespawning animals where females contained much less (0.015–0.04 μmol/ml pcv) than males. GTP levels (Table 2) ranged between 0.02 and 0.15 μmol/ml pcv, with no apparent change or gender-specific differences. Only at the end of the maturation period did the GTP concentration drop to levels between 0.002 and 0.015 μmol/ml pcv; this decrease took place earlier in males (sample 4) than in females (sample 6). For samples 4–6, the GTP/GDP ratios (Fig. 2) in each sample group were significantly higher (unpaired *t*-test; $P < 0.025$) in female compared to male eleocytes. At the end of sexual maturation of *N. virens* (data for sample groups 5 and 6 combined), the GTP/GDP ratios were significantly lower (unpaired *t*-test; $P < 0.025$) than in either of the preceding sample groups (1–4) in male and female eleocytes, respectively.

Table 2. Concentrations of guanine nucleotides (given as mmol/ml of packed cell volume; means ± SD) in eleocytes of *Nereis virens* isolated from immature and female (a) and from male animals (b). Number of determinations is given in parentheses. For small sample size ($n = 3$), the range of determinations is given. For explanation of maturation stages, see Table 1

a Immature (sample 1) and female (samples 2–6) animals

	Sample no.					
	1(Immature)	2	3	4	5	6
GMP	0.058 ± 0.012 (7)	0.066 ± 0.024 (10)	0.031 (3)0.019–0.039	0.069 ± 0.035 (6)	0.068 ± 0.029 (8)	0.024 ± 0.008 (6)
GDP	0.011 ± 0.002 (7)	0.007 ± 0.003 (10)	0.011 (3)0.009–0.012	0.011 ± 0.006 (6)	0.018 ± 0.006 (8)	0.002 (3)0.001–0.002
GDP+GMP	0.069 ± 0,013 (7)	0.074 ± 0.022 (10)	0.042 (3)0.031–0.048	0.084 ± 0.031 (6)	0.087 ± 0.031 (8)	0.026 (3)0.013–0.037
GTP	0.069 ± 0.019 (7)	0.057 ± 0.028 (7)	0.094 (3)0.064–0.125	0.079 ± 0.032 (6)	0.060 ± 0.042 (6)	0.005 ± 0.003 (4)
GTP+GDP +GMP	0.138 ± 0.029 (7)	0.139 ± 0.042 (7)	0.136 (3)0.112–0.155	0.162 ± 0.056 (5)	0.150 ± 0.049 (7)	0.032 (3)0.013–0.045

b Male eleocytes

	Sample no.				
	2	3	4	5	6
GMP	0.078 ± 0.058 (12)	0.074 ± 0.024 (5)	0.056 ± 0.021 (6)	0.105 ± 0.033 (8)	0.132 (3)0.086–0.185
GDP	0.018 ± 0.013 (12)	0.020 ± 0.005 (4)	0.013 ± 0.007 (6)	0.016 ± 0.005 (6)	0.012 (3)0.008–0.015
GDP+GMP	.0.096 ± 0.053 (12)	0.092 ± 0.026 (4)	0.069 ± 0.019 (6)	0.118 ± 0.033 (6)	0.144 (3)0.094–0.198
GTP	0.069 ± 0.037 (12)	0.069 ± 0.023 (5)	0.038 ± 0.013 (6)	0.013 ± 0.011 (7)	0.013 (3)0.010–0.015
GTP+GDP +GMP	0.165 ± 0.046 (12)	0.169 ± 0.032 (5)	0.106 ± 0.024 (6)	0.130 ± 0.035 (5)	0.157 (3)0.110–0.212

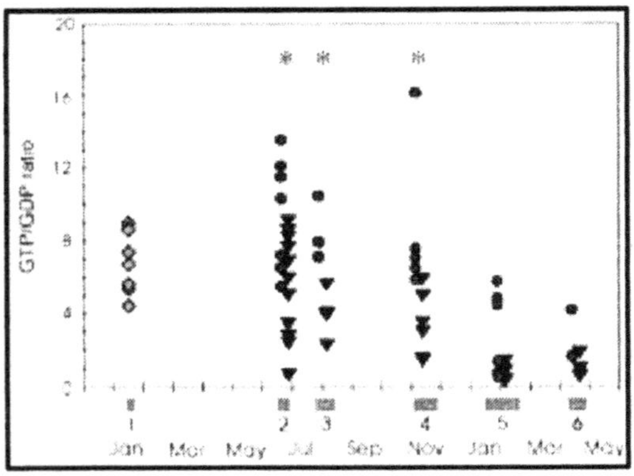

Fig. 2 Seasonal changes in the GTP/GDP ratios in eleocytes from immature (*green symbols*), female (*red symbols*) and male (*blue symbols*) *Nereis virens* isolated at different stages of sexual maturation. For explanation of the maturation stages see Table 1. *Asterisks* indicate significant differences between male and female eleocyte samples (unpaired *t*-test; $P < 0.025$). *Hatched squares* below the *abscissa* designate the time at which eleocyte samples were prepared. *Numbers 1–6* correspond to the samples as characterized in Table 1.

Since GDP acts as a competitive inhibitor of GTP binding to several factors involved in protein synthesis (Pall 1985), the cellular GTP/GDP ratio is considered to be a regulatory factor in protein synthesis (Pall 1985; Land *et al.* 1993). In this light, the higher GTP/GDP ratios in female eleocytes would favour the rate of protein synthesis compared to males, which is consistent with the capacity of female eleocytes to synthesize vitellogenin. GTP and ATP concentrations were also much more closely correlated in female (Spearman's rank correlation; r_s=0.95; $P < 0.001$) than in male eleocytes (r_s=0.71; $P < 0.001$; see Fig. 3), indicating a tighter coupling of GTP regeneration by ATP, a reaction mediated by the enzyme nucleoside diphosphate kinase. Activity levels of this enzyme are similar in eleocytes of either sex (U. Hoeger, unpublished). However, the overall decline in the GTP/GDP ratios, suggestive of a gradual decrease in protein synthesis, is much in contrast to our expectations since female eleocytes continuously increase their rate of both vitellogenin synthesis and secretion into the medium during the course of gametogenesis in *N. virens* (Heil 1995). On the other hand, nucleotide pools are compartmentalized in the cell and may differ in their composition depending on the cellular compartment, as was found for rat liver Golgi vesicles (Fleischer 1981). Some of the GDP (as well as GMP) may be generated inside the Golgi apparatus by the breakdown of sugar nucleotides involved in glycosylation reactions (Capasso and Hirschberg 1984, Abeijon *et al.* 1993). Thus, the observed changes in the apparent GTP/GDP ratios could also reflect differences in the degree of GDP sequestration in these subcellular compartments and do not necessarily reflect the cytosolic GTP/GDP ratios for protein synthesis. The range of GTP/GDP ratios measured in this study (0.5–16) compares well with the GTP/GDP ratios found for different mammalian cell types (0.5–12; calculated from Traut 1994).The possible localization of some of the GDP (and GMP) in the eleocyte vacuole seems less likely since no correlation with the high AMP+ADP levels was found.

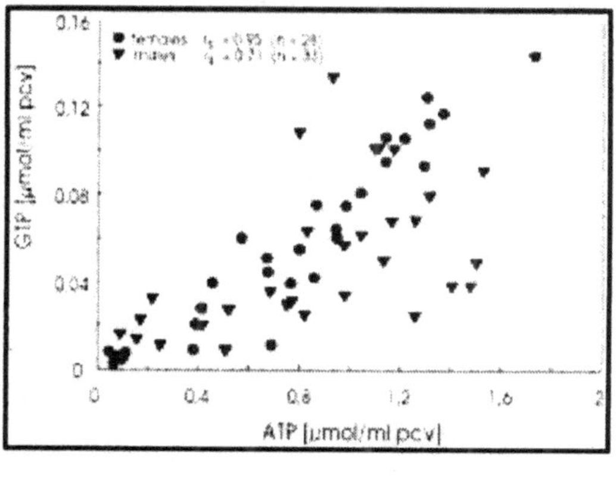

Fig. 3 Correlation between the concentrations of ATP and GTP (expressed as µmol/ml of packed cell volume) in eleocytes of female (red symbols) and male (blue symbols) eleocytes. Data from all sample groups were combined. The correlations were significant at the p = 0.01 level. r_s; spearman's rank correlation coefficient; n; number of samples.

The concentrations of nucleotides determined in this study fluctuated considerably within the different sample groups. Nucleotide concentrations in eleocyte samples isolated from animals at the same stages of germ cell development but collected in different years (see Table 1) were similar and did not significantly add to the observed sample variability. Likewise, differences in the time intervals between capture of the worms and eleocyte isolation (i.e. 2–20 days) were not found to contribute to the variation in the nucleotide levels.

Apparently, the nucleotide levels are not tightly regulated in eleocytes. For comparison, the ATP content of the body wall musculature of another polychaete, *Arenicola marina*, is in a much narrower range (2.1–2.4 µmol/g tissue) and these values do not change during several hours of metabolic stress (i.e. anoxia; Schöttler *et al.* 1984). Eleocytes are not directly involved in vital functions such as movement and the functioning of the nervous system, and in such cell types the organism may allow a much greater tolerance for metabolically unfavourable situations such as low ATP levels and low GTP/GDP ratios, since this may not be a factor critical for the immediate survival of the animal.

Acknowledgements. This study was supported by the grants Ho 889/3-1 and Ho 889/4-1 from the Deutsche Forschungsgemeinschaft, the Naturwissenschaftlich-Medizinisches Forschungszentrum and the Feldbausch Foundation at the University of Mainz. We thank Dr. G. Wegener and Dr. A. Fischer for helpful comments on the manuscript.

References

Abe, H., Okuma, E. (1991) Rigor-mortis progress of carp acclimated to different water temperatures. *Nippon Suisan Gakkai Shi.* **57**:2095–2100

Abeijon, C., Yanagisawa, K., Mandon, E.C., Hausler, A., Moremen, K., Hirschberg, C.B., Robbins, P.W. (1993) Guanosine diphosphatase is required for protein and sphingolipid glycosylation in the Golgi lumen of *Saccharomyces cerevisiae. J. Cell. Biol.* **122**:307–323

Beis, I., Newsholme, E.A. (1975) The contents of adenine nucleotides, phosphagens, and some glycolytic intermediates in resting muscles from vertebrates and invertebrates. *Biochem. J.* **152**:23–32

Capasso, J.M., Hirschberg, C.B. (1984) Mechanisms of glycosylation and sulfation in the Golgi apparatus: evidence for nucleotide sugar/nucleoside monophosphate and nucleotide sulfate/nucleoside monophosphate antiports in the Golgi apparatus membrane. *Proc. Natl. Acad. Sci. USA* **81**:70517055

Dobson, G.P., Veech, R.L., Passoneau, J.V. (1991) NMR and enzymatic analysis of cytosolic phosphocreatine. ATP, Pi and intracellular pH in the isolated working perfused rat heart. *NMR Biomed.* **4**:1–9

Fischer, A., Hoeger, U. (1993) Metabolic links between somatic sexual maturation and oogenesis in nereid annelids. *Adv. Invertebrate Reprod.* **23**:131–138

Fischer, A., Dorresteijn, A.W.C., Hoeger, U. (1996) Metabolism of oocyte construction and the origin of histospecificity in the cleaving egg. Lessons from nereid annelids. *Int. J. Dev. Biol.* **40**:421–430

Fleischer, B. (1981) The nucleotide content of rat liver golgi vesicles. *Arch. Biochem. Biophys.* **212**:602–610

Geier, G., Hoeger, U. (1997) Nucleoside uptake in male germ cells of the polychaete *Nereis virens. Invertebrate Reprod. Dev.* **31**:75–78

Heil, P. (1995) Die Bereitstellung von Vitellogenin durch die Elaeozyten im Wachstumsverlauf der Oocyten bei *Nereis virens*. PhD thesis, University of Mainz, pp 1134

Hoeger, U. (1991) Hydrolytic activities in eleocytes of *Nereis virens* (Annelida, Polychaeta), during sexual maturation. *Mar. Biol.* **110**:7–12

Hoeger, U., Dunn, J.F., Märker, C. (1995) Adenylate compartmentation and storage in coelomic cells of the polychaete, *Nereis virens. J. Exp. Biol.* **198**:2079–2085

Hoeger, U., Märker, C., Geier, G. (1996) Adenylate storage, metabolism and utilization in coelomic cells of the polychaete, *Nereis virens. Experientia* **52**:481–486

Land, S.C., Buck, L.T., Hochachka, P.W. (1993) Response of protein synthesis to anoxia and recovery in anoxia-tolerant hepatocytes. *Am. J. Physiol.* **265**:R41–R48

Pall, M.L. (1985) GTP, a central regulator of cellular anabolism. *Curr. Top. Cell. Regul.* **25**:1–20

Schöttler, U., Surholt, B., Zebe, E. (1984) Anaerobic metabolism in *Arenicola marina* and *Nereis diversicolor* during low tide. *Mar. Biol.* **81**:69–73

Traut, T.W. (1994) Physiological concentrations of purines and pyrimidines. *Mol. Cell. Biochem.* 141:1–22

Exp. Biol. Online 2:13 (1997)

An improved method for screening *Frankia* viability in strain CcI3

N.R. McEwan[1,2] and C.T. Wheeler[2]

[1]Land Resources Department, Scottish Agricultural College, Craibstone Estate, Aberdeen, AB21 9TN, Scotland
[2]Division of Biochemistry and Molecular Biology, Bower Building, University of Glasgow, Glasgow G12 8QQ, Scotland

Correspondence to: N.R. McEwan (e-mail: n.mcewan@ab.sac.ac.uk)

Received: 16 June 1997 / Accepted: 6 October 1997
Published: October 20, 1997

Abstract. An improved method for determining the viability of the nitrogen-fixing actinomycete *Frankia* is presented. This method uses tetrazolium red as a vital stain, which proved more effective than a previously used method of acridine orange staining.

Key words: *Frankia* – Tetrazolium red – Vital staining

Introduction

Frankia is a nitrogen-fixing actinomycete, which fixes nitrogen in symbiotic association with the roots of over 200 dicotyledonous species of plants, spanning 25 genera and 8 different families (Baker and Schwintzer, 1990). These plants are generally woody shrubs or trees and are used, among other things, as fuel woods and fruits, as well as in sustainable agricultural practices such as prevention of soil erosion and creation of windbreaks (Benoit and Berry, 1990).

The first *Frankia* was isolated in 1978 (Callaham *et al.*, 1978). The organism is characterised by a very slow growth rate *in vitro*, with an approximate doubling time of 18–36 h or more. Isolations performed on material extracted from plant nodules normally require a minimum of 3–6 weeks before colonies can be viewed by eye. Cultures grown *in vitro* can survive even after storage for several years. It is thought that this may be due to survival of dormant spores, which germinate following transfer to fresh culture medium, rather than the re-initiation of fresh growth by hyphae. Assessment of cell viability in this slow-growing filamentous, multicellular actinomycete is of particular importance in studies designed to improve cell culture methods, or after treatments which attempt to genetically transform the organisms (e.g. electroporation). At present, there has only been one method reported for assessing cell viability in *Frankia* (Schwenke, 1991) a method that uses acridine orange as a vital stain. However, acridine orange staining detects levels of RNA or DNA in cells, thus indicating active transcription, rather than cell viability.

This paper presents an alternative approach to assessing viability in *Frankia*, based on the use of tetrazolium red (2,3,5-triphenyltetrazolium chloride) as a vital stain. Tetrazolium salts are low-redox-potential indicators which on reduction form insoluble, coloured formazans (Smith, 1951).

Materials and Methods

Growth of cells

Cultures of *Frankia* strain CcI3, an isolate from *Casuarina cunninghamiana* originally supplied from Harvard Forest, Harvard University, USA, were grown in BuCT medium as previously described (Hooker and Wheeler, 1987) for 7 days. Cells were harvested by centrifugation at 13,000 *g* for 5 min. Cells were either killed by autoclaving at 121°C for 20 min prior to staining, or were stained immediately after harvesting.

Staining with acridine orange

Cells were stained with 0.05% (w/v) acridine orange in 50 mM potassium phosphate buffer (pH 7.0) for 10 min, followed by two rinses in 50 mM potassium phosphate buffer solution. Staining was assessed by viewing preparations with fluorescein filters and a fluorescence microscope. Red or orange staining is indicative of RNA. Green or yellow is indicative of the presence of DNA, but can normally be visualised only when levels of RNA are low.

Staining with tetrazolium red

Cells were stained using 0.15% (w/v) tetrazolium red salt in 50 mM potassium phosphate buffer (pH 7.0) for 2 h in the dark. Cells which actively take up the salt have a deep red or pink colour when viewed by bright field microscopy. Cells which are dead, and unable to take up the salt actively, remain colourless (Patriquin and Dobereiner, 1978).

Results and Discussion

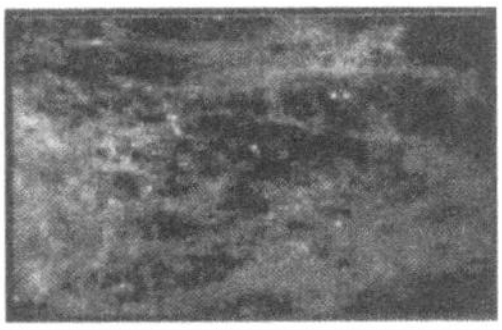

Fig. 1 Acridine orange staining of viable mycelium. The cells stain primarily bright orange, indicating the presence of RNA. Dimensions of figure = 1.04 mm × 0.69 mm

Cells which were stained with acridine orange immediately after harvesting gave a bright orange colour, as seen in Fig. 1, indicative of the presence of RNA and active transcription of DNA.

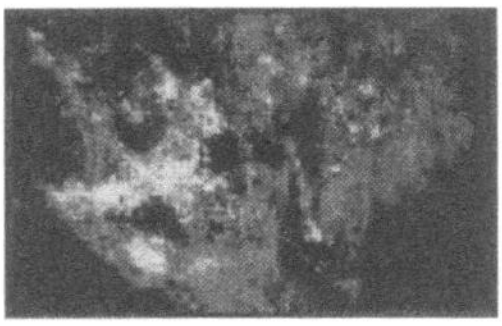

Fig. 2 Acridine orange staining of autoclaved mycelium. The cells still stain primarily bright orange, indicating the presence of RNA. Dimensions of figure = 0.33 mm × 0.22 mm

However, when cells which had been killed by autoclaving for 20 min were stained in a similar manner, they also carried substantial amounts of red staining when viewed by fluorescence microscopy, as seen in Fig. 2. This suggests that although the cells had been killed, there was still sufficient RNA present in the cells to react with the acridine orange dye.

When viable cells were stained with the tetrazolium salt, they appeared dark red in bright field microscopy, indicating active uptake of the salt (Fig. 3).

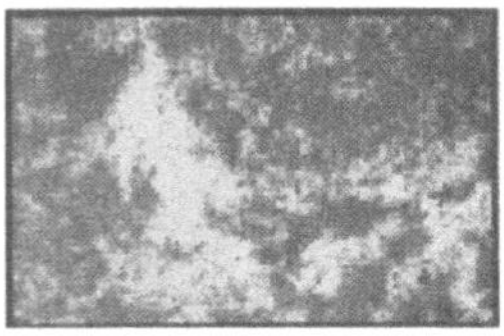

Fig. 3 Tetrazolium salt staining of viable mycelium. The cells actively take up the salts and stain primarily deep red. Dimensions of figure = 1.04 mm × 0.69 mm

Cells which had been autoclaved prior to incubation showed no red colouring when stained with tetrazolium (Fig. 4).

Fig. 4 Tetrazolium salt staining of autoclaved mycelium. The cells are unable to actively take up the salts and therefore are unstained. Dimensions of figure = 1.04 mm × 0.69 mm

While acridine orange staining is indicative of cell viability and may be useful for routine assessment, for example of cell culture systems, it is not an appropriate technique for more definitive assessment of cell viability following more specific experimental treatments (e.g. electroporation) where it is vital to assess rapidly the presence of living cells in the mycelium. Tetrazolium staining is a more effective technique for this purpose.

References

Baker, D.D., Schwintzer, C.R. (1990) Introduction. In: *The biology of* Frankia *and actinorhizal plants.* Academic, New York, pp.1–13

Benoit, L.F., Berry, A.M. (1990) Methods for production and use of actinorhizal plants in forestry, low maintenance landscape, and revegetation. In: *The biology of* Frankia *and actinorhizal plants*. Academic, New York, pp. 281–297

Callaham, D., Del Tredici, P., Torrey, J.G. (1978) Isolation and cultivation in vitro of the actinomycete causing root nodulation in Comptonia. *Science* **199**: 899–902

Hooker, J.E., Wheeler, C.T. (1987) The effectivity of *Frankia* for nodulation and nitrogen fixation in *Alnus rubra* and *A. glutinosa. Physiologia Plantarum* **70**: 333–341

Patriquin, D.G, Doberiner, J. (1978) Light microscopy observations of tetrazolium-reducing bacteria in the endorhizosphere of maize and other grasses in Brazil. *Can. J. Microbiol.* **24**: 734–742

Schwenke, J. (1991) Rapid, exponential growth and increased biomass yield of some *Frankia* strains in buffered and stirred mineral medium (BAP) with phosphatidyl choline. *Plant Soil* **137**: 37–41

Smith, F.E. (1951) Tetrazolium salt. *Science* **113**: 751–754

Exp. Biol. Online 2:14 (1997)

Metamorphic development of locusta-tachykinin immunoreactive neurons of the antennal lobes of the beetle *Tenebrio molitor* and the effect of fenvalerate application

Rainer Wegerhoff

Institute of Biochemistry, Olshausenstr. 40, University of Kiel, D-24098 Kiel, Germany.

Correspondence to: Rainer Wegerhoff (email.: rwegerhoff@biochem.uni-kiel.de)

Received: 28 July 1997 / Accepted: 15 October 1997
Published: October 24, 1997

Abstract. The effects of the widely used neurotoxic pyrethroid insecticides on neuronal development or plasticity are unclear. To expand knowledge about the influence of the pyrethroid fenvalerate on neuronal development, metamorphic remodelling of the primary olfactory neuropil of the beetle *Tenebrio molitor* has been studied. The antennal lobe is subdivided into distinct glomeruli before metamorphosis. This is in contrast to that which occurs in other well-studied holometabolous insects such as the moth *Manduca sexta* and the honeybee. As an indicator of antennal lobe interneurons, locusta-tachykinin immunoreactive neurons have been used. They project into the antennal lobes and form tufted arbors in larval and adult stages within glomeruli throughout the neuropil. These glomerular structures are invaded by glomerular sensory afferent axons and are surrounded by processes of glia cells. With pupation, the glomerulization is lost and no locusta-tachykinin or substance P immunoreactivity is visible in the antennal lobe. The immunoreactivity reappears during metamorphosis, starting with diffusely branched arbors that later become tufted. Application of the neurotoxic insecticide fenvalerate at pupation in sublethal concentrations resulted in a loss or reduction of glomerular pattern formation by neurons and glia cells during metamorphosis. Labelling of antennal sensory axons revealed that the olfactory neuropil was not deafferented, and also that the sensory axons were not organized into a normal glomerular pattern. In addition to the morphological differences, fenvalerate treatment caused locusta-tachykinin immunoreactivity to reappear prematurely during metamorphosis. Possible reasons for fenvalerate-induced alterations in antennal lobe development and their implications for normal development are discussed.

Key words: Glia cells – Olfactory lobe – Peptidergic interneurons – Pyrethroid insecticide – Tachykinin

Introduction

Pyrethroids are a widely used and chemically diverse class of neurotoxic insecticides. Many have a low toxicity against mammals, as indicated by high LD_{50} values, and a high toxicity against most arthropods (Aldridge 1990; Naumann 1990; Vijverberg and Van den Berken 1990; Wegerhoff 1996). The primary biological target site of the lipophilic pyrethroids is thought to be the neuronal voltage-gated sodium channels, where the effect results in repetitive firing or a stable depolarization (Vijverberg and Van den Berken 1990; Brown and Narahashi 1992; Narahashi 1996a,b). For some pyrethroids, such as fenvalerate, a block in gap-junction communication has also been shown (Hemming *et al.*1993; Tateno *et al.* 1993). However, little is known about the effects of pyrethroids on the developing nervous system and their possible impact on neuronal plasticity (Hartzell and Scudder 1942; Eriksson and Nordberg 1990; Eriksson and Fredriksson 1991; Sheets *et al.* 1994).

Following embryogenesis, the brains of holometabolous insects exhibit dramatic changes in neuronal structure and function during metamorphosis. Persisting and newly developed neurons establish new neuropil regions or remodel already existing parts of the nervous system (Wegerhoff *et al.* 1996b). Thus, the metamorphosing insect brain represents a good model for studying the formation and plasticity of nervous tissue (Technau and Heisenberg 1982; Hildebrand 1985; Wegerhoff and Breidbach 1992; Breidbach and Wegerhoff 1994) and could be useful for analysing the effects of pyrethroids on neuronal development and plasticity.

The current study has focused on the metamorphic remodelling of the first-order neuropil of the insect olfactory pathway, the antennal lobe (AL), in *Tenebrio molitor*. During metamorphosis, the small antennae of the larvae (also called mealworm) are replaced by complex antennae of the adult beetle. The peripheral changes are accompanied by changes in the AL during metamorphosis. Several newly developed sensory neurons and interneurons also innervate this neuropil with glomerular terminations (Breidbach and Marx 1990). Glomeruli are thought to be fundamental functional subunits in olfactory information processing and share a basic plan of organization with the olfactory brain centres in vertebrates. For some insects, such as the moth *Manduca sexta*, the adult AL arises in a part of a larval lobe that is not glomerular (Oland and Tolbert 1996). In *Manduca*, glial cells play an essential role in the development of glomeruli, which is triggered by the ingrowth of adult olfactory receptor axons (Oland and Tolbert 1987, 1996; Baumann *et al.* 1996). In other insects with a pupal stage, including the beetle *Tenebrio molitor*, the larva already has a glomerular AL, which is remodelled during metamorphosis (Nordlander and Edwards 1970; Levine 1984; Wegerhoff and Breidbach 1991; Becker and Breidbach 1993).

As has been shown for the adult cockroach and the blowfly, certain neurons in the AL exhibit strong locusta-tachykinin immunoreactivity (Nässel 1993; Muren *et al.* 1995). In *Tenebrio*, however, the AL is also immunopositive in larval stages (Wegerhoff *et al.* 1996b). Therefore, the immunohistochemical staining of locusta-tachykinin immunoreactive (LomTK-IR) neurons was used as a marker of developmental processes. About 190 LomTK-IR neurons, forming 17 clusters, are found in each hemisphere of an adult *Tenebrio* brain. Neurons from three distinct clusters project into the AL where they form dense, tufted arborizations (Wegerhoff *et al.* 1996b).

Materials and Methods

Staging of pupal development

Meal beetles, *Tenebrio molitor* L., were kept under crowded conditions in a laboratory culture at room temperature with a light cycle of 12 h light and 12 h darkness. For staging the 14 days of pupal development, the ongoing differentiation of the compound eye was observed. Rows of ommatidia (OR) were counted in a dorsal to ventral lane and only individuals of the same number of OR and the same day of ongoing pupal development were compared. The duration of pupal development was the same for fenvalerate-treated or unmanipulated beetles kept under identical conditions.

For immunocytochemistry at least ten unmanipulated and at least ten manipulated individuals (males and females) of each developmental stage were used. For the labelling of sensory neurons of unmanipulated, fenvalerate-treated or tetrodotoxin- (TTX-) treated antennae from six individuals were used in each group.

Immunohistochemistry

To analyse the metamorphic development of the AL in *Tenebrio molitor*, brains were dissected out and fixed in 4% paraformaldehyde for 4 h and embedded in gelatin ovalbumin. Serial sections of 40–60 μm were obtained using a Vibratome (Bachhofer, Reutlingen). To reduce background staining the sections of the brain were incubated in PBT-NG [0.1 M phosphate-buffered saline (PBS), pH 7.4, containing 2% normal goat serum, 1% bovine serum albumin (BSA) and 1% Triton-X-100] for 3 h. Then the sections were incubated in the primary polyclonal antibody (anti-locusta-tachykinin II, 1:15000 in PBS containing 2% normal goat serum and 1% Triton X-100) for 48 h at room temperature. Detection was performed using a TRITC-, Cy-3- or DTAF-coupled secondary antibody (Dianova) and occasionally an additional peroxidase anti-peroxidase conjugate for DAB staining. Finally, the sections were dehydrated and embedded in DePex via Xylol.

The production and purification of the locusta-tachykinin II antiserum have been described before (Veenstra *et al.* 1995). As a control, the primary locusta-tachykinin II antiserum was preabsorbed with 20 nmol locusta-tachykinin II (Bachem, Heidelberg, Germany) per ml diluted antiserum (LomTK II 1:15000), overnight at 4°C, and applied as primary antibody to the sections as described above. Also double staining with another tachykinin-specific antibody (anti-substance P, monoclonal, rat, Serva, Heidelberg, Germany) was done (for details see Wegerhoff I 1996a).

Experimental manipulations

The synthetic pyrethroid fenvalerate was applied to freshly hatched pupae by dropping 1–5 µl ethanol containing different amounts of fenvalerate, on the head capsule of the insects. For controls, only pure ethanol was applied. All pupae were kept separately in 35-mm tissue chambers for different times, and the brains of pupae or adults were removed. Also, the antenna from one side was removed at the first proximal segment, from several freshly hatched pupae, to show the effect of deantennation. To reduce the neuronal activity of antennal sensory neurons, 1 µM TTX in PBS was applied by a microcapillary into one antenna of freshly hatched pupae. As a control pure PBS was applied. Innervation of the AL was detected by nickel-chloride labelling of antennal sensory axons. A glass capillary filled with 5% $NiCl_2$ was inserted into the antenna, or the distal two-thirds of the antenna was removed and the cut end of the remaining antenna was inserted into a capillary with 5% $NiCl_2$. After 4 h the capillary was removed and the dissected brain was immersed in 200 µl of 0.1 M PBS containing 15 µl saturated dithiooxamide in pure ethanol to precipitate the $NiCl_2$. After several washes and fixation in 4% paraformaldehyde the nickel deposits were silver enhanced as previously described (Kutsch and Schneider 1987).

Nuclear staining and staining of glia cells

For nuclear staining previously immunostained sections were incubated with a 10^{-6} M solution of 4,6-diamidino-2-phenyl-indol (DAPI, Molecular Probes) in PBS for 5–10 min. In some cases propidium iodide staining was used as previously described by Baumann *et al.* (1996). The sections were incubated for 5–10 min in 0.1mg RNase/ml PBS (pH 6.5), washed in PBS and incubated for 15 min in 25 µM propidium iodide (Molecular Probes) and washed again in PBS. For glial cell staining biotinylated *Datura stramonium* lectin (DSL; Oxford GlycoSciences) was used. The brains were fixed in 4% paraformaldehyde and embedded in bovine serum albumin (BSA) / gelatin. After postfixation and cutting on a Vibratome the 30- to 50-µm-thick sections were incubated with DSL at a concentration of 10 µg/ml Tris-buffered saline, 0.1 M, pH 7.4 (TBS) containing 1% Triton-X-100 and BSA for 24 h. After washing with TBS the sections were incubated with streptavidin-Cy2 (Dianova) in TBS overnight. Sections were finally dehydrated, cleared and embedded in Depex via Xylol.

All preparations were viewed using an Olympus BX-50 microscope with a PM-30 exposure control unit for photography.

Results

In general, the immunohistochemical results were uniform within each developmental stage for either unmanipulated or fenvalerate-treated insects. There was no obvious difference between female and male beetles.

Locusta-tachykinin immunoreactivity in the antennal lobe during metamorphosis

Unmanipulated development

In the last larval instar the AL was innervated by projections of about 22 bilaterally symmetrical LomTK-IR neurons. The side branches in the AL were organized in a glomerular fashion (Fig. 2A). With the onset of metamorphosis the immunostaining in the AL decreased but remained glomerular (0–4 OR). After 2 days of metamorphic development (7 OR), no immunoreactive side branches could be detected in the AL (Fig. 2B). The respective somata and primary projections also were no longer immunopositive. LomTK immunoreactivity could be displayed again in a pupa of 14 OR (Fig. 2C). This weak immunoreactivity was confined to diffuse arbors throughout the AL. The staining began to be glomerular again in the 16-OR-old pupa (75% of metamorphic development). Older pupae had a densely innervated glomerular neuropil with adult-like morphology (Fig. 1, Fig. 2D, Fig. 3A). The number of immunoreactive AL somata had increased to 75 per adult hemisphere (Fig. 5). The preabsorption with locusta-tachykinin II prevented immunolabelling.

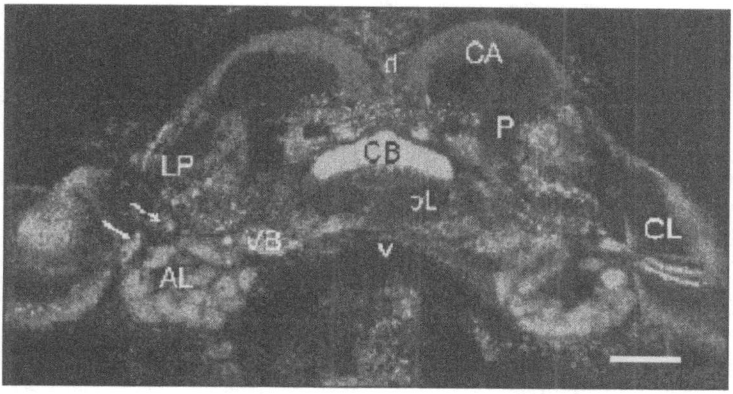

Fig. 1. Frontal Vibratome section of an adult beetle brain at the level of the central body (*CB*). Locusta-tachykinin immunopositive stainings (*green*) are found in various parts of the brain. Red propidium iodide staining displays somata. *Arrows* point to immunopositive somata at the antennal lobe (*AL*). (*bL* Beta lobe, *CA* calyx, *d* dorsal, *LP* lateral protocerebrum, *OL* optic lobe, *v* ventral, *VB* ventral body). *Scale bar* = 100 μm

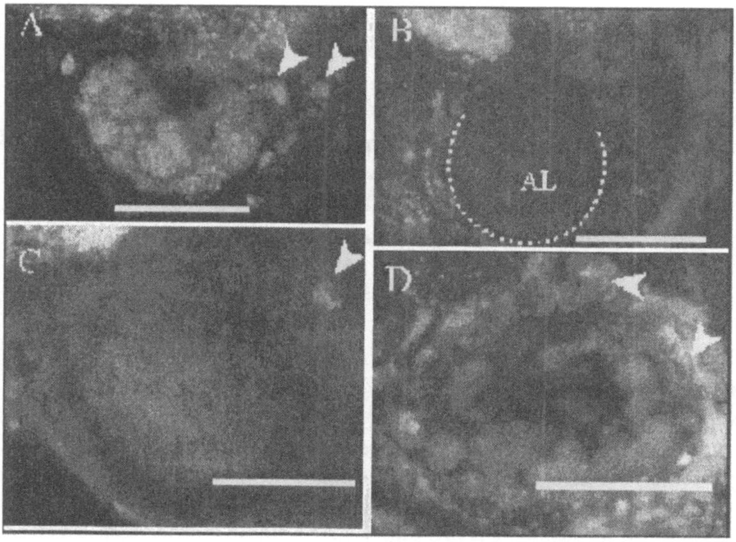

Fig. 2. Locusta-tachykinin immunoreactivity (*LomTK-IR*) in the antennal lobe (*AL*) of the beetle *Tenebrio molitor*. Frontal Vibratome sections showing different developmental stages. *Arrowheads* point to LomTK-IP somata. **A** Prepupa; the AL is densely innervated by LomTK-IP fibres with tufted arborizations. **B** Pupa, eight rows of ommatidia old (*8-OR-old*); note that the AL is free of immunoreactivity. **C** 14-OR-old pupa. First diffusely arranged immunopositive fibres are visible. **D** 18-OR-old pupa, the LomTK-IP fibres are restricted to the glomerulus structures (*green*). Note that the propidium iodide staining (*red*) outlines the arrangement of all somata in the rind around the glomerulus neuropil. *Scale bars* = 100 μm

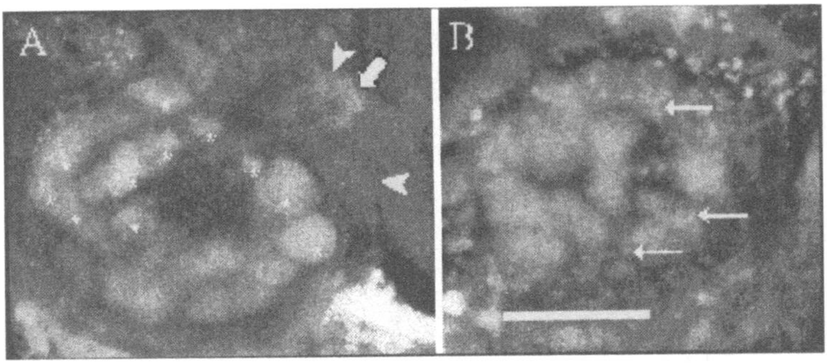

Fig. 3. Adult antennal lobe, frontal Vibratome sections. **A** LomTK-IR neurons (*green*) and substance-P-like immunoreactive neurons (*red*). *Arrowheads* point to LomTK-IR somata, *arrows* point to substance-P-like immunoreactive somata. Note that both cell groups are separated and that substance-P-immunopositive fibres are more concentrated to the dorsal glomeruli. *Stars* point to glomeruli with immunoreactivity from both stainings. **B** Anterior surface of the antennal lobe Lom-TK-IR neurons (*green*) and propidium iodide counterstaining (*red*). Note that the surrounding somata (most probably glial cells) did not enwrap the glomerular neuropil pattern. *Scale bars* = 100 µm

Pyrethroid application

Different amounts of the pyrethroid fenvalerate were applied topically to newly hatched pupae. Following the application of 24 nmol fenvalerate, the pupae developed a normal adult morphology including compound eyes, adult antennae and legs during metamorphosis. However, the wings remained in the pupal orientation on the ventral side of the body and, therefore, the beetles were not able to walk. After application of more than 100 nmol fenvalerate, 20–40 % of the pupae died during metamorphosis. At amounts below 24 pmol the beetles developed a normal, adult gross morphology and were able to walk.

Three days after 24 nmol fenvalerate was applied to freshly hatched pupae, LomTK immunoreactivity was detected in the AL (Fig. 4C). This was earlier than could found at this stage (8 OR) during normal metamorphic development (Fig. 2B) or in controls. The intensity of immunoreactivity within the AL neuropil increased during ongoing pupation (Fig. 4B,C). The number of LomTK-IR neurons increased to approximately 75. The location of their cell bodies and their primary projection pattern were not altered in comparison to those in untreated adult individuals (Fig. 5). However, the immunoreactive branches created an aglomerular pattern (Fig. 4C, Fig. 6). If 240 pmol fenvalerate was applied instead, LomTK-IR neurites were partially glomerular (Fig. 7A). If only 24 pmol fevalerate was applied, glomerular structures were only slightly reduced (Fig. 7C). Glomeruli were most well developed in the anterior–ventral region of the AL (Fig. 7).

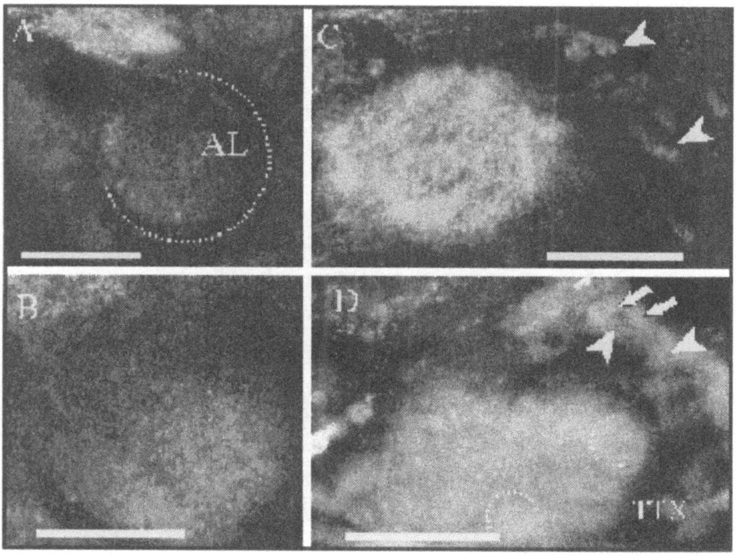

Fig. 4. Fenvalerate and tetrodotoxin (*TTX*) treatment of freshly hatched pupae. **A–C** Fenvalerate application (24 nmol). **A** 8-OR-old pupa, first premature LomTK immunoreactive staining in the antennal lobe (*AL*). **B** 12-OR-old pupa, increased immunoreactive staining in the AL. **C** Adult AL, *arrowheads* point to LomTK-IR somata, note that the staining in the AL remains aglomerular. **D** LomTK-IR neurons (*green*) and substance-P-like immunoreactive neurons (*red*) after TTX treatment. *Arrows* point to substance-P-like immunoreactive somata and *arrowheads* to LomTK-IR somata. Note that a glomerular structure is slightly visible in the AL. *Scale bars* = 100 μm

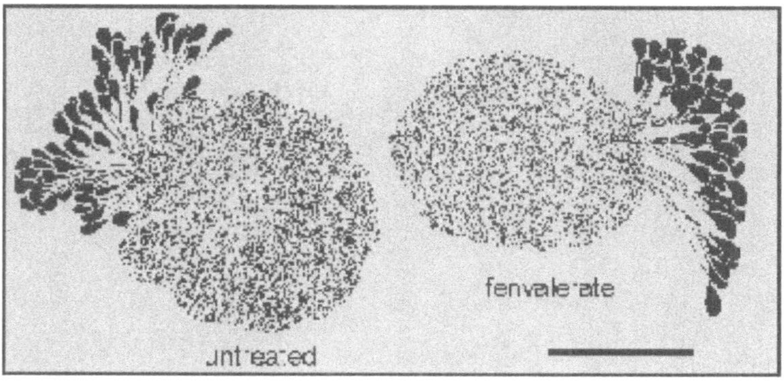

Fig. 5. Reconstruction of LomTK-IR neurons of the adult antennal lobe from serial Vibratome sections. Untreated and treated (2.4 nmol fenvalerate) situation. *Scale bar* = 100 μm

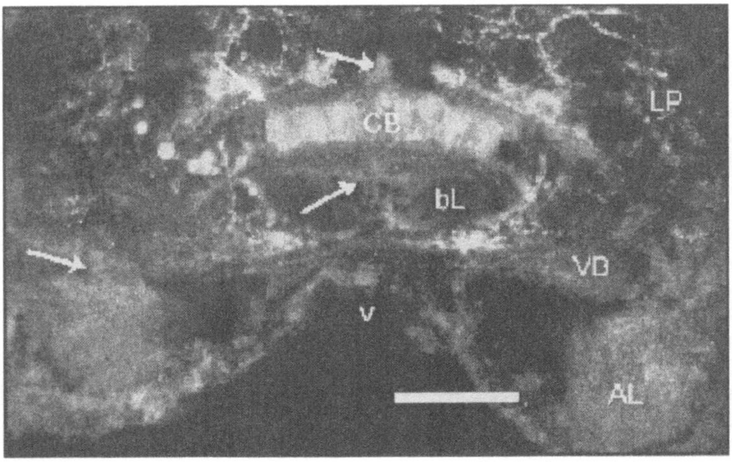

Fig. 6. Adult brain after fenvalerate treatment (2.4 nmol) with the onset of pupation. Triple staining of a Vibratome section. Locusta-tachykinin immunoreactive cells are labelled *green*, substance-P-like immunopositive cells are labelled *red*, and somata *blue* (DAPI-staining). Note the aglomerular red and green staining in the antennal lobe (*AL*). *Arrows* point to glia cells which ensheath the central body (*CB*), the beta lobes (*bL*) and the AL. *Scale bar* = 100 μm

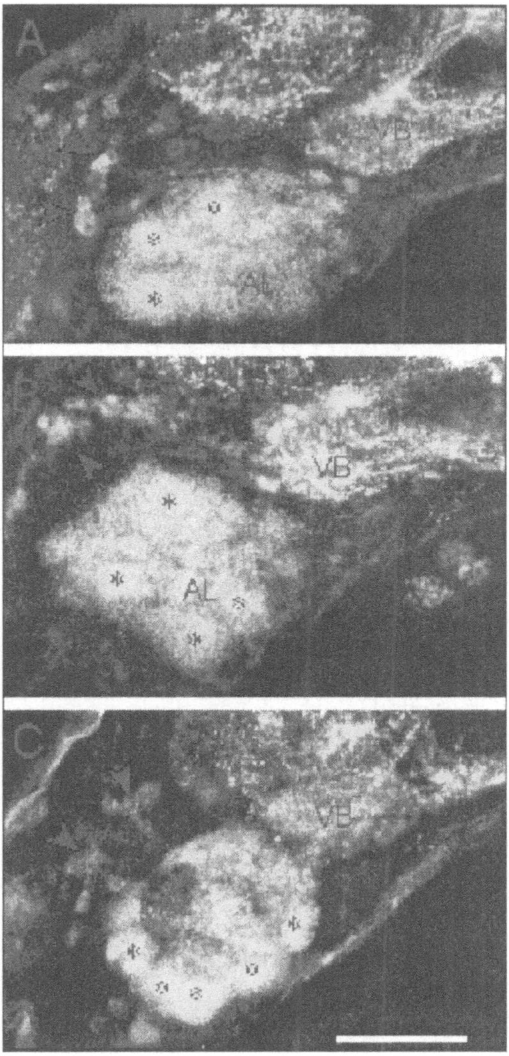

Fig. 7. LomTK-IR neurons in the adult antennal lobe (*AL*) after treatment of freshly hatched pupae with from 24 pmol to 240 pmol fenvalerate; frontal Vibratome sections. *Arrowheads* point to LomTK-IR somata at the AL. **A** Effect of 240 pmol fenvalerate; note that in the AL first gloemerular-like structures are labelled (*stars*). **B** Effect of 70 pmol fenvalerate; note the increased glomerulization compared to **A** (*stars*). **C** Effect of 24 pmol fenvalerate; glomerulization (*stars*) is still reduced as compared to controls (Fig. 3). (*VB* Ventral body). *Scale bar* = 100 μm

Substance-P-like immunoreactivity

Double staining with an anti-substance P antibody and the LomTK-II antibody revealed different groups of tachykin-immunoreactive neurons in the adult AL. Both groups showed a glomerular arborization pattern (Fig. 3A, Fig. 4D). The substance-P-like immunopositive neurons were immunostained less at the prepupal stage and the immunostaining reappeared at the end of metamorphosis. The dorsal glomeruli of the AL are strongly immunoreactive whereas the ventral ones are only weakly stained (Fig. 3A).

After application of 2.4 nmol fenvalerate, the immunoreactive neurons showed a non-tufted pattern within the AL without premature reappearance of immunoreactivity (Fig. 6). The staining was concentrated in the dorsal part of the AL.

Glia cells of the antennal lobe

The neuropil of the larval, pupal and adult AL was free of any somata (Fig. 2D). All somata that were stained by propidium iodide or DAPI procedures were scattered outside the neuropil of the AL. Even during metamorphosis no invasion of somata was detected (Fig. 1, Fig. 3B). As a marker of glial cells the *Datura stramonium* lectin (DSL) was used, since this has been shown previously to stain these cells specifically (Hähnlein *et al.* 1996). The glomeruli are ensheathed by small processes of DSL-labelled fibres (Fig. 8A). Mainly the outer caps of the glomeruli are labelled. The DSL staining is also prominently found around the central body, corresponding to known glial processes found in immunohistochemical and ultrastuctural studies (Breidbach *et al.* 1992, Wegerhoff and Breidbach 1992). After pyrethroid application (24–2.4 nmol) no DSL-stained processes formed such boundaries around AL glomeruli (Fig. 8B), although stained processes around the central body were not influenced in their morphological arrangement (Fig. 8).

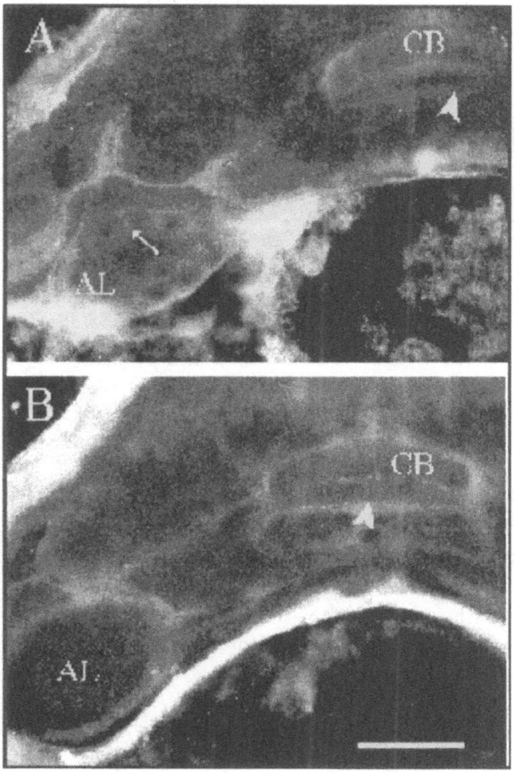

Fig. 8. *Datura stramonium* lectin staining of processes at the antennal lobe (*AL*) and the central body (*CB*). **A** Frontal section through the brain of an adult untreated beetle. The *arrow* indicates the ensheathing of the glomeruli by lectin-labelling. The *arrowhead* points to the ellipsoid body of the central complex, where intense labelling is present. **B** Frontal section of a fenvalerate-treated adult brain (24 nmol with the onset of pupation). Note that no labelling is present in the AL neuropil; however, the outer perineural sheath exhibits strong staining. The *arrowhead* points to the ellipsoid body, compare with Fig. 8A. *Scale bar* = 100 μm

Tetrodotoxin treatment and deantennation

The injection of TTX into the antennae of freshly hatched pupae (0–2 OR) resulted in a reduction of the glomerular immunoreactive side branches of LomTK-IR neurons after metamorphosis (Fig. 4D), in contrast to the PBS controls. After deafferentation (deantennation) the LomTK-IR neurons of the manipulated side invaded the AL but did not show any glomerular tufting during metamorphosis (Fig. 9). The number and location of immunoreactive somata in the adult AL and in the unmanipulated adult beetle were not different.

Fig. 9. Adult brain, one antenna was removed at the onset of metamorphosis (*deant.* Deantennated, *n* normal side), frontal Vibratome section. Note that on the deantennated side the LomTK-IR neurons form an aglomerular pattern, whereas those cells on the contralateral side (*control*) form the normal adult pattern. *Scale bar* = 100 µm

Sensory innervation of the antennal lobe

At the end of metamorphosis, labelling of sensory axons from the antenna showed that all glomeruli in the AL were innervated (Fig. 10A). In fenvalerate-treated pupae (24 nmol), the sensory axons displayed a diffusely branching pattern within the AL (Fig. 10B). These arborizations were more concentrated at the surface but remained within this neuropil. In TTX-treated pupae also, sensory axons terminated diffusely in the AL neuropil.

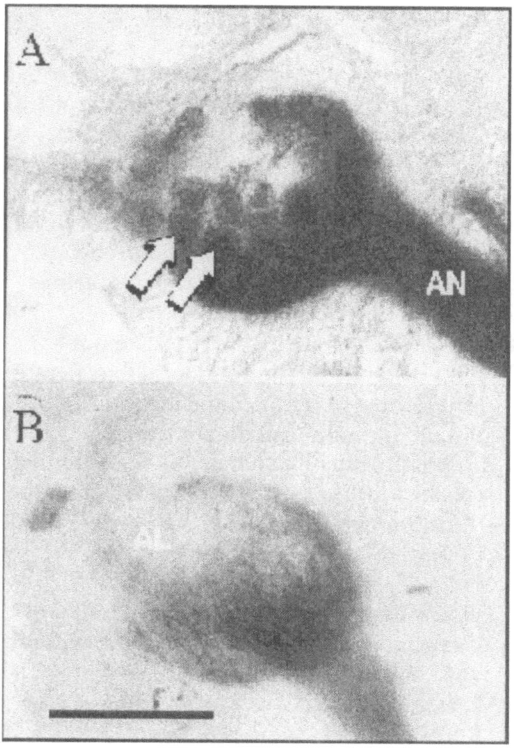

Fig. 10. Sensory axonal innervation of the antennal lobe in normal development (**A**) and after fenvalerate treatment (**B**). Nickel chloride backfills of one antenna of adult beetles, Vibratome sections. **A** Untreated antennal lobe; note the tufted arborizations of axonal terminals within the glomeruli (*small arrows*). **B** Fenvalerate treatment (24 nmol); note the diffuse branching of the sensory projections. (*AN* Antennal nerve). *Scale bars* = 100 μm

Discussion

The results presented show that the metamorphic reorganization of the AL is affected by the application of fenvalerate. The reestablishment of glomeruli in the AL is reduced or inhibited by the application of this pyrethroid at the onset of metamorphosis. The morphologies of antennal sensory axons, AL interneurons and AL glial processes are affected.

In the pupa LomTK-IR interneurons with projections to the AL are only visible at the beginning and the end of the metamorphosis. The reduction of immunoreactivity during metamorphosis is mainly restricted to the AL, whereas other neuropil areas, such as the central body, are not affected (Wegerhoff *et al.* 1996b). The immunoreactivity reappears at the end of metamorphosis, starting with a diffuse staining that becomes glomerular before reaching the adult stage. It will be interesting to investigate whether the larval LomTK-IR neurons of the AL persist during metamorphosis and become immunoreactive again at the end of the pupation, or if all adult LomTK-IR neurons are newly formed.

The application of fenvalerate (24 nmol) results in an earlier reappearance of LomTK immunoreactivity in AL neurons during metamorphosis than in control animals. These neurons show the same somata location and primary projection of their neurites as do the immunoreactive neurons in untreated adult beetles. This indicates that the treatment with fenvalerate does not change the type of LomTK-IR neuron that invades the AL or their primary target finding. However, after projecting into the AL, the neurites do not develop glomerular branches. A similar aglomerular morphology of neuronal branches in the AL also develops during metamorphosis from gamma-amino-butyric acid (GABA) immunoreactive neurons of *Tenebrio* after fenvalerate treatment (data not shown). The degree of deglomerulization depends on the amount of fenvalerate applied: 24 pmol fenvalerate is sufficient to cause a significant effect on AL morphology. These beetles are able to walk normally, but do not show a discriminative olfactory behaviour in a pitfall-trap olfactometer as do untreated ones (R. Wegerhoff and Gäthje, unpublished results). The effects of fenvalerate applications are not restricted to the AL, although the induced changes are most obvious in this area.

The influence of antennal-sensory neurons on the formation of glomeruli in the AL has been investigated in studies of several insect species, especially the moth *Manduca sexta* (Oland and Tolbert 1996). A deantennation, that removes the antenno-sensory projection to the AL, results in the AL having an aglomerular adult morphology (Sanes *et al.* 1977; Rodrigues and Pinto 1988; Salecker and Boeckh 1996), similar to the results in this study. However, deafferantation could not be the reason for the aglomerular structure of the adult AL in *Tenebrio* after fenvalerate treatment, since the sensory axonal innervation of the AL of *Tenebrio* is not blocked after fenvalerate treatment. The branching pattern formed by the axon terminals is also diffuse, not glomerular.

For the moth *Manduca sexta* it has been shown that glial-neuronal interactions are necessary for glomerulus formation, which occurs during metamorphosis (Oland *et al.* 1988; Baumann *et al.* 1996). In this insect sensory axons grow into the AL and form protoglomeruli which become templates for further glomerulus development. Glial cells migrate into the AL neuropil to surround the protoglomeruli. After protoglomeruli have been formed and been enveloped by glial cells, the AL interneurons grow out to meet and synapse with the sensory axon terminals in a glomerular fashion (Oland and Tolbert 1987; Malun *et al.* 1994; Baumann *et al.* 1996). An invasion of glial cell somata to form borders around each glomerulus has no analogy in *Tenebrio*. Glial cells do, however, extend processes that form glomerular boundaries in normal development, a morphological situation similar to that in the cockroach or bee brain (Salecker and Boeckh 1995, 1996; Hähnlein and Bicker 1996). Therefore, a comparable influence of glial cells in *Tenebrio* would have to be restricted to the processes of the glial cells. Such glial processes do not develop upon fenvalerate treatment. If the metamorphic development of glomeruli in *Tenebrio* also requires protoglomeruli formed by sensory axons and enveloped by glia cells, then the primary cellular effect of fenvalerate might be on ingrowing sensory axons and on glial processes.

The primary biological effect of pyrethroids is thought to be on the voltage-gated sodium channels. This would suggest that increased neuronal activity is the reason for the developmental changes observed. According to the findings in studies of vertebrates, both enhanced and reduced neuronal activities can change an activity-dependent pattern formation (Schatz 1990, 1995). The sodium channel blocker TTX was used in the current study to reduce neuronal activity. TTX application resulted in less glomerular patterns of LomTK-IR in the AL, suggesting that appropriate electical activity is important for the re-formation of the AL in *Tenebrio*. In the moth *Manduca sexta* the primary formation of glomeruli is shown to be an activity-independent process (Oland and Tolbert 1996; Oland *et al.* 1996). AL development starts during metamorphosis and after TTX application into the hemolymph the formation of glomeruli is not reduced. In *Tenebrio* the developmental situation is different. The AL is already a glomerular neuropil in larval stages (Breidbach and Marx 1990; Breidbach and Wegerhoff 1994); therefore, newly invading sensory neurons or brain interneurons project into a preestablished neuropil. Electrophysiological studies should reveal more about the role of activity-dependent processes in the metamorphic remodelling of the AL in the beetle brain.

The premature appearance of LomTK-IR neurons in the AL of *Tenebrio* is possibly an effect of a change in gene expression upon pyrethroid treatment. An example of such an effect of pyrethroids has been shown in a study of the adult rat brain, in which c-Fos and c-Jun transcription-factor protein expression is induced after cypermetrin or permethrin application (Hassouna *et al.* 1996). However, so far studies of gene expression in the insect brain have not been reported.

In conclusion, this study provides the first evidence that pyrethroids can induce changes in neuronal and glial fine structure at a defined cellular level during insect brain development. The gradual increase of deglomerularization correlating with the amount of fenvalerate applied indicates that a plastic mechanism of cell–cell interaction in the AL pattern formation is affected. Further analysis of the pyrethroid effects reported here will allow a better understanding of the developmental processes involved.

Acknowledgements. The author would like to thank Ms Marlies Rusch for her expert technical assistance, Martin Lobemeier for preparing Fig. 4C, Leslie Tolbert, Olaf Breidbach and Soerge Kelm for helpful comments on the manuscript, Hans Agricola (Jena) for his generous gift of the anti-locusta-tachykinin II antibody, Ingo Jensen (Olympus Akademie, Hamburg) for technical support and Professors B. Havsteen and H. Busse for general support.

References

Aldridge, W. N. (1990) An assessment of the toxicological properties of pyrethroids and their neurotoxicity. *Crit. Rev. Toxicol.* **21**: 89–104

Baumann, P.M., Oland, L.A., Tolbert, L.P. (1996) Glial cells stabilize axonal protoglomeruli in the developing olfactory lobe of the moth *Manduca sexta*. *J. Comp. Neurol.* **373**: 118–128

Becker, M., Breidbach, O. (1993) Distribution of GABA-like immunoreactivity throughout metamorphosis of the supraoesophageal ganglion of the beetle *Tenebrio molitor* L. (Coleoptera, Tenebrinoidae). In: Elsner, N., Heisenberg, M. (eds)*Proc. Gött. Neurobiol. Conf.* Thieme, Stuttgart, p. 738

Breidbach, O., Marx, J.M. (1990) Antennale sensorische Projektionen im Zentralnervensystem der Larve des Mehlkäfers *Tenebrio molitor* L. (Insecta, Coleoptera). *Z. Ang. Zool.* **77**: 291–310

Breidbach, O., Wegerhoff, R. (1994) FMRFamide-like immunoreactive neurons in the brain of the beetle, *Tenebrio molitor* L. 1758 (Coleoptera: Tenebrinoidae): constancies and variations in development from embryo to the adult. *Int. J. Insect Morphol. Embryol.* **23**: 383–404

Breidbach, O., Dennis, R., Marx, J., Görlach, C., Wiegand, H., Wegerhoff, R. (1992) Insect glial cells show differential expression of a glycolipid-derived, glucuronic acid-containing epitope throughout neurogenesis: detection during postembyogenesis and regeneration in the central nervous system of *Tenebrio molitor* L. *Neurosci. Lett.* **147**: 5–8

Brown, L.D., Narahashi, T. (1992) Modulation of nerve membrane sodium channel activation by deltamethrin. *Brain Res.* **584**: 71–76

Eriksson, P., Fredriksson, A. (1991) Neurotoxic effects of two different pyrethroids. Bioallethrin and Deltamethrin, on immature and adult mice: changes in behavioural and muscarinic receptor variables. *Toxicol. Appl. Pharmacol.* **108**: 78–85

Eriksson, P., Nordberg, A. (1990) Effects of two pyrethroids, bioallethrin and deltamethrin, on subpopulations muscarinic and nicotinic receptors in the mouse brain. *Toxicol. Appl. Pharmacol.* **102**: 446–463

Hähnlein, I., Bicker, G. (1996) Morphology of neuroglia in the antennal lobes and mushroom bodies of the brain of the honeybee. *J. Comp. Neurol.* **367**: 235–245

Hähnlein, I., Härtig, W., Bicker, G. (1996) *Datura stramonium* lectin staining of glial associated extracellular material in insect brains. *J. Comp. Neurol.* **376**: 175–187

Hartzell, A., Scudder, H. I. (1942) Histological effects of pyrethrum and an activator on the central nervous system of the housefly. *J. Econ. Entomol.* **35**: 428–433

Hassouna, I., Wickert, H., El-Elaimy, I., Zimmermann, M., Herdegen, T. (1996) Systemic application of pyrethroid insecticides evokes differential expression of c-Fos and c-Jun proteins in rat brain. *Neurol. Toxicol.* **17**: 415–432

Hemming, H., Flodstrom, S., Warngard, L. (1993) Enhancement of altered hepatic foci in rat liver and inhibition of intercellular communication in vitro by the pyrethroid insecticides fenvalerate, flucythrinate and cypermethrin. *Carcinogenesis* **14**: 2531–2535

Hildebrand, J.G. (1985) Metamorphosis of the insect nervous system: influence of the periphery on the postembryonic development of the antennal sensory pathway in the brain of *Manduca sexta*. In: Selverstone, A.I. (ed)*Model neuronal networks and behaviour*. Plenum, New York, pp. 124–148

Kutsch, W., Schneider, H. (1987) Histological characterization of neurons innervating functionally different muscles of *Locusta*. *J. Comp. Neurol.* **261**: 515-528

Levine, R.B. (1984) Changes in neuronal circuits during insect metamorphosis. *J. Exp. Biol.* **112**: 27–44

Malun, D., Oland, L.A., Tolbert, L.P. (1994) Uniglomerular projection neurons participate in early developmet of olfactory glomeruli in the moth *Manduca sexta*. *J. Comp. Neurol.*350: 1–22

Muren, J.E., Lundquist, C.T., Nässel, D.R. (1995) Abundant distribution of locustatachykinin-like peptide in the nervous system and intestine of the cockroach *Leucophaea maderae*. *Philos. Trans. R. Soc. Lond. [Biol]*. **348**: 423–444

Narahashi, T. (1996a) Toxins that modulate the sodium channel gating mechanism. *Ann. N. Y. Acad. Sci.* **479**: 133–151

Narahashi, T. (1996b) Transmitter-activated ion channels as the target of chemical agents. *Adv. Exp. Med. Biol.* **287**: 61–73

Nässel, D.R. (1993) Neuropeptides in the insect brain: a review. *Cell Tissue Res.* **273**: 1–29

Naumann, K. (1990) Synthetic pyrethroid insecticides. In:*Structure and properties*. Springer, Berlin Heidelberg New York

Nordlander, R.H., Edwards, J.S. (1970) Postembryonic brain development in the monarch butterfly, *Danaus plexippus plexippus* L. III. Morphogenesis of centers other than the optic lobes. *Wilhelm Roux. Arch.* **164**: 247–260

Oland, L.A., Tolbert, L.P. (1987) Glial patterns during early development of antennal lobes of *Manduca sexta*: a comparison between normal lobes and lobes of deprived of antennal axons. *J. Comp. Neurol.* **255**: 196–207

Oland, L.A., Tolbert, L.P. (1996) Multiple factors shape development of olfactory glomeruli: insights from an insect model system. *J. Neurobiol.* **30**: 92–109

Oland, L.A., Pott, W.M., Bukhman, G., Sun, X.J., Tolbert, L. P. (1996) Activity blockade does not prevent the construction of olfactory glomeruli in the moth *Manduca sexta. Int. J. Dev. Neurosci.* **14**: 983–996

Rodrigues, V., Pinto, L. (1988) The antennal glomerulus as a functional unit of odor coding in *Drosophila melanogaster*. In: Singh, R.N., Strausfeld, N.J. (eds) *Neurobiology of sensory systems*. Plenum, New York, pp. 387–396

Salecker, I., Boeckh, J. (1995) Embryonic development of the antennal lobes of a hemimetabolous insect, the cockroach *Periplaneta americana*: light and electron microscopic observations. *J. Comp. Neurol.* **352**: 33–54

Salecker, I., Boeckh, J. (1996) Influence of receptor axons on the formation of olfactory glomeruli in a hemimetabolous insect, the cockroach *Periplaneta americana. J. Comp. Neurol.* **370**: 262–279

Sanes, J.R., Prescott, D.J., Hildebrand, J.G. (1977) Cholinergic neurochemical development of normal and deafferented antennal lobes during metamorphosis of the moth *Manduca sexta. Brain Res.* **119**: 389–402

Schatz, C.J. (1990) Impulse activity and the patterning of connections during CNS development. *Neuron* **5**: 745–756

Schatz, C.J. (1995) Brain waves and brain wiring. In: Elsner, N., Menzel, R. (eds) *Gött. Neurobiol. Rep.* Thieme, Stuttgart, pp.159–187

Sheets, L.P., Doherty, J.D., Law, M.W., Reiter, L.W., Crofton, K.M. (1994) Age-dependent differences in the susceptibility of rats to deltamethrin. *Toxicol. Appl. Pharmacol.* **126**: 186–190

Tateno, C., Ito, S., Tanaka, M.,Yoshitake, A. (1993) Effects of pyrethroid insecticides on gap junctional intercellular communications in Balb/c3T3 cells by dye-transfer assay. *Cell Biol. Toxiol.* **9**: 215–221

Technau, G. M., Heisenberg, M. (1982) Neuronal reorganization during metamorphosis of the corpora pedunculata in*Drosophila melanogaster. Nature* **295**: 405–407

Veenstra, J.A., Lau, G.W., Agricola, H.J., Petzel, D.H. (1995) Immunohistochemical localization of regulatory peptides in the midgut of the female mosquito *Aedes aegypti. Histochem. Cell Biol.* **104**: 337–347

Vijverberg, H.P.M., Van den Berken, J. (1990) Neurotoxicological effects and the mode of action of pyrethroid insecticides. *Crit. Rev. Toxicol.* **21**: 105–126

Wegerhoff, R. (1996) Pyrethroide: Unbedenkliche Insektizide?. *PHi* **3**: 94–97

Wegerhoff, R., Breidbach, O. (1991) Distribution of GABA like immunoreactivity throughout postembryogenesis of the midbrain in the beetle *Tenebrio molitor* L. (Insecta, Coleoptera). In: Elsner, N., Penzlin, H. (eds) *Proc. Gött. Neurobiol. Conf.* Thieme, Stuttgart, p.511

Wegerhoff, R., Breidbach, O. (1992) Structure and development of the larval central complex in a holometabolous insect, the beetle *Tenebrio molitor. Cell Tissue Res.* **268**: 341–358

Wegerhoff, R., Breidbach, O., Lobemeier, M. (1996a) Tachykinin and FMRFamide like immunopositive neurons in the developing central complex of the beetle *Tenebrio molitor*. In: Krisch, B., Mentlein, R.(eds) *The peptidergic neuron.* Birkhäuser, Basel, pp. 205–211

Wegerhoff, R., Breidbach, O., Lobemeier, M. (1996b) Development of Locustatachykinin immunopositive neurons in the central complex of the beetle *Tenebrio molitor. J. Comp. Neurol.* **375**: 157–166

Exp. Biol. Online 2:15 (1997)

Effect of azadirachtin on the nutrition, development and biogenic amine levels in the Eastern Death's Head hawk moth, *Acherontia styx* (lepidoptera: sphingidae)

Edward William Awad*, Fabienne Eugénie Saadé and Mohammed Hadi Amiri

Department of Biology, Faculty of Science, United Arab Emirates University

Present address:
*Département de sciences biologiques, Université de Montréal, C.P. 6128, Succursale "Centre ville", Montréal, Québec, Canada, H3C 3J7

Correspondence to: Edward William Awad (e-mail: awadew@montrealnet.net)

Received: 20 May 1997 / Accepted: 6 October 1997
Published: October 29, 1997

Abstract. Previous studies have suggested the involvement of biogenic amines in insect metamorphic events and post-embryonic development. The effect of azadirachtin (AZ), a natural antifeedant and growth-disrupting compound, on the nutrition, development, and biogenic amine contents of the last instar larvae of the Eastern Death's Head hawk moth, *Acherontia styx*, was examined. Single doses of AZ, injected into the haemolymph at day 1 post-ecdysis, inhibited food consumption in a dose-dependent manner (ED_{50} = 0.65±0.08 µg AZ/g body weight), and was found to be highly effective at producing pupal deformities and inhibiting larval growth (0.1–0.2 µg AZ/g body weight range). Biogenic amine contents, namely octopamine (OA), dopamine (DA) and serotonin (5-HT), in the brain and the haemolymph of 4-day- and 8 day-old larvae were analysed using high-performance liquid chromatography (HPLC) with an electrochemical detector (ECD). A dose-response relationship between AZ and biogenic amine contents in the brain and the haemolymph was also established. Low doses of AZ (0.1–0.2 µg AZ/g body weight) caused a dramatic reduction in OA and 5-HT levels in both the brain and the haemolymph. However, higher doses (0.9–1.2 µg AZ/g body weight) were needed to induce a significant reduction in DA levels. The significance of these findings in relation to the possibility of the involvement of biogenic amines in regulating metamorphic events in insects through mediation of juvenile hormone synthesis and/or release is discussed.

Key words: *Acherontia styx* – Azadirachtin – Biogenic amines – Development –Nutrition

Introduction

Azadirachtin (AZ), a tetranorterpenoid first isolated from the seeds of the neem tree, *Azadirachta indica* A. Juss. (Meliaceae) (Butterworth and Morgan, 1971), is known for its potent insecticidal effects. It acts as a phagodeterrent, antifeedant and growth-disrupting substance and is most effective against Lepidoptera [see Mordue (Luntz) and Blackwell, 1993]. AZ is an example of natural chemical defence by plants, affecting feeding primarily through chemoreception (deterrence) and secondarily through toxic effects [Mordue (Luntz) and Blackwell, 1993]. The growth-regulatory effects of AZ are due to the disruption of the hormonal control of metamorphosis and moulting. These effects are manifested by changes in haemolymph ecdysteroid and juvenile hormone titres due to a blockage and/or delay in the release of those hormones from neurohaemal organs (Barnby and Klocke, 1990; Beckage *et al.*, 1988; Garcia *et al.*, 1986; Reynolds and Wing, 1986; Schlüter *et al.*, 1985; Subrahmanyan *et al.*, 1989).

Biogenic amines are known to play important physiological roles as neurotransmitters, neuromodulators and/or neurohormones in a large variety of insects (see Downer and Hiripi, 1993). It is now established that these neuroactive substances act as systems that regulate many essential processes, including: octopaminergic (OA-ergic) modulation of carbohydrate metabolism (Downer *et al.*, 1984), muscle contraction (Ramirez and Pearson, 1991), and firefly bioluminescent organ activity (Nathanson, 1979); OA-ergic and dopaminergic (DA-ergic) regulation of flight motor activity (Classen and Kammer, 1986); serotonergic (5-HT-ergic) modulation of feeding (Lange *et al.*, 1989); 5-HT-ergic and DA-ergic modulation of salivary gland secretion (Gifford *et al.*, 1991); and OA-ergic enhancement of short-term memory and sensitivity to sensory stimuli in honeybees (Mercer and Menzel, 1982; Menzel, 1983). A role for biogenic amines in insect development and metamorphosis was also suggested. There have been a number of reports of changes of biogenic amines in the central nervous system during metamorphic development (Awad *et al.*, 1997; Bodnaryk, 1980; Cantera and Carlberg, 1988; Fuzeau-Braesch *et al.*, 1979; Geng *et al.*, 1993; Takeda *et al.*, 1991; Woodring *et al.*, 1988). OA, DA and 5-HT levels in the brain of last larval instars reached a peak in the wandering phase of two sphingid species, *Manduca sexta* (Geng *et al.*, 1993) and *Acherontia styx* (Awad *et al.*, 1997), suggesting the occurrence of neuronal processes, prior to and/or during the wandering phase, that are modulated by biogenic amines and are involved in the metamorphic events exhibited in the life cycle of these insects. In addition, OA was reported to modulate juvenile hormone synthesis and release by the corpus allatum in *Locusta migratoria* (Lafon-Cazal and Baehr, 1988) and *Diploptera punctata* (Thompson *et al.*, 1990).

Several studies have reported the effect of AZ on nutrition, growth, and development in Lepidoptera (e.g. Barnby and Klocke, 1987; Isman, 1993; Koul *et al.*, 1987; Osman, 1993). The growth-inhibitory effect of AZ and its interference with endocrine events have been extensively studied [see Mordue (Luntz) and Blackwell, 1993]. However, there is no information about the effect of AZ on biogenic amine systems in insects. Banerjee and Rembold (1992) reported that AZ causes an elevation of 5-HT levels in the central nervous system of both 5th instar nymphs and adults of *Locusta migratoria* and *Schistocerca gregaria*. The presumed involvement of biogenic amines in metamorphic development and the interference of AZ with 5-HT pools and the mode of release of growth- and development-regulating hormones argue in favour of the possible interference of AZ with biogenic amine pools in the central nervous and neurohaemal systems in insects. Therefore, this study was undertaken to examine the effect of various doses of AZ on the development and nutrition the of last instar larvae of the Eastern Death's Head hawk moth, *Acherontia styx*, and its possible interference with biogenic amine pools, namely, octopamine (OA), dopamine (DA) and serotonin (5-HT), in the brain and the haemolymph of this insect.

Materials and Methods

Experimental insects

Acherontia styx eggs were obtained from the field and reared on fresh *Clerodendrum inerme* (Verbenaceae) leaves at 20–25°C, 70% relative humidity (RH), and photoperiod 14 h light:10 h dark. Under these conditions, the life cycle of this insect comprises five larval instars. The instar stage of each larva was determined by counting the number of head capsules shed. The 5th instar lasts for 8–9 days and ends with a short-lived wandering stage during which the insect actively searches for a suitable pupation site in the soil. Experimental animals consisted solely of 5th instar larvae. Samples for biogenic amine analysis were prepared from middle and late 5th instar larvae corresponding to day 4 (V4; ecdysis day was considered as day 0) and day 8 (V8), respectively.

Azadirachtin injections

Pure AZ was dissolved in 90% ethanol and diluted with distilled water. Then, 25 µl AZ solution, at a dose ranging between 0.1 and 2.4 µg AZ/g of larval body weight, was injected into the haemolymph at the base of the right 4th abdominal larvapod 24 h after day 0. Larvae were injected only once during the 5th instar stage. Controls were injected with the corresponding ethanol and water dilutions.

Food utilization, growth and developmental bioassays

To evaluate the effect of injected AZ on *A. styx* nutrition and to assess any possible AZ-induced changes in biogenic amine levels due to antifeedancy, individual 5th instar larvae were injected with varying doses of AZ and placed in plastic jars containing a known weight of leaf diet. Food was changed daily and the remaining uneaten leaves and faeces were collected, dried and weighed. Larvae were weighed daily, from day 0 to day 6 inclusively. To estimate larval dry weight, 5th instar larvae at different developmental stages were dried and weighed and the water content was determined. This was found to account for 89–91% of body weight, irrespective of the developmental stage of 5th instar larvae. The following nutritional parameters were then measured, based on Reese and Beck (1976): amount of food ingested, approximate digestibility, efficiency of conversion of digested food and efficiency of conversion of ingested food.

To assess any changes in biogenic amine levels that occur as a consequence of starvation, a group of ten insects was semi-starved by alternating feeding days. Samples for biogenic amine analysis from those insects were prepared on day 4 and day 8 post-ecdysis corresponding to those days when the insects were starved for 24 h.

To monitor developmental responses of *A. styx* 5th instar larvae to different doses of injected AZ, the duration of the 5th instar, the onset of the wandering stage, the duration of the buried stage, the slipping of the head capsule, pupal transformations as well as mortality and net gain in dry weight were noted.

Sample preparation

Haemolymph samples (100–300 µl) from V4 and V8 larvae were mixed with 0.1 N HCl (1:1) solution and centrifuged at 15,000 *g* for 20 min at 4°C to precipitate cell debris and large proteins. The supernatant was collected and stored at -80°C until needed. Prior to analysis, haemolymph samples were thawed, mixed with an equal volume of an extraction solution containing 1% glacial acetic acid, 0.5% sodium disulphide, 0.5% ethylenediamine tetraacetic acid (EDTA) and 50 mg/ml sodium metabisulphite (pH 3.5; adapted from Waters, Millipore, Massachusetts, USA), diluted 4 times and filtered.

Brains of V4 and V8 larvae were dissected in cold saline (0.02% KCl, 0.02% $CaCl_2$, 0.4% sucrose and 0.9% NaCl; pH 7.5), homogenized in 50 µl of the extraction solution and centrifuged at 15,000 *g* for 20 min at 4°C and then processed as for the haemolymph.

HPLC with electrochemical detector (ECD)

Samples (25 µl) were manually injected on the HPLC column through an injector valve fitted with a 50-µl loop. OA, DA and 5-HT were separated through a C_{18}

reverse-phase column (4.6 × 250 mm; BioRad, California, USA) maintained at 25°C. They were detected simultaneously using an ECD (Waters 460, Millipore) with a graphite electrode set at an oxidizing potential of 600 mV. Signals from the ECD were recorded and integrated using a Waters 746 data processor module. A running buffer (adapted from Shimadzu, Japan) consisting of 20 mM sodium citrate (pH 6.2–3.5), 5 mM octane sulphonic acid sodium salt, 50 μM EDTA and acetonitrile (10:1) was run isocratically at a rate of 1 ml/min. Under these conditions, the detection limits for OA, DA and 5-HT were 1000 pg/ml, 60 pg/ml and 150 pg/ml, respectively.

Standards and samples were injected alternately for identification and quantification analyses. Each peak was identified by comparison of the retention times, by spiking and by comparison of relative retention time ratios between compounds and the internal standard, 3,4-dihydroxy-benzylamine (DHBA). Biogenic amine contents were determined by comparing areas of the peaks with known external standards. Samples were spiked with external standards.

To determine the rate of recovery of the analysed amines in the tissues, 100 pg/μl standards for each compound was added to samples from V8 larvae before processing for HPLC analysis. Percentage recovery rate was calculated by subtracting the values of the control (endogenous amines in samples prepared from V8 larvae) from corresponding values of samples with added standards. The mean recovery of DHBA was 93%. The recovery of DA was 91%, while that of OA and 5-HT was 96%.

Data analysis

Paired data were analysed for significance by Student's *t*-test. Multiple comparisons among groups of means were performed using one-way analysis of variance (ANOVA, Tukey-Kramer method) and Duncan's multiple range test. Probability values of $P<0.05$ were taken to indicate statistical significance. Results are presented as means ± standard deviation (sd) for each variable presented from each experiment.

Estimates of ED_{50} values (median effective dose of AZ inducing 50% of the maximal effect) of the dose-response effect of AZ on growth and biogenic amine levels were calculated using log-dose regression analysis.

Chemicals

All chemicals were of analytical reagent grade. The following compounds were purchased from Sigma (St. Louis, Mo., USA): azadirachtin (95% purity), sodium disulphide, EDTA, sodium metabisulphite, OA, DA, 5-HT, sodium citrate, octane sulphonic acid, DHBA, NaCl, $CaCl_2$, KCl and sucrose. HPLC-grade HCl, glacial acetic acid and acetonitrile were obtained from Aldrich (Wisconsin, USA).

Results

Effect of AZ on food utilization by *A. styx* 5th instar larvae

Data on the effect of AZ on food utilization by *A. styx* larvae are shown in Table 1. AZ, at 0.4 µg/g larval body weight or higher, significantly reduced the amount of food ingested by *A. styx* larvae (ED_{50} = 0.65±0.08 µg AZ/g body weight) compared to the controls. However, approximate digestibility values of larvae treated with 1 µg AZ/g larval body weight were significantly higher than those of controls. AZ, on the other hand, reduced the efficiency of conversion of ingested and digested food into biomass by *A. styx* larvae. These reductions in efficiency occurred at doses of AZ that were higher (0.8 µg/g body weight for efficiency of conversion of both ingested and digested food) than those that reduced food consumption and net gain in weight (0.4 µg/g body weight) (Table 2).

Table 1. Nutritional parameters of *Acherontia styx* 5th instar larvae injected with different doses of azadirachtin. [*AI* Amount of food ingested (food consumption), *AD* approximate digestibility, *ECI* efficiency of conversion of digested food, *ECD* efficiency of conversion of ingested food]. Means within *columns* followed by the same *letter* are not significantly different (p>0.05). n=30 per dose

AZ dose (µg/larval body weight)	AI (g)	AD (%)	ECI (%)	ECD (%)
0.0 (control)	7.51±1.19a	57.32±5.79a	12.10±1.16a	23.00±3.93a
0.1	7.21±1.04a	58.25±5.37a	11.78±1.63b	24.21±3.67b
0.2	6.22±1.16a	59.46±5.67a	11.27±1.12b	24.16±3.27b
0.4	5.78±1.02a	59.28±6.12a	10.57±1.16b	19.73±3.12b
0.8	3.60±0.93b	61.48±6.05a	7.28±1.54c	10.42±2.67c
1.0	2.33±0.54b	76.23±6.55b	5.61±1.26c	6.54±2.16d
1.4	1.86±0.36c	74.95±6.98b	4.04±1.10d	3.71±1.87e

Effect of AZ on growth and development of *A. styx* 5th instar larvae

AZ was found to disrupt the development of 5th instar larvae in a dose-dependent manner (Table 2). At low doses (0.1–0.2 μg AZ/g of body weight), the rate of pupation was similar to that of the controls, but with a high rate (30–60%) of deformed pupae (larval-pupal intermediates). At 0.4 μg AZ/g larval body weight, pupation rate was reduced by almost half (ED_{50} = 0.32±0.04 μg/g body weight). A low pupation rate (26.7%) associated with a high rate of deformed pupae (100%) was recorded for larvae injected with 0.8 μg AZ/g larval body weight (ED_{50} = 0.79±0.09 μg/g body weight). All larvae injected with doses greater than 0.8 μg AZ/g body weight did not survive past day 6 post-injection. The ED_{50} value for mortality was 0.51 μg AZ/g body weight. AZ was also found to prolong the duration of the 5th instar. This was associated with a decrease in the rate of onset of the wandering stage in over-aged 5th instar larvae. Larvae exhibiting the wandering behaviour did not delay entering the prepupal stage (buried stage). However, AZ (at doses between 0.4 and 0.8 μg/g body weight) induced a prolongation of the prepupal stage.

Table 2. Developmental responses of *Acherontia styx* 5th instar larvae to different doses of azadirachtin (*AZ*). *n* = 30 per dose

AZ dose[a]	Percent pupation	Percent deformed pupae	Mean number of days to wandering stage	Percent entering the wandering stage	Mean duration of prepupal stage (days)	Net gain in weight (g dry weight)
0.0 (control)	100.0	0.0	8.5±0.7	100.0	4.0±0.9	0.92±0.23
0.1	93.3	30.0	9.6±0.8	93.3	4.5±0.8	0.85±0.26
0.2	86.7	60.0	12.4±1.1	86.7	5.4±1.2	0.76±0.20
0.4	56.7	80.0	22.0±1.4	60.0	8.4±0.9	0.42±0.16
0.8	26.7	100.0	24.1±2.3	33.3	9.2±1.0	0.32±0.12[b]

[a]: (μg/g larval body weight)

[b]: No significant change in dry weight was observed with larvae treated with AZ doses higher than 0.8 μg/g body weight over a period of 6 days (see Results for details)

AZ, at a dose above 0.2 µg/g body weight, reduced the net gain in weight of 5th instar *A. styx* larvae (Table 2). The calculated ED_{50} of AZ in terms of its effect on net weight gain was 0.35±0.04 µg/g larval body weight. Larvae injected with doses greater than 0.8 µg AZ/g body weight did not show any significant changes in total body weight over a period of 4 days post-ecdysis. These individuals then lost weight and died on day 6 post-ecdysis.

Effect of AZ on biogenic amine contents in the brain of 5th instar *A. styx* larvae

OA, DA and 5-HT levels in the brain of 5th instar larvae are shown in Fig. 1, corresponding to those of our previous report (Awad *et al.*, 1997). OA levels were the highest (Fig. 1B). No significant differences in all biogenic amine levels between V4 and V8 in the control were observed. AZ caused a general dose-dependent reduction in all biogenic amine levels tested in the brain. AZ was most potent at reducing 5-HT levels in the brain with an ED_{50} (V4) = 0.10±0.03 µg AZ/g body weight (Fig. 1C). Slightly higher doses of AZ were needed to reduce brain OA levels with an ED_{50} (V4) = 0.19±0.04 µg AZ/g body weight (Fig. 1B). However, DA levels were the least affected by AZ injections with an ED_{50} (V4) = 0.91±0.13 µg AZ/g body weight (Fig. 1A). No significant differences were obtained for AZ ED_{50} values in terms of biogenic amine levels in the brain of V8 larvae as compared to those of V4 larvae. 5-HT in V4 and V8 was not detected with AZ doses greater than 0.4 µg/g larval body weight.

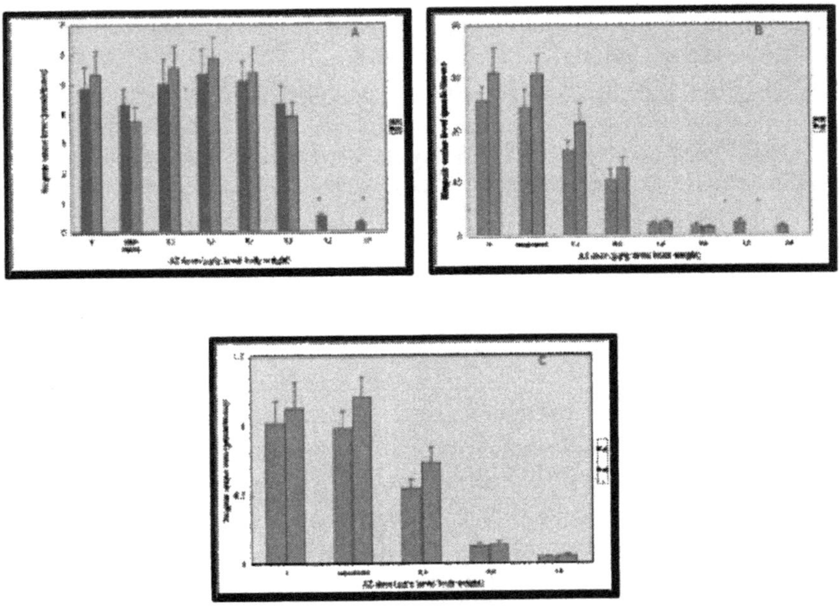

Fig. 1A–C Effect of several doses of AZ on dopamine (*DA*) (**A**), octopamine (*OA*) (**B**) and serotonin (*5-HT*) (**C**) contents in the brain of last instar larvae of the Eastern Death's Head hawk moth, *Acherontia styx*. (*) Larvae injected with a dose greater than 0.8 µg AZ/g body weight did not survive past day 6 post-injection. (Left column - blue- *V4* 4-day-old larvae, Right column-red- *V8* 8-day-old larvae). n=10 per dose

OA and 5-HT levels in the brain of semi-starved larvae were not significantly different from those of the controls (Fig. 1B, C). A slight reduction, however, in DA levels of semi-starved larvae was observed (Fig. 1A).

Effect of AZ on biogenic amine concentrations in the haemolymph of 5th instar *A. styx* larvae

Biogenic amine concentrations in the haemolymph of 5th instar larvae are shown in Fig. 2, corresponding to those of our previously published report (Awad *et al.*, 1997). OA was the most abundant biogenic amine, whereas 5-HT was not detected. A 4.8- and a 1.5-fold increase in DA (Fig. 2A) and OA (Fig. 2B) concentrations, respectively, were observed to occur with the development of control 5th instar larvae. There was a dose-dependent decrease in DA and OA concentrations in the haemolymph of V4 and V8 larvae treated with AZ, with ED_{50} values for V4 larvae being 0.73±0.08 µg/g (DA) and 0.24±0.03 µg/g (OA) body weight. No significant differences were obtained for AZ ED_{50} values in terms of biogenic amine concentrations in the haemolymph of V8 larvae as compared to those of V4 larvae.

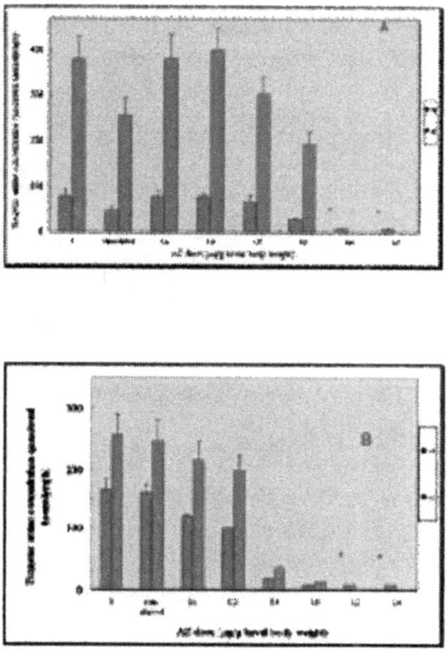

Fig. 2A,B Effect of several doses of AZ on DA (**A**) and OA (**B**) concentrations in the haemolymph of the last instar larvae of the Eastern Death's Head hawk moth, *Acherontia styx*. (*) Larvae injected with a dose greater than 0.8 μg AZ/g body weight did not survive past day 6 post-injection. (Left column - blue- *V4* 4-day-old larvae, Right column-red- *V8* 8-day-old larvae). n=10 per dose

There was a reduction in the DA concentration in the haemolymph of semi-starved larvae as compared to the controls (Fig. 2A). On the other hand, OA concentrations in the haemolymph of semi-starved larvae were not significantly different to those of the controls (Fig. 2B).

Discussion

In addition to reporting findings on the effect of AZ on nutrition and development of the last instar larvae of the Eastern Death's Head hawk moth, *A. styx*, this study presents, for the first time, data on the effect of this natural insecticide on biogenic amine levels in the central nervous system and the haemolymph of a lepidopteran species.

Fifth instar larvae injected with AZ consumed less food, gained less weight and were less efficient at converting ingested and digested food into biomass. These results are similar to those reported for larvae of the European corn borer *Ostrinia nubilalis* (Arnason *et al.*, 1985), the cabbage webworm *Crocidolomia binotalis* (Fagoonee, 1984) and several noctuid species (Barnby and Klocke, 1987; Isman, 1993). The fact that approximate digestibility increased with increasing AZ dose may well reflect a decreased rate of passage of food through the gut, as for example in *Locusta migratoria* [Mordue (Luntz) *et al.*, 1985]. This prolongation of food in the gut might explain the increase approximate digestibility in AZ-treated larvae, probably due to an increase in exposure to digestive enzymes as proposed by Barnby and Klocke (1987).

AZ was also found to be extremely effective at producing pupal deformities and larval growth inhibition at doses as low as 0.2 µg AZ/g body weight. The dose-dependent effects of AZ on the development of the last instar larvae of *A. styx* parallel those reported for other insect species (Barnby and Klocke, 1987; Dorn *et al.*, 1986; Isman, 1993; Jaganadh and Nair, 1992; Koul *et al.*, 1987; Osman, 1993; Schlüter *et al.*, 1985; Simmonds *et al.*, 1990). Treatment of 1-day-old *A. styx* 5th instar larvae with low doses of AZ resulted in the production of permanent larvae or deformed pupae. Over-aged larvae or nymphs treated with AZ were also reported in relation to *Spodoptera mausita* (Jaganadh and Nair, 1992), *Manduca sexta* (Schlüter *et al.*, 1985), *Locusta migratoria* [Mordue (Luntz) *et al.*, 1985], *Rhodnius prolixus* (Garcia and Rembold, 1984) and *Bombyx mori* (Koul *et al.*, 1987). A single dose of AZ was enough to cause the observed morphogenetic effects in *A. styx*. *Rhodnius prolixus* nymphs treated with a single dose of AZ (ng AZ/nymph range) were reported to be unable to moult for several months (Garcia and Rembold, 1984).

Dramatic changes in OA and 5-HT levels were observed in the brain of *A. styx* larvae treated with small single doses of AZ. There seems to be a correlation between AZ doses, larval development and OA and 5-HT contents, judging from the ED_{50} values of AZ on last larval instar development and biogenic amine contents in the brain. In addition, the release of OA from neurohaemal organs into the haemolymph seems to be affected by low doses of AZ, correlating with AZ doses needed to inhibit larval development. The dependency of the reduction of OA concentration in the haemolymph on the dose of AZ and the dramatic reduction in OA concentrations (9 times) at low AZ doses (0.4 µg/g body weight) make it unlikely that this reduction occurred as a result of haemolymph dilution. It was reported that neurosecretory cells in the central nervous system of insects are affected by different insecticides (Jankovic-Hlandni *et al.*, 1983; Ivanovic et al., 1985). Neurosecretory cells or other cerebral regions may respond to AZ by selectively affecting the biosynthesis of biogenic amines. Therefore, some of the actions of AZ on larval growth development may result from an imbalance in the metabolism, mode of inactivation and/or mode of release of these neuroactive amines and related substances.

It was reported that water and food deprivation for a period of 10–30 days decreases the levels of biogenic amines in *Blaberus craniifer* (Barreteau *et al.*, 1993) and *Tribolium castaneum* (Hirashima *et al.*, 1994). AZ induced a reduction in food consumption, but not complete starvation, in *A. styx* larvae, even with doses as high as 1.4 µg/g body weight. Therefore, feeding inhibition does not appear to be the primary cause of the observed effects on growth and biogenic amine levels. This is further supported by the fact that OA and 5-HT concentrations in semi-starved larvae were not significantly different from those of the controls. In addition, relatively higher doses (ED_{50} = 0.65±0.08 μg AZ/g body weight) were needed to reduce food consumption, whereas metamorphic effects of AZ were observed at lower doses (3–6 times lower). Several studies have suggested that insect growth-inhibitory and antifeedant effects are independent of each other and that AZ-induced effects on insect development are due to interference with development hormone titres (Garcia and Rembold, 1984; Malczewska *et al.*, 1988; Sieber and Rembold, 1983; Smith and Mitchell, 1988). Such developmental responses to AZ are due to a disruption of the neuroendocrine control of moulting and metamorphosis [see Mordue (Luntz) and Blackwell, 1993]. Therefore, the results of the present study suggest that the primary cause of reduction in OA and 5-HT synthesis and/or release into the haemolymph is AZ, but not feeding inhibition. On the other hand, the reduction in DA levels in the brain and the haemolymph of AZ-treated and semi-starved larvae may be related to inhibition of food consumption. Higher doses of AZ, corresponding to those needed to induce 50% feeding inhibition, were necessary to get a significant reduction in DA levels in the brain and the haemolymph.

The developmental effects of AZ are known to be attributed to a disruption of endocrine events through blockade of the release of juvenile hormones and ecdysteroids. In several AZ-treated insect species, the peaks of those titres were delayed and shifted to a longer time period, with or without a decrease in their intensity [Beckage *et al.*, 1988; Mordue (Luntz) *et al.*, 1985; Redfern *et al.*, 1982; Rembold *et al.*, 1987; Schlüter *et al.*, 1985; Sieber and Rembold, 1983]. OA was reported to mediate hormone neuropeptide release in insects (Orchard and Loughton, 1981; Pannabecker and Orchard, 1986). Thompson *et al.* (1990) have suggested that OA acts as a natural neuromodulator of juvenile hormone synthesis, possibly through regulating the release of the inhibitory neuropeptide allatostatin. In addition, treatment with OA and cAMP resulted in a dose-dependent inhibition of juvenile hormone synthesis by the corpora allata of the adult cockroach, *Diploptera punctata* (Thompson *et al.*, 1990). Therefore, this study suggests that AZ-induced-overageing may be caused by a delay in the peak of juvenile hormone levels so that the presence of this hormone extends into the critical period for commitment to larval rather than pupal characteristics. Presumably, this might occur through a reduction in the rate of synthesis of OA in the central nervous system and/or its release into the haemolymph, thus maintaining a high juvenile hormone titre. This argument is further supported by previous results which have shown that OA synthesis and/or release in *A. styx* increases with development of the last larval instar and peaks prior to the onset of the wandering stage (Awad *et al.*, 1997). However, it remains to be demonstrated whether injected OA induces inhibition of juvenile hormone titre in *A. styx* last instar larvae and whether AZ-treated larvae can be rescued from the effects of AZ by the injection of OA.

Our findings on the effect of AZ on 5-HT contents in the brain contradict those of Banerjee and Rembold (1992) and Banerjee (1994). In their immunohistochemical studies, AZ (at doses of 1.8–4.0 μg/g insect) was found to cause a general accumulation of 5-HT in the brain and other neuroendocrine organs of both last instar nymphs and adults of *Locusta migratoria* and

Schistocerca gregaria. This contradiction may be attributed to species differences. In addition, AZ has direct effects on a whole variety of tissues and organs. This suggests that AZ has a number of different modes of action or that it causes a specific toxic lesion to all types of cells, which is manifested differently in different tissues [Mordue (Luntz) and Blackwell, 1993]. The mode of action of AZ-induced 5-HT accumulation in locusts might be different from that causing a reduction in 5-HT contents in *A. styx*. The former could have resulted from the exposure to relatively high doses of AZ, which then interferes with several systems controlling 5-HT pools in locusts.

Several physiological processes are influenced by 5-HT in insects. Of direct relevance to the effect of AZ on food consumption is the stimulation of midgut muscle contractions in cockroaches and silkworms (Cook and Holman, 1978; Hukuhara *et al.*, 1981), the increase in amplitude and rate of contraction of mandibular closer muscles in locusts (Baines *et al.*, 1990; Baines and Downer, 1991) and the modulation of olfactory information in *Manduca sexta* (Kent *et al.*, 1987; Sun *et al.*, 1993). The secondary antifeedant effects of AZ could result from toxic lesions in the serotonergic system of *A. styx* affecting midgut motility, food manipulation and processing of olfactory signals.

The present findings shed new light on the aminergic control of metamorphic events in insects. The differential and selective responses in biogenic amine levels to different doses of AZ demonstrate the usefulness of AZ as a chemical probe with which to study the proposed involvement of biogenic amines in the process of insect development. However, many questions remain unanswered and deserve to be investigated in future works. These include the mechanism by which AZ induces a reduction in biogenic amine contents and the effect of AZ on OA synthesis and release from neurohaemal organs.

Acknowledgements. We thank Professor Dean Harron and Mr. Rajan S. Stephen of the Pharmacology Department of the United Arab Emirates University for their advice. This work was supported by a grant from the Research Council of the United Arab Emirates University (grant to E.W.A. and M.H.A.). Other support was provided by the Department of Pharmacology, Faculty of Medicine and Health Sciences, United Arab Emirates University.

References

Arnason, J.T., Philogene, B.J.R., Donskov, N., Hudon, M., McDougall, C., Fortier, G., Morand, P., Gardner, D., Lambert, J., Morris, C., Nozzollilo, C. (1985) Antifeedant and insecticidal properties of azadirachtin to the European Corn Borer, *Ostrinia nubilalis. Entomolog. Exp. Appl.* **38**: 29–34

Awad, E.W., Amiri, M.H., Stephen, R.S. (1997) Developmental changes in biogenic amine levels in the central nervous system and the haemolymph of the Eastern Death's Head hawk Moth, *Acherontia styx* (Lepidoptera: Sphingidae). *Comp. Biochem. Physiol.* **116C**: 219–225

Baines, R.A., Downer, R.G.H. (1991) Pharmacological characterization of a 5-hydroxytryptamine-sensitive receptor/adenylate cyclase complex in the mandibular closer muscles of the cricket, *Gryllus domestica. Arch. Insect Biochem. Physiol.* **16**: 153–163

Baines, R.A., Tyrer, N.H., Downer, R.G.H. (1990) Serotonergic innervation of the locust mandibular closer muscle modulates contractions through the elevation of cyclic adenosine monophosphate. *J. Comp. Neurol.* **294**: 623–632

Banerjee, S. (1994) Serotonin immunoreactivity and its content in azadirachtin treated locusts. Symposium on Toxicology, Conservation and Biodiversity. Proc. Acad. Environ. Biol. **3**: 25–31

Banerjee, S., Rembold, H. (1992) Azadirachtin A interferes with control of serotonin pools in the neuroendocrine system of locusts. *Naturwissenschaften* **79**: 81–84

Barnby, M.A., Klocke, J.A. (1987) Effects of azadirachtin on the nutrition and development of the tobacco budworm, *Heliothis virescens* (Fabr.) (Noctuidae). *J. Insect Physiol.* **33**: 69–75

Barnby, M.A., Klocke, J.A. (1990) Effects of azadirachtin on levels of ecdysteroids and prothoracicotropic hormone-like activity in *Heliothis virescens* (Fabr.) larvae. *J. Insect Physiol.* **36**: 125–131

Barreteau, H., Goudey-Perrière, F., Perrière, C., Jacquot, C., Gayal, P., Grosclaude, J.-M., Brousse-Gaury, P. (1993) Influence of fasting and isolation on biogenic amine levels in the nervous system of the male cockroach *Blaberus craniifer* Burm. (Dictyoptera, Blaberidae). *Comp. Biochem. Physiol.* **105C**: 11–16

Beckage, N.E., Metcalf, J.S., Nielson, B.D., Nesbit, D.J. (1988) Disruptive effects of azadirachtin on development of *Cotesia congregata* in host tobacco hornworm larvae. *Arch. Insect Biochem. Physiol.* **9**: 47–65

Bodnaryk, R.P. (1980) Changes in brain octopamine levels during metamorphosis of the moth, *Mamestra configurata* Wlk. *Insect Biochem.* **10**: 169–173

Butterworth, J.H., Morgan, E.D. (1971) Investigation of the locust feeding inhibition of the seeds of the neem tree, *Azadirachta indica. J. Insect Physiol.* **17**: 969–977

Cantera, R., Carlberg, M. (1988) Serotonin levels in the central nervous system of the blow fly *Calliphora erythrocephala* during post-embryonic development. *Insect Biochem.* **18**: 507–509

Classen, D.E., Kammer, A.E. (1986) Effects of octopamine, dopamine, and serotonin on production of flight motor output by thoracic ganglia of *Manduca sexta*. *J. Neurobiol.* **17**: 1–14

Cook, B.J., Holman, G.M. (1978) Comparative pharmacological properties of muscle function in the foregut and the hindgut of the cockroach, *Leucophaea maderae*. *Comp. Biochem. Physiol.* **61C**: 291–295

Dorn, A., Rademacher, J.M., Sehn, E. (1986) Effects of Azadirachtin on the moulting cycle, endocrine system, and ovaries in last-instar larvae of the milkweed bug, *Oncopeltus fasciatus*. *J. Insect Physiol.* **32**: 231–238

Downer, R.G.H., Hiripi, L. (1993) Biogenic amines in insects. In: Borkovec, A.B., Loeb, M.J. (eds) *Insect neurochemistry and neuropharmacology 1993*. CRC, Boca Raton, Fla., pp. 23–38

Downer, R.G.H., Orr, G., Cole, G., Orchard, I. (1984) The role of octopamine and cyclic AMP in regulating hormone release from corpora cardiaca of the American cockroach. *J. Insect Physiol.* **30**: 457–462

Fagoonee, I. (1984) Effect of azadirachtin and of a neem extract on food utilization by *Crocidolomia binotalis*. In: Schmutterer, H., Ascher, K.R.S. (eds) Natural pesticides from the neem tree (*Azadirachta indica* A. Juss.) and other tropical plants. Proceedings of the 2nd International Neem Conference. German Agency for Technical Cooperation, Eschborn, Germany, pp. 211–223

Fuzeau-Braesch, S., Coulon, J.F., David, J.C. (1979) Octopamine levels during the molt cycle and adult development in the migratory locust, *Locusta migratoria*. *Experentia* **35**: 1349–1350

Garcia, E.S., Rembold, H. (1984) Effects of azadirachtin on ecdysis of *Rhodnius prolixus*. *J. Insect Physiol.* **30**: 939–941

Garcia, E.S., Uhl, M., Rembold, H. (1986) Azadirachtin, a chemical probe for the study of moulting process in *Rhodnius prolixus*. *Z. Naturforsch* **41C**: 771–775

Geng, C., Sparks, T.C., Skomp, J.R., Gajewski, R.P. (1993) Biogenic amines in the brain of *Manduca sexta* during larval-pupal metamorphosis. *Comp. Biochem. Physiol.* **106C**: 275–284

Gifford, A.N., Nicholson, R.A., Pitman, R.M. (1991) The dopamine and 5-hydroxytryptamine content of locust and cockroach salivary neurons. *J. Exp. Biol.* **161**: 405–414

Hirashima, A., Nagano, T., Eto, M. (1994) Effect of various insecticides on the larval growth and biogenic amine levels of *Tribolium castaneum* Herbst. *Comp. Biochem. Physiol*. **107C**: 393–398

Hukuhara, T., Satake, S., Sato, Y. (1981) Rhythmic contractile movements of the larval midgut of the silkworm, *Bombyx mori. J. Insect Physiol.* **27**: 469–473

Isman, M.B. (1993) Growth inhibitory and antifeedant effects of azadirachtin on six noctuids of regional economic importance. *Pesticide Sci.* **38**: 57–63

Ivanovic, J., Jankovic-Hlandni, M., Stani, V., Kalafati, V. (1985) Differences in the sensitivity of protocerebral neurosecretory cells arising from the effect of different factors in *Morimus funereus* larvae. *Comp. Biochem. Physiol.* **80A**: 107–113

Jaganadh, V., Nair, V.S.K. (1992) Azadirachtin-induced effects on larval-pupal transformation of *Spodoptera mauritia. Physiol. Entomol.* **17**: 56–61

Jankovic-Hlandni, M., Ivanovi, J., Nenadovi, V., Stani, V. (1983) The selective response of the protocerebral neurosecretory cells of the *Cerambyx cerdo* larvae to the effect of different factors. *Comp. Biochem. Physiol.* **74A**: 131–136

Kent, K.S., Hoskins, S.G., Hildebrand, J.G. (1987) A novel serotonin-immunoreactive neuron in the antennal lobe of the sphinx moth *Manduca sexta* persists throughout postembryonic life. *J. Neurobiol.* **18**: 451–465

Koul, O., Amanai, K., Ohtaki, T. (1987) Effect of azadirachtin on the endocrine events of *Bombyx mori. J. Insect Physiol.* **33**: 103–108

Lafon-Cazal, M., Baehr, J.C. (1988) Octopaminergic control of corpora allata activity in an insect. *Experentia* **44**: 895–896

Lange, A.B., Orchard, I., Barrett, F.M. (1989) Changes in hemolymph serotonin levels associated with feeding in the blood-sucking bug, *Rhodnius prolixus. J. Insect Physiol.* **35**: 393–399

Malczewska, M., Gelman, D.B., Cymborowski, B. (1988) Effects of azadirachtin on development, juvenile hormone and ecdysteroid titres in chilled *Galleria mellonella* larvae. *J. Insect Physiol.* **34**: 725–732

Menzel, R. (1983) Neurobiology of learning and memory: the honeybee as a model system. *Naturwissenschaften* **70**: 504–511

Mercer, A., Menzel, R. (1982) The effects of biogenic amines on conditioned and unconditioned responses to olfactory stimuli in the honeybee *Apis mellifera. J. Comp. Physiol.* **145**: 363–368

Mordue (Luntz), A.J., Blackwell, A. (1993) Azadirachtin: an update. *J. Insect Physiol.* **39**: 903–924

Mordue (Luntz), A.J., Cottee, P.K., Evans, K.A. (1985) Azadirachtin: its effect on gut motility, growth and moulting in *Locusta. Physiol. Entomol.* **10**: 431–437

Nathanson, J.A. (1979) Octopamine receptors, adenosine 3',5'-monophosphate and neural control of firefly flashing. *Science* **203**: 65–68

Orchard, I., Loughton, B.G. (1981) Is octopamine a transmitter mediating hormone release in insects? *J. Neurobiol.* **12**: 143–153

Osman, M.Z. (1993) Effects of neem seed extracts on growth and development of larvae of *Pieris brassicae* L. (Lepidopetra, Pieridae). *J. Appl. Entomol.* **115**: 254–258

Pannabecker, T., Orchard, I. (1986) Octopamine and cAMP mediate release of adipokinetic hormone I and II from isolated locust neuroendocrine tissue. Mol. *Cell Endocrinol.* **48**: 153–159

Ramirez, J.-M., Pearson, K.G. (1991) Octopaminergic modulation of interneurons in the flight system of the locust. *J. Neurophysiol.* **66**: 1522–1537

Redfern, R.E., Kelly, T.J., Borkove, A.B., Hayes, D.K. (1982) Ecdysteroid titers and molting aberrations in last stage *Oncopeltus* nymphs treated with insect growth regulators. *Pest. Biochem. Physiol.* **18**: 351–356

Reese, J.C., Beck, S.D. (1976) Effects of allelochemics on Black Cutworm, *Agrotis ipsilon*: effects of *p*-benzoquinone, hydroquinone and duroquinone on larval growth, development and utilization of food. *Ann. Ent. Soc. Am.* **69**: 59–67

Rembold, H., Uhl, M., Müller, T. (1987) Effect of azadirachtin A on hormone titres during the gonadotrophic cycle of *Locusta migratoria*. In: Schmutterer, H., Ascher, K.R.S. (eds) Proceedings of the 3rd International Neem Conference. German Agency for Technical Cooperation, Eschborn, Germany, pp. 289–298

Reynolds, S.E., Wing, K.D. (1986) Interactions between azadirachtin and ecdysteroid-dependent systems in the tobacco hornworm, *Manduca sexta*. Abstracts Vth International Congress on Pesticide Chemicals. IUPAC, Ottawa, Canada, P.2 D/E–08

Schlüter, U., Bidmon, H.J., Grewe, S. (1985) Azadirachtin affects growth and endocrine events in larvae of the tobacco hornworm *Manduca sexta*. *J. Insect Physiol.* **31**: 773–777

Sieber, K.-P., Rembold, H. (1983) The effects of azadirachtin on the endocrine control of moulting in *Locusta migratoria*. *J. Insect Physiol.* **29**: 523–527

Simmonds, M.S.J., Blaney, W.M., Ley, S.V., Anderson, J.C., Toogood, P.L. (1990) Azadirachtin: structural requirements for reducing growth and increasing mortality in lepidopterous larvae. *Entomologia Exp. Appl.* **55**: 169–181

Smith, S.L., Mitchell, M.J. (1988) Effects of azadirachtin on insect cytochrome P-450 dependent ecdysone 20-mono oxygenase activity. *Biochem. Biophys. Res. Commun.* **154**: 559–563

Subrahmanyan, B., Müller, T., Rembold, H. (1989) Inhibition of turnover of neurosecretion by azadirachtin in *Locusta migratoria*. *J. Insect Physiol.* **35**: 493–500

Sun, X.J., Tolbert, L.P., Hildebrand, J.G. (1993) Ramification pattern and ultrastructure characteristics of the serotonin-immunoreactive neuron in the antennal lobe of the moth *Manduca sexta*: a laser scanning confocal and electron microscope study. *J. Comp. Neurol.* **338**: 5–16

Takeda, N., Takaoka, H., Shimizu, T., Yazawa, M., Yagi, S. (1991) Biogenic amine levels in the central nervous system and haemolymph of the silkworm, *Bombyx mori*. *Comp. Biochem. Physiol.* **100C**: 667–682

Thompson, C.S., Yagi, K.J., Chen, Z.F., Tobe, S.S. (1990) The effects of octopamine on juvenile hormone biosynthesis, electrophysiology and cAMP content of the corpora allata of the cockroach *Diploptera punctata*. *J. Comp. Physiol.* **160B**: 241–249

Woodring, J.P., Meier, O.W., Rose, R. (1988) Effect of development, photoperiod and stress on octopamine levels in the house cricket *Acheta domestica*. *J. Insect Physiol.* **34**: 759–765

Exp. Biol. Online 2:16 (1997)

Experimental simulation in behavioral ecology: a multimedia approach with the spatial searching simulation RattleSnake©

D. Duvall[1], D. Chiszar[2], R.A. Mintzer[3], and N.O. Roster[1]

[1]Department of Zoology, Oklahoma State University, Stillwater, OK 74078, USA
[2]Department of Psychology, University of Colorado, Boulder, CO 80309, USA
[3]Franklin McLean Memorial Research Institute, University of Chicago, 5801 South Ellis, Chicago, IL 60637, USA

Correspondence to: D. Duvall (e-mail: duvalld@okway.okstate.edu)

Received: 26 March 1997 / Accepted: 6 October 1997
Published: October 30, 1997

Abstract. Searchers in nature often have accurate knowledge of the spatial location of the resource targets they seek, though in many other cases they have none. For example, the spatial distribution of targets such as food patches or potential mates may shift or change unpredictably from season to season. Searchers encountering circumstances of these sorts may be said to be "naive". This problem is compounded by the fact that spatial distributions of targets may vary statistically as well: they may be distributed randomly, uniformly, or they may be clustered. Accordingly, since we study an animal system in nature that encounters such challenges (i.e., free-ranging rattlesnakes in many parts of their range), we wrote a comprehensive spatial searching program for Macintosh systems that simulates this problem thoroughly, RattleSnake©. In a large series of experimental simulations using this software, we found that search paths of high vector magnitude (approaching 1.0), or those that approached straight lines, generated large numbers of collisions in large, clustered worlds. No search path was any better than any other in large, randomly or uniformly distributed worlds. Zig-zag paths of low vector magnitude (approaching zero) in small worlds of all types and of all densities were efficacious, due to continuous turning which prevented searchers from moving out of or exiting patches. Thus it appears that there are design rules in nature governing target collision probabilities in some but not all two-dimensional spatial worlds. Search paths of high vector magnitude, or those approaching straight lines, generate high collision frequencies in statistically clustered spatial worlds, for example.

RattleSnake© thus may be useful in programs of basic and/or applied behavioral ecology, including conservation, as well as in laboratory and multimedia classroom education.

Key words: Macintosh simulation – Rattlesnakes – Searching behavior

Introduction

Prairie rattlesnakes in Wyoming searching for important resource units or "targets" are like many other kinds of organisms in that the location of the targets is often unknown. However, the mechanistic design of searchers should reflect current selection pressures for search paths that maximize the location of targets whose position is unknown. Some examples of these types of searchers are male rattlesnakes in Wyoming searching for spatially sparse, widely distributed, and unpredictable receptive females, shore birds searching at sea for constantly shifting and moving kelp beds or schools of fish, or even the gametic propagules of many kinds of wind-pollinated plants. Searchers of this kind are "naïve". These are not the omniscient searchers typically modeled and studied in optimality theory. As discussed by Pyke (1978), Stephens and Krebs (1986), Smallwood and Cartar (1996) and many others, most optimality models assume that searchers are all-knowing, though this condition is unlikely to be met in many cases in nature. This is not to say that these models are not incredibly useful; they are. It is simply that more attention is needed to be paid to the problems that naivesearchers face as they likewise attempt to optimize the costs and benefits of searching for key resource and habitat units. Bell (1991) provides an excellent text on searching behavior. Lima and Zollner (1996) also discuss this matter.

Returning to the case of male prairie rattlesnakes searching for female mates in Wyoming, courtships and copulations typically occur several kilometers away from the population's communal over-wintering den. Females become receptive later in the summer season as they search for widely dispersed and spatially unpredictable patches, or clusters of deer mice (Duvall *et al.*, 1990; King and Duvall, 1990). Most males fail to locate even a single receptive female in most seasons (Duvall and Schuett, 1997), so the problem they face is considerable.

We have developed and now conducted experiments with the general, though precise, Macintosh spatial searching simulation RattleSnake©. The simulation considers completely the problem of how a naivesearcher should best travel in two dimensions when seeking targets of unknown location. QuickTime Demonstration (935 KB) shows how this software works. Herein we introduce and describe the operation of RattleSnake point-and-click simulation software, as well as its application. We designed several experiments to consider (1) the efficacy of some distinctive search patterns exhibited by prairie rattlesnakes in Wyoming, a system we have studied thoroughly, and (2) the general problem of how to search for targets in uniform, random, or clustered, two-dimensional spatial distributions, where target locations are unknown or when the searcher is naïve. Uniform, random, and clustered distributions in two-dimensional space are defined by the following relationships in target occurrence per unit space: $\mu>\sigma^2$, $\mu=\sigma^2$, $\mu<\sigma^2$, respectively (Duvall and Beaupre, 1997). Many introductory statistics texts address this issue (see Zar, 1996).

Aside from variation in density, population and sample statistical relationships between μ and σ^2 can describe all possible two-dimensional relationships in target distributions in space and time (Zar, 1996). Accordingly, five different sets of experimental simulations were run to assess the efficacy of variation in preprogrammed search geometry in several large samples of individual, naivevirtual rattlesnakes. These virtual rattlesnakes were run through worlds characterized by uniform, random, or clustered spatial distributions of targets. The experiments examined:

1. "Small worlds", with random, uniform, and clustered resource distributions. In this spatial array we predicted that searchers with low vector magnitude (low **r**), exhibiting zig-zag random walks, would be the most successful at colliding with targets. The variable "**r**" is a geometric vector with values ranging from 0.0 to 1.0.
2. "Large worlds", with random, uniform, and clustered resource distributions. It was assumed that searchers traveling in the random and clustered distributions would not be made more or less effective by following any search vector. Conversely, in a clustered distribution, searchers with a high **r** (approaching a straight-line bearing) would be most successful at hitting discrete, dispersed targets.
3. "A closer look at vector magnitude", where we considered in more detail the effects of a wider range of **r** values on collision success in large random or clustered worlds, given the results of experiments 1 and 2 above.
4. "The ratio of step length to season length", where we asked how variation in the ratio of these two parameters affects searcher success in large worlds with clustered resources, again given the anticipated results of previous experiments.
5. "Target radius", which emerged in the course of our experiments as the key general parameter affecting collision success of naivesearchers in clustered worlds.

General rationale and methods

All experiments were done with RattleSnake©, a comprehensive 68K Macintosh program designed to assess the efficacy of variation in: (1) two-dimensional geometric movement paths of naivevirtual rattlesnake searchers; and (2) in target distributions. Though not optimized for it, RattleSnake© will run in the PowerPC environment as well. Newer versions are being developed. Written in floating point computation in "C" and Apple "SANE" (Standard Apple Numerics Environment, Apple Computer, 1988), RattleSnake© first sets out the specified world size, as well as target size, distribution, and density. Next, any number of individual independent searchers are sent out into the world along vector paths of any desired step-length, distance, speed, and two-dimensional geometry, the latter ranging from 0.0 (random walk) to 1.0 (a perfectly straight line). Finally, virtual searcher performance then is scored on many different dependent variables and compiled (Fig. 1). The simulation and pseudocode can be supplied at nominal costs, to established investigators and educators, to D. Duvall and R.A. Mintzer.

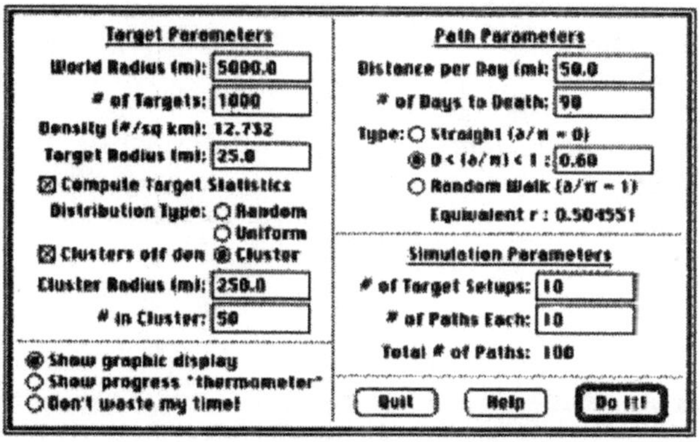

Fig. 1 RattleSnake© control screen including variables, constraints, and output. The program is designed to analyze target, path and bearing, and distribution parameters for naivesearchers in two-dimensional space.

When running RattleSnake©, continuous but precisely specifiable parameters such as world, target and cluster size, search geometry, and the like are entered in the Control Screen (Fig. 1). In all cases, unless otherwise noted, searchers were programmed to move a distance of 50 m per day in a 270-day season, and along the following preprogrammed vector bearings (r): 0.0, 0.2, 0.4, 0.6, 0.8, and 1.0. These values derive from 11 years of fieldwork on prairie rattlesnakes in Wyoming. The number of target setups, or independent replicates, was set at 10, and the number of individual searchers per setup was also 10. This resulted in samples of 100 statistically independent, virtual rattlesnake searchers per simulation. Depending upon the experiment, (see below) worlds were either small (1000 m radius) or large (5000 m radius). If a searcher reached the world boundary, the searcher "fell off" and died. World boundaries are thus nonreflective.

Given the complete and general structure of our simulation, a searcher bouncing off reflective world boundaries is no different than simply running an additional virtual searcher. All designs in all experiments were multi-factor in construction, and completely randomized and orthogonal. All data were analyzed using χ^2 (Li, 1964) as an exact equivalent of analysis of variance and all p values are reported.

Experiment 1 (small worlds)

Rationale and methods

In a world that is small, or one in which resources are rich or dense, a searcher should never travel far away. Searchers should follow vectors of low magnitude to maximize collision success. This would maximize turning behavior and reduce the likelihood of exiting the patch (see Pyke, 1978).

A world with a radius of 1000 m was used, and random, uniform, and clustered target distributions were examined. In the random case (simulation 1), targets had a radius of 25 m and the number of targets was 1, 10, 25, 50, 75, 100, or 1000. The uniform case (simulation 2), utilized targets that were 25 m in radius and the number of targets was 10, 25, 50, 75, 100, or 1000. In the clustered case (simulation 3), target radius was 25 m and cluster radius was 250 m with five targets per cluster. The number of clusters was 2, 5, 10, 15, or 20 in as many different groups.

Results and discussion

The simulation shows that in a small world with resources that are distributed randomly (simulation 1), a searcher with an r value close to 0 (randomly walking) is more successful at hitting targets compared with one following a vector bearing approaching 1.0 ($p<0.001$; Fig. 2). Not surprisingly, searchers are also more successful at hitting targets when the number of targets increases ($p<0.001$). In a small world with uniformly distributed resources (simulation 2), randomly walking searchers (r approaching 0) were significantly more successful at hitting targets ($p<0.001$; Fig. 3). Finally, in a small world where resources are clustered (simulation 3; Fig. 4), searchers were more successful at hitting targets when traveling along random walks versus straight-line paths ($p<0.001$). Searchers were also more successful as the number of targets increased ($p<0.001$), but this, again, is not surprising.

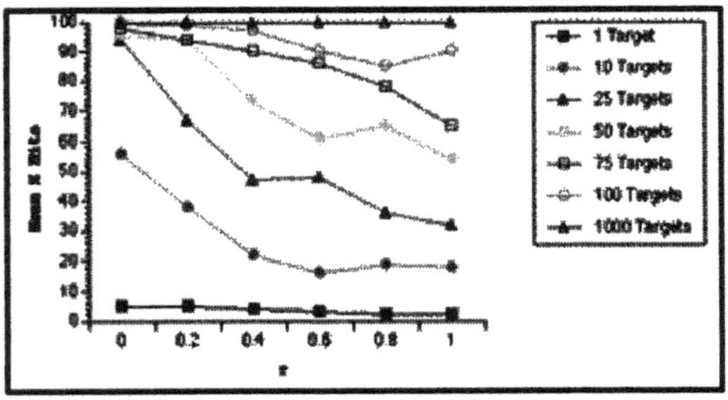

Fig. 2 Portrayed here are mean per cent hits, or collisions for naivesearchers traveling along six different vector bearings (r) in a small world with different numbers of randomly distributed

resource targets. See text for additional discussion and definition of terms.

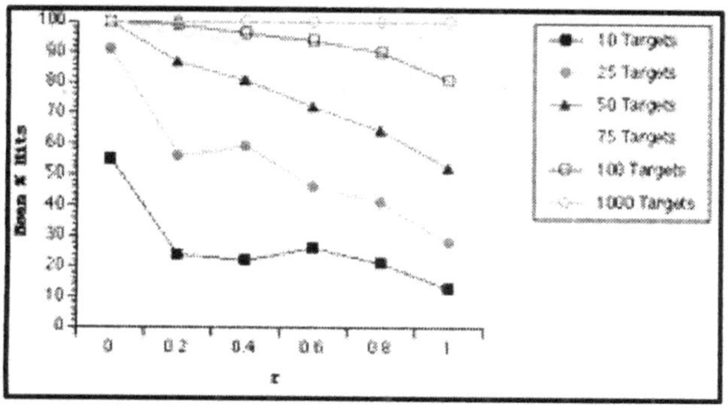

Fig. 3 Mean per cent hits, or collisions, of searchers assigned to differing vectors in a small world with uniformly distributed resources. See Figs. 1 and 2 and text for additional discussion.

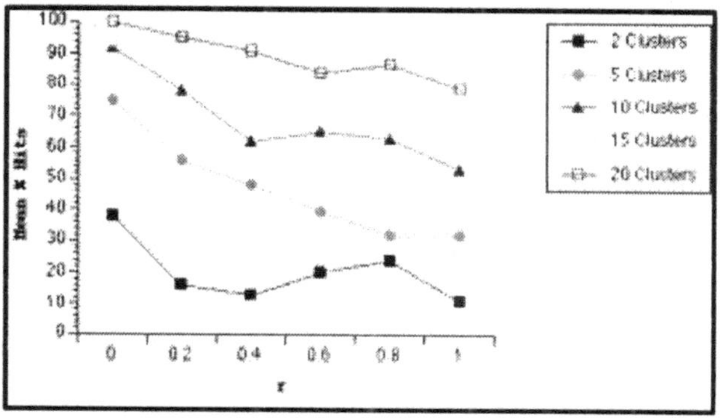

Fig. 4 Mean per cent hits, or collisions, of searchers assigned to differing vectors in a small world with clustered resources. See Figs. 1–3 and text for additional discussion.

RattleSnake© thus shows that when the world is effectively small, search vectors approaching 0 (random walks) represent the best tactic whether resources are distributed randomly, uniformly or are clustered. Frequent turning movements also keep searchers from leaving a patch (or "falling off" the world).

Experiment 2 (large worlds)

Rationale and methods

In large worlds, or ones in which resources are more scarce or spread out, searchers will have to explore a larger area to find the same number of resource units that they would encounter in a small world. An efficient searcher, thus, will be one that can find resources quickly and with some consistency. Increasing the number of targets in the three distributions will have less of an effect since the targets will be spread out over a larger area.

Large worlds were similar to those in Experiment 1, but with some key differences. World radius, for example, was increased to 5000 m. The tests of clustered distributions in large worlds were also altered slightly. Clustered tests were comprised of two densities. The first (simulation 6a), had 10 targets per cluster and the number of clusters was set at 1, 10 and 50, and the second (simulation 6b) had 50 targets per cluster, with the number of clusters the same as in simulation 6a.

Results and discussion

In large worlds, RattleSnake© shows that in random (simulation 4; Fig. 5) or uniform (simulation 5; Fig. 6) distributions, a searcher is equally likely to hit targets when moving along paths of any vector. There is no benefit to randomly walking or following search paths of high vector magnitude. As the number of targets reaches 1000 there is an unsurprising significant increase in searcher success (p<0.001).

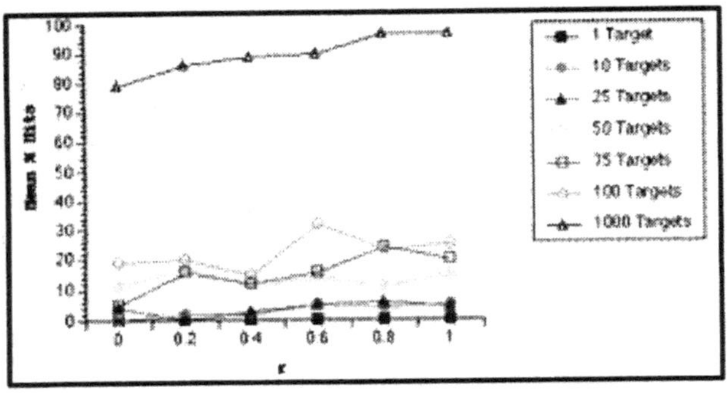

Fig. 5 Mean per cent hits, or collisions, of searchers assigned to differing vectors in a large world with randomly distributed resources. See Figs. 1–4 and text for additional discussion.

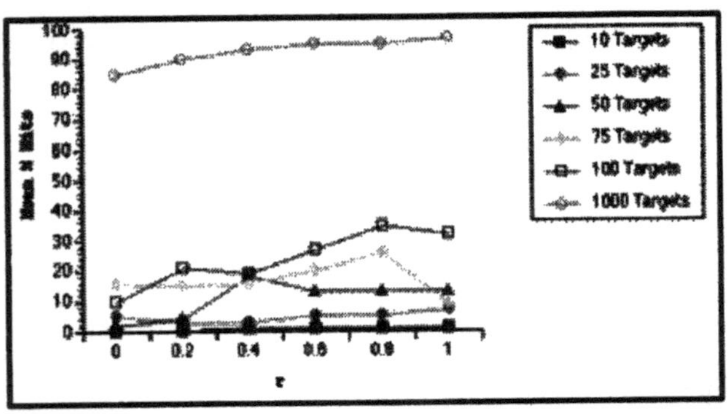

Fig. 6 Mean per cent hits, or collisions, of searchers assigned to differing vectors in a large world with uniformly distributed resources. See Figs. 1–5 and text for additional discussion.

This significant increase in searcher success as the number of targets increases is true also in large worlds where the resources are clustered (simulations 6a and

6b). Likewise, there is a significant increase in searcher success (Figs. 7 and 8) as r approaches 1 ($p<0.001$). In both simulation 6a and 6b, where there are 10and 50 targets per cluster, respectively, searcher success increased significantly when (1) the number of targets ($p<0.001$), (2) the number of clusters ($p<0.001$), and (3) the number of targets per cluster increased ($p<0.05$).

It was found that straight-line travelers in clustered large worlds, however, were uniquely and significantly more able to find targets of unknown location regardless of target density ($p<0.001$; Figs. 7 and 8). These results corroborate what has been observed in studies of prairie rattlesnakes in Wyoming: both (1) males and females; and (2) males alone showed lengthy straight-line movements when searching for widely distributed and spatially unpredictable deer mice and receptive females (King and Duvall, 1990; Duvall and Schuett, 1997), respectively. It appears as though the tactic of searching naïvely along a bearing of high vector magnitude may be an effective plan or rule of thumb in nature, given particle collision dynamics in clustered two-dimensional space.

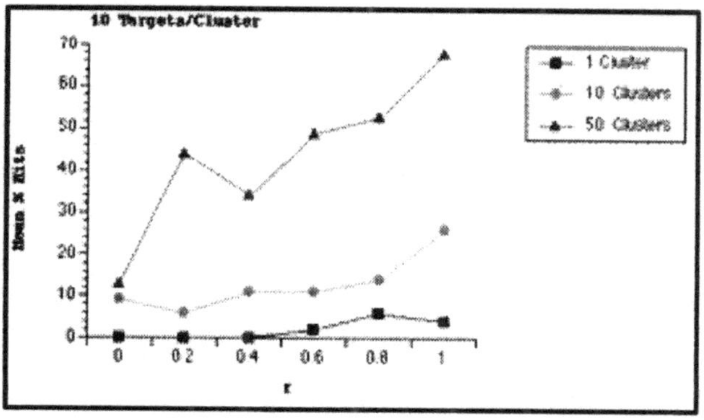

Fig. 7 Mean per cent hits, or collisions, of searchers assigned to differing vectors in a large world of clustered resources with ten targets per cluster. See Figs. 1–6 and text for additional discussion.

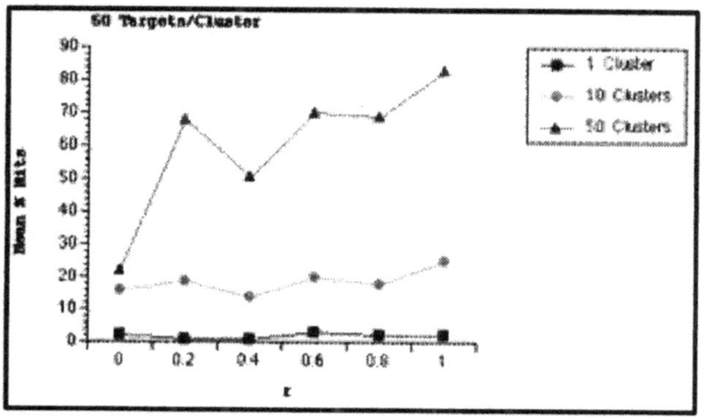

Fig. 8 Mean per cent hits, or collisions, of searchers assigned to differing vectors in a large world of clustered resources with 50 targets per cluster. See Figs. 1–7 and text for additional discussion.

Experiment 3 (a closer look at r)

Rationale and methods

Experiment 3 was intended to take a closer look at increased resolution in **r**, by increasing the number of vector bearing levels studied. It is important to know if vector bearing (**r**) and average collisions co-vary positively and in a linear manner. Moreover, this potential co-variance was considered in both random and clustered worlds (simulations 7a and 7b). The results in experiments 6a and 6b (see Figs. 7 and 8) hinted at a second-order polynomial function at **r** = 0.4 in the higher-density world. This needed to be checked and an increased resolution in **r** seemed a good way to proceed.

Simulations 7a and 7b examined random and clustered distributions, respectively. In each simulation, **r** varied by increments of one-tenth, generating 11 bearing values from 0.0 to 1.0. The season length was lowered to 90 days and the number of searchers per replicate was set to 36 with 10 target setups for a total of 360 searchers. Additionally in simulations employing random distributions, 25-m targets were arrayed in densities of 50 or 500 targets. The clustered trial had 25-m targets with a cluster radius of 250 m. Finally, the number of clusters was set to 5 or 50.

Results and discussion

In the random simulation, and again not surprisingly, searcher success increased when the number of targets increased from 50 to 500 (Fig. 9). Searcher success, however, did not change as r varied. However, in large worlds with clustered resources, the number of collisions increased as **r** increased ($p<0.01$). Even as the searcher moved only slightly away from a random walk (r=0), hits increased (Fig. 10).

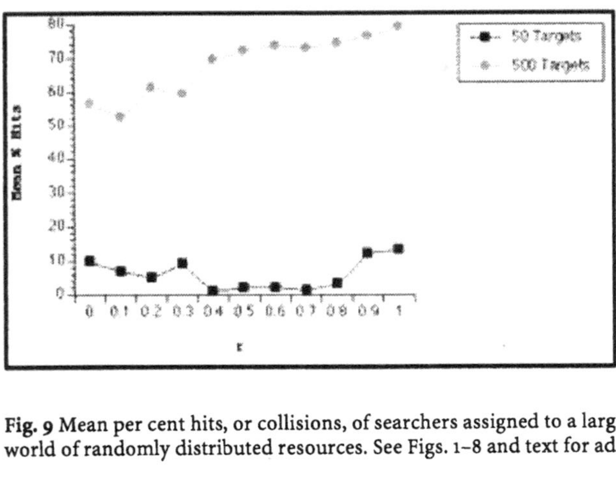

Fig. 9 Mean per cent hits, or collisions, of searchers assigned to a larger range of vectors in a large world of randomly distributed resources. See Figs. 1–8 and text for additional discussion.

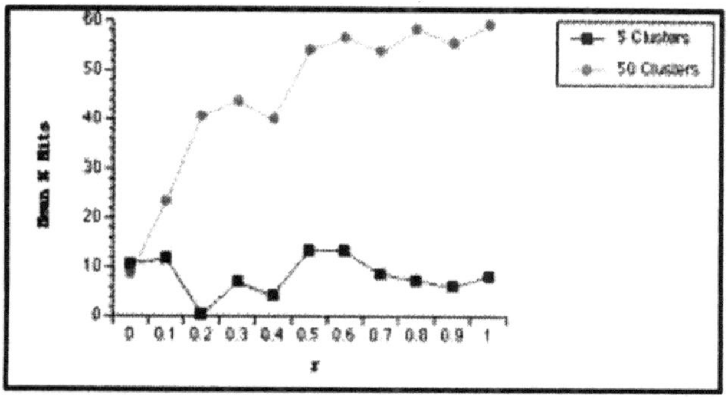

Fig. 10 Mean per cent hits, or collisions, of searchers assigned to a larger range of vectors in a large world with clustered resources. See Figs. 1–9 and text for additional discussion.

Experiment 3 shows that, in large worlds with randomly distributed resources, increased vector magnitude does not generate increased collisions. However, in the clustered case they did. It appears that an almost one-to-one, positive, linear association exists between increasing vector magnitude and increasing hit rates (Fig. 10). Likewise, the dip at $r = 0.4$ was dampened, suggesting that no systematic effects exist here.

Experiment 4 (the ratio of step length to season length)

Rationale and methods

Experiment 4 was designed to consider the effect of the ratio of step length (SL), or the distance crossed per movement, to season length (SeL) on collision numbers. We predicted that whenever SL:SeL ratio values amounted to a lengthy straight-line movement, this would increase searcher success by artificially straightening out a searcher's path. That is, vector bearing is effectively high when animals travel great distances per movement regardless of the preprogrammed r to which they are assigned. Pyke (1978) imagined other connections between these variables.

All other conditions in this experiment were similar to those employed above. However, the key difference is the use of the SL:SeL ratio. The SL:SeL ratios studied included: 1:4500, 25:180, 50:90, 100:45, 500:9. Thus all searchers covered the same distance, but in varying numbers of movements. Additionally, the use of 50 clusters with 50 targets per cluster for a total of 2500 targets rendered densities the same as those in previous experiments.

Results and discussion

RattleSnake© demonstrates that fixed bearing paths will always generate the highest collision rates, whether these paths result from many short movements along a fixed vector, or small numbers of lengthy movements of no particular vector. These results thus corroborate and virtually replicate experiments 2 and

4 above (Fig. 11).

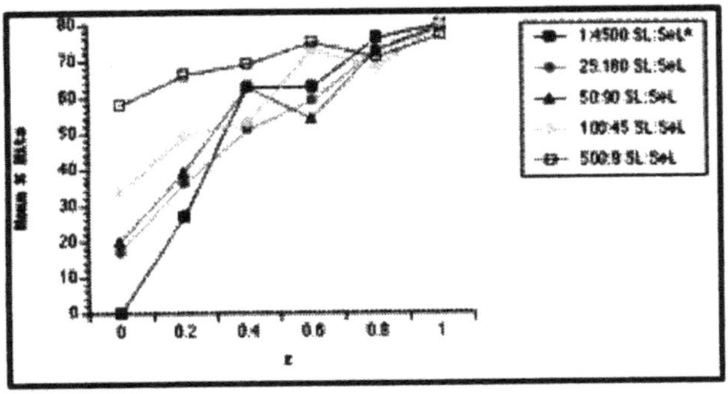

Fig. 11 Mean per cent hits, or collisions, of searchers assigned to differing vectors and differing ratios of step length (*SL*) to season length (*SeL*)* in a large world with clustered resources. See Figs. 1–10 and text for additional discussion.

Experiment 5 (target size)

Rationale and methods

Experiment 5 was designed to consider the effects of variation in target size on collision success. As targets become larger they become equivalent to clusters of the same diameter, thus, should be similarly easier to hit. This may be the general mechanism underlying the positive effect of searching along increasingly fixed vector paths in clustered worlds. In experiment 5 searchers were tested in worlds with targets of three different radii: 2.5 m, 25 m and 250 m. Otherwise this set of simulations was similar to others described above.

Results and discussion

Increasing target radius generated a corresponding and significant effect on collision success ($p<0.01$; Fig. 12). This suggests that at least one general mechanism behind the overall positive association between increasing vector magnitude and collision success in clustered worlds is based more upon variation in cluster or target radius than on variation in target density. However, a significant interaction between target radius and increasing vector magnitude was also detected ($p<0.01$), indicating that the positive effects of higher vector magnitude were most pronounced in worlds where target radius was 250 m (Fig. 12).

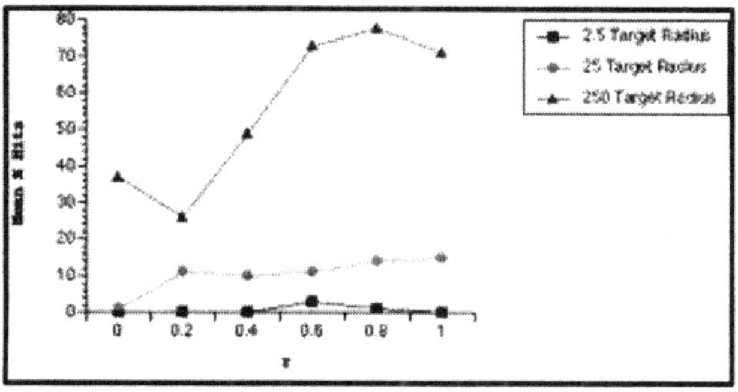

Fig. 12 Mean per cent hits, or collisions, of searchers assigned to differing vectors in a large world with randomly distributed resources of differing radii. See Figs. 1–11 and text for additional discussion.

These findings are not surprising. Simulation 4 is essentially identical to this experiment. Specifically, variation in r in large worlds with randomly distributed targets had no effect on collision success. However, when targets had a radius of 250 m, searcher success increased significantly as r approached 1 (Fig. 12). Since these targets are extremely large, they become equivalent to clusters of similar diameter. As observed in simulation 6, in a large world with clustered resources, a naivesearcher that exhibits a higher r will be more effective at hitting targets.

Discussion and prospects

RattleSnake© has been a useful tool in modeling the efficacy of variation in search paths and bearings, when the position of targets or resource units is unknown. Lack of omniscience may arise from little or no knowledge of target location, shifting or changing spatio-temporal location of targets, or a variety of other causes. Studies like ours, that assume naivesearching, represent a qualitative and useful departure from many other such models in which perfect knowledge is assumed. As discussed above, we have found predictions derived from RattleSnake© to be helpful in our empirical studies of rattlesnakes in nature (see King and Duvall, 1990; Duvall and Beaupre, 1997; Duvall and Schuett, 1997). Resource unit clustering in Wyoming has been confirmed, including both food patches and potential mates, and snakes there follow search paths of high vector magnitude (King and Duvall, 1990). Moreover, the sexes have even been found to differ predictably in these respects (Duvall and Beaupre, 1997), in accordance with differing selective forces for searching abilities. It also helps that computer simulation approaches pure, brute-force mathematical simulation in power (see Duvall and Beaupre, 1997). These sorts of claims are supported by our demonstration herein that at least one kind of search path (straight lines), in at least one kind of world (clustered, and of various densities), generates more collisions with targets of unknown location than any other. This is interesting for several reasons. First, many natural biological communities evidence spatial distributions of key resource targets of this sort. They are not at all uncommon. The rattlesnake communities we have studied in Wyoming seem to exhibit some of these properties and, likewise, many of their searching and orientation tendencies are consistent with our models. Second, it suggests that some sort of deterministic collision geometry must exist for potentially interacting particles in most and perhaps all clustered distributions (those in two dimensions). Though we demonstrate its general existence herein in a way that is potentially useful to ecologists and biologists of many stripes, we make no other claims and nor do we derive it mathematically. For our purposes, it is useful simply to demonstrate that threshold values that may amount to selection gradients occur for collision dynamics in two dimensions. This may lead, in many cases, to various sorts of animal taxis, kinesis, orientation patterns, navigational capacities, and the like. Thus, both the ease of simulation and comprehensive as well as the general scope of RattleSnake© suggest that others might also find this simulation useful. For example, Lima and Zollner (1996) advocate application of related simulations in landscape ecology.

As biology teachers and conservationists, we also see potential use for software like ours in the teaching laboratory and in management and conservation planning. For example, for a variety of reasons, it is increasingly the case that sufficient numbers of wild animals for study are not available. This sometimes derives from matters of feasibility, ethos and mores, and/or economic necessity. Large and powerful simulations, such as RattleSnake©, that allow studies of any number of virtual searchers, that are exceptionally easy to operate and are comprised of intuitive multimedia user interfaces may find broad and general application.
Finally, the graphical nature of our simulation as well as the point-and-click Apple Macintosh environment lend themselves nicely to publication in multimedia venues such as *Experimental Biology Online*.

Acknowledgements. We thank the University of Wyoming, Arizona State University, and Oklahoma State University for financial support.

References

Apple Computer (1988) Standard Apple Numerics Environment, 2nd edn. Addison-Wesley, Reading, Mass.

Bell, W.J. (1991) In: Broom, D.M., Colgan, P.W. (eds) *Searching behaviour: the behavioural ecology of finding resources.* Chapman and Hall, London.

Duvall, D., Beaupre, S.J. (1997) Sexual strategy and size dimorphism in rattlesnakes: integrating proximate and ultimate causation. *Am. Zoologist* (in press)

Duvall, D., Schuett, G.W. (1997) Straight-line movement and competitive mate searching in prairie rattlesnakes, *Crotalus viridis viridis. Anim. Behav.* **54**: 329–334

Duvall, D., Goode, M.J., Hayes, W.K., Leonhardt, J.K., Brown, D.G. (1990) Prairie rattlesnake vernal migration: field experimental analysis and survival value. *Natl Geog Res* **6**: 457–469

King, M.B., Duvall, D. (1990) Prairie rattlesnake seasonal migrations: episodes of movement, vernal foraging and sex differences. *Anim. Behav.* **39**: 924–935

Li, J.C.R. (1964) Analysis of variance versus Chi-square test. In : *Statistical inference*. Edwards Brothers, Ann Arbor, pp. 471–484

Lima, S.L., Zollner, P.A. (1996) Towards a behavioral ecology of ecological landscapes. *Tree* **11**: 131–135

Pyke, G.H. (1978) Are animals efficient harvesters? *Anim. Behav.* **26**: 241–250

Smallwood, P.D., Cartar, R.V. (1996) Risk sensitivity in behavioral ecology: where are we now? Introduction to the symposium. *Am. Zool.* **36**: 389–391

Stephens, D.W., Krebs, J.R. (1986) *Foraging theory*. Princeton University Press, New Jersey

Zar, J.H. (1996) *Biostatistical analysis*. Prentice Hall, New Jersey

Exp. Biol. Online 2:17 (1997)

Effects of Diazinon on bluegill sunfish, *Lepomis macrochirus*, gills: scanning electron microscope observations

H. M. Dutta[1], J. S. D. Munshi[2], P. K. Roy[2], N. K. Singh[2], L. Motz[1] and S. Adhikari[2]

[1]Department of Biological Sciences, Kent State University, Kent OH 44242 USA
[2]Postgraduate Department of Zoology, Bhagalpur University, Bhagalpur, India

Correspondence to: H.M. Dutta (email: hdutta@kentvm.kent.edu)

Received: 29 April 1997 / Accepted: 13 November 1997
Published: November 19, 1997

Abstract. Gills of bluegill sunfish, *Lepomis macrochirus*, exhibited varied degrees of structural damage following a 24-h exposure to sublethal concentrations (15 µg/l, 30 µg/l, 45 µg/l, 60 µg/l and 75 µg/l) of Diazinon [*O,O*-diethyl-*O*-(2-isopropyl-6-methyl-4 pyrimidinyl ester or phosphorothioate]. Exposure to 15 µg/l and 30 µg/l resulted in exocytosis of some material to the cell surface and perforations of the microridges. At higher doses (above 45 µg/l), the extrusion was reduced and the cells were swollen. Compared to control values, the thickness of the microridge on the gill arch and on the gill filament generally increased with exposure to Diazinon. Also, the distance between microridges decreased with increased exposure concentrations. At 60 µg/l, gill arch microridges fused and some ridges of gill filaments disappeared. At 75 µg/l exposure, epithelial cells of the gill arch became obscured with severe cellular extrusions and the lamellar surfaces swelled. The mucus extrusion, lamellar swelling and reduced microridges may be related to a defence mechanism which reduces the water surface around the gill and increases the barrier distance for diffusion of toxicants from outside to the blood capillaries. Although this mechanism protects the fish from toxicants, it also reduces the oxygen supply which leads to suffocation of the fish.

Key words: Bluegills – Diazinon – Gills – *Lepomis macrochirus* – Scanning electron microscopy

Introduction

Diazinon (*O*, *O*-diethyl-*O*-(2-isopropyl-6-methyl-4-pyrimidinyl ester, or phosphorothioate), a commonly used organophosphate pesticide, may damage or kill non-target organisms, especially fish. Gills of fish are potentially vulnerable to waterborne toxicants, and respiratory distress is one of the earliest indications of pesticide poisoning (Murthy 1986). There are some histopathological (light microscopy) studies about the effects of pesticides on fish gills (Munshi and Singh 1971; Reddy *et al.* 1977; Richmonds and Dutta 1989; Dutta *et al.* 1992, 1993), and some electron microscopy studies of fish gill surface morphology following their exposure to crude oil (Engelhardt *et al.*, 1981; Prasad 1989, 1991), zinc sulphate (Crespo 1982), cadmium (Karlsson *et al.* 1985), acid water and aluminum (Karlsson *et al.* 1986; Munshi and Singh 1992), acid and aluminum (Tietge *et al.* 1988), plant extracts (Roy *et al.* 1986, 1990), and exposure of tight junctions in gills to acid (Freda *et al.*1991). Few studies are available on the effects of pesticides on gill surface morphology (Hart and Oglesby 1979; Roy and Munshi 1991). Histopathology appears to be a very sensitive parameter and is crucial for determining the cellular damage that occurs in the target organs. Scanning electron microscopy (SEM) is an important tool for revealing the surface ultrastructure of various organs, including gills. Damage to surface ultrastructures cannot be identified by light or transmission electron microscopy; therefore, it was necessary to use SEM to reveal changes and damage that may occur in microridged epithelial, chloride and mucous cells of the gill. This study investigated changes to the surfaces of the gill of bluegill sunfish, *Lepomis macrochirus*, following exposure to Diazinon.

Materials and methods

Bluegills, *Lepomis macrochirus*, were obtained from a fish hatchery near Baltic, Ohio, USA. This source was selected for the supply of fish, since there is no agricultural land nearby. In this study, five mature female bluegills (total length 12–13.5 cm; weight 32–45 g) for each exposure concentration were used. A control group of five fish was also kept under similar conditions without the Diazinon. Fish were acclimatized to the laboratory condition in several 100-gallon glass aquaria for 7–8 weeks. Fish were well fed twice daily with live guppies and commercially balanced fish foodsticks prior to experiments.

City water was used in the aquaria, having been dechlorinated in a reservoir. All aquaria were fitted with large-sized Whisper Power filters (Whisper Power Tetera Second Nature, Willinger Brothers, USA). Water temperature and dissolved oxygen concentrations were determined using a YSI model 50 oxygen meter. Total hardness and alkalinity were estimated with a digital titrator and Hach's water analysis kit. Water temperature was 21 ± 1°C and a 14:10 h light:dark cycle was maintained by using auto-control fluorescent lighting for the hulding and experimental periods. Water quality data were as follows: pH 7.25 ± 0.16, dissolved oxygen 6.91 ± 0.38 mg/ l, total hardness 125.55 ± 3.35 µg/l as

$CaCO_3$, alkalinity 34.57 ± 1.63 µg/l as $CaCO_3$.

Commercial Diazinon produced by Ortho Chevron Chemical (California, USA) was used for this study. Commercial-grade Diazinon has 25% active Diazinon and 57% aromatic petroleum derivative solvent and 18% inert ingredients. A test solution with the desired concentrations of Diazinon was calculated using a 25% active gradient and was prepared with a previously stocked solution of Diazinon within the expiry date, according to the appropriate tank capacity. The experimental water was kept in the test tank for at least 24 h without any large-sized Whisper Power filters. Only an air tube was placed in the water during the experiment.

On the basis of the LC_{50} (120–220 for a 96-h exposure) of the Diazinon for bluegills (Johnson and Finley 1980), the fish were exposed to different sublethal concentrations of Diazinon, i.e.15 µg/l, 30 µg/l, 45 µg/l, 60 µg/l and 75 µg/l for 24 h. The exposure water from the tank was analysed for its Diazinon concentration at the beginning, middle and end of the experiment by high-performance liquid chromatography (HPLC). After the exposure period, the fish were anaesthetized with 100 mg/l MS222 and 200 mg/l sodium bicarbonate (Sigma) and gills from five control and five experimental fish were excised. Samples of the second gill arch from the right side were immediately excised. Gill tissues were fixed in 2.5% glutaraldehyde in 0.1 M phosphate buffer at pH 7.4 for 24 h. The samples were washed in 0.1 M phosphate buffer (pH 7.4), and dehydrated in ethanol and acetone/anhydrous acetone. The dehydrated tissue was critical-point dried, gold coated and examined under a Cambridge Stereoscan 100 scanning electron microscope. Microridge thickness and distance between microridges of the gill arch and gill filament were taken automatically (randomly) from the pictures on the stereoscan screen. All measurements were taken in a horizontal direction. The images were rotated to align the feature of interest in the horizontal direction. Ten measurements from each tissue of five exposed fish samples per concentration and five control fish were taken. Measurements of the microridges (thickness and distances between them) of the gills exposed to 15 µg/l and 30 µg/l Diazinon could not be taken because of their corrugated appearances.

Statistical comparisons were carried out using Student's *t*-test (Table 1).

Table 1 Morphometric differences between control and experimental measurements between control and exposed fish. (*RGA* Thickness of the microridges on gill arch, *DBGRA* distance between the microridges on gill arch)

	Variables	No. of samples	Mean ± SD (µm)	*t*-value	Two-tail probability
	Control versus different concentrations:				
1.	RGA control	50	0.315 ± 0.008		
	RGA 45 µg/l	50	0.290 ± 005	2.92	0.017
	RGA 60 µg/l	50	0.372 ± 0.015	−3.28*	0.010
2.	DBRGA control	50	0.388 ± 0.018		
	DBRGA 45 µg/l	50	0.291 ± 0.005	5.04	0.001
	DBRGA 60 µg/l	50	0.256 ± 0.010	6.21	0.000
3.	Control	50	0.181 ± 0.004		
	RGF 45 µg/l	50	0.167 ± 0.012	1.19	0.266
	RGF 60 µg/l	50	0.249 ± 0.005	−8.93*	0.000
	RGF 75 µg/l	50	0.234 ± 0.006	−8.03*	0.000
4.	Control	50	0.336 ± 0.007		
	DBRGF 45 µg/l	50	0.484 ± 0.021	−7.18*	0.000
	DBRGF 60 µg/l	50	0.604 ± 0.050	−5.71*	0.000
	DBRGF 75 µg/l	50	0.249	10.87	0.000

*Negative *t*-values indicate that mean values of exposed fish exceed mean values of control group

Results

Control gill morphology

The bluegill sunfish has four pairs of gills. Each gill arch has: (1) an inner row of pyramid-like gill rakers and (2) two rows of outer gill filaments with lamellae.

The surface of the gill arch is covered by microridged epithelial cells among which mucus gland openings can be found (Plate 1, Fig. 1). Epithelial cells are polygonal in shape. Ridges are in a maze-like pattern (Plate 1, Figs. 1, 2).

Filaments are covered by microridged epithelial cells with a cup-like depressed central portion (Plate 1, Fig. 3). The ridges are not as prominent as those on the arch. Mucus gland openings are present among them (Plate 1, Fig. 4). Each lamella is transversely arranged on the filament surface (Plate 1, Fig. 3). Margins of lamellae are covered by microridged cells while the central part bears microvilli (Plate 1, Fig. 3). One peculiarity in the epithelial cells of the gill is the presence of many small projections of plasma membrane in the inter-ridge area, and these are more numerous in gill filaments than in the gill arch.

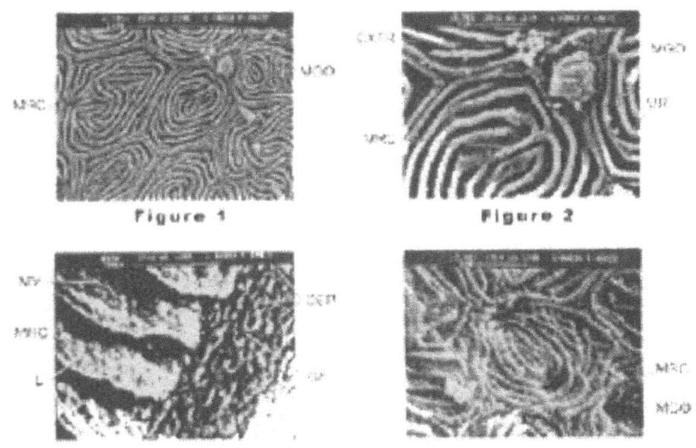

Plate 1 The gill of a control fish. Fig. 1 Surface of a gill arch with microridged epithelial cells (*MRC*) and mucus gland openings (*MGO*). × 4790. Fig. 2 Surface of a gill arch with extrusion (*EXTR*); microridge (*MR*); microridged epithelial cell (*MRC*); mucus gland opening (*MGO*). × 1270. Fig. 3 Gill filament (*GF*) with microridged epithelial cells (*MRC*) and cup-like depressions (*DEP*) and lamellae (*L*) with microvilli (*MV*). × 857. Fig. 4 Gill filament showing the microridges of the microridged epithelial cells (*MRC*); mucus gland opening (*MGO*). × 1240

Treated gill morphology

Following exposure to 15 µg/l Diazinon, the epithelial cell boundaries were not prominent and the ridges had lost their normal pattern in gill arches. Red blood cells were occasionally seen on the surface, suggesting some vascular damage. Cellular debris is suggestive of necrosis of the epithelial surface. A granular deposit on the surface of the cells was also observed (Plate 2, Fig. 1). A mucus film, which was lost during tissue preparation, was found over the filament and lamellae. Several droplets of oozing material were seen on the cell surface, giving the epithelial cells a papulate appearance. This material accumulated in the furrow between the ridges (Plate 2, Fig. 3). The lamellar surface was also covered with some secretory substance (Plate 2, Fig. 4).

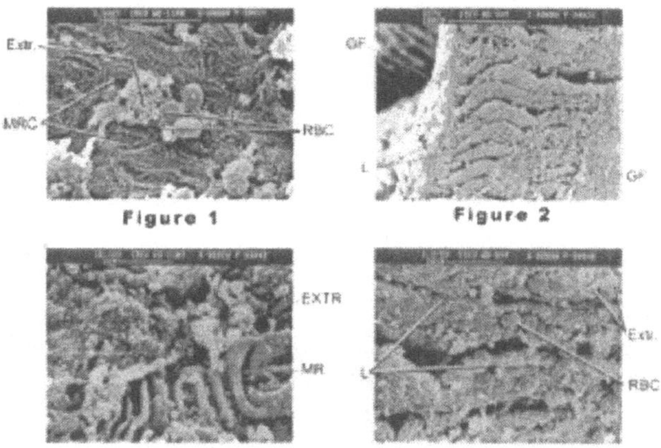

Plate 2 Gill of a fish exposed to 15 µg/l Diazinon. Fig. 1 Gill arch with (not prominent) epithelial cell boundaries; red blood cells (*RBC*). × 1840. Fig. 2 Gill filaments (*GF*) with lamellae (*L*). × 357. Fig. 3 Gill filament showing the droplets of oozing material. × 1210. Fig. 4 Lamellar surface (*L*) showing extrusion (*EXTR*) and red blood cells (*RBC*). × 1270

Following exposure to 30 µg/l Diazinon, only a few ridges on the gill arch epithelium were seen. The ridges had a pimpled appearance due to extrusion of cytoplasmic material (Plate 3, Fig. 1, Fig. 2). Ridges were also broken with perforations (Plate 3, Fig. 2). The entire filament surface was covered by extruding droplets and broken parts of the ridges (Plate 3, Fig. 3). Most of the lamellae were swollen, causing hypertrophism of the lamellar cells (Plate 3, Fig. 4). Lamellar cells were blebbed at many places.

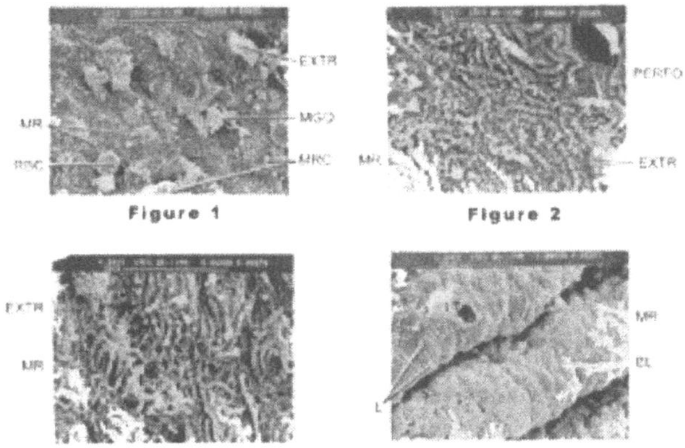

Plate 3 Gill of a fish exposed to 30 µg/l Diazinon. Fig. 1 Gill arch with a few visible microridges and more extrusions. × 1820. Fig. 2 Gill arch with broken and perforated microridges and lot of extrusions. × 5670. Fig. 3 Gill filament with microridges and extruding droplets. × 4800. Fig. 4 Lamellae (*L*) with swollen celis. Blebbed (*BL*). × 1240

Following exposure to 45 µg/l Diazinon, gill arch surfaces had normal microridged epithelial cells (Plate 4, Fig. 1, Fig. 2). The surface in the depressions had an undulating appearance. Mucus gland openings were also seen (Plate 4, Fig. 2). Compared to the gill arch, the gill filaments retained microridged epithelial cells, but they were comparatively thinner with wider inter-ridge spaces. It can be seen that the central portion of the cell has smaller elevations where the ridges are fused (Plate 4, Fig. 3). Few blebs were present and lamellae were swollen with flake-like cells. Cells closer to filaments and at the margin were microridged (Plate 4, Fig. 4).

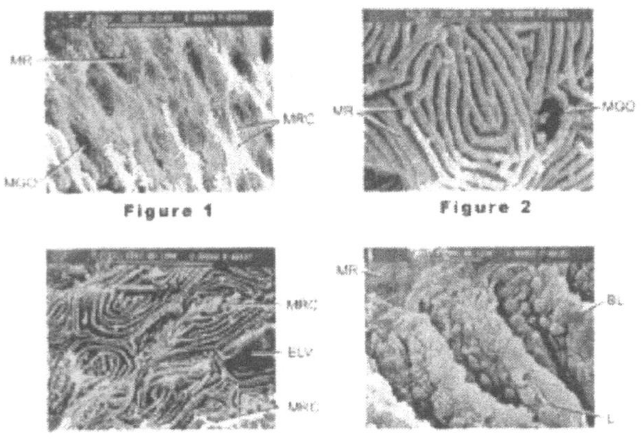

Plate 4 The gill of a fish exposed to 45 µg/l Diazinon. Fig. 1 Gill arch surface showing normal microridged epithelial cells with undulating appearance in the depressions. × 3600. Fig. 2 Gill arch surface with microridges and mucus gland openings. × 1230. Fig. 3 Gill filament with centrally

located elevations (*ELV*). × 4710. Fig. 4 Swollen lamellae with blebbed cells (*BL*). × 1722

Following exposure to 60 µg/l Diazinon, the gill arch was covered by microridged epithelial cells. The ridges were swollen with reduced inter-ridge spaces. In many places, adjacent ridges were so swollen that they appeared to fuse, and this swelling gave different elevations to the cells (Plate 5, Fig. 1). The gill filament surface was covered by microridged epithelial cells, but the density of ridges was very low. The cell surface had some blebs. The centres of many cells were elevated in such a manner that the ridges in those areas were not seen (Plate 5, Fig. 2, Fig. 3). Lamellae were swollen with blebbed microridged cells (Plate 5, Fig. 4).

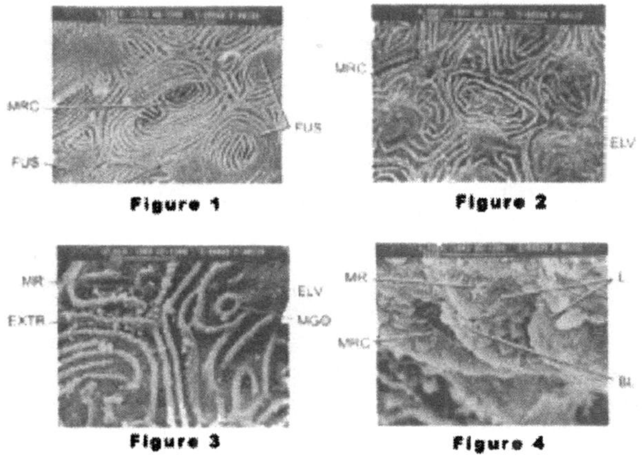

Plate 5 Gill of a fish exposed to 60 µg/l Diazinon. Fig. 1 Gill arch showing swollen and fused microridges (*FUS*) with reduced inter-ridge spaces. × 4600. Fig. 2 Gill filament with several elevations (*ELV*) due to the fusion of microridges. × 4880. Fig. 3 Gill filament with a low density of microridges and some blebs on the cell surface. × 12,400. Fig. 4 Swollen lamellae (*L*) with blebbed (*BL*) microridged cells. × 1710

Following exposure to 75 µg/l Diazinon, the microridged epithelial cells of the gill arch surface were covered with extensive cellular extrusions (Plate 6, Fig. 1). The microridged epithelial cells had a slightly convex appearance (Plate 6, Fig. 2, Fig. 3). The microridged epithelial cells on the margins of the lamellar surface were extremely swollen (Plate 6, Fig. 4).

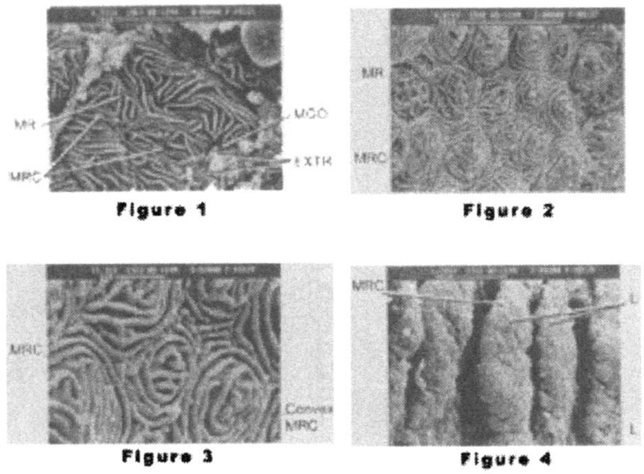

Plate 6 The gill of a fish exposed to 75 μg/l Diazinon. Fig. 1 Gill arch with microridged epithelial cell and severe cellular extrusion. × 4,710. Fig. 2 Gill filament with microridged cells with a convex appearance. × 4870. Fig. 3 Same as in Fig. 2 at a higher magnification. × 1230. Fig. 4 Swollen lamellar surface (*L*) with microridged epithelial cells on their margins. × 1260

Nearly all the morphometric differences between control and experimental measurements were highly significant (see Table 1) with *P* values below or considerably below 1%. The only exception to this was the difference between the thickness of microridges exposed to 45 μg/ml concentrations and that of their control.

Discussion

Exposure to sublethal concentrations of Diazinon had a variety of effects on gill morphology. Commonly reported effects of toxicants on gills fall into two categories: lesions and reactions. Lesions include necrosis and rupture of respiratory epithelium, which are direct and deleterious effects (Temmink *et al.* 1983). Lesions are dose dependent and develop by two mechanisms: autolysis induced by the cell's own enzymes and rapid lysis caused by direct action of a toxicant (Mallatt 1985). Reactions that are also dose dependent include epithelial lifting, fusion, hypertrophy, hyperplasia, mucus secretion, vascular stasis, mucus cell proliferation, chloride cell proliferation and leucocyte infiltration (Morgan and Tovell 1973; Mallatt 1985; Dutta *et al.* 1996). These are defence responses of the fish.

Organic xenobiotics such as pesticides pass rapidly through the gills to enter the blood stream (Rankin *et al.* 1983), which facilitates their distribution to the branchial cells and other organs. Gills also take an active part in metabolism and elimination of the xenobiotics (Dutta *et al.* 1996). Oozing of intracellular material from the cell surface after exposure to low doses of Diazinon may be a way of eliminating the absorbed and metabolized toxicants. This type of reaction occurred more frequently on gill filaments than on the gill arch, possibly because filaments are more exposed than the arch and are also more vascularized. Mucus hypersecretion after exposure to different toxicants has been reported frequently (Burton *et al.*1972; Dalela *et al.* 1979; Crespo 1982; Richmonds and Dutta 1989; Dutta *et al.* 1993). Mucus restricts the toxicant entry. The role of the chloride cell in excretion and neutralization of toxicants has also been studied (Karnaky 1980; Crespo *et al.* 1981; Oronsaye and Brafield 1984). Chloride cell proliferation may occur with concomitant excretion of the toxicants (Dutta *et al.* 1996), but in the present study the extrusion was so extensive that it was difficult to differentiate chloride cells from other epithelial cells. According to Dutta *et al.* (1996), following 48 h of exposure of *H. fossilis* to malathion, the chloride cells became swollen and traversed to whole lamellar epithelia, and were in direct contact with the lymphoid space and exterior to the epithelium. They proposed that the chloride cells play a role in the excretion of circulating secondary metabolites generated from the pesticide and some accumulated ions.

In this study, at higher doses of Diazinon (45 μg/l upward) the extrusion lessened and the cells were found to be swollen. The ridges of the gill arch in the centre of some microridge cells fused (at 60 μg/l) and ridges of gill filaments in each cell began to disappear. At lower doses, the ridges looked pimpled and perforated, which gave the cell a bizarre structure. The pimples of the ridge may be vesiculation of the cell plasma membranes, which break apart to give the perforated appearance. Similar changes in the microridges were observed by Kudo and Kimura (1984) to occur in bacterial gill disease. However, the loss of microridges after exposure to different toxicants has been reported (Hart and Oglesby 1979; Jacobs *et al.* 1981; Jagoe and Haines 1983; Roy *et al.* 1986, 1990; Roy and Munshi 1991). The loss may be a defence mechanism, as loss of microridges results in a reduction of the cellular area. The thickness of the microridges on both the gill arch and the gill filament increased with different exposures, except for the group exposed to 45 μg/l Diazinon. An increase in this thickness is probably a defence mechanism that increases the unstirred boundary layer effect and thereby slows delivery of Diazinon to the epithelial surface. Cell swellings occur more on the lamellae (visually perceived on the micrographs), but the filament epithelium bulges, forming a convex surface. Lamellar swelling is also a generalized response to toxicants (Jacobs *et al.* 1981; Kudo and Kimura 1984; Roy *et al.* 1986, 1990). This is also a defence mechanism since it reduces the water space (interlamellar space). This swelling may lead to disruption of gas exchange and osmoregulation, which are branchial responses. Lamellar swelling may be due to epithelial lifting, hyperplasia, or formation of lymphatic spaces.

The distance between the microridges on the gill arch for all exposures and on the gill filaments at 75 μg/l Diazinon decreased (Table 1), possibly because the distance between the microridges is inversely related to the swelling of the ridges, which reduces the area of gill surface exposed to the contaminated water.

In the case of gill filaments, the thickness of the microridges increased, as did the distance between the microridges, following exposure to 45 μg/l and 60 μg/l Diazinon. Such a situation may be caused by the bulging of gill filament epithelial cells, forming a convex surface. The bulging should increase the surface area, but fusion between microridges on the gill filament following exposure to 45 µg/l and 60 µg/l Diazinon may reduce the surface area exposed to contaminated water.

It is clear that fish gills possess several protective structures and mechanisms. If one cellular barrier is damaged by cellular excretion of a toxicant or its metabolic product, then other defensive responses, such as a hyperplastic reaction, lifting of the cells and a thickening of the basal lamina of the blood capillaries, may also occur. In spite of these actions, toxicants may still enter the blood stream. Branchial responses that serve to slow the entry of toxicants may have the undesirable effect of threatening to suffocate the fish (Richmonds and Dutta 1989; Dutta *et al.* 1993). Severe damage in the form of necrosis and rupture of the gill epithelium and microridges results in hypoxia and respiratory failure. In addition, the fish encounters problems with ionic and acid-base balance (Richmonds and Dutta 1989).

Conclusion

The gill is a structure that is essential for the survival of fish. This study confirms that sublethal doses of Diazinon (a pesticide) have an adverse effect on the surface ultrastructures of fish gills. Though some damage occurred at 15 µg/l, 30 µg/l and 45 µg/l exposures, the most significant damage was observed in the gill filament and gill arch at 60 µg/l and 75 µg/l Diazinon. Defensive reactions predominate, except for isolated cellular rupture. Defensive reactions include loss of microridges, fusion and changes in the thickness of the microridges as well as lamellar swelling. Lamellar swelling may result from epithelial lifting, hyperplasia or formation of lymphatic spaces. Fusion and reduction in the microridges lead to a reduction in the respiratory area, while swelling may lead to disruption of gas exchange and osmoregulation, which will definitely harm the fish. At lower doses of Diazinon, the microridges look pimpled and perforated, which may be due to vesiculation of the cells' plasma membranes. Oozing of the intracellular materials from perforated microridges may be a way of eliminating absorbed and metabolized toxicants.

Structural changes in Diazinon-exposed gills, whether lesions or defence reactions, are detrimental.

Acknowledgements. The authors express sincere thanks to Dr. A. Graham for helping us take the SEM photographs in his laboratory (Department of Biological Sciences, Kent State University, Kent, Ohio, USA). The authors also thank Mrs. Linda Matz of the Department of Biological Sciences, Kent State University, for typing several versions of this manuscript.

References

Burton, D.T., Jones, A.H., Cairns, J. (1972) Acute zinc toxicity to rainbow trout (*Salmo gairdneri*) Confirmation of the hypothesis that death is related to tissue hypoxia. *J. Fish. Board Can.* **29**: 1463–1466

Crespo, S. (1982) Surface morphology of dogfish (*Scyliorhinus canicula*) gill epithelium and surface morphological changes following treatment with zinc sulphate: a scanning electron microscope study. *Mar. Biol.* **67**: 159–166

Crespo, S., Soriano, E., Sampera, C., Balasch, J. (1981) Zinc and copper distribution in excretory organs of the dogfish, *Scyliorhinus canicula* and chloride cell response following treatment with zinc sulphate. *Mar. Biol.* **65**: 117–123

Dalela, R.C., Bhatnagar, M.C., Tyagi, A.K., Verma, S.R. (1979) Histological damage of gills in *Channa gachua* after acute and subacute exposure to endosulfan and rogor. *Mikroskopie (Vienna)* **35**: 301–307

Dutta, H.M., Zeno, T., Richmonds, C. (1992) Toxic action of Diazinon on the gills of bluegill sunfish *Lepomis macrochirus*. *Ohio J. Sci.* **92**: 09

Dutta, H.M., Richmonds, C., Zeno, T. (1993) Effects of Diazinon on the gills of bluegill sunfish, *Lepomis macrochirus*. *J. Environ. Pathol. Toxicol. Oncol.* **12**: 219–227

Dutta, H.M., Munshi, J.S.D., Roy, P.K., Singh, N.K., Adhikari, S., Killius, J. (1996) Ultrastructural changes in the respiratory lamellae of the catfish, *Heteropneustes fossilis*, after sublethal exposure to malathion. *Environ. Pollut.* **9**: 329–341

Engelhardt, F.R., Wong, M.P., Dvey, M.E. (1981) Hydromineral balance and gill morphology in rainbow trout, *Salmo gairdneri*, acclimated to fresh and sea water, as affected by petroleum exposure. *Aqua. Toxicol.* **1**: 175–186

Freda, J., Sanchez, D.A., Bergman, H.L. (1991) Shortening of branchial tight junctions in acid-exposed rainbow trout (*Oncorhynchus mykiss*). *Can. J. Fish. Aquat. Sci.* **48**: 2028–2033

Hart, R.K., Oglesby, G.B. (1979) Toxin induced changes in the gill epithelium of rainbow trout (*Salmo gairdneri*). *Scanning Electron Microsc.* **3**: 355–361

Jacobs, D., Exmond, E.F., Melisky, E.L., Hocutt, C.H. (1981) Morphological changes in gill epithelium of heat stressed rainbow trout, *Salmo gairdneri*: evidence in support of a temperature induced surface area hypothesis. *Can. J. Fish. Aquat. Sci.* **38**: 16–22

Jagoe, C.H., Haines, T.A. (1983) Alterations in gill epithelial morphology of yearling sunapee trout exposed to acute acid stress. *Trans. Am. Fish. Soc.* **112**: 689–695

Johnson, W., Finley, M. (1980) Handbook of acute toxicity of chemical to fish and aquatic invertebrates. USDA Fish and Wildlife Service/Resource Publication 137, Washington, DC

Karlsson, L., Runn, D., Haux, C., Forlin, L. (1985) Cadmium induced changes in gill morphology of Zebrafish, *Brachydanio rerio* (Ham.) and rainbow trout, *Salmo gairdneri* (Rich.). *J. Fish Biol.* **27**: 81–95

Karlsson, L., Bjorklund, I., Ljungberg, O., Runn, P. (1986) Acid water and aluminum exposure: experimentally induced gill lesions in brown trout, *Salmo trutta* L. *J. Fish Dis.* **91**: 1–25

Karnaky, K.J. (1980) Ion-secreting epithelia: chloride cells in the head region of *Fundulus heteroclitus*. *Am. J. Physiol.* **238**: 185–198

Kudo, S., Kimura, N. (1984) Scanning electron microscopic studies on bacterial gill disease in rainbow trout fingerlings. *Jpn. J. Ichthyol.* **30**: 393–403

Lindstrom, P., Koivusaari, U., Hanninen, O. (1981) Metabolism of xenobiotics by Vendace (*Coregonus albula*). *Comp. Biochem. Physiol.* **68C**: 121–126

Mallatt, J. (1985) Fish gill structural changes induced by toxicants and other irritants: a statistical review. *Can. J. Fish. Aquat. Sci.* **42**: 630–648

Morgan, M., Tovell, P.W.A. (1973) The structure of the gill of the trout, *Salmo gairdneri* (Richardson). *Z. Zellfo. Mikros. Anat.* **142**: 147–162

Munshi, J.S.D., Singh, R.K. (1971) Investigation on the effect of insecticides and other chemical substances on the respiratory epithelium of the predatory and weed fishes. *Ind. J. Zool.* **22**: 127–134

Munshi, J.S., Singh, D.H. (1992) Scanning electron microscopic evaluation of effects of low pH on gills of *Channa punctata* (Bloch). *J. Fish Biol.* **41**: 83–89

Murthy, A.S. (1986) Toxicity of pesticides to fish, Vol. 2. CRC, Boca Raton, Fla.

Oronsaye, J.A.O., Brafield, A.E. (1984) The effect of dissolved cadmium on the chloride cells of the gills of the stickleback, *Gasterosteus aculeatus* L. *J. Fish Biol.* **25**: 253–258

Prasad, M.S. (1989) Effects of crude oil on the air breathing organs of the striped gourami, *Colisa fasciatus*: a SEM study. *Ecotoxicol. Environ. Safr.* **18**: 211 –218

Prasad, M.S. (1991) SEM study on the effects of crude oil on the gills and air-breathing organis of climbing perch, *Anabas testudineus*. *Bull. Environ. Contam. Toxicol.* 47: 882–889

Rankin, J.C., Stagg, R.M., Bolis, L. (1983) Effect of pollutant on gills. In: Houlihan, D.F., Rankin, J.C., Shuttleworth, T.J. (eds) *Gills* (SEB Seminar Series No.16). Cambridge University Press, Cambridge, pp. 207–219

Reddy, T.G., Palaniappan, S., Pillay, K.B.P. (1977) Respiratory and histopathologicaleffects of the pesticide Disyston on the fish, *Colisa latia*. All India Symposium on Environmental Biology, University Kerala, pp. 57

Richmonds, C., Dutta, H.M. (1989) Histopathological changes induced by malathion in the gills of bluegill, *Lepomis macrochirus*. *Bull. Environ. Contam. Toxicol.* 43: 123–130

Roy, P.K., Munshi, J.S.D. (1991) Malathion induced structural and morphometric changes of gills of a freshwater major carp,*Cirrhinus mrigala* (Ham.). *J. Environ. Biol.* 12: 79–87

Roy, P.K, Munshi, J.S.D., Munshi, J.D. (1986) Scanning electron microscopy evaluation of effects of saponin on the gills of the climbing perch, *Anabas testudineus* (Bloch) (Anabantidae: Pisces). *Ind. J. Exp. Biol.* 24: 511–516

Roy, P.K., Munshi, J.D., Dutta, H.M. (1990) Effect of saponin extracts on morpho histology and respiratory physiology of an air-breathing fish, *Heteropneustes fossilis* (Bloch).*J. Freshwater Biol.* 2: 135–145

Temmink, J., Bouweister, P., deJong, P., Vander Berg, J. (1983) An ultrastructural study of chromate induced hyperplasia in the gill of rainbow trout (*Salmo gairdneri*). *Aquat. Toxicol.* 4: 165–179

Tietge, J.E., Johnson, R.D., Bergman, H.L. (1988) Morphometric changes in gill secondary lamellae of brook trout (*Salvelinus fontinalis*) after long-term exposure to acid and aluminum. *Can. J. Fish. Aquat. Sci.* 45: 1643–1648

Exp. Biol. Online 2:18 (1997)

The multiple soluble malate dehydrogenase of *Hoplias malabaricus* (Erythrinidae, Characiformes)

M.R. Aquino-Silva, M.L.B. Schwantes and A.R. Schwantes

Department of Genetics and Evolution, Laboratory for Molecular Evolution, Universidade Federal de São Carlos. C. Postal 676. Zip 13565-905, S. Carlos, SP, Brazil

Correspondence to: M.R. Aquino-Silva (email: dars@power.ufscar.br)

Received: 11 June 1997 / Accepted: 18 November 1997
Published: November 21, 1997

Abstract. To explain the six-banded pattern obtained upon electrophoresis of the soluble form of malate dehydrogenase (sMDH, EC 1.1.1.37) from the characiform *Hoplias malabaricus*, a recent locus duplication of its A isoform (*sMDH-A**), in addition to its *sMDH-B** isoform, was proposed. Klebe's serial dilutions carried out using skeletal muscle, heart and liver extracts showed that the A1 and A2 subunits have the same visual end-points, indicating that these A-duplicated genes have a nondivergent pattern. Since there is no evidence of polyploidy in the Erythrinidae family, the *MDH-A** loci have probably evolved from regional gene duplication. While these *sMDH-A** loci encode nondivergent thermostable isoforms, the *sMDH-B** encodes a thermolabile one. Thermostable sMDHs differ from the thermolabile sMDHs in that they have a higher K_m of oxaloacetate. Liver, muscle and heart unfractionated sMDH levels at three different temperature and two pH regimens were analysed and the results showed that, in the adaptative temperature range of *Hoplias*, the variation in K_m under conditions of constant pH (imidazole buffer) was less (approximately threefold) than that measured in the presence of temperature-dependent pH imidazole buffer (sevenfold). Estimation of the ratio of both isoforms in these tissues by Klebe's method showed that, in unfractionated liver – where K_m values were the highest and the minimum K_m was obtained at 30°C (both for temperature-dependent pH and constant-pH imidazole buffer) – the duplicate A (thermostable, A1 and A2) and B (thermolabile) subunits were detected in a ratio of 2:1. On the other hand, in muscle extracts – in which the lowest K_m values were measured, with the minimum K_m at 10–20°C (temperature-dependent pH and constant-pH imidazole buffer, respectively) – a ratio of two thermolabile to one thermostable subunits was observed.

Key words: Characiformes – *Hoplias malabaricus* – K_m – Recent locus duplication – Soluble malate dehydrogenase – Thermolabile isoform – Thermostable isoform

Introduction

A measure of the specialization of enzyme function is the extent to which it is expressed in a number of tissue types. Generally, less specialized enzymes are expressed more widely. In lower vertebrates, the distribution of creatine kinase (Fisher and Whitt, 1978), lactate dehydrogenase (Markert *et al.*, 1975) and other multi-locus enzymes (Fisher *et al.*, 1980) reveals a marked proliferation of isozyme loci accompanied by increasing restriction of their expression in various differentiated tissues. The malate dehydrogenase system (MDH, EC 1.1.1.37) is represented by two major forms in vertebrates and invertebrates: soluble (sMDH) and mitochondrial (mMDH). The two forms differ in their electrophoretic mobility (Markert and Moller, 1959), kinetic behaviour (Kitto and Lewis, 1967), amino acid composition (Thorner and Cooper, 1963), antigenic properties (Kitto and Lewis, 1967), and are under the control of separate gene loci (Davidson and Cortner, 1967; Bailey *et al.*, 1969, 1970; Whitt, 1970). Two gene loci, *sMDH-A** and *sMDH-B**, encode sMDH in most fish and amphibian species that have been studied (Bailey *et al.*, 1969, 1970; Whitt, 1970; Wheat *et al.*, 1971; Schwantes and Schwantes, 1977, 1982a,b; De Luca *et al.*, 1983; Coppes *et al.*, 1987; Fenerich-Verani *et al.*, 1990; Farias and Almeida-Val, 1992; Lin and Somero, 1995a,b; Caraciolo *et al.*,1996). As in other multi-locus systems, throughout the primitive fishes both *sMDH-A** and *sMDH-B** loci have generalized expression, and so tend to be similiarly expressed in all tissues (Fisher *et al.*, 1980; Schwantes and Schwantes, 1982a). On the other hand, in advanced teleosts the expression of these loci is progressivley restricted as the tissue becomes more differentiated. Three categories describe the types of gene expression: nondivergent, unidirectionally divergent and bidirectionally divergent (Ferris and Whitt, 1979). Utilizing the serial dilutions of Klebe's (1975) method, the relative contributions of duplicate genes to the isozymic activity in certain tissues is obtained and the category to which each pair of duplicate genes belongs is determined. In the nondivergent category, duplicate genes are expressed equally in all tissues in which the enzyme is present. This pattern could result from a recent gene duplication that has not yet functionally diverged. In the unidirectionally divergent category, which includes most of the divergent patterns observed (Ferris and Whitt, 1979), the duplicate genes are expressed unequally in one or more tissues. In the bidirectionally divergent category there is no consistent predominance of one locus product over another among the various tissues. This pattern of expression reflects the predominance of different genes in different tissues and was probably developed through alterations in the regulatory genes (Ferris and Whitt, 1979).

Graves and Somero (1982), studying A4-lactate dehydrogenases of congeneric barracudas from the Eastern Pacific, showed that differences in the average or maximal habitat temperature of the order of a few degrees Celsius appear to be sufficient to favour selection of adaptatively different orthologous homologues (i.e. homologues encoded by a single gene locus common to different species) of enzymes. Also, differential expression of paralogous homologues (i.e. homologues encoded by different gene loci in a species) of proteins during thermal acclimation or acclimatization has been reported (Hochachka, 1965; Hochachka and Somero, 1973, 1984; Tsukuda and Ohsawa, 1974; Tsugawa, 1976; Schwantes and Schwantes, 1982a,b; De Luca *et al.*, 1983; Coppes *et al.*, 1987a–c; Coppes and Somero, 1990; Lin and Somero, 1995a,b). Adaptation to temperature by the two soluble MDH gene loci of teleost fish, where *sMDH-A** encodes a thermostable isoform and *sMDH-B** encodes a thermolabile one, was first shown by Schwantes and Schwantes (1982a,b). These authors, studying the sMDH of several vertebrates, showed that in perciform *Leiostomus xanthurus* (Sciaenidae) tissue extracts, the paralogous isoforms were present at differing quantitative levels and their activities could be modified by changes in the temperature of the environment. Furthermore, thermostability and thermal dependency tests showed that, similar to that which occurs during acclimatization and acclimation, the A isoform was more stable than the B in the presence of heat (Schwantes and Schwantes, 1982b). The B isoform appeared to be activated by low temperatures and to be more thermolabile at high ones. De Luca *et al.* (1983), studying the subtropical fish *Astyanax fasciatus* (Characidae), also reported that the A and B subunits were present at differing quantitative levels but that their activities were almost independent of the season. Lin and Somero (1995b), comparing the sMDH of Eastern Pacific barracudas from different latitudes, suggested that the variation in therelative quantities of paralogous homologues and the absence of a thermolabile isoform in warm-adapted species may be a specific adaptation to high temperature. However, Farias and Almeida-Val (1992) and Caraciolo *et al.* (1996), studying the sMDH of Amazon fishes, detected a unidirectionally divergent pattern of expression of both isoforms, with the thermostable isoform being predominant.

Monteiro *et al.* (1991) demonstrated the presence of the paralogous isoforms of sMDH in most of 22 species of subtropical fish belonging to the orders Characiformes, Siluriformes and Perciformes. However, as well as other teleosts (Whitt, 1970; Bailey *et al.*, 1970; Wheat *et al.*, 1971; Farias amd Almeida-Val, 1992), one of these species showed a much more complex electrophoreticpattern of sMDH. To explain the six-banded pattern detected in the characiform *Hoplias malabaricus*, a recent locus duplication hypothesis for its *sMDH-A** was proposed in addition to its *sMDH-B**. In order to gain insight into this possible recent gene duplication in *H. malabaricus*, the present paper describes tissue specificities, thermostability and kinetic properties of its sMDH.

Materials and methods

Fifty specimens of *H. malabaricus* were collected from Monjolinho Reservoir at the Federal University of São Carlos, São Paulo State, Brazil, where the annual temperature range was 12.3–23.0°C. White muscle, heart and liver from each individual were dissected immediately after capture and kept at -20°C. A small piece of each tissue was homogenized (w/v) in 50 mM phosphate buffer (pH 7.0), using a Potter-Elvejhem tissue grinder, and then centrifuged at 19,000 *g* for 30 min at 4°C in a Sorvall RC 5B centrifuge. The resulting crude extracts were used for electrophoretic and spectrophotometric analyses.

Electrophoreses were carried out using horizontal gels containing 14% (w/v) corn starch prepared according to Val *et al.* (1981), using the pH 6.9 Whitt (1970) buffer system. A voltage gradient of 5 V/cm was applied for 14–17 h at 4°C. After electrophoreses, the starch gels were sliced lengthwise and the lower halves incubated in an MDH staining solution described by Monteiro *et al.* (1991). Contributions of duplicate genes to the isozyme activities in each tissue were quantified by Klebe's (1975) method. The thermal stability of the isozymes was tested by subjecting each tissue extract to a temperature of 55°C for 1–20 min. Afterwards, these extracts were cooled on ice, centrifuged at 19,000 *g* for 30 min at 4°C, and electrophoresed together with the control (kept in an ice and water mixture). After electrophoreses, the gels were stained as described above. Nomenclature of sMDH gene loci, subunits and iso/allozymes is taken from Shaklee *et al.* (1989).

The effect of temperature on *Hoplias* sMDH was measured in the direction of oxaloacetate reduction by the change in absorbance at 340 nm using a Beckman 25K spectrophotometer at three different temperatures (10, 20 and 30°C) and two pH regimens, i.e. constant pH (7.00) buffer at each of the three temperatures, and temperature-dependent pH buffer at both 10°C (pH 7.08) and 30°C (pH 6.90). The assays were carried out using a 50 mmol l^{-1} imidazole chloride buffer containing 0.2 mmol l^{-1} reduced nicotinamide adenine dinucleotide (NADH) and different concentrations of oxaloacetate. Oxamate, at a concentration of 10 mM, was added to inhibit lactate dehydrogenase (LDH) activity, which could result from any pyruvate present in the assay medium. Temperature was controlled using a Lauda K-2/R circulating bath. All reactions were done at least in triplicate and initiated by adding 10 μl of enzyme to 1.0 ml of assay medium. Apparent Michaelis-Menten constants (K_m) were calculated by the Lineweaver-Burk method, using double-reciprocal plots of velocity versus substrate concentrations. When saturation curves were sigmoidal the enzyme-substrate (E-S) affinities were determined by the Hill plot and determining the half-maximal saturation, $S_{0.5}$.

Results

A six-banded electrophoretic pattern for sMDH was observed following analysis of 50 *H. malabaricus* specimens (Fig.1). It suggests the existence of three gene loci: *sMDH-B**, *sMDH-A1** and *sMDH-A2**,the latter being the result of a recent locus duplication since both homodimers exhibited a nondivergent pattern of tissue expression. Probably the band nearest the anode represents the B isoform encoded by the *MDH-B**, while the bands furthest away from the anode represent the A isoforms produced by the association of the A1 and A2 subunits.

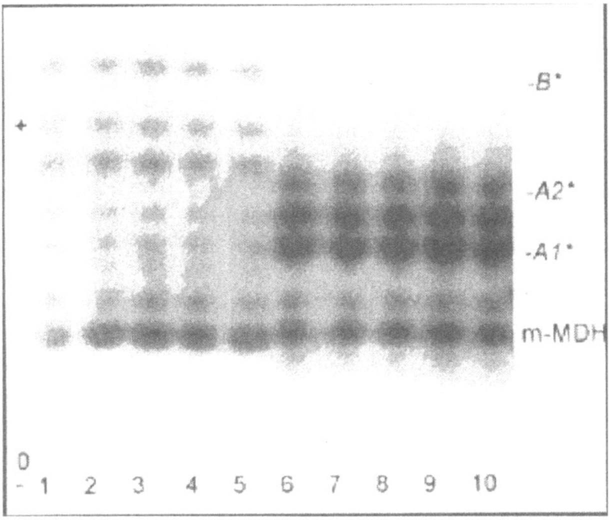

Fig. 1 Malate dehydrogenase from skeletal muscle (1–5), and liver (6–10) extracts of *Hoplias malabaricus*

Klebe's serial dilutions carried out on skeletal muscle, heart and liver extracts showed the same visual end-points for A1 and A2 subunits, indicating the same molar ratio as expected for nondivergent genes. In skeletal muscle extracts, A (A1 and A2) and B subunits were present in a proportion of approximately 1:4 and their contribution to the present isozymes was one A (A1 and A2) to two B subunits. In heart and liver extracts, these subunits were present in an approximate proportion of 1:1 and 4:1, respectively. Thus, the contribution of each subunit to all the present isozymes varied from 1A:1B in heart to 2A:1B in liver.

Also thermal inactivation tests applied to the products of the *sMDH-A1** and *sMDH-A2** loci showed a nondivergent pattern of thermostability between A isoforms (Fig. 2). These least-migrating bands represented the thermostable sMDH isoforms and the fastest-migrating bands, the thermolabile.

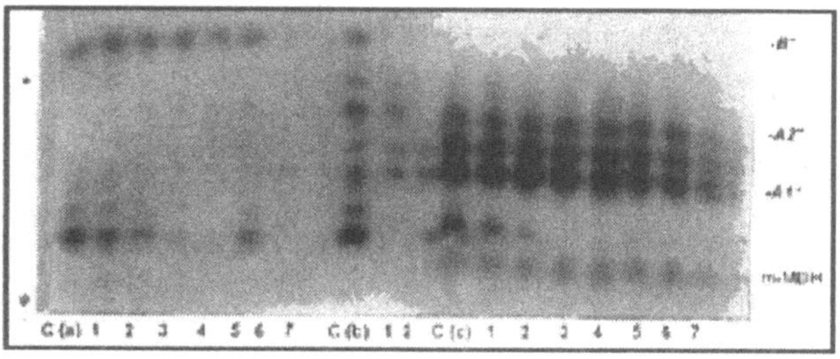

Fig. 2a–c The effects of heat treatment on malate dehydrogenase (MDH) from **a** skeletal muscle, **b** heart, and **c** liver extracts of *Hoplias malabaricus*. Samples of unfractionated MDH were incubated at 55°C for (*1*) 1, (*2*) 2, (*3*) 3, (*4*) 5, (*5*) 10, (*6*) 15, and (*7*) 20 min, electrophoresed and stained for residual activity.(*C* Control)

To compare the effect of temperature and pH on sMDH function of *H. malabaricus*, we examined the response of its apparent K_m of oxaloacetate to measurement temperatures under two pH regimens, both in an imidazole buffer system: temperature-dependent pH (10 and 30°C) and a constant pH (10, 20 and 30°C). MDH of *H. malabaricus* showed hyperbolic kinetics under both pH regimens and the three temperatures tested, with the exception of white-muscle MDH which displayed sigmoidal kinetics at 30°C under temperature-dependent pH (Fig. 3). Table 1 and Figure 4 illustrate the variation in *Hoplias* apparent $K_m/S_{0.5}$ for oxaloacetate as a function of assay temperature and pH regimen. K_m values were higher for liver extracts under both pH regimens (27–54 μmol l^{-1} at temperature-dependent pH and 45–54 μmol l^{-1} at constant pH) than for heart (19–28 μmol l^{-1} and 14–3 μmol l^{-1}) and skeletal muscle (6–45 μmol l^{-1} and 14–25 μmol $^{-1}$) extracts. Minimum K_m values for liver extracts were obtained at 30°C (27 μmol l^{-1} and 45 μmol l^{-1}) under both pH regimens, for heart extracts at 10°C (19 μmol l^{-1} at temperature-dependent pH) and 30°C (14 μmol l^{-1} at constant pH) and for white muscle at 10°C (6 μmol l^{-1} at temperature-dependent pH) and 20°C (14 μmol l^{-1} at constant pH).

Table 1 Effects of the temperature and pH regimen on the apparent K_m ($\mu mol\ l^{-1}$) values for oxalacetate with *H. malabaricus* tissues

		Tissue		
Temp. (°C)	pH regimens	muscle	heart	liver
10	[a]7.08	6	19	54
	[b]7.00	23	37	54
20	[b]7.00	14	16	50
30	[a]6.90	45 ($S_{0.5}$)	28	27
	[b]7.00	25	14	45

[a]Temperature-dependent pH
[b]Constant pH, both imidazole buffer system

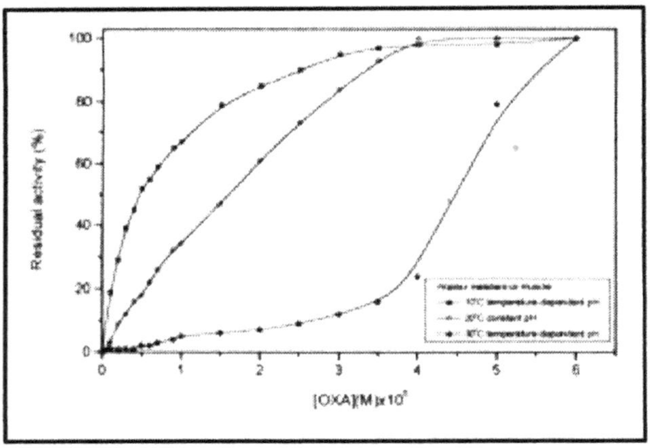

Fig. 3 MDH saturating curves (oxaloacetate reduction) of muscle from *H. malabaricus* as affected by temperature and pH regimen. Note that at 30°C the enzyme assumes a sigmoidal kinetics

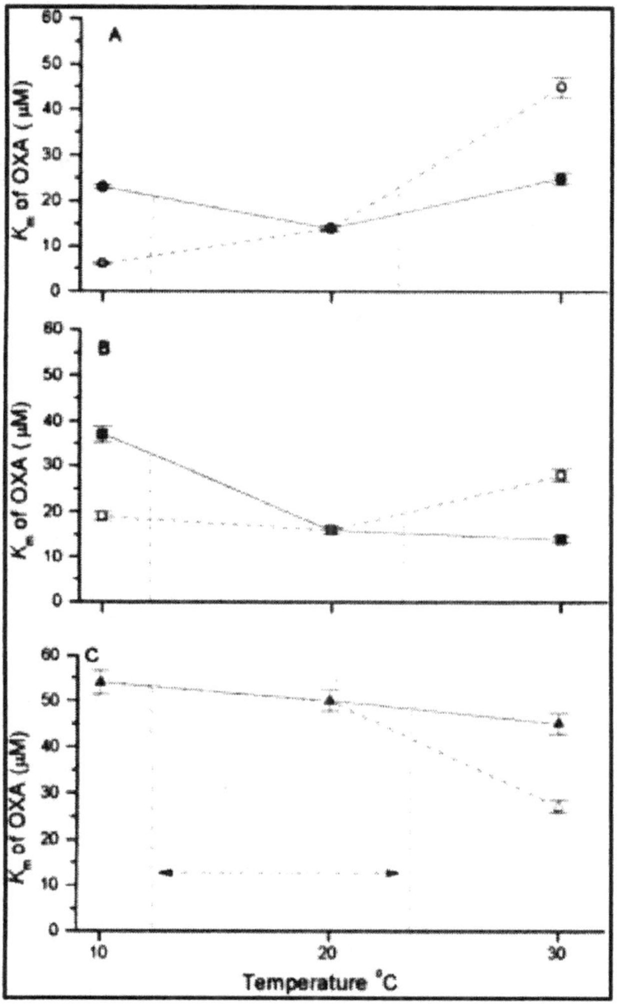

Fig. 4 The effect of temperature on the MDH $K_{m(oxa)}$ in muscle (A), heart (B) and liver (C) of *H. malabaricus* under two pH regimens: temperature-dependent pH (*open symbols*) and constant pH (*closed symbols*), both in an imidazole buffer system. <--> Species physiological temperature (12.3–23.0°C). *Error bars* are 95 % confidence intervals for each K_m determination

Discussion

The sMDH of tissue extracts from 50 *H. malabaricus* individuals analysed here confirms the six-banded pattern obtained previously by Monteiro *et al.* (1991) in a study of 44 other specimens of this species and reinforces the recent locus duplication hypothesis for its *sMDH-A** isoform. These *H. malabaricus sMDH-A** loci were hypothesized since both homodimers codified by them exhibited a nondivergent pattern of expression. In addition, the nondivergent pattern of thermostability of these locus products contributedto the acceptance of the recent locus duplication hypothesis. Duplicated isozyme loci can be formed by different kinds of gene and genome amplifications such as aneuploidy, polyploidization, and regional gene or in-tandem duplication. Additional copies of structural genes and regulatory elements are then free to accept mutations and to evolve to new functions, facilitating radiation of the vertebrates. Ohno (1974) has argued that in-tandem duplications frequently produce additional copies of structural genes that are under the control of the same regulatory element, which makes divergence in these duplicate loci more difficult. According to Bertollo *et al.* (1986), the Erythrinidae family has shown very variable karyotypes, with a diploid number ranging from 39 to 42 chromosomes and variable chromosomal rearrangements during its evolutionary history. Thus, the *MDH-A** loci that are apparently characteristic of *H. malabaricus* would have evolved from in-tandem duplication, which agrees with suggestions made by Whitt (1987), namely that almost all isozymes in diploid fish have arisen from regional gene duplication. On the other hand, according to Ferris and Whitt (1978), the rate of accepted in-tandem duplications appears to be rather low.

Among the categories that describe the types of gene expression (Ferris and Whitt, 1979) found using Klebe's method, the *sMDH-A** isozyme pattern of *H. malabaricus* might be referred to as nondivergent. In this same species, sMDH loci might be included in the bidirectionally divergent category, where the B isoform predominates in skeletal muscle extracts and the A isoforms in liver. This pattern of expression was also described by other authors to occur in teleost species (Bailey *et al.*, 1969, 1970; Clayton *et al.*, 1973; Aspinwall, 1974; Fisher *et al.*, 1980; De Luca *et al.*, 1983; Coppes *et al.*, 1987a; Monteiro *et al.*, 1991). Otherwise, Schwantes and Schwantes (1982a) suggested that sMDH loci expression in several fish and amphibian species follows a unidirectionally divergent pattern, with *sMDH-A** being expressed to a greater extent than *sMDH-B** in all tissues.

Adaptation to temperature by the two soluble MDH gene loci of teleost fish, where *sMDH-A** encodes a thermostable isoform and *sMDH-B** encodes a thermolabile one (as shown by Schwantes and Schwantes, 1982a,b; De Luca *et al.*, 1983; Coppes *et al.*, 1987a; Farias and Almeida-Val, 1992; Lin and Somero, 1995a,b; Caraciolo *et al.*, 1996), was also detected here for *H. malabaricus*. According to Shaklee *et al.* (1977), the metabolic reorganization that occurs during thermal acclimation can be, and usually is, a tissue-specific process. An increased expression of the slowest-migrating isoforms and thermostable sMDH is a better adaptive response to an increase in the temperature of the environment and the concomitant increase in oxidative metabolism compared with an increase in the expression of the fastest-migrating and thermolabile isoforms.

Comparative studies of orthologous homologues of several enzymes from animals adapted to different temperatures have revealed strong conservation of certain kinetic properties at the physiological temperatures of the species (Somero, 1978; Somero *et al.*, 1983; Hochachka and Somero, 1984; Somero, 1986; Coppes and Somero, 1990). Although K_m values tend to be strongly perturbed bychanges in temperature, highly similar values are found at normal body temperatures among species that have widely different average body temperatures. According to Hochachka and Somero (1984), the biological significance of this marked conservation of K_m values for A4-LDH is indicated by similar pyruvate concentrations among different vertebrates. If this biological significance is valid for unfractionated sMDH, then, in the physiological temperature range of *H. malabaricus* analysed here, the oxaloacetate concentrations in their cellular microenvironment must be around the K_m values obtained. In addition, the liver, where contributions of duplicate A (A1 and A2) and B subunits to the present isozymes obtained by Klebe's method were 2:1, the K_m values would be two to sevenfold higher than in muscle extracts and two to threefold than in heart.

Conservation of K_m values is interpreted as reflecting selection for the retention of the catalytic and regulatory capacities of enzymes and is manifested by adaptations to changes in pressure, osmotic conditions, pH as well as temperature (Hochachka and Somero, 1984). According to Reeves (1977), the pH values of blood and cytosol (pH_i) decrease with rising temperature with an approximate slope of 0.017 pH units per °C rise in temperature.For this reason, Hochachka and Somero (1984) suggest that this pH-temperature relationship has conditioned the short-term and evolutionary time course of changes in body temperature. In nonphysiological conditions of constant pH (phosphate buffer), the effect of a change in experimental temperature on the apparent K_m of pyruvate among interspecific homologues of A4-LDH was approximately tenfold whereas in a biologically realistic pH regimen (range of physiological temperature) this effect was twofold (Somero, 1981). However, our results obtained with unfractionated sMDH from liver, muscle and heart at three temperature assays and two pH regimens showed that, in the adaptative temperature range of *Hoplias*, the variation in K_m under conditions of constant pH (imidazole buffer) was smaller (approximately threefold) than that measured in the presence of a temperature-dependent pH buffer (i.e. sevenfold).

However, according to Lin and Somero (1995b), while the conservation of K_m at physiological temperatures for barracuda A4-LDH was achieved by an evolutionary change in the amino acid sequence of orthologous homologues, the barracuda's cytosolic form of MDH may also be achieved by altering the ratio between paralogous thermostable and thermolabile isozymes. As thermostable sMDH differs from the thermolabile form in that it has a higher K_m of oxaloacetate (Bailey *et al.*, 1970; De Luca *et al.*, 1983; Lin and Somero, 1995b, Aquino-Silva *et al.*, in preparation), the contribution of this high-K_m thermostable isoform predominated when present in approximately twofold excess relative to the thermolabile isoform. In agreement with these authors, estimation of the ratio of both isoforms in unfractionated liver, heart and muscle homogenates of *Hoplias* by Klebe's (1975) method showed that in the liver, where the highest K_m values were measured, duplicate A (thermostable, A1 and A2) and B (thermolabile) subunits were present in the isozymes in a ratio of

2:1. On the other hand, muscle extracts, where the lowest K_m values were recorded, contained the two thermolabile subunits for every thermostable subunit.

According to Hochachka and Somero (1968), the enzymes show minimum K_m values at environmental temperatures close to those encountered by the fish in their normal habitat, i.e. near their acclimatization temperatures. Also, Hochachka and Somero (1973) have proposed that there are certain generalized strategies of biochemical adaptation employed by organisms to meet varying environmental conditions. If adjustments in the rates of enzymatic activity are produced by changing the types of enzymes present in the system, then the strategy employed is a qualitative one, resulting in the "warm" isozymes, which display high affinity for substrates at temperatures near the organism's upper thermal range, and the "cold" isozymes, near the organism's lower thermal range. In agreement with these authors, the minimum K_m for unfractionated muscle extracts of *H. malabaricus* was obtained at 10–20°C (temperature-dependent pH and constant-pH using imidazole buffer) where a ratio of two paralogous thermolabile isoforms to one thermostable isoform was observed. Also, for liver homogenates, the minimum K_m was obtained at 30°C (both temperature-dependent pH and constant-pH, using imidazole buffer)and a ratio of two duplicate A (thermostable, A1 and A2) isoforms to one B (thermolabile) was observed.

Three different physiological roles have been proposed for sMDH: gluconeogenesis, lipogenesis and malate-aspartate shuttle during aerobic glycolysis (Zink and Shaw, 1968), where probably both A and B isoforms have different roles (Bailey *et al.*, 1970). According to Coppes *et al.* (1987), in white muscle malate is oxidized by the B isoform to yield oxaloacetate with reduction of NAD. When temperature decreases, the metabolic activities are reduced, and oxaloacetate would be utilized in hexose synthesis or in transaminations. When temperature increases, metabolic activity increases too, and oxaloacetate would be reduced to malate by the A isoform, yielding NAD which would be used by glyceraldehyde-3-phosphate dehydrogenase, favouring the glycolytic direction. However, according to Lin and Somero (1995b), available kinetic data do not clearly establish what the different metabolic roles for the two teleost sMDHs might be.

References

Aspinwal, N. (1974) Genetic analysis of duplicate malate dehydrogenase in the pink salmon, *Oncorhynchus gorbuscha*. *Genetics* **76**: 65–72

Bailey, G.J., Cocks, G.T., Wilson, A.C. (1969) Gene duplication in fishes: malate dehydrogenase of salmon and trout. *Biochem. Biophys. Res. Commun.* **34**: 605–612

Bailey, G.S., Wilson, A.C., Halver, J.E., Johnson,C.L. (1970) Multiple forms of supernatant malate dehydrogenase in salmonid fishes: biochemical, immunological and genetic studies. *J. Biol. Chem.* **245**: 5927–5940

Bertollo, L.A.C., Moreira-Filho, O., Galetti, P.M. Jr. (1986) Cytogenetics and taxonomy considerations based on chromosome studies of freshwater fish. *J. Fish Biol.* **28**: 153–159

Caraciolo, M.C., Val, A.L., Almeida-Val, V.M.F. (1996) Malate dehydrogenase polymorfism in amazon curimatids (Teleostei: Curimatidae): evidence of an ancient mutational evente. *Rev. Bras. Genet.* **19**: 57–64

Clayton, J.W., Harris, R.E.K., Tretiak, D.N. (1973) Identification of supernatant and mitochondrial isozymes of malate dehydrogenase on electropherograms applied to the taxonomic discrimination of walleye (*Stizolection vitreum vitreum*), sanger) and suspected interspecific hybrid fishes.*J. Fish Res. Board Can.* **30**: 927–938

Coppes, Z.L., Somero, G.N. (1990) Temperature-adaptative differences between the A4 lactate dehydrogenases of stenothermal and eurythermal sciaenid fishes.*J. Exp. Zool.* **254**: 121–131

Coppes, Z.L., Schwantes, M.L.B., Schwantes, A.R. (1987a) Adaptative features of enzymes from family Scianidae (Perciformes) - I. Studies on soluble malate dehydrogenase (sMDH) and creatine kinase (CK) of fishes from the South Coast of Uruguay. *Comp. Biochem. Physiol. B* **88**: 203–210

Coppes, Z.L., Schwantes, M.L.B., Schwantes, A.R. (1987c) Adaptative features of enzymes from family Scianidae (Perciformes) - III. Studies on lactate dehydrogenase (LDH) of fishes from the South Coast of Uruguay. *Comp. Biochem. Physiol. B* **88**: 1005–1012

Davidson, R.G., Cortner, J.A. (1967) Genetic variant of human erythrocyte malate dehydrogenase. *Nature* **215**: 761–762

Coppes, Z.L., Schwantes, M.L.B., Schwantes, A.R. (1987b) Adaptative features of enzymes from family Scianidae (Perciformes) - II. Studies on phosphoglucose isomerase (PGI) of fishes from the South Coast of Uruguay. *Comp. Biochem. Physiol. B* **88**: 211–218

De Luca, P.H., Schwantes, M.L.B., Schwantes, A.R. (1983) Adaptative features of ectothermic enzymes. IV. Studies on malate dehydrogenase of *Astyanax fasciatus* (Characidae) from Lobo Reservoir (São Carlos, São Paulo, Brasil). *Comp. Biochem. Physiol. B* 47: 315–324

Farias, I.P., Almeida-Val, V.M.F. (1992) Malate dehydrogenase (sMDH) in amazon cichlid fishes: evolutionary features. *Comp. Biochem. Physiol. B* **103**: 939–943

Fenerich-Verani, N., Schwantes, M.L.B., Schwantes, A.R. (1990) Patterns of gene expression during *Prochilodus scrofa* (Characiformes: Prochilodontidae) embryogenesis – II. Soluble malate dehydrogenase.*Comp. Biochem. Physiol. B* **97** : 247–255

Ferris, S.D.,Whitt, G.S. (1978) Philogeny of tetraploid Catostomid fishes based on the loss of duplicate gene expression. *Syst. Zool.* **27**: 189–206

Ferris, S.D., Whitt, G.S. (1979) Evolution of the differential regulation of duplicate genes after polyploidization. *J. Mol. Evol.* **12**: 267–317

Fisher, S.F.,Whitt, G.S. (1978) Evolution of isozyme loci and their differential tissue expression: creatine kinase as a model system. *J. Mol. Evol.* **12**: 25–55

Fisher, S.F., Shaklee, J.B., Ferris, S.D., Whitt, G.S. (1980) Evolution of five multi-locus isozymes systems in the chordates. *Genetics* **52/53**: 73–85

Graves, J.E., Somero, G.N. (1982) Electrophoretic and functional enzymic evolution in four species of eastern Pacific barracudas from different thermal environments. *Evolution* **36**: 97–106

Hochachka, P.W. (1965) Isoenzymes in metabolic adaptation of a poikilotherm: Subunit relationships in lactic dehydrogenase of goldfish. *Arch. Biochem. Biophys.* **111**: 96g

Hochachka, P.W., Somero, G.N. (1968) The adaptation of enzymes to temperature. *Comp. Biochem. Physiol.* **27**: 659–668

Hochachka, P.W., Somero, G.N. (1973) *Strategies of biochemical adaptation.* Saunders, Philadelphia

Hochachka, P.W., Somero, G.N. (1984) *Biochemical adaptation.* Princeton University Press, Princeton, New Jersey

Kitto, G.B., Lewis, R.G. (1967) Purification and properties of tuna supernatant and mitochondrial malate dehydrogenase. *Biochim. Biophys. Acta* **139**: 1–15

Klebe, R.J. (1975) A simple method for the quantification of isozyme patterns. *Biochem. Gen.* **13**: 805–812

Lin, J.J., Somero, G.N. (1995a) Temperature-dependent changes in expression of thermostable and thermolabile isozymes of cytosolic malate dehydrogenase in the eurythermal goby fish *Gillchthys mirabilis. Physiol. Zool.***68**: 114–128

Lin, J.J., Somero, G.N. (1995b) Termal adaptation of cytoplasmic malate dehydrogenase of Eastern Pacific barracuda (*Sphyraena* ssp): the role of differential isoenzyme expression. *J. Exp. Biol.* **198**: 551–560

Markert, C.L., Moller, F. (1959) Multiple forms of enzyme: tissue, ontogenic and species specific patterns. *Proc. Natl. Acad. Sci. USA* **45**: 753–763

Markert, C.L., Shaklee, J.B., Whitt, G.S. (1975) Evolution of a gene. *Science* **188**: 102–114

Monteiro, M.C., Schwantes, M.L.B., Schwantes, A.R. (1991) Malate dehydrogenase in subtropical fish belonging to the orders Characiformes, Siluriformes and Perciformes. I. Duplicate gene expression and polymorfism. *Comp. Biochem. Physiol. B* **100**: 381–390

Ohno, S. (1974) Protochordata, cyclostomata and pisces. In: John, B. (ed.) *Animal cytogenetics*. Gebrii der Borntraerger, Berlin

Reeves, R.B. (1977) The interaction of body temperature and acid-base balance in ectothermic vertebrates. *Annu. Rev. Physiol.* **39**: 559–586

Schwantes, M.L.B., Schwantes, A.R. (1977) Eletrophoretic studies on polyploid amphibians. III. Lack of locus duplication evidences through tetraploidization. *Comp. Biochem. Physiol. B* **57**: 341–351

Schwantes, M.L.B., Schwantes, A.R. (1982a) Adaptative features of ectothermic enzymes. I. Temperature effects on the malate dehydrogenase from a temperate fish, *Leiostomus xanthurus*. *Comp. Biochem. Physiol. B* **72**: 49–58

Schwantes, M.L.B., Schwantes, A.R. (1982b) Adaptative features of ectothermic enzymes. II. The effects of acclimation temperature on the malate dehydrogenase of the spot, *Leiostomus xanthurus*. *Comp. Biochem. Physiol. B* **72**: 59–64

Shaklee, J.B., Christiansen, J.A., Sidell, B.D., Prosser, C.L., Whitt, G.S. (1977) Molecular aspects of temperature acclimation in fish: contributions of changes in enzyme activities and isozyme patterns to metabolic reorganization in the green sunfish. *J. Exp. Zool.* **201**: 1–20

Shaklee, J.B., Allendorf, F.W., Morizot, D.C.F., Whitt, G.S. (1989) Genetic nomenclature fot protein - coding loci in fish: proposed guildelines. *Trans. Am. Fish. Soc.* **118**: 218–227

Somero, G.N. (1978) Temperature adaptation of enzymes: biological optimization through structure-function compromises. *Annu. Rev. Ecol. Syst.* **9**: 1–29

Somero, G.N. (1981) pH temperature interaction on proteins. Principles of optimal pH and buffer system design. *Marine Biol. Lett.* **2**: 163–178

Somero, G.N. (1986) Protein adaptation and biogeography: threshold effects on molecular evolution. *Trends Ecol. Evol.* 1: 124–127

Somero, G.N., Siebenaller, J.F., Hochachka, P.W. (1983) Biochemical and physiological adaptations of deep-sea animals. In: Rowe, G. (ed.). *The sea.* Wiley, New York

Thorner, C.J.R., Cooper, P.M. (1963) Preparation of pig heart supernatant malate dehydrogenase. *Biochim. Biophys. Acta* **81**: 397–399

Tsugawa, K. (1976) Direct adaptation of cells to temperature: similarchanges of LDH isozyme pattern by in vitro and in situ adaptation in *Xenopus laevis.Comp. Biochem. Physiol. B* 55: 259–261

Tsukuda, H., Ohsawa, W. (1974) Effect of temperature acclimation on the isozyme pattern of liver lactate dehydrogenase in the goldfish, *Carasssius auratus* (L.). *Ann. Zool. Jpn.* **182**: 59–68

Val, A.L., Schwantes, A.R., Schwantes, M.L.B., De Luca, P.H. (1981) Amido hidrolisado de milho com suporte eletroforético. *Ciênc Cult.* **33**: 992–996

Wheat, T.E., Childers, W.F., Miller, E.T., Whitt, G.S. (1971) Genetic and in vitro molecular hybridization of malate dehydrogenase isozymes in interspecific bass (*Micropterus*) hybrids. *Anim. Blood Grps. Biochem. Genet.* **2**: 3–14

Whitt, G.S. (1970) Genetic variation of supernatant and mitochondrial malate dehydrogenase isozymes in the Teleosts *Fundulus heteroclitus. Experientia* **26**: 734–736

Zink, M.W., Shaw, D.A. (1968) Regulation of malic isozymes and malic dehydrogenase in *Neurospora crassa.Can. J. Microbiol.* **14**: 907–912

Exp. Biol. Online 2:19 (1997)

METHODS REVIEW

Quantitative imaging of intact cells and tissues by multi-dimensional confocal fluorescence microscopy

M.D. Fricker[1], C.-M. Chow[2], R.J. Errington[3], M. May[4], J. Mellor[2], A.J. Meyer[1], M. Tlalka[1], D.J. Vaux[5], J. Wood[1] and N.S. White[1]

[1]Department of Plant Sciences, University of Oxford, South Parks Road, Oxford, OX1 3RB, UK
[2]Microbiology Unit, Department of Biochemistry, University of Oxford, South Parks Road, Oxford, OX1 3QU, UK
[3]162 Cell Physiology, University of Nijmegen, PO Box 9101, 6500 HB Nijmegen, The Netherlands
[4]Laboratorium voor Genetica, Universiteit Gent, KL Ledeganckstraat 35, B-9000, Gent, Belgium
[5]Sir William Dunn School of Pathology, South Parks Road, Oxford, OX1 3RE, UK

Correspondence to: M.D. Fricker (e-mail: mark.fricker@plant-sciences.oxford.ac.uk)

Received: 18 September 1997 / Accepted: 18 November 1997
Published: December 4, 1997

Abstract. Confocal fluorescence microscopy enables visualisation and quantitation of fluorescent probes at high resolution deep within intact tissues, with minimal disturbance both of cell–cell interactions and the mechanical, ionic and physiological effects of the extracellular matrix. We illustrate the principles of multiple-parameter 3-D (*x*,*y*,*z*) imaging using reconstruction of nuclear channels in mammalian cells. Repeated sampling in time generates 4-D (*x*,*y*,*z*,*t*) images which can be used to follow dynamic changes, such as blue-light-dependent chloroplast re-orientation, in intact tissues. Quantitative measurements from multi-dimensional images require calibration of the spatial dimensions of the image and the fluorescence intensity response. This must be determined throughout the volume, which must be sampled to correct for geometric distortion as well as photometric errors arising from the complete optical system, including the specimen. The effects of specimen calibration are illustrated for morphological analysis of stomatal closing responses to abscisic acid in *Commelina* from 4-D images. Calibrated 4-D imaging allows direct volume measurements and we have followed volume regulation of chondrocytes in cartilage explants during osmotic perturbation. In intact cartilage, unlike in isolated cells, the chondrocytes exhibit volume regulatory mechanisms. In other cases, the fluorescence intensity of the probe may be related to a physiological parameter of interest and changes in its distribution within the cell. Optical sectioning permits discrimination of signal in separate compartments within the cell and can be used to follow transport events between different organelles. We illustrate 3-D (*x*,*y*,*t*) measurements of vacuolar glutathione conjugate pump activity in intact roots of *Arabidopsis* by following the sequestration of a fluorescent conjugate between glutathione and monochlorobimane. Dynamic measurements of protein localisation are now possible following the introduction of chimeric fusion proteins with green fluorescent protein (GFP) from *Aequoria victoria*. We have analysed the disposition of heterochromatin in nuclei of living *Schizosaccharomyces pombe* cells expressing a chimeric construct between Swi6 and GFP. Heterochromatin dynamics can be followed throughout mitosis in 4-D (*x*,*y*,*z*,*t*) images. Statistical analysis of the fluorescence histograms from each nucleus over time provides quantitative support for aggregation and dispersion of Swi6-GFP clusters during mitosis, rather than dissociation of Swi6 from the heterochromatin. A wide range of single-wavelength and ratio probes are available for imaging different ion activities. We compare 3-D (*x*,*y*,*t*) measurements of ion activities made using single-wavelength (Fluo-3 for calcium) and ratio (BCECF for pH) measurements, using stomatal responses in *Vicia faba* to peptides from the auxin-binding protein of maize and tip growth in pollen tubes of *Lilium longiflorum* as examples. Ratioing techniques have many advantages for quantitative fluorescence measurements and we conclude with a discussion of techniques to develop ratioing of single-wavelength probes against alternative references, such as DNA, protein or cell wall material.

Key words: Chondrocyte volume measurements – Chromodomain proteins – Confocal fluorescence microscopy – Glutathione conjugate pump – Nuclear channels – Quantitative imaging

Abbreviations: *ABP1* Auxin-binding protein from *Zea mays*, *BCECF* [2',7'- bis -(2- carboxyethyl)- 5-(and-6)- carboxyfluorescein], *[Ca²⁺]* cytoplasmic free calcium concentration, *CLSM* confocal laser scanning microscopy, *CMFDA* 5-chloromethylfluorescein diacetate, *Con-A* concanavalin-A, *ER* endoplasmic reticulum, *FC* fusicoccin, *Fluo-3* {1- [2-amino-5- (2,7- dichloro-6- hydroxy-3- oxy-9- xanthenyl)- phenoxy]-2- [2-amino-5- methylphenoxy] ethane-*N*,*N*,*N'*,*N'*-tetraacetic acid}, *FWHM* full-width at half maximum, *GFP* green fluorescent protein, *GS- B* glutathione-bimane conjugate, *GSH* glutathione, *GS-X* any glutathione conjugate, *NA* numerical aperture, *PI* propidium iodide, *PSF* point spread function, *r.m.s.* root mean square, *SBL* strong blue light, *s.d.* standard deviation, *S/N* signal-to-noise ratio, *WBL* weak blue light

Introduction

Fluorescent probes provide powerful tools with which to visualise and quantify dynamic events within single living cells with a relatively low degree of perturbation. Many techniques are now available for quantifying fluorescence signals from reporter molecules within biological specimens. These include fluorimetry, flow cytometry and microscope photometry (all non-imaging methods) as well as wide-field, two-photon and confocal microscopies. Several recent volumes cover many of the technical details and practical applications of these techniques (e.g. Wang and Taylor, 1989; Taylor and Wang, 1990; Mason, 1993; Matsumoto, 1993; Nuccitelli, 1994; Pawley, 1995; Gilroy, 1997; Van Duijn and Wiltink, 1997). The choice of measurement system depends on the specific range of sampling conditions and specimens, and several techniques are often used in combination to provide the required range of spatial, temporal and spectral resolution. All of the above methods can be used with single cells or monolayers maintained in culture. However, measurements are often needed from cells within intact, living tissues that are functioning within an environment more closely resembling their normal physiological context. Most notably this includes the effects of cell–cell interactions, mechanical interactions and the ionic physiology of the extracellular matrix (e.g. Errington *et al.*, 1997). Confocal microscopy is particularly useful for such preparations, as it provides high-contrast images, largely free from out-of-focus blur, and can be used directly for non-invasive serial optical sectioning deep within intact tissues. In this article, we illustrate the principle of fluorescence quantitation by confocal microscopy of single cells and cells in intact tissues, with examples from our own work. We describe errors typically encountered during optical sectioning of thick material and present pragmatic solutions. We also discuss some of the data-presentation strategies used to visualise multi-dimensional confocal microscopy images.

Optical sectioning and 3-D reconstruction

In conventional fluorescence microscopy, the excitation light continuously illuminates a large volume in the specimen to allow simultaneous measurement of the fluorescence emission from all points in the 3-D field. Fluorescence is unavoidably excited throughout the entire volume that extends above and below the plane of focus of the objective. Thus, the fluorescence image is not sharply focused at the microscope image plane and is spread out or blurred at the camera. If the specimen is very thin, in the order of a few micrometres, this often presents few problems. However, in larger cells or cells that form part of a multicellular structure, the out-of-focus blur significantly degrades the contrast and distorts any attempts at high-resolution measurements. In a confocal laser scanning microscope (CLSM), the excitation light behaves as a "near-point" source which is focused into the specimen by a high NA lens. Fluorescence is still excited in the cones of illumination above and below the focus. However, this out-of-focus fluorescence is prevented from contributing to the final image by a small physical aperture in the imaging plane near the detector. In-focus fluorescence from the focal point passes through this aperture and reaches the detector. Thus, the illumination and detection are co-aligned or "confocal". A simple 2-D optical section is generated by scanning the confocal point in the (x,y) plane. A series of optical sections collected at different focus levels constitutes a three-dimensional (x,y,z) (3-D) image. The removal of out-of-focus blur allows discrimination of fluorescence from subcellular domains, whilst any contaminating signal from outside the focal plane is rejected.

When combined with 3-D visualisation techniques, optical sectioning may be sufficient to reveal novel information previously obscured by the lack of both contrast and z-discrimination. For example, serial optical sections collected at 0.5-μm intervals from the base of a human cervical epithelial tumour cell (HeLa cell) labelled with fluorescently tagged Concanavalin A (Con-A) are shown as an animation in Fig. 1A (taken from Fricker *et al.*, 1997b). Only the nucleus and perinuclear regions are shown. Fluorescence from labelled Con-A binding to high-mannose glycoproteins present in the lumen of the endoplasmic reticulum (ER) and the lumen between the inner and outer membranes of the nuclear envelope is visible. Fluorescently labelled structures are also present within the nucleus. As these features run through several sections, it is difficult to determine their extent and morphology in a single optical section. Smooth animation of the z-series of images assists in interpretation of the 3-D organisation, capitalising on depth-cues and the visual memory of the viewer. The morphology of these structures can be seen more clearly in projected views (White, 1995), after removal of the sections containing the thin layer of cytoplasm underlying the nucleus (Fig. 1B). In these height-coded reconstructions, structures above a threshold intensity nearest the viewer are coded as bright, masking underlying structures, to provide a 3-D impression not available from single sections (White, 1995). It is now clear that the structures within the nucleus form continuous tubes, the most prominent of which traverses the nucleoplasm, contacting the nuclear envelope at two locations. A second shorter tube is visible that terminates within the nucleus. This can be seen more clearly if a rotation series of projected views is animated (Fig. 1B). Since the pixel value now represents "height" or z-position, then the (x,y) and (z) co-ordinates can be interactively measured from an on-screen (x,y) cursor position and pixel (z) value.

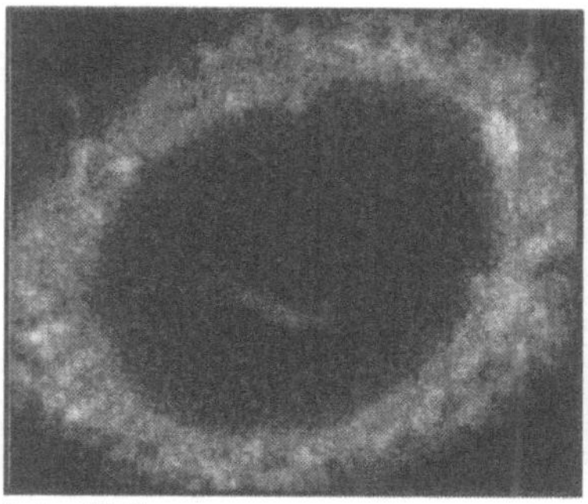

(384 KB)

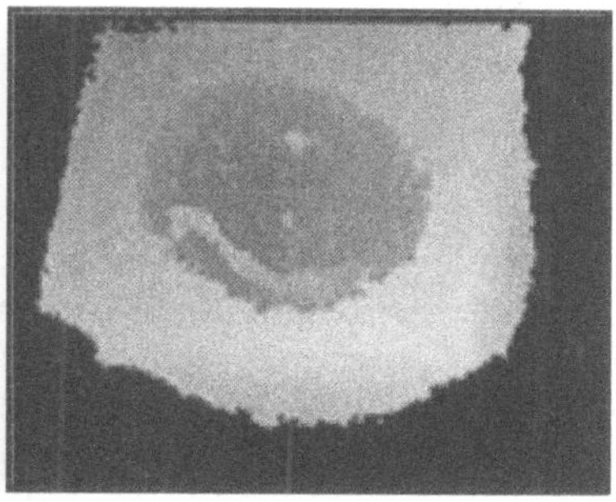

(288 KB)

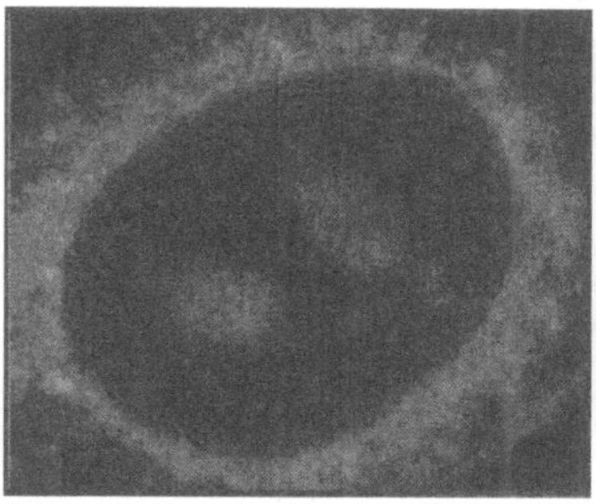

(928 KB)

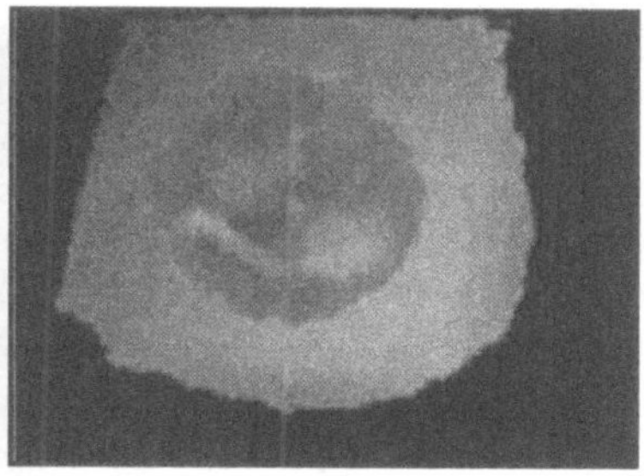

(576 KB)

Fig. 1A–D Optical sectioning and 3-D visualisation of nuclear channels in mammalian cells. **A** Serial optical sections at 0.5-μm intervals of the nucleus and peri-nuclear region of a HeLa cell labelled with fluorescein-tagged concanavalin A (*Con-A*). Label is visible in the endoplasmic reticulum, nuclear envelope and within the nucleus. The morphology of the channels can be visualised in height-coded reconstructions, viewed from the base of the cell (**B**). Structures above a threshold intensity are coded as bright nearest the viewer and mask underlying structures. Double labelling with propidium iodide (*PI*) to stain nucleic acids (**C** – *red*) reveals the relative juxtaposition of the nucleoli and the channels (**C** – *green*). In a merged image of the height-coded Con-A (**D** – *green*) and a maximum brightness projection of the PI signal (**D** – *red*), the major channel runs underneath one nucleolus and adjacent to the second, whilst the shorter channel terminates close to the second nucleolus. Images were collected with a modified Bio-Rad MRC600 (Bio-Rad Microscience,Hemel Hempsted, Herts., UK) CLSM attached to a Nikon Diaphot TMD inverted m icroscope using a Nikon 60 × 1.4 NA Plan-Apochromat oil immersion lens. Excitation was at 488 nm, using an argon-ion laser (25 mW ILT,Salt Lake City, Utah, USA). Emission was at 540 ±15 nm for Con-A and >600 nm for PI. Image processing and 3-D reconstruction were performed using COMOS, MPL and ThruView PLUS (Bio-Rad Microscience) on a PC workstation. Video animation was performed using Confocal Assistant v. 4.02 (T.C. Berlje)

Multi-channel 3-D microscopy

In comparison to generating views of a single label in a 3-D image, it is more difficult to achieve a satisfactory projection algorithm that combines the information from two 3-D images of different labels in the same specimen. Double labelling of HeLa cells with fluorescent Con-A (shown in green in Fig. 1C) and with the nucleic acid stain propidium iodide (PI) (shown in red in Fig.

1C) reveals that the major tube passes underneath one PI-positive domain and adjacent to the second. The shorter channel also terminates close to the second PI-positive region (Fig. 1C). The PI-positive regions are likely to represent the nucleoli. In Fig. 1D, a rotation series of the height-coded Con-A image (green) is combined with a maximum brightness rotation series of the PI image (red). This approach assists in the understanding of the 3-D juxtaposition of the two structures. However, there are still some visual ambiguities, as the height-coded image appears "solid", whilst the transparency of the merge operation makes the PI view appear "through" the Con-A-stained structure.

At the light microscope level, the channels show a cytoplasmic core accessible to high-molecular-weight markers introduced into the cytoplasm by scrape loading The channels can also be labelled with a range of other markers including antibodies to luminal ER markers, or to rough ER membrane proteins (Fricker *et al.*, 1997b) or with anti-lamin antibodies. Simple "merging" by combining coloured channels together can be effective for visualisation of the co-localisation of several labels. However, it is often difficult for the viewer to separate the individual components when their colour alters depending on the degree of overlap between other signals. An alternative approach is to present a combination of (x,y), (x,z) and/or (y,z) orthogonal views derived from projections over selected subregions of the image to highlight specific structures. For example, in Fig. 2, HeLa cells were triple labelled with Con-A (Fig. 2A), anti-lamin antibodies (Fig. 2B) and the DNA dye Yo-Pro 1 (Molecular Probes, Eugene, Ore., USA) (Fig. 2C). Maximum projections of six optical sections in the (x,y) plane are shown for each label individually and as a three-colour image. In addition, ten-pixel-deep (x,z) and (y,z) projections are shown along the lines indicated by the black arrowheads. Each of the Con-A sites in the nucleus is associated with regions of lamin immuno-reactivity and a reduced level of DNA label. However, there are also additional intranuclear lamin-reactive sites that do not have associated Con-A labelling. Coloured images can provide useful information on co-localisation (e.g. Hassan *et al.*, 1994), but the fine details of the labelling are much clearer in the individual monochrome views. This example illustrates the importance of using the computer display resources (colour, pixel resolution, etc.) efficiently to aid presentation and interpretation of multi-dimensional images (see White, 1995).

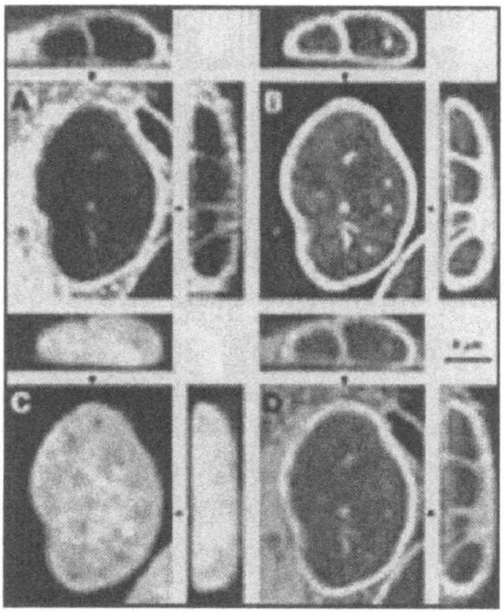

Fig. 2A–D Con-A-labelled channels are associated with lamin immuno-reactivity in HeLa cells. Maximum projections of six optical sections at 0.3-μm (z) intervals from a triple-labelled HeLa nucleus are shown for Con-A (**A**), polyclonal anti-lamin reactivity (**B**) and nucleic acid labelling with To-Pro-3 (**C**). Associated (x,z) *above* and (y,z) *right* 10-pixel-wide projections along the lines indicated by the *black arrowheads* are also shown. In the merged image (**D**) each of the Con-A sites co-localizes with sites of lamin reactivity, but there are additional sites of lamin reactivity that do not have associated Con-A signals. *Scale bar* = 5 μm. Images were collected with a BioRad MRC1000 CLSM (Bio-Rad Microscience) attached to a Nikon Diaphot 300 inverted microscope using a Nikon 60 ×1.4 NA Plan-Apochromat oil-immersion objective with excitation at 488 nm, 568 nm, and 647 nm (15 mW Argon-Krypton; ILT). Fluorescence emissions were collected at 552±35 nm for Yo-Pro-1 (DNA), 605±32 for TRITC (anti-lamin) and 680±32 for Cy-5 (Con-A). Image processing and 3-D reconstruction were performed using COMOS, ThruView PLUS and Lasersharp (Bio-Rad Microscience) on a PC workstation. Montages were assembled using Adobe Photoshop v.4.0 (Adobe Systems)

Using this type of 3-D CLSM technique in conjunction with serial section transmission electron microscopy (TEM), we have found evidence for the presence of long, branching intranuclear membrane channels that are derived from the ER, as deep, narrow invaginations of the nuclear envelope in all tissue culture cell types examined and in primary cells in culture. The number of channels and their morphological complexity vary widely but remain characteristic for a given cell type (Fricker *et al.*, 1997b). Whilst the association between channels and nucleoli sugests a possible role in nucleocytoplasmic transport, we are currently trying to ascertain the function of these intranuclear channels.

Time-lapse 3-D (4-D) microscopy

Whilst optical sectioning and 3-D reconstruction reveal the detail of structures within a short period of time, repeated 3-D sampling [4-D (x,y,z,t) microscopy] provides a means to follow dynamic changes in cell and organelle morphology, distribution or physiology in living tissue. For all measurements and particularly in 4-D microscopy, the signal-to-noise (S/N) ratio must be sufficient to record the signal of interest with minimal disruption to the cell. Extended imaging time, high laser intensities and high dye loading are not compatible with cell viability and can severely disrupt many of the cell's physiological processes (Terasaki and Dailey, 1995). Thus, for 4-D imaging it is essential to optimise the data-aquisition regime (e.g. Errington *et al.*, 1997).

As an example, we have been using 4-D imaging to record the position of chloroplasts in leaves from *Lemna trisulca* during re-orientation induced by blue light. In this and many other species, the position of the chloroplasts continuously alters in response to the ambient light intensity to maintain a balance between light capture for photosynthesis and light avoidance to reduce photodamage. The arrangement of the chloroplasts was imaged in 4-D using chlorophyll autofluorescence excited at 543 nm to avoid stimulating the blue-light-sensitive transduction system. Chloroplast positions were recorded throughout the entire cell volume from each 3-D image with z-sampling of 1- to 2-μm steps at depths of up to 50 μm from the surface. Chloroplasts were tracked during exposure to specific light regimes at 1- to 10-min intervals and visualised as height-coded images (Fig. 3). The chloroplasts initially moved from a random arrangement in darkness to a profile position clustered adjacent to the anticlinal cell walls in response to strong blue light (SBL, ~20 W m^{-2}). In weak blue light (WBL, ~0.6 W m^{-2}), the chloroplasts moved from the sides of the cells to a full-face position adjacent to the periclinal cell walls. The height-coded views (Fig. 3) can be used to extract the three spatial co-ordinates, (x, y) and (z), directly from the screen. Thus, the position, distance moved, direction and average velocity of each chloroplast can be readily extracted from the 4-D image. When different parts of individual cells were illuminated with different intensities or wavelengths of light, the individual chloroplasts responded according to the local light intensity that they experienced (e.g. Fig. 4). This suggests that the signal transduction system has sufficient spatial resolution to allow perception and transduction of opposite responses within the same cell.

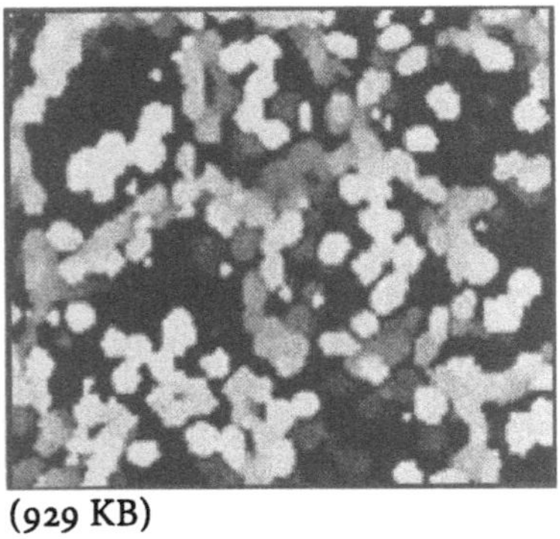

(929 KB)

Fig. 3 Chloroplast re-orientation in *Lemna trisulca* in response to blue light. Chloroplasts are initially distributed randomly around the periphery of the cells, but move rapidly to a band adjacent to the anti-clinal walls in response to strong blue light. They subsequently move to the periclinal walls in response to a weak blue-light stimulus. Images were collected with a modified Bio-Rad MRC600 (Bio-Rad Microsciences) CLSM with excitation at 543 nm (1.3 mW Gre-Ne; Melles-Griot, Rochester, N.Y., USA) and emission at >600 nm using a Zeiss 25 × 0.8 NA Plan-Neofluar multi-immersion objective. Views are presented as height-coded reconstructions of 32 optical sections at 1-μm intervals. Image processing and 3-D reconstruction were performed using MPL and

Lasersharp (Bio-Rad Microscience) on a PC workstation. Video animation was performed using Confocal Assistant v. 4.02 (T.C. Berlje)

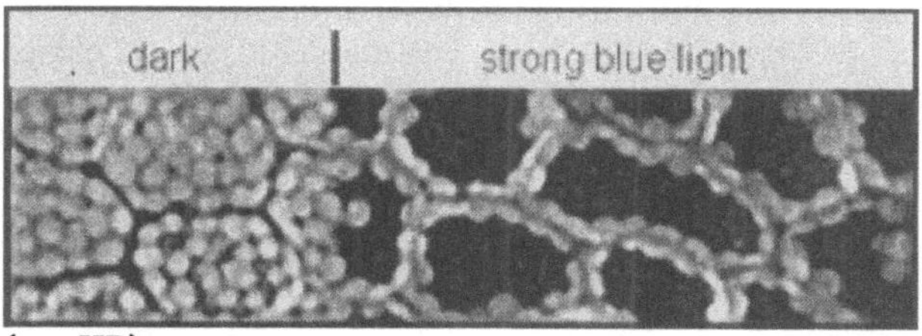

(352 KB)

Fig. 4 Chloroplast re-orientation in response to differential illumination of adjacent regions. Maximum projections of 30 images at 1-μm sections at approximately 10-min time intervals. Within individual cells, the chloroplasts respond according to the local light regime they experience. Thus, in the dark the chloroplasts are distributed throughout the cell and only move to a limited degree. In weak blue light the chloroplasts move to the anti-clinal surface. In strong blue light they migrate to a band around the periclinal walls. Imaging and processing conditions as for Fig. 3

Sources of error in multi-dimensional microscopy

The views shown in Figs. 1–3 have already been processed significantly from the original 3-D and 4-D images to allow presentation on a 2-D display device. If the visualisation and analysis techniques are to produce reproducible and reliable data three criteria must be fulfilled: first, the dimensions of the original images must be correctly calibrated in (x, y, z) and (t); second, the intensity response must be uniform throughout the volume sampled; third, objective criteria must be applied at each stage of the analysis. Each voxel (3-D pixel) of a digitally stored image represents an approximation to the intensity emitted from a corresponding volume within the specimen. This volume element represents the confocal "optical probe" or point spread function (PSF). Aberrations in the optical path of the microscope and physical interactions between the specimen and both the illumination and fluorescence light significantly affect the position and shape of the confocal probe (Carlsson, 1991; Visser *et al.*, 1991; Hell and Stelzer, 1993; Cogswell and Larkin, 1995; White *et al.*, 1996; Sheppard and Török, 1997). Increased scattering is associated with local refractive index inhomogeneities in the specimen and results in some loss of signal. However, the most important of these interactions are refractions, which occur at all non-opaque boundaries through the optical path of the microscope and the specimen. Aberrations introduced by the sample are particularly difficult to measure and control, as there may be many refractive index boundaries with varying orientation and transparency. If the illumination or fluorescence light is refracted so that it no longer appears to emanate from a "point source" the effect is known as spherical aberration. As refractive index varies with wavelength in most materials (dispersion), the magnitude of these errors will also vary with wavelength (chromatic aberration). The result is a scaling change in (x, y) and (z), and broadening of the imaging probe, which may happen to different degrees around the specimen volume. Blurring of the probe results in a decrease in fluorescence signal passing through the confocal detector aperture, leading to significant attenuation of intensity. Thus, in confocal microscopy, spatial errors also give rise to photometric errors (Hell and Stelzer, 1995; White *et al.*, 1996). We can simulate these errors using simple Gaussian models of the PSF after application of appropriate geometric and chromatic errors (Fig. 5 – see White *et al.*, 1996). The chromatic shifts have also been measured directly using reflection imaging (e.g. Fricker and White, 1992; Cogswell and Larkin, 1995) or multi-labelled fluorescent microbeads (e.g. Manders, 1997).

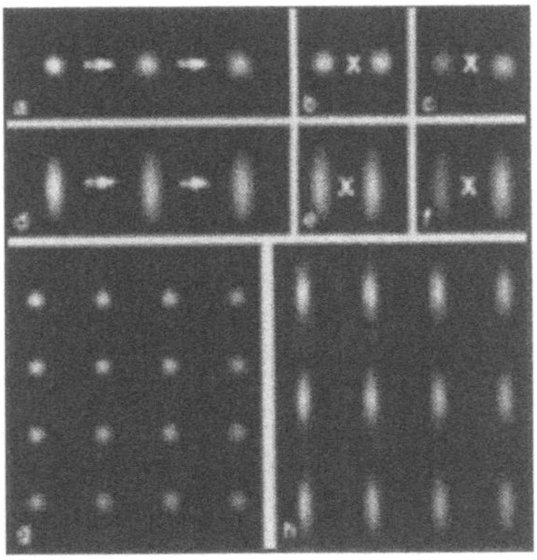

Fig. 5A–H Three-dimensional Gaussian point spread functions (*PSFs*) crudely model the image intensity from a theoretical point object. **A** The effects of blurring the PSF by increasing the Gaussian rms radius to simulate monochromatic error. Subsequent chromatic shifts are represented by a wavelength-dependent change in magnification. The reponse of the complete microscope is represented as the illumination response × the detection response with blurring for both illumination and detection (**B**) and chromatic shifts for the detection (**C**). The corresponding axial effects are shown in *panels* **D–F**. The effects of these errors across the field of view in (x,y) and with depth (z) are shown in *panels* **G** and **H**, respectively. The Gaussian PSFs are repeated across one quadrant (origin top left) of a 15-mm primary image plane (**G**) and 30 µm in depth (**H**). Model PSFs at each position are enlarged for clarity. Spherical blurring to one-third the PSF diameter over the field was used with chromatic error up to half the PSF. The attenuation arising from blurring and chromatic shifts is clearly visible at the extremes of the field and with increasing depth. Gaussian PSF models were generated with Semper (Synoptics), montages were assembled using Adobe Photoshop v. 4.0 (Adobe Systems)

The extent of blurring in a biological specimen can be appreciated from measurement of the PSF at different points in the sample. Fig. 6 shows 2-D (x,z) sections of measured confocal PSFs from 3-D images of 100-nm fluorescent beads located above and below an epidermal strip from *Commelina communis* (White *et al.*, 1996). The response above the epidermal strip is sharper than that below. The blurring is predominantly axial and asymmetric, with only a small lateral component. The full-width at half the maximum peak height (FWHM) of the PSF above the epidermis is 200 nm in (x,y) and 600 nm in z. The corresponding values below the epidermis are 250 nm in (x,y) and 1200 nm in (z). Separate calibration of these errors for a given specimen is required to convert the raw digital data to a true rectilinear array of corrected intensities (White *et al.*, 1996; Errington *et al.*, 1997). Spatial errors in (x,y) lead to changes in magnification across the field. These are typically small, with less than a 10% difference in linear scaling at the field edge. In practice, the geometric consequences of magnification distortion are minimised by only sampling the central half to one-third of the field of view. However, due to the need for accurate confocal alignment of illumination and fluorescence, the resulting attenuation across the entire field of view may be as high as 50%. Axial (z) focus errors are far more significant and yield spatial errors of up to 33% depending on the combination of lens, immersion medium and sample (White *et al.*, 1996). An approximate value for the z-magnification error is $\eta_{immersion} / \eta_{medium}$. The corresponding z-attenuation varies from less than a small percentage for $\eta_{immersion} = \eta_{medium}$ for a transparent sample, to greater than 90% over 100 μm for oil into water and plant specimens.

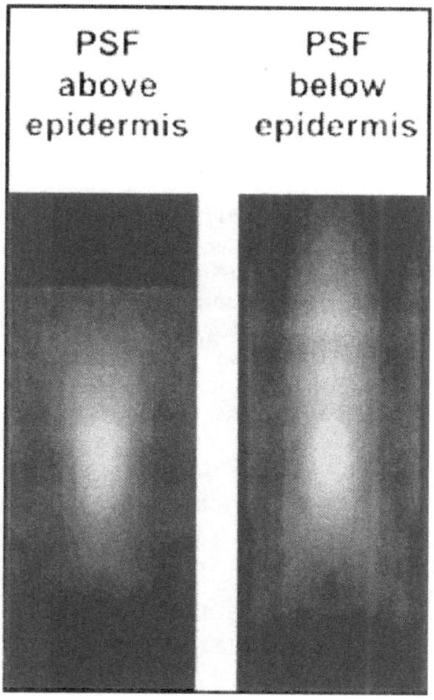

Fig. 6 Measurement of aberrated PSFs *in situ*. The PSF was measured *in situ* from 100-nm fluorescent beads located above and below an epidermal peel from leaf tissue of *Commelina communis* using a 60× 1.4 NA Plan-Apochromat oil-immersion lens with excitation st 488 nm and emission at >520 nm.. The PSF is more asymmetrical and blurred in a predominantly axial direction when measured below the epidermis. The full-width at half the maximum peak (*FWHM*) increases from 200 nm to 250 nm in (x,y) and 600 nm to 1200 nm in (z)

To calibrate more accurately the distortion of axial focus position due to mismatched immersion and mountant, the distance between glass/aqueous-medium/glass boundaries in a slide-coverslip sandwich can be measured and compared with the "correct" distance determined with a matched immersion system, usually water immersion into an aqueous medium (Fig. 7) (Hell *et al.*, 1993; Hell and Stelzer, 1995; White *et al.*, 1996; Errington *et al.*, 1997; Sheppard and Török, 1997). This problem is reduced if water-immersion lenses can be used for the experiment and the specimen is largely aqueous. Appropriate high-NA water-immersion lenses have recently become available from several manufacturers to help address this issue (Keller, 1995). The correction for z-axis focus error can be applied directly to the images to view correctly scaled (*x*,*z*) sections, but for some measurements it is more efficient to apply the correction to the numerical values extracted during the analysis (White, 1995; White *et al.*, 1996; Errington *et al.*, 1997).

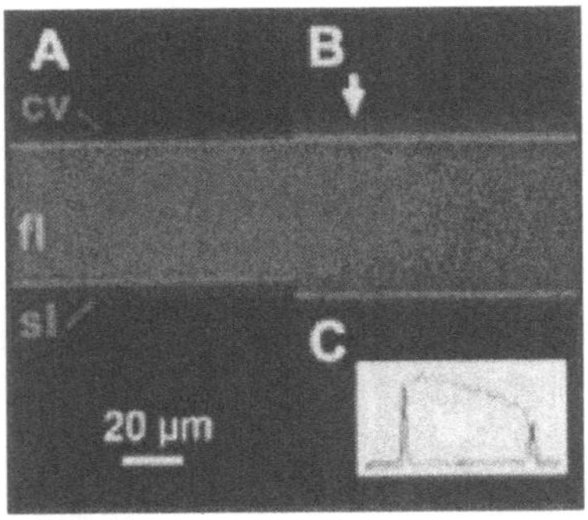

Fig. 7A–C Empirical determination of (*z*) distortion. Reflection (*red*) and fluorescence (*green*) (*x*,*z*) profiles through a sandwich of 53 µM fluorescein (*fl*) in buffer between a coverslip (*cv*) and slide (*sl*). A (*x*,*y*) image collected with a 25× 0.8 NA multi-immersion lens in water-immersion mode. **B** Image collected with the same lens in oil-immersion mode. The increase in separation of the reflection peaks in the oil-immersion mode arises from the *z*-focus distortion due to mismatched immersion and mountant. A (*z*)-scaling factor of 0.88 should be applied to images collected under these conditions to give correctly calibrated (*z*)-spacing. The inset graph (**C**) show an (*x*,*z*) transect at the point indicated by the *arrow*. The reflection peaks define the glass/medium boundaries and correspond to the 50% fluorescence intensity. The decrease in intensity with depth arises as a consequence of increased probe blurring

Partial correction for signal attenuation with depth into biological specimens can be achieved by a pragmatic approach (Errington *et al.*, 1997; White *et al.*, 1996), based on measurement of the loss in signal with depth in a model system derived from the specimen after permeabilisation and infiltration with a uniform "sea" of fluorochrome (Fig. 8). The resultant axial intensity response combines the effects of the depth-dependent "sea" response and the additional contribution of the permeabilised tissue.

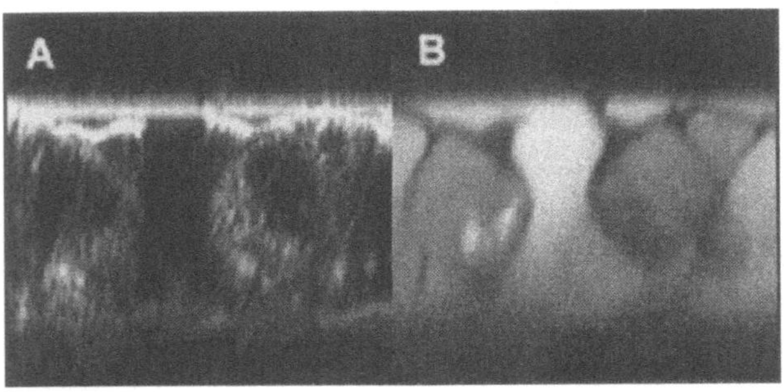

Fig. 8A,B Pragmatic measurement of sample-dependent axial attenuation of fluorescence. Fixed, permeabilised epidermal strip from *Commelina communis* were infiltrated with a "sea" of 53 µM fluorescein and imaged with excitation at 488 nm and emission at 535±25 nm using a Nikon 60× 1.4 NA Plan-Apochromat oil-immersion objective. Axial (*x*,*z*) sections in reflection (**A**) and fluorescence (**B**) are shown. The reflection image (**A**) shows the highly refractive (and reflective) cuticle/cell wall on the outer surface adjacent to the coverslip. The fluorescence signal decreases with depth throughout the epidermis as a consequence of sample-dependent abberations. The difference in intensity between the pore and the guard cell lumen arises from incomplete equilibration of internal and external solutions

The attenuation profile through permeabilised epidermal tissue from *Commelina communis* is shown in Fig. 9. A pronounced decrease in intensity with depth through the specimen can be observed from (*x*,*z*) transects through the pore region (Fig. 9). The level of attenuation through the permeabilised vacuole of the guard cells is similar after normalisation to account for the absolute difference in intensities between the pore and the vacuole (Fig. 10). The level of attenuation is substantial, even over a relatively small depth into the specimen, probably as a result of the complex wall architecture and the cytoplasmic density of cells in the stomatal complex. *In vivo*, the effects are likely to be marginally worse even than this case, as a number of refractive boundaries, particularly membranes, will be distorted or extracted during the fixation/permeabilisation procedure.

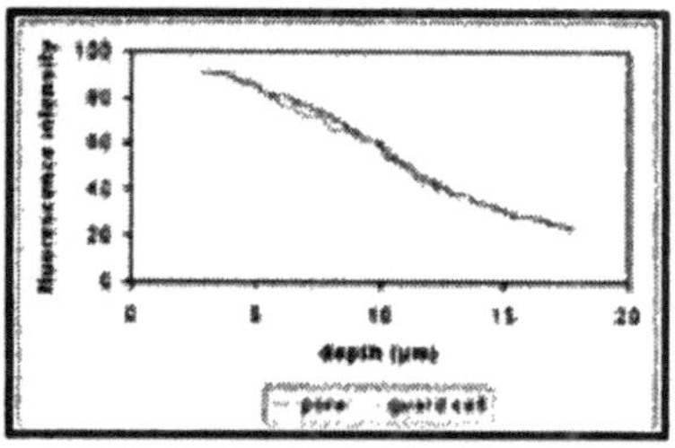

Fig. 9 sample-dependent axial attenuation profiles for guard cells. Axial intensity profile through the pore region (*red*) and guard cell vacuole (*green*) of a permeabilised and infiltrated epidermal srip from *Commelina communis*. The start of the vacuolar trace has been normalised to the corresponding pore profile at the same depth for comparison. Graphical analysis was performed using Excel (Microsoft)

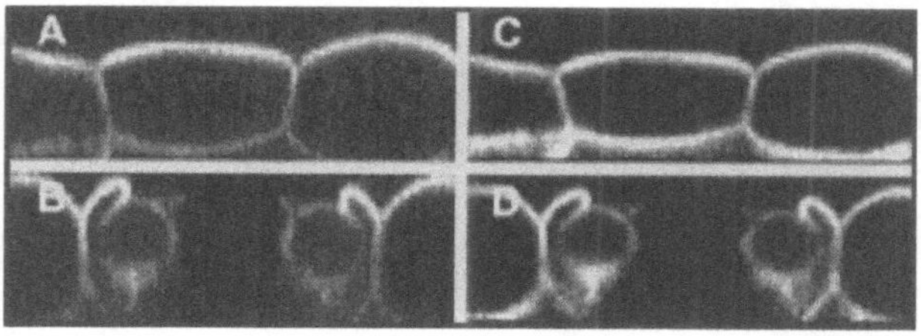

Fig. 10A–D Digital correction of axial distortion and attenuation. Cell walls in the guard cell complex from *Commelina communis* were stained with primulin and imaged with excitation at 442 nm (15 mW He-Cd, Lambda Photometrics) and emission at >520 nm using a Nikon 60× 1.4 NA Plan-Apochromat objective. Median vertical (*x*,*z*) sections through uncorrected images of the terminal subsidiary cell (**A**) and the stomatal pore (**B**) show axial stretching and axial attenuation. The corresponding images in **C** and **D** are correctly scaled in (*z*) and the intensities have been restored using a parametised model derived from the permeabilised in situ "sea" response (see Fig. 9). Graphical analysis was performed in Excel (Microsoft) and correction factors applied using MPL (Bio-Rad Microscience)

A parametised model of the attenuation function can be fitted to the attenuation profile. Inversion of the function gives the correction factor needed to restore the fluorescence intensity at a given depth. The effects of attenuation correction can be seen after labelling the cell walls of the stomatal complex with primulin. Without any correction, there is a substantial reduction in intensity and also progressive blurring of cell-wall fluorescence in (*x*,*z*) sections (Fig. 10A,B). Application of a depth-dependent correction factor, derived from inversion of the attenuation profile, is sufficient to restore the apparent intensity from the cell walls deeper into the tissue, although the increase in blurring of the confocal probe is still apparent in the rescaled images (Fig. 10C,D). After correction of each 3-D image in the time series, changes in the relationship between cells in the stomatal complex can be followed during physiological responses. Images are shown for the median (*x*,*y*) plane (Fig. 11A), median (*x*,*z*) plane (Fig. 11B) and as a height-coded reconstruction of half the data set (Fig. 11C) for a stomatal complex from *Commelina communis* during abscisic-acid-induced pore closure. In the corrected data, cells in the stomatal complex show the correct 3-D proportions and there are clearer details of the interactions with the neighbouring subsidiary cells.

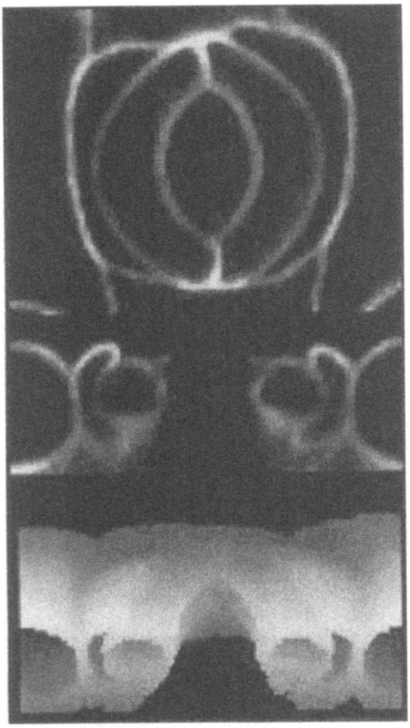

(640 KB)

Fig. 11A–C Four-dimensional imaging of stomatal responses to abscisic acid in epidermal strips from *Commelina communis* stained with primulin. Median (*x*,*y*) (**A**) and (*x*,*z*) (**B**) sections are shown, along with tilted height-coded reconstructions of half the data set (**C**) after correction for (*z*)-distortion and (*z*)-attenuation. Three-dimensional images of 35 sections at 1-μm intervals were collected at 10-min time intervals using a modified Bio-Rad MRC600 CLSM with a Nikon 60× 1.4 NA oil-immersion lens. Excitation at 442 nm (15 mW He-Cd, Lambda Photometrics) and emission at >520 nm. Seven time points out of ten are shown after application of 1 μM abscisic acid

Multi-dimensional measurements: volume

In addition to allowing correct visualisation, a fully calibrated image can provide a wealth of quantitative data. For measurements of lengths, areas and volumes, the object of interest is usually segmented from the image. The simplest form of segmentation relies on a single-intensity threshold to define the boundary of the object. As the accuracy of the subsequent measurements is critically dependent on the position of the boundary, it is essential to have an objective means to determine the threshold value. For large, relatively homogeneous objects, the 50% intensity between the background and the centre of the object is straightforward to determine and relatively robust (Errington *et al.*, 1997), although there is a systematic error using this method for objects similar in size to the PSF (Van Vliet, 1993). Pixel connectivity provides an additional criterion to a simple threshold segmentation. Using this method, voxels above the threshold and with a defined degree of connectivity with neighbours are included. This "seed fill" function can be used to semi-automatically segment individual objects even if there are other objects above the same threshold in the image.

As an example of dynamic volume measurement, we have been studying volume regulation of chondrocytes in explants of porcine articular cartilage. Porcine chondrocytes occur in clusters of single or paired cells, surrounded by an extensive extracellular matrix. Cells were labelled with 5-chloromethylfluorescein diacetate (CMFDA) and 4-D confocal fluorescence images were collected during osmotic treatments. The individual chondrocytes were segmented using a 50% threshold and their volumes calculated from summation of the number of pixels remaining in the binarised images. A composite image is shown in Fig. 12 where the individual chondrocytes have been "seed-filled" from the original 3-D images, colour coded and re-assembled in a rotation series of projected views.

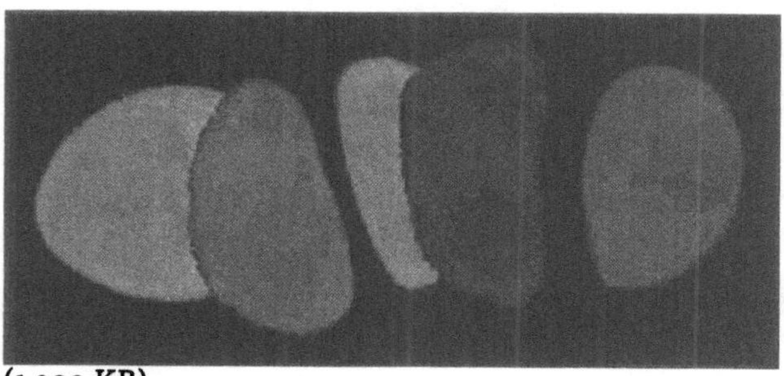

(1.200 KB)

Fig. 12 Postacquisition digital correction and reconstruction of porcine chondrocytes in cartilage explants. Chondrocytes in cartilage explants from the metacarpophalangeal joint of ~8-month-old pigs were labelled with CMFDA and imaged using a Zeiss 25× 0.8 NA multi-immersion objective with excitation at 488 nm and emission at 535±25 nm. Three-dimensional images were corrected for axial distortion using a (z)-scaling factor of 0.88, and axial attenuation using an empirically determined correction equation. Each chondrocyte was then segmented using a 50% threshold between the cell and the background. The views presented have been colour-coded to indicate the cell identity and re-assembled as a rotation series

We have found it particularly useful to visualise changes in volume directly in reconstructed views from 4-D images, as a significant change in volume is often not readily apparent from the small changes in linear dimensions. In the same way that it is customary to code other derived parameters - such as [Ca^{2+}], pH, etc., using pseudo-colour to enhance perception of small changes - volumes can also be coded relative to the starting volume of each cell. The volume of individual chondrocytes in cartilage explants was measured. A composite image was then reconstructed, in which the morphology and shape of the cells is preserved and the relative volume is coded in colour (Errington *et al.*, 1997). Thus, Fig. 13 shows the results of a typical time-course experiment following cell volume changes during exposure of porcine chondrocytes *in situ* in a cartilage explant to anisotonic conditions. The sequence shows reconstructions of 3-D images at six time points that follow, in particular, cell volume recovery in response to shrinkage induced by hypertonicity after pre-swelling. The colour-coded views show relative volume in colour with respect to their initial volume (green). The pixel brightness represents the fluorescence intensity of the cells labelled with CMFDA. On transition from 280 mosmol kg^{-1} H_2O to 140 mosmol kg^{-1} H_2O, the cells enlarged asynchronously by varying amounts, ranging from 10% to 30% (coded as green to red in the images). After this hypotonic swelling, all the cells shrank rapidly in response to a subsequent exposure to a hypertonic solution (380 mosmol kg^{-1} H_2O) but then slowly regained their initial volume. The average change in volume for a group of eight cells is shown in Fig. 14.

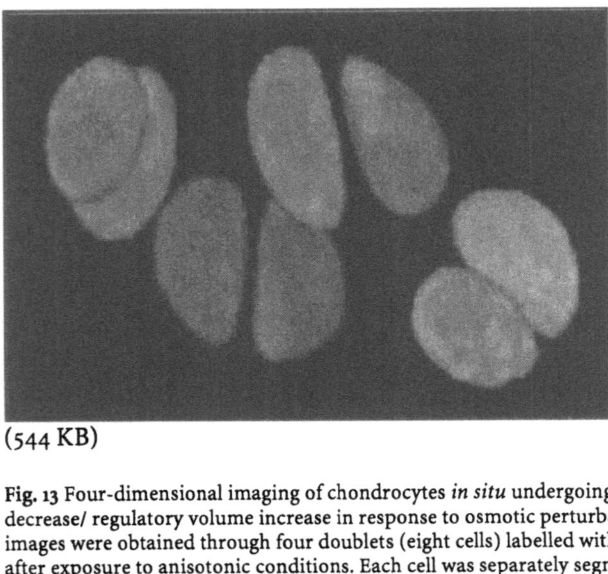

(544 KB)

Fig. 13 Four-dimensional imaging of chondrocytes *in situ* undergoing a post-regulatory volume decrease/ regulatory volume increase in response to osmotic perturbation. Three-dimensional images were obtained through four doublets (eight cells) labelled with CMFDA before, during and after exposure to anisotonic conditions. Each cell was separately segmented, the volume determined and this parameter used to colour-code the cell in the reconstructed views relative to its starting volume (*green* = 1). These cells swelled (colour tends towards *red*) when exposed to a hypotonic shock from 280 to 140 mosmol/l but to different extents and at varying rates (frames 1–3). After this pre-swelling, cells all shrank rapidly and co-ordinately (*frame 4*) upon perfusion with hypertonic buffer (140–380 mosmol/l). The cells then regained most of their initial volume after 60 min (*frame 6*) despite the continued presence of the 380 mosmol/l buffer. Imaging conditions as for Fig. 12

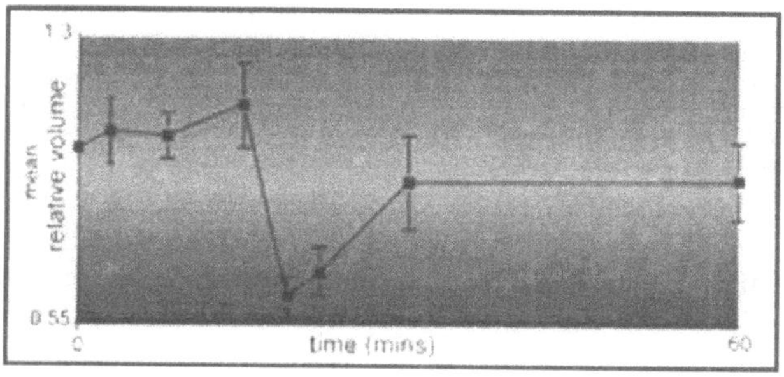

Fig. 14 Graphical representation of the mean volume change (± 1 s.d. unit) for the cells shown in Fig. 13. The calibrated colour wedge corresponds to the coding used in Fig. 13

Non-invasive imaging of living chondrocytes within intact cartilage explants avoids disturbance of the 3-D physical and chemical environment and progressive de-differentiation observed with isolated chondrocytes. Such *in situ* volume measurements indicate that, when surrounded by extracellular matrix, chondrocytes remain osmotically sensitive and exhibit volume-regulatory behaviour when subjected to osmotic perturbation. In contrast, a regulatory volume increase (RVI) in chondrocytes that have been isolated from cartilage has not been observed.

Multi-dimensional measurements: intracellular compartmentalisation

In the examples above, a single-intensity threshold was used to segment the object of interest and to subsequently extract the volume of the cells or organelles. In many cases the fluorescence intensity is related to the concentration of the probe and can be used to follow the concentration and distribution of a molecule of interest. For reliable measurements, it is important that the complete intensity distribution, including the background, is within the useable range of the detector system. Any clipping of the histogram will cause a systematic error in the mean intensity (Fricker *et al.*, 1997a). If the level of sample-dependent attenuation is relatively low, as with weakly refractive cells adjacent to the coverslip for example, then the intensity can be calibrated directly against standard solutions. The defined volume sampled in the confocal microscope makes this much more straightforward than calibration using wide-field microscopies, as there is no requirement to correct for a varying pathlength. In addition, optical sectioning also allows measurements of the distribution of the molecule in adjacent compartments and can be used to follow dynamic transport events *in situ.*

As an example, we have measured the distribution of glutathione (GSH), in primary roots of *Arabidopsis thaliana* (May *et al.*, 1997; Sanchez-Fernandez *et al.*, 1997) following conjugation to the membrane-permeant dye monochlorobimane (MCB) in the cytoplasm to give a fluorescent glutathione-bimane (GS-B) adduct (Coleman *et al.*, 1997a,b). In this tissue GS-B conjugate is rapidly transported into the vacuole (Fig. 15), probably through the action of a glutathione conjugate (GS-X) pump (Coleman *et al.*, 1997b; Ishikawa *et al.*, 1997) that has recently been cloned from *Arabidopsis* (Lu *et al.*, 1997). The kinetics of the labelling reaction and transport to the vacuole can be followed for individual cells in the root. In Fig. 16, the labelling kinetics for a cell in a trichoblast cell file and an atrichoblast cell file are shown. The signal from the nucleus in both cell types (blue) increased to a peak after about 1000 s and then declined to a steady-state plateau after ~ 2000 s. The signal from the cytoplasm followed the same pattern, but at a slightly lower level. The signal in the vacuole increased with time in the cell from the trichoblast file. The rate of vacuolar accumulation increased as the concentration of GS-B increased in the cytoplasm. The total vacuolar fluorescence continued to increase when the cytoplasmic signal had plateaued, but at a slightly slower rate. In the cell from the atrichoblast cell file, the signal from the vacuole began to plateau after 2000 s. The difference in the absolute concentration in the vacuole between the two cell types reflects their differing cytoplasm-to-vacuole ratios to a large extent. This type of measurement would be exceedingly difficult to make with conventional imaging systems, as the dye labels all the cells, giving a large and varying amount of out-of-focus blur, and it would not be possible to separate the vacuole and cytoplasmic signals.

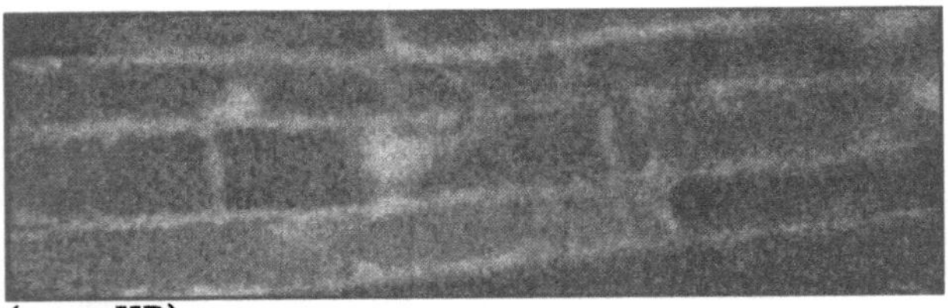

(1.500 KB)

Fig. 15 Labelling of root epidermal cells in an intact root of *Arabidopsis thaliana* with monochlorobimane. Single (*x*,*y*) optical sections through the epidermal tissue of a primary root of a 6-day-old *Arabidopsis* seedling were collected at 60-s intervals after exposure to 100 μM monochlorobimane. Fluorescence from the fluorescent glutathione-bimane (*GS-B*) conjugate appears initially in the cytoplasm and nucleus but is rapidly transported to the vacuole. Excitation was at 422 nm (15 mW He-Cd, Lambda Photometrics) and emission at >470 nm

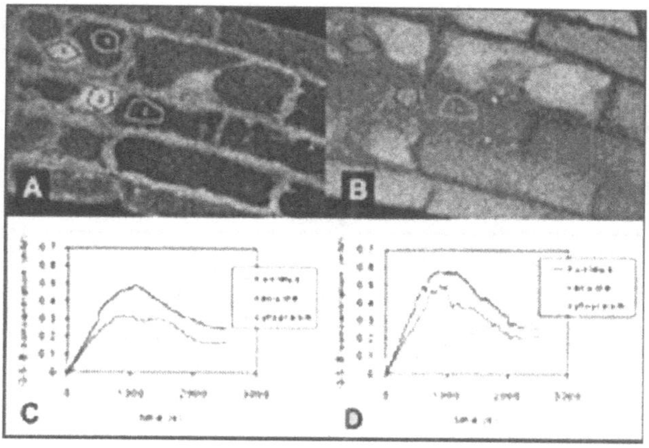

Fig. 16 Analysis of GS-B labelling kinetics for trichoblast and atrichoblast cell types in the epidermis of an intact *Arabidopsis thaliana* root. The average intensity from the nucleus (*blue*), cytoplasm (*red*) and vacuole (*green*) was followed over time as the GS-B conjugate was formed and then transported into the vacuole. The calibration was derived from the fluorescence intensity of a series of GS-B standards measured *in vitro* under the same imaging conditions. Graphical analysis was performed with Excel (Microsoft)

Multi-dimensional measurements: dynamic 3-D distribution of proteins

The wide availability of coding sequences and expression cassettes for GFP and its various spectral variants (Chalfie *et al.* 1994; Heim *et al.*, 1994;Heim and Tsien, 1996; Ormö *et al.*, 1996; Yang *et al.*, 1996; Brejc *et al.*, 1997) has stimulated considerable interest in the development of fluorescently tagged chimeric reporters to follow the *in vivo* dynamics of a target protein (e.g. Cheng *et al.*, 1996; Rizzuto *et al.*, 1996). We have been investigating techniques to visualise the distribution of heterochromatin inside living cells of *Schizosaccharomyces pombe*, using expression of chimeric genes encoding a fusion between the chromodomain protein Swi6 and GFP from *Aequorea victoria*. In interphase cells expressing Swi6-GFP, the fluorescence was restricted to the nucleus and was localised as a punctate distribution adjacent to the nuclear membrane in both the wild-type and a *swi6* null strain (Fig. 17). This distribution pattern is similar to the distribution of endogenous Swi6 determined using immuno-fluorescence in fixed cells (Ekwall *et al.*, 1995). The bright spots are thought to correspond to clusters of centromeres, telomeres and the mating type loci (Ekwall *et al.*, 1995) that all contain heterochromatin.

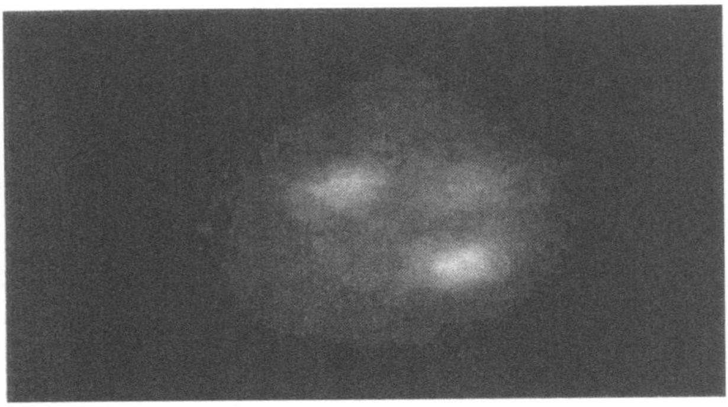

(160 KB)

Fig. 17 Rotation series of maximum projections of an interphase nucleus in living *Schizosaccharomyces pombe* expressing Swi6-GFP. Living cells of *Schizosaccharomyces pombe* expressing Swi6-GFP were imaged with excitation at 488 nm and emission at 535±25 nm using a Nikon 60× 1.4 NA oil-immersion objective. 3-D images were collected with a *z*-focus increment of 0.3 μm and displayed as maximum brightness projections. Two bright spots and a third weaker one are visible at the periphery of the nucleus in the rotation sequence. The diameter of the nucleus is about 1.8 μm

By using the GFP-tagged protein, it was possible to follow the dynamics of these heterochromatic regions throughout the cell cycle. In Fig. 18 a group of dividing cells of *S. pombe* is shown as a bright-field view and as a maximum projection of the 3-D images collected at 0.3-μm intervals at successive 10-min time points. Whilst all cells show a punctate distribution of Swi6-GFP in interphase, the relative disposition of the spots is quite dynamic. Several of the cells underwent mitosis and the punctate pattern disappeared during this phase of the cell cycle (Fig. 18). As the cells progressed through the subsequent G1 and S phases, the punctate pattern was re-established. This indicates that there was either a change in chromatin structure or its higher-level organisation at mitosis. To analyse the dynamic changes in the Swi6-GFP distribution in more detail, intensity-weighted histograms of the fluorescence intensity within the entire nucleus were derived. The bright features in the interphase cells are represented by a few voxels at high intensities (Fig. 19A). However, as the overall fluorescence intensity within the nuclei changed during the cell cycle, i.e. as the DNA was replicated and then separated during mitosis, the histograms for each time point were normalised and expressed in standard deviation (s.d.) units from the median (Fig. 19B). The bright features were objectively segmented from the interphase cells as intensities above 2 s.d. units from the median (Fig. 19B). This approach provides an opportunity to standardise comparisons at different time points and between different cells and treatments. Interestingly, analysis of the intensity weighted fluorescence histograms throughout mitosis indicated that roughly the same amount of material remained above 2 s.d. units from the median throughout division (Fig. 19C), although spots were no longer clearly discernible (Fig. 20B). These data provide quantitative support for a model whereby the change in the spots during mitosis results from dispersion of Swi6-GFP clusters, giving rise to a speckled distribution in the images, rather than dissociation of Swi6-GFP from the heterochromatin to give a homogeneous distribution in the nucleus.

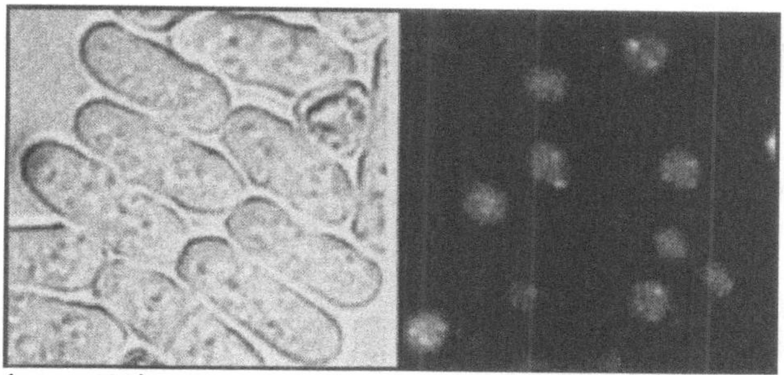

(1.600 KB)

Fig. 18 Distribution of Swi6-GFP fluorescence in cells of a *swi6* null strain of *S. pombe*. Maximum projections of 3-D confocal fluorescence images (*right panel*) and non-confocal bright field images (*left panel*) of *S. pombe* expressing Swi6-GFP. Imaging and visualisation conditions as for Fig. 17. Cells were followed over 120 min and 3-D fluorescence images and corresponding non-confocal bright field images were collected at 5-min intervals

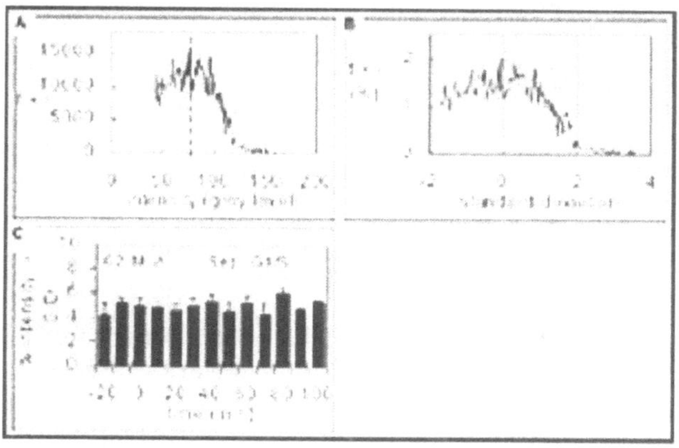

Fig. 19A–C Analysis of Swi6-GFP fluorescence histograms. The intensity-weighted fluorescence histogram is shown for a nucleus in an individual cell of a *swi6* null strain expressing Swi6-GFP (**A**). The median intensity is indicated by a *broken vertical line*. After normalisation, the histogram is expressed in s.d. units from the median (**B**). The shape of the histogram was similar to that in **B** at all stages of the cell cycle after normalisation, in terms of the spread of values about the median and the presence of a tail extending over 2 s.d. units above the median. The amount of signal above 2 s.d. units from the median was averaged from 6 cells (12 daughters) (**C**). Image processing and analysis routines were written in MPL (Bio-Rad Microscience). Graphical analysis was performed using Excel (Microsoft)

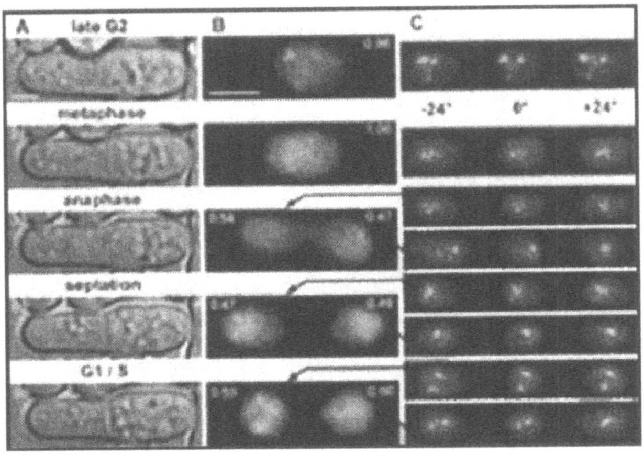

Fig. 20A–C Dynamics of Swi6-GFP fluorescence in an individual cell of a *swi6* null strain during mitosis and septation. Dividing cells were imaged at 10-min intervals over 2 h. Five time points at key stages in the cell cycle of a representative cell are shown as a single non-confocal bright-field images with the time inset (**A**) and a 3-D maximum projection confocal fluorescence image (**B**). The total fluorescence intensity of the nucleus (expressed as a fraction of the value at metaphase) is indicated in each image. **C** Normalised intensities greater than 2 s.d. from the median are coded in *green* and merged with the total Swi6-GFP fluorescence in *red*. Three rotated views at 24° increments are shown. The green image correlated well with the discrete spots at the G2 stage (***B**,first panel*). The distribution became more speckled at mitosis (**B**,*second and third panels*) but began to re-appear as discrete spots after septation (**B**,*last two panels*). Scale bar 2 µm

Multi-dimensional measurements: ion concentrations

The basic principle of intracellular ion measurement involves introduction of a chelating agent whose fluorescence properties alter with the ion activity (e.g. Bright *et al.*, 1989). This may be a simple quantitative change in intensity (single-wavelength dyes) or a shift in either the excitation or emission spectrum (ratio dyes) upon binding to the ion of interest. Removal of the out-of-focus information by confocal optical sectioning provides a measurement from a defined sampling volume (voxel) in the specimen. However, the limited depth of focus in a single optical section makes measurements very sensitive to small changes in specimen or organelle position and the *S*/*N* ratio is usually worse in comparison to wide-field imaging (see Pawley, 1995, pp. 19–38; Sandison *et al.*, 1995 for a discussion of the *S*/*N* ratio). Single-wavelength measurements of any physiological parameter cannot readily discriminate between genuine changes in fluorescence that reflect a change in the physiological process and changes in fluorescence due to dye leakage, photobleaching, re-distribution, compartmentalisation or degradation. This difficulty can be illustrated for a single-wavelength measurement of cytoplasmic Ca^{2+} ($[Ca^{2+}]_{cyt}$) levels using Fluo-3 iontophoretically microinjected into stomatal guard cells from *Vicia faba* (Fricker *et al.*, 1994). Signal was present in the thin peripheral cytoplasm and within the nucleus (Fig. 21A). In addition, the position of the chloroplasts was visible from their autofluorescence. Even with no additional stimuli, the level of fluorescence gradually fell during the experiment (Fig. 21B). Upon exposure to a peptide derived from the C-terminus of the auxin-binding protein (ABP) from *Zea mays* (reviewed by Napier and Venis, 1995) there was a rapid increase in fluorescence, corresponding to an increase in $[Ca^{2+}]_{cyt}$. Upon wash-out of the peptide the signal decreased again, to a level lower than the pre-stimulus value. The rate of decay of the basal fluorescence was estimated to be around 10% over 80–150 s in unstimulated cells. The average percentage increase in response to the peptide was 43% from the pre-stimulus level (Fricker *et al.*, 1994). Calibration of the changes in fluorescence in terms of calcium would require both a reliable *in situ* calibration at the end of the experiment (which is difficult to achieve for these cells – Fricker *et al.*, 1994) and a series of assumptions about the background fluorescence decay rate.

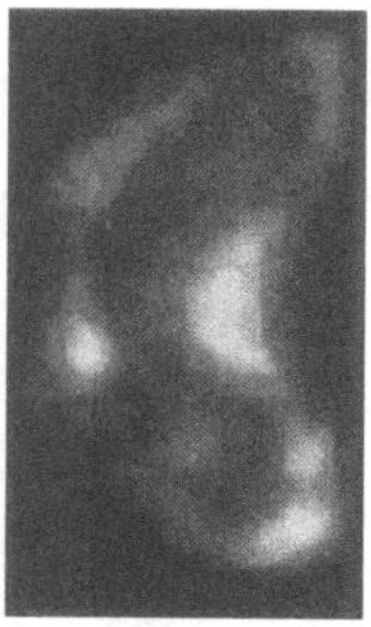

(960 KB)

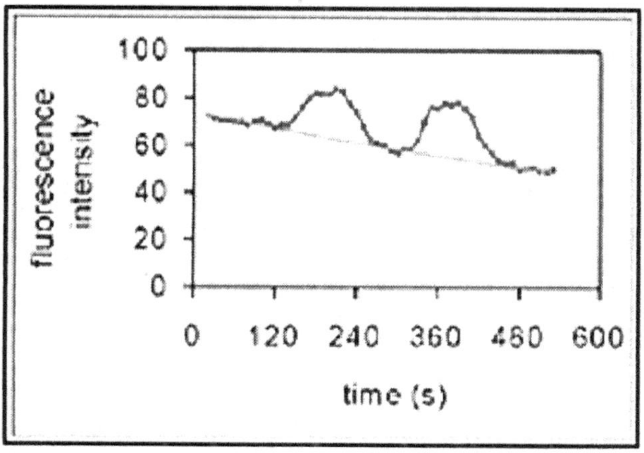

Fig. 21 Imaging of calcium dynamics in stomatal guard cells of *Vicia faba*. **A** Time course of fluorescence changes in a single optical section through a guard cell iontophoretically loaded with the calcium indicator Fluo-3, and imaged with excitation at 488 nm and emission at >520 nm. Fluorescence was visible in the nucleus, in the cytoplasm at the periphery of the cell and in strands crossing the vacuole, but was excluded from the vacuole itself. Strong autofluorescence from the chloroplasts was also visible. Fluorescence slowly declined in the un-stimulated cell. Addition of the C-terminal peptide from the auxin binding protein (*ABP1*) from *Zea mays* caused an increase in fluorescence that was rapidly reversed on wash-out of the peptide. **B** Graph of Fluo-3 fluorescence changes with time for an averaged area of cytoplasm. The C-terminal peptide caused a reproducible increase in fluorescence in two successive applications that was rapidly reversed on wash-out of the peptide (*red trace*). However, these changes were superimposed on a decay in the resting level of

Fluo-3 fluorescence (indicated by the *green line*) making it difficult to calibrate the changes in fluorescence caused by addition of the pepti

The ratio of fluorescence intensities determined at two excitation or emission wavelengths at or near the spectral peak of the free and bound forms can automatically correct for local variations in dye concentration, cell thickness, dye bleaching or leakage (Bright *et al.*, 1989). Implementation of ratio measurements using CLSM instruments requires laser excitation lines of appropriate wavelengths for the fluorescent probes (e.g. Kurtz and Emmons, 1993). As with all quantitative confocal fluorescence measurements, rigorous attention must be paid to problems of chromatic abberation to ensure correct recording of images in (*x*,*y*) and (*z*) (Fricker and White, 1992).

The intensity of the fluorescence for the pH indicator dye BCECF is not dependent on pH when excited at 442 nm, but increases dramatically with increasing pH at 488 nm. Thus the 442 nm image (Fig. 22A, *left*) only reflects the dye distribution in the cytoplasm (levels of autofluorescence were negligible), whilst the 488 nm image (Fig. 22A, *centre*) represents a combination of dye distribution and the local pH. Ratio images were calculated by dividing the 488 nm image by the 442 nm image (Fig. 22A, *right*) and indicates a uniform pH over most of the cytoplasm. The ratio image was related to cytoplasmic pH after appropriate calibration. The unstimulated cells also showed a steady decline in fluorescence (Fig. 22B). However, expressing the pH-dependent wavelength as a ratio of the pH-independent wavelength compensated for these changes and revealed that the resting pH alters little during the experiment (Fig. 22B). There was a substantial and rapid alkalinisation in response to the C-terminal peptide, which returned to resting levels after wash-out of the peptide (Thiel *et al.*, 1993; Fricker *et al.*, 1994).

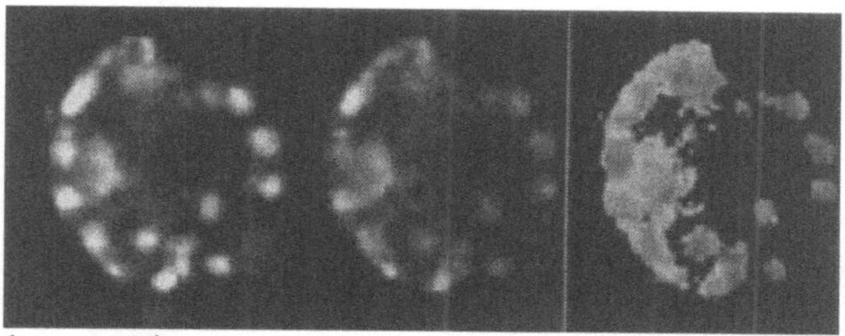

(2.000 KB)

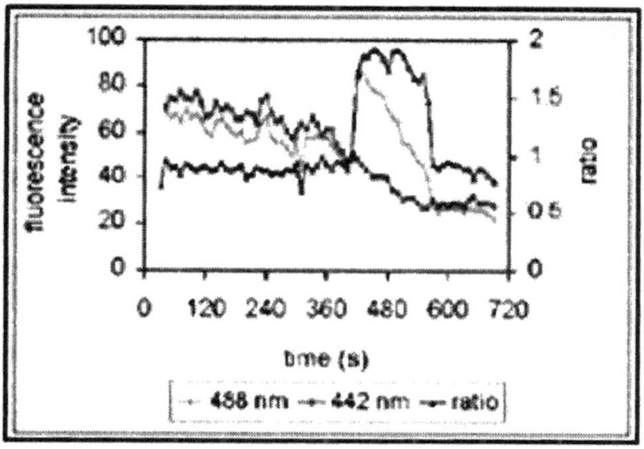

Fig. 22 Imaging of pH dynamics in stomatal guard cells of *Vicia faba*. **B** Time course of fluorescence changes in a single optical section through a guard cell iontophoretically loaded with the pH indicator BCECF with excitation at 442nm (*left panel*) and 488 nm (*centre panel*) and emission at >520 nm. The ratio image (488 nm/442 nm) (*right panel*) is pseudocolour coded from blue (pH 6.5) to red (pH 8.0) and has been masked to exclude regions of low dye concentration determined from the 442 nm image. The chloroplasts are visible from their autofluorescence in both guard cells. Addition of the C-terminal peptide from ABP1 caused an increase in ratio that was rapidly reversed on washout of the peptide. **B** The graphs from a region of cytoplasm in the 442 nm image showed a gradual decline in intensity during the experiment. However, ratioing the 488 nm signal against the 442 nm signal effectively compensated for the change in the amount of dye present, and the ratio trace indicated that the resting pH was maintained essentially constant. Addition of the C-terminal peptide caused a pronounced alkalinisation that was rapidly reversed on wash-out of the peptide to a similar resting level. The rate of decay of the 442 nm signal increased during exposure to the peptide, suggesting that the bleaching or leakage rate had also altered. Calibration using the ionophore nigericin and a high K^+ concentration indicate the resting pH was pH 7.2 and the peak increase about pH 7.8

Ratioing is particularly important for measurements on cells that alter their shape or organisation with time, as the resulting changes in dye concentration are likely to be more complex than a simple exponential decay due to dye leakage or bleaching. In addition, measurements of spatial gradients are critically dependent on techniques that compensate for uneven dye distribution. Tip–acid pH gradients have been reported to be associated with developing rhizoids of *Pelvetia fastigiata* (Gibbon and Kropf, 1994) and appear to be critical for tip growth in this system. We have investigated whether similar tip–acid pH gradients are associated with tip growth in pollen tubes of *Lilium longiflorum*

(Fricker *et al.*, 1997c). However, we found no evidence for a central role of tip-acid cytoplasmic pH gradients in the organisation or maintenance of pollen tube tip growth. Similar conclusions were reached by Parton *et al.* (1997) in a survey of several tip-growing systems. As an example, in Fig. 23 a tube that has temporarily stopped growing was induced to grow rapidly by addition of the fungal toxin fusicoccin (FC), to activate the plasma membrane proton pump. In the non-growing tube and during the early stages of elongation, no pH gradients were apparent. After about 4 min, an alkaline pH gradient towards the tip developed when the tube was growing vigorously.

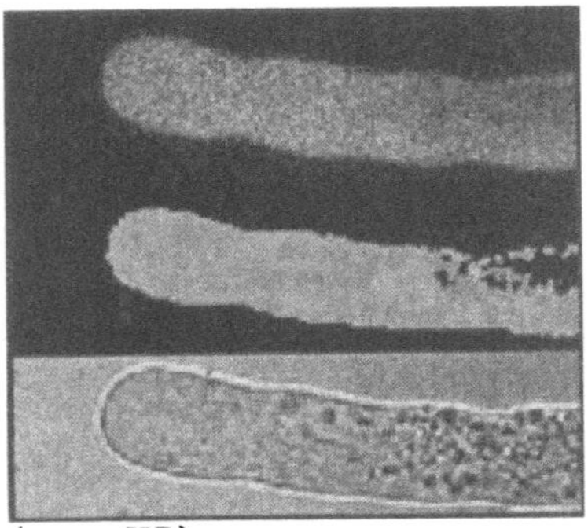

(1.000 KB)

Fig. 23 Confocal ratio imaging of pH in pollen tubes of *Lilium longiflorum*. BCECF was loaded as the acetoxy methyl ester and showed no obvious punctate staining in the cytoplasm (442 nm excitation - *top panel*). There was no marked pH gradient along the tube when it was initially not growing or during the early stages of fusicoccin-stimulated growth (ratio image - *centre panel*). After about 4 min the tip had become more alkaline by about 0.16 pH units, based on the *in situ* calibration to pH 6.7 and pH 7.3 shown at the end of the sequence. The non-confocal bright field image shows that the tube was actively streaming throughout the experiment until the calibration

Future perspectives

In principle, ratio techniques also provide automatic correction for depth- and sample-dependent attenuation in thick tissues, provided that the excitation and emission wavelengths all experience similar levels of aberration. The recent synthesis of genetically encoded and ratioable Ca^{2+} indicators based on GFP and calmodulin (Miyawaki *et al.*, 1997) should greatly facilitate calcium measurements from intact tissues. Unfortunately, many of the probes that are of interest, such as the GS-B conjugate, do not have spectral attributes that allow ratioing. One approach we are investigating is to correlate the parameter of interest with a second label, such as DNA, protein or cell wall material, that has a well-defined distribution (Fig. 24). This approach has the added advantage that it is possible to directly compare measurements obtained using an imaging approach with results from biochemical assays that are routinely expressed in units of protein or nucleic acid. Even with such ratioing techniques, the optical aberrations in biological specimens, and hence the level of attenuation, severely limit the depth over which reliable measurements can be made. Techniques that increase the penetration depth include the use of adaptive optics to continuously optimise the data collection and two-photon imaging which uses long-red or far-red illumination wavelengths that suffer less aberration in most systems. Application of two-photon imaging to intact biological specimens is not without drawbacks, however (see discussions in Denk *et al.*,1995; Jacobsen *et al.*, 1994; Brakenhoff *et al.*, 1996). In addition to improving the data collection, a number of post-collection image restoration techniques have been developed (e.g. Holmes *et al.*, 1995; Van der Voort and Strasters, 1995; Margadant *et al.*, 1996; Van Kempen *et al.*, 1997) that may improve the accuracy of quantitation from aberrated images.

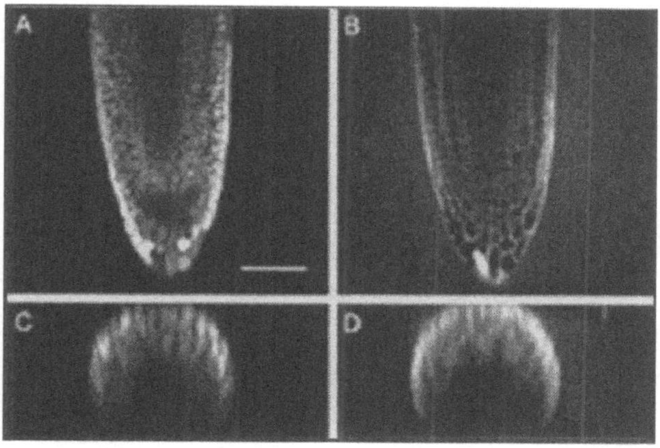

Fig. 24 *In vivo* labelling of glutathione (*GSH*) with monochlorobimane in intact roots of *Arabidopsis thaliana.* **A** An (*x,y*) image of the root tip after 20 min of labelling. Most of the GS-B has been transported to the vacuole. *Scale bar* = 50 µm. **B** An (*x,y*) image of the cell walls labelled with propidium iodide and imaged simultaneously. **C, D** Corresponding (*x,z*) images showing the attenuation of the fluorescence intensity with increasing depth through the root

Acknowledgements. We would like to thank the following for financial support: AFRC (M.D.F.); BBSRC (M.D.F., J.M., C.M.C); INTAS (M.D.F. and M.M.); Medical Research Council (D.J.V.); Nuffield Foundation (M.D.F.); Rhône-Poulenc and the Bio Avenir Program (M.D.F., A.J.M., N.S.W.); Science and Engineering Research Council (N.S.W. and M.D.F.; J.W.); Royal Society (M.D.F.); Wellcome Trust (D.J.V.; R.J.E.). N.S.W. is currently a Royal Society Industry Fellow.

References

Brakenhoff, G.J., Müller, M., Ghauharali, R.I. (1996) Analysis of efficiency of two-photon versus single-photon absorption for fluorescence generation in biological objects. *J. Microsc.* **183**: 140–144

Brejc, K., Sixma, T.K., Kitts, P.A., Kain, S.R., Tsien, R.Y., Ormö, M., Remington, S.J. (1997) Structural basis for dual excitation and photoisomerization of the *Aequorea victoria* green fluorescent protein. *Proc. Natl. Acad. Sci. USA* **94**:2306–2311

Bright, G.R., Fisher, G.W., Rogowska, J., Taylor, D.L. (1989) Fluorescence ratio imaging microscopy. In: *Methods in cell biology*, vol. 30. Academic Press, London

Carlsson, K. (1991) The influence of specimen refractive index, detector signal integration, and non-uniform scan speed on the imaging properties in confocal microscopy. *J. Microsc.* **163**: 167–178

Chalfie, M., Tu, Y., Euskirchen, G., Ward, W.W., Prasher, D.C. (1994) Green fluorescent protein as a marker for gene expression. *Science* **263**: 802–805

Cheng, L.Z., Fu, J., Tsukamoto, A., Hawley, R.G. (1996) Use of green fluorescent protein variants to monitor gene-transfer and expression in mammalian cells. *Nature Biotechnology* **14**: 606–609

Cogswell, C.J., Larkin, K.G. (1995) The specimen illumination path and its effect on image quality. In: Pawley, J.B. (ed) *Handbook of confocal microscopy*, 2nd edn. Plenum, New York, pp. 127–138

Coleman, J.O.D., Blake-Kalff, M.M.A., Davies, T.G.E. (1997a) Detoxification of xenobiotics by plants: chemical modification and vacuolar compartmentation. *Trends Plant Sci.* **2**: 144–151

Coleman, J.O.D., Randall, R., Blake-Kalff, M.M.A. (1997b) Detoxification of xenobiotics in plant cells by glutathione conjugation and vacuolar compartmentation: a fluorescent assay using monochlorobimane. *Plant Cell Environ.* **20**: 449–460

Denk, W., Piston, D.W., Webb, W.W. (1995) Two-photon molecular excitation in laser-scanning microscopy. In: Pawley, J.B. (ed)*Handbook of confocal microscopy*, 2nd edn. Plenum, New York, pp 445–458

Ekwall, K., Javerzat, J.-P., Lorentz, A., Schmidt, H., Cranston, G., Allshire, R. (1995) The chromodomain protein Swi6: a key component at fission yeast centromeres. *Science* **269**: 1429–1431

Errington, R.J., Fricker, M.D., Wood, J.L., Hall, A.C., White, N.S. (1997) Four-dimensional imaging of living chondrocytes in cartilage using confocal microscopy: a pragmatic approach. *Am. J. Physiol.* **272**: C1040–C1051

Fricker, M.D., White, N.S. (1992) Wavelength considerations in confocal microscopy of botanical specimens. *J. Microsc.* **166**: 29–42

Fricker, M.D., White, N.S., Thiel, G., Millner, P., Blatt, M.R. (1994) Peptides derived from the auxin binding protein elevate Ca^{2+} and pH in stomatal guard cells of *Vicia*: a confocal fluorescence ratio imaging study. In: Blatt, M.R., Leigh,

R.A., Sanders, D. (eds) *Membrane transport in plants and fungi: molecular mechanisms and contol. SEB Symp. Ser.* **48**: 215–228

Fricker, M.D., Errington, R.J., Wood, J.L., Tlalka, M., May, M., White, N.S. (1997a) Quantitative confocal fluorescence measurements in living tissues. In: Van Duijn, B. Wiltnik, A. (eds) *Signal transduction – single cell research.* Springer, Berlin, Heidelberg, New York (in press)

Fricker, M.D., Hollinshead, M., White, N.S., Vaux, D. (1997b) Interphase nuclei of many mammalian cell types contain deep, dynamic, tubular membrane-bound invaginations of the nuclear envelope. *J. Cell Biol.* **136:** 531–544

Fricker, M.D., White, N.S., Obermeyer, G. (1997c) pH gradients are not associated with tip growth in pollen tubes of *Lilium longiflorum. J. Cell Sci.* **110:** 1729–1740

Gibbon, B.C., Kropf, D.L. (1994) Cytosolic pH gradients are associated with tip growth. *Science* **263:** 1419–1421

Gilroy, S. (1997) Fluorescence microscopy of living plant cells. *Annu. Rev. Plant. Physiol. Plant Mol. Biol.* **48:** 165–190

Hassan, A.B., Errington, R.J., White, N.S., Jackson, D.A., Cook, P.R. (1994) Replication and transcription sites are colocalized in human cells. *J. Cell Sci.* **107:** 425–434

Heim, R., Tsien, R.Y. (1996) Engineering green fluorescent protein for improved brightness, longer wavelengths and fluorescence resonance energy transfer. *Curr. Biol.* **6:** 178–182

Hell, S.W., Reiner, G., Cremer, C., Stelzer, E.H.K. (1993) Aberrations in confocal fluorescence microscopy induced by mismatches in the refractive index. *J. Microsc.* **169:** 391–405

Hell, S.W., Stelzer, E.H.K. (1995) Lens aberrations in confocal fluorescence microscopy. In: Pawley, J.B. (ed) *Handbook of confocal microscopy,* 2nd edn. Plenum, New York, pp. 347–354

Holmes, T.J., Bhattacharyya, S., Cooper, J.A., Hanzel, D., Krishnamurthi, V., Lin, W.-C., Roysam, B., Szarowski, D.H., Turner, J.N. (1995) Light microscopic images reconstructed by maximum likelihood deconvolution. In: Pawley, J.B. (ed) *Handbook of confocal microscopy,* 2nd edn. Plenum, New York, pp. 389–402

Ishikawa, T., Li, Z.S., Lu, Y.P., Rea, P.A. (1997) The GS-X pump in plant, yeast, and animal cells: structure, function and gene expression. *Biosci. Rep.* **17:** 189–207

Jacobsen, H., Hänninen, P., Soini, E., Hell, S.W. (1994) Refractive-index-induced aberrations in two-photon confocal fluorescence microscopy. *J. Microsc.* **176**: 226–230

Keller, H.E. (1995) Objective lenses for confocal microscopy. In: Pawley, J.B. (ed) *Handbook of confocal microscopy*, 2nd edn. Plenum, New York, pp. 111–126

Kurtz, I., Emmons, C. (1993) Measurement of intracellular pH with a laser scanning confocal microscope. In: *Methods in cell biology*, vol. 38. Academic Press, London

Lu, Y.-P., Li, Z.-S., Rea, P.A. (1997) *AtMRP1* gene of *Arabidopsis* encodes a glutathione-*S*-conjugate pump: isolation and functional definition of a plant ABC transporter gene. *Proc. Natl. Acad. Sci. USA* **94**: 8243–8248

Manders, E.M.M. (1997) Chromatic shift in multicolour confocal microscopy. *J. Microsc.* **185**: 321–328

Margadant, F., Leemann, T., Niederer, P. (1996) A precise light attenuation correction for confocal scanning microscopy with $O(n^{4/3})$ computing time and $O(n)$ memory requirements for n voxels. *J. Microsc.* **182**: 121–132

Mason, W.T. (1993) *Fluorescent and luminescent probes for biological activity.* Academic Press, London

Matsumoto, B. (1993) Cell biological applications of confocal microscopy. In: *Methods in cell biology*, vol. 38. Academic Press, London

May, M.J., Van Montagu, M., Inze, D. (1997) Recent advances in glutathione metabolism and function in plants. *J. Exp. Bot.* (in press)

Miyawaki, A., Llopis, J., Heim, R., McCaffery, Adams, J.A., Ikura, M., Tsien, R.Y. (1997) Fluorescent indicators for Ca^{2+} based on green fluorescent proteins and calmodulin. *Nature* **388**: 882–887

Napier, R.M., Venis, M.A. (1995) Auxin action and auxin-binding proteins. *New Phytol.* **129**: 167–201

Nuccitelli, R. (1994) A practical guide to the study of calcium in living cells. In: *Methods in cell biology*, vol. 40. Academic Press, London

Ormö, M., Cubitt, A.B., Kallio, K., Gross, L.A., Tsien, R.Y., Remington, S.J. (1996) Crystal structure of the *Aequoria victoria* green fluorescent protein. *Science* **273**: 1392–1395

Parton, R.M., Fischer, S., Malhó, R., Papasouliotis, O., Jelitto, T.C., Leonard, T., Read, N.D. (1997) Pronounced cytoplasmic pH gradients are not required for tip growth in plant and fungal cells. *J. Cell Sci.* **110**: 1187–1198

Pawley, J. (ed) (1995) *Handbook of confocal microscopy*, 2nd edn. Plenum, New York

Rizzuto, R., Brini, M., De Giogi, F., Rossi, R., Heim, R., Tsien, R.Y., Pozzan, T. (1996) Double labelling of subcellular structures with organelle-targeted GFP mutants *in vivo*. *Curr. Biol.* **6**: 183–188

Sanchez-Fernandez, R., Fricker, M., Corben, L.B., White, N.S., Sheard, N., Leaver, C.J., Van Montagu, M., Inze, D., May, M.J. (1997) Cell proliferation and hair tip growth in the *Arabidopsis* root are under mechanistically different forms of redox control. *Proc. Natl. Acad. Sci. USA* **94**: 2745–2750

Sandison, D.R., Williams, R.M., Wells, K.S., Strickler, J., Webb, W.W. (1995) Quantitative fluorescence confocal laser scanning microscopy (CLSM). In: Pawley, J.B. (ed) *Handbook of confocal microscopy*, 2nd edn. Plenum, New York, pp 39–54

Sheppard, C.J.R., Török, P. (1997) Effects of specimen refractive index on confocal imaging. *J. Microsc.* **185**: 366–374

Taylor, L.D., Wang, Y.-L. (1990) Fluorescence microscopy of living cells in culture. In: *Methods in cell biology*, vol. 30, part B. Academic Press, London

Terasaki, M., Dailey, M.E. (1995) Confocal microscopy of living cells. In: Pawley, J.B. (ed) *Handbook of confocal microscopy*, 2nd edn. Plenum, New York, pp 327–346

Thiel, G., Blatt, M.R., Fricker, M.D., White, I.R., Millner, P. (1993) Modulation of K^+ channels in *Vicia* stomatal guard cells by peptide homologs to the auxin-binding protein C-terminus. *Proc. Natl. Acad. Sci. USA* **90**: 11493–11497

Van der Voort, H.T.M., Strasters, K.C. (1995) Restoration of confocal images for quantitative image analysis. *J. Microsc.* **178**: 165–181

Van Duijn, B., Wiltink, A. (1997) *Signal transduction –single cell techniques.* Springer, Berlin Heidelberg New York

Van Kempen, G.M.P., Van Vliet, L.J., Verveer, P.J., Van der Voort, H.T.M. (1997) A quantitative comparison of image restoration methods for confocal microscopy. *J. Microsc.* **185**: 354–365

Van Vliet, L.J. (1993) *Grey-scale measurements in multi-dimensional digitised images.* Delft University Press, Delft

Visser, T.D., Groen, F.C.A., Brakenhoff, G.J. (1991) Absorption and scattering correction in fluorescence confocal microscopy. *J. Microsc.* **163**: 189–200

Wang, Y.-L., Taylor, L.D. (1989) Fluorescence microscopy of living cells in culture. In: *Methods in cell biology*, vol. 29, part A. Academic Press, London

White, N.S. (1995) Visualisation systems for multi-dimensional CLSM images. In: Pawley, J.B. (ed) *Handbook of confocal microscopy*, 2nd edn. Plenum, New York, pp 211–254

White, N.S., Errington, R.J., Fricker, M.D., Wood, J.L. (1996) Aberration control in quantitative imaging of botanical specimens by multidimensional fluorescence microscopy. *J. Microsc.* **181**: 99–116

Yang, F., Moss, L.G., Phillips, G.N. (1996) The molecular structure of green fluorescent protein. *Nature Biotechnology* **14**: 1246–1251

Author index

F

G

H

J

K

Y

Keyword index

S

T

V

W

X

Z

SPRINGER NATURE

GPSR Compliance

The European Union's (EU) General Product Safety Regulation (GPSR) is a set of rules that requires consumer products to be safe and our obligations to ensure this.

If you have any concerns about our products, you can contact us on ProductSafety@springernature.com

In case Publisher is established outside the EU, the EU authorized representative is:

Springer Nature Customer Service Center GmbH
Europaplatz 3
69115 Heidelberg, Germany

FSC
www.fsc.org
MIX
Papier | Fördert gute Waldnutzung
FSC® C083411

Kolibri 360 GmbH
Ferdinand-Jühlke-Straße 7
99095 Erfurt, Deutschland
produktsicherheit@kolibri360.de